STUDENT'S SOLUTIONS MANUAL

RANDY GALLAHER
Lewis and Clark Community College

KEVIN BODDEN
Lewis and Clark Community College

STATISTICS: INFORMED DECISIONS USING DATA

THIRD EDITION

Michael Sullivan, III
Joliet Junior College

D1401411

Prentice Hall
is an imprint of

Reproduced by Pearson Prentice Hall from electronic files supplied by the author.

ISBN-13: 978-0-321-57751-1
ISBN-10: 0-321-57751-5

1 2 3 4 5 6 BB 10 09 08

Prentice Hall
is an imprint of

www.pearsonhighered.com

Table of Contents

Preface

Preface

This solutions manual accompanies *Statistics: Informed Decisions Using Data, 3e* by Michael Sullivan, III. The Instructor's Solutions Manual contains detailed solutions to all exercises in the text as well as the Consumer Reports® projects and the case studies. The Student's Solutions Manual contains detailed solutions to all odd exercises in the text and all solutions to chapter reviews and tests. A concerted effort has been made to make this manual as user-friendly and error free as possible. We offer a special note of thanks to Sarah Streett who carefully checked this entire manual for accuracy. Please feel free to send us any suggestions or corrections.

Randy Gallaher and Kevin Bodden
Department of Mathematics
Lewis & Clark Community College
5800 Godfrey Road
Godfrey, IL 62035
rgallahe@lc.edu   kbodden@lc.edu

Chapter 1

Data Collection

Section 1.1

1. Statistics is the science of collecting, organizing, summarizing and analyzing information in order to draw conclusions and answer questions. In addition, statistics is about providing a measure of confidence in any conclusions.

3. individual

5. statistic; parameter

7. A qualitative variable describes or classifies individuals based on some attribute or characteristic. Some examples are gender, zip codes, class (freshman, sophomore, etc.) and ethnicity. A quantitative variable is a numerical measure of the individuals on which arithmetic operations can be sensibly performed. Some examples are temperature, height, blood pressure, and life expectancy.

9. A variable is at the nominal level of measurement if the values of the variable categorize but cannot be ranked or put in a specific order (e.g. gender). In addition, arithmetic operations have no sensible interpretation at the nominal level.

   A variable is at the ordinal level of measurement if it has the characteristics of the nominal level, but its values can be ranked or placed in a specific order (e.g. sport team rank). However, arithmetic operations still have no sensible interpretation at the ordinal level.

   A variable is at the interval level of measurement if it has the properties of the ordinal level and the difference in the values of the variable has meaning (e.g. Fahrenheit temperature). Addition and subtraction can be performed on the values of the variable and have meaningful results at the interval level. At the interval level, we still lack a 'true zero'. That is, a value of 0 does not mean the absence of the quantity.

   A variable is at the ratio level of measurement if it has the properties of the interval level and has a true zero so that ratios of values have meaning (e.g. weight). A value of 0 means the absence of the quantity. Multiplication and division can be performed on the values of the variable and have meaningful results.

11. The process of statistics is our approach to performing statistical analyses:
    (1) Identify the research objective
    (2) Collect the data needed to answer the question(s) posed in the research objective.
    (3) Describe the data (graphically and numerically).
    (4) Perform inference (extend the results of the sample to the population and report a level of confidence in the results).

13. 18% is a parameter because it describes a population (all of the governors).

15. 25% is a statistic because it describes a sample (the high school students surveyed).

17. 0.366 is a parameter because it describes a population (all of Ty Cobb's at-bats).

19. 23% is a statistic because it describes a sample (the 6,076 adults studied).

21. Qualitative   23. Quantitative

25. Quantitative   27. Qualititative

29. Discrete   31. Continuous

33. Continuous   35. Discrete

37. Nominal   39. Ratio

41. Ordinal   43. Ratio

45. The population consists of all teenagers 13 to 17 years old who live in the United States. The sample consists of the 1,028 teenagers 13 to 17 years old who were contacted by the Gallup Organization.

47. The population consists of all of the soybean plants in this farmer's crop. The sample consists of the 100 soybean plants that were selected by the farmer.

49. The population consists of all women 27 to 44 years of age with hypertension. The sample consists of the 7,373 women 27 to 44 years of age with hypertension who were included in the study.

51. Individuals: Hitachi #P50X901, Mitsubishi #WD-73833, Sony #KDF-50E3000, Panasonic #TH-65PZ750U, Phillips #60PP9200D37, Samsung #FP-T5884, LG #52LB5D.
Variables: Size (in.), Screen Type, Price ($).
Data for "size": 50, 73, 50, 65, 60, 58, 52; data for "screen type": plasma, projection, projection, plasma, projection, plasma, plasma; data for "price": $4,000, $4,300, $1,500, $9,000, $1,600, $4,200, $3,500.
The variable "size" is continuous; the variable "screen type" is qualitative; the variable "price" is discrete.

53. Individuals: Alabama, Colorado, Indiana, North Carolina, Wisconsin.
Variables: Minimum age for Driver's License (unrestricted); mandatory belt-use seating positions, maximum allowable speed limit (rural interstate) in 2007.
Data for "minimum age for driver's license": 17, 17, 18, 16, 18; data for "mandatory belt-use seating positions": front, front, all, all, all; data for "maximum allowable speed limit (rural interstate) 2007": 70, 75, 70, 70, 65.
The variable "minimum age for driver's license" is continuous; the variable "mandatory belt-use seating positions" is qualitative; the variable "maximum allowable speed limit (rural interstate) 2007" is continuous.

55. (a) The research objective was to determine if the application of duct tape is as effective as cryotherapy in the treatment of common warts.

(b) The population is all people with warts. The sample consisted of 51 patients with warts.

(c) Descriptive statistics: 85% of patients in group 1 and 60% of patients in group 2 had complete resolution of their warts.

(d) The conclusion was that duct tape is significantly more effective in treating warts than cryotherapy.

57. (a) The research objective was to determine the hour of the day when adults feel at their best.

(b) The population is adult Americans aged 18 years or older.

(c) The sample consisted of the 1,019 adults surveyed.

(d) Descriptive statistic: 55% felt they were at their best in the morning.

(e) Inference: Gallup is 95% certain that the percentage of all adult Americans ages 18 years or older who feel they are at their best in the morning is between 52% and 58%.

59. *Jersey number* is nominal (the numbers generally indicate a type of position played). However, if the researcher feels that lower caliber players received higher numbers, then *jersey number* would be ordinal since players could be ranked by their number.

61. (a) The research question is to determine the role that TV watching by children younger than 3 plays in future attention problems for the children.

(b) The population of interest is all children under the age of 3 years.

(c) The sample consisted of the 967 children whose parents answered questions about TV habits and behavior issues.

(d) Descriptive statistic: The risk of attention problems five years later doubled for each hour per day that kids under 3 watched violent child-oriented programs.

(e) Inference: Children under the age of 3 years should not watch television. If they do watch, it should be educational and not violent child-oriented entertainment. Shows that are violent double the risk of attention problems for each additional hour watched each day. Even educational programs can result in a substantial risk for attention problems.

Section 1.2

1. The response variable is the variable of interest in a research study. An explanatory variable is a variable that affects (or explains) the value of the response variable. In research, we want to see how changes in the value of the explanatory variable affect the value of the response variable.

3. Confounding exists in a study when the effects of two or more explanatory variables are not separated. So any relation that appears to exist between a certain explanatory variable and the response variable may be due to some other variable or variables not accounted for in the study. A lurking variable is a variable not accounted for in a study, but one that affects the value of the response variable.

5. Cross-sectional studies collect information at a specific point in time (or over a very short period of time). Case-control studies are retrospective (they look back in time). Also, individuals that have a certain characteristic (such as cancer) in a case-control study are matched with those that do not have the characteristic. Case-control studies are typically superior to cross-sectional studies. They are relatively inexpensive, provide individual level data, and give longitudinal information not available in a cross-sectional study.

7. There is a perceived benefit to obtaining a flu shot, so there are ethical issues in intentionally denying certain seniors access to the treatment.

9. This is an observational study because the researchers merely observed existing data. There was no attempt by the researchers to manipulate or influence the variable(s) of interest.

11. This is an experiment because the explanatory variable (teaching method) was intentionally varied to see how it affected the response variable (score on proficiency test).

13. This is an observational study because the survey only observed preference of Coke or Pepsi. No attempt was made to manipulate or influence the variable of interest.

15. This is an experiment because the explanatory variable (carpal tunnel treatment regimen) was intentionally manipulated in order to observe potential effects on the response variable (level of pain).

17. (a) This is a cross-sectional study because the researchers collected information about the individuals at a specific point in time.

    (b) The response variable is whether the woman has nonmelanoma skin cancer or not. The explanatory variable is the daily amount of caffeinated coffee consumed.

    (c) It was necessary to account for these variables to avoid confounding due to lurking variables.

19. (a) This is an observational study because the researchers simply administered a questionnaire to obtain their data. No attempt was made to manipulate or influence the variable(s) of interest. This is a cross-sectional study because the researchers are observing participants at a single point in time.

    (b) The response variable is body mass index. The explanatory variable is whether a TV is in the bedroom or not.

    (c) Answers will vary. Some lurking variables might be the amount of exercise per week and eating habits. Both of these variables can affect the body mass index of an individual.

    (d) The researchers attempted to avoid confounding due to lurking variables by taking into account such variables as 'socioeconomic status'.

    (e) No. Since this was an observational study, we can only say that a television in the bedroom is associated with a higher body mass index.

21. Answers will vary. This is a prospective, cohort observational study. The response variable is whether the worker had cancer or not and the explanatory variable is the amount of electromagnetic field exposure. Some possible lurking variables include eating habits, exercise habits, and other health related variables such as smoking habits. Genetics (family history) could also be a lurking variable. Because this was an observational study, and not an experiment, the study only concludes that high electromagnetic field exposure is associated with higher cancer rates.

    The author reminds us that this is an observational study, so there is no direct control over the variables that may affect cancer rates. He also points out that while we should not simply dismiss such reports, we should consider the results in conjunction with results from future studies. The author concludes by mentioning known ways (based on extensive study) of reducing cancer risks that can currently be done in our lives.

23. (a) The research objective is to determine whether lung cancer is associated with exposure to tobacco smoke within the household.

    (b) This is a case-controlled study because there is a group of individuals with a certain characteristic (lung cancer but never smoked) being compared to a similar group without the characteristic (no lung cancer and never smoked). The study is retrospective because lifetime residential histories were compiled and analyzed.

    (c) The response variable is whether the individual has lung cancer or not. This is a qualitative variable.

    (d) The explanatory variable is the number of "smoker years". This is a quantitative variable.

    (e) Answers will vary. Some possible lurking variables are household income, exercise routine, and exposure to tobacco smoke outside the home.

    (f) The conclusion of the study is that approximately 17% of lung cancer cases among nonsmokers can be attributed to high levels of exposure to tobacco smoke during childhood and adolescence. No, we cannot say that exposure to household tobacco smoke causes lung cancer since this is only an observational study. We can, however, conclude that lung cancer is associated with exposure to tobacco smoke in the home.

    (g) An experiment involving human subjects is not possible for ethical reasons. Researchers would be able to conduct an experiment using laboratory animals, such as rats.

Section 1.3

1. The frame is necessary because it is the list we use to assign numbers to the individuals in the population.

3. Sampling without replacement means that no individual may be selected more than once as a member of the sample.

5. Answers will vary. We will use one-digit labels and assign the labels across each row (i.e. *Pride and Prejudice* – 0, *The Sun Also Rises* – 1, and so on). Starting at row 5, column 11, and proceeding downward, we obtain the following labels: 8, 4, 3 In this case, the 3 books in the sample would be *As I Lay Dying*, *A Tale of Two Cities*, and *Crime and Punishment*. Different labeling order, different starting points in Table I in Appendix A, or use of technology will likely yield different samples.

7. (a) {616, 630}, {616, 631}, {616, 632}, {616, 645}, {616, 649}, {616, 650}, {630, 631}, {630, 632}, {630, 645}, {630, 649}, {630, 650}, {631, 632}, {631, 645}, {631, 649}, {631, 650}, {632, 645}, {632, 649}, {632, 650}, {645, 649}, {645, 650}, {649, 650}

    (b) There is a 1 in 21 chance that the pair of courses will be EPR 630 and EPR 645.

9. (a) Starting at row 5, column 22, using two-digit numbers, and proceeding downward, we obtain the following values: 83, 94, 67, 84, 38, 22, 96, 24, 36, 36, 58, 34,.... We must disregard 94 and 96 because there are only 87 faculty members in the population. We must also disregard the second 36 because we are sampling without replacement. Thus, the 9 faculty members included in the sample are those numbered 83, 67, 84, 38, 22, 24, 36, 58, and 34.

   (b) Answers will vary depending on the type of technology used. If using a TI-84 Plus, the sample will be: 4, 20, 52, 5, 24, 87, 67, 86, and 39.

   Note: We must disregard the second 20 because we are sampling without replacement.

11. (a) Answers will vary depending on the technology used (including a table of random digits). Using a TI-84 Plus graphing calculator with a seed of 17 and the labels provided, our sample would be North Dakota, Nevada, Tennessee, Wisconsin, Minnesota, Maine, New Hampshire, Florida, Missouri, and Mississippi.

   (b) Repeating part (a) with a seed of 18, our sample would be Michigan, Massachusetts, Arizona, Minnesota, Maine, Nebraska, Georgia, Iowa, Rhode Island, Indiana.

13. (a) The list provided by the administration serves as the frame. Number each student in the list of registered students, from 1 to 19,935. Generate 25 random numbers, without repetition, between 1 and 19,935 using a random number generator or table. Select the 25 students with these numbers.

   (b) Answers will vary.

15. Answers will vary. Members should be numbered 1 – 32, though other numbering schemes are possible (e.g. 0 – 31). Using a table of random digits or a random-number generator, four different numbers (labels) should be selected. The names corresponding to these numbers form the sample.

Section 1.4

1. Stratified random sampling may be appropriate if the population of interest can be divided into groups (or strata) that are homogeneous and non-overlapping.

3. Convenience samples are typically selected in a nonrandom manner. This means the results are not likely to represent the population. Convenience samples may also be self-selected, which will frequently result in small portions of the population being overrepresented.

5. stratified sample

7. False. In many cases, other sampling techniques may provide equivalent or more information about the population with less "cost" than simple random sampling.

9. True. Because the individuals in a convenience sample are not selected using chance, it is likely that the sample is not representative of the population.

11. Systematic sampling. The quality-control manager is sampling every 8th chip.

13. Cluster sampling. The airline surveys all passengers on selected flights (clusters).

15. Simple random sampling. Each known user of the product has the same chance of being included in the sample.

17. Cluster sampling. The farmer samples all trees within the selected subsections (clusters).

19. Convenience sampling. The research firm is relying on voluntary response to obtain the sample data.

21. Stratified sampling. Shawn takes a sample of measurements during each of the four time intervals (strata).

23. The numbers corresponding to the 20 clients selected are 16, $16+25=41$, $41+25=66$, $66+25=91$, $91+25=116$, 141, 166, 191, $216, 241, 266, 291, 316, 341, 366, 391, 416,$ $441, 466, 491.$

25. Answers will vary. To obtain the sample, number the Democrats 1 to 16 and obtain a simple random sample of size 2. Then number the Republicans 1 to 16 and obtain a simple random sample of size 2. Be sure to use a different starting point in Table I or a different seed for each stratum.

For example, using a TI-84 Plus graphing calculator with a seed of 38 for the Democrats and 40 for the Republicans, the numbers selected would be 6, 9 for the Democrats and 14, 4 for the Republicans. If we had numbered the individuals down each column, the sample would consist of Haydra, Motola, Engler, and Thompson.

```
38→rand
              38
randInt(1,16)
               6
               9
```
```
40→rand
              40
randInt(1,16)
              14
               4
```

27. (a) $\dfrac{N}{n}=\dfrac{4502}{50}=90.04 \to 90$; Thus, $k=90$.

  (b) Randomly select a number between 1 and 90. Suppose that we select 15. Then the individuals to be surveyed will be the 15th, 105th, 195th, 285th, and so on up to the 4425th employee on the company list.

29. Simple Random Sample:
  Number the students from 1 to 1280. Use a table of random digits or a random-number generator to randomly select 128 students to survey.

  Stratified Sample:
  Since class sizes are similar, we would want to randomly select $\dfrac{128}{32}=4$ students from each class to be included in the sample.

  Cluster Sample:
  Since classes are similar in size and makeup, we would want to randomly select $\dfrac{128}{32}=4$ classes and include all the students from those classes in the sample.

31. Answers will vary. One design would be a stratified random sample, with two strata being commuters and noncommuters, as these two groups each might be fairly homogeneous in their reactions to the proposal.

33. Answers will vary. One design would be a cluster sample, with the clusters being city blocks. Randomly select city blocks and survey every household in the selected blocks.

35. Answers will vary. Since the company already has a list (frame) of 6,600 individuals with high cholesterol, a simple random sample would be an appropriate design.

37. (a) For a political poll, a good frame would be all registered voters who have voted in the past few elections since they are more likely to vote in upcoming elections.

  (b) Because each individual from the frame has the same chance of being selected, there is a possibility that one group may be over- or underrepresented.

  (c) By using a stratified sample, the strategist can obtain a simple random sample within each strata (political party) so that the number of individuals in the sample is proportionate to the number of individuals in the population.

39. Answers will vary.

Section 1.5

1. It is rare for frames to be completely accurate because the population may change frequently (e.g. the voter roll), making it difficult to keep the frame up to date. Also, the population may be very large (e.g. the population of the United States), making it difficult to obtain a complete frame.

3. A closed question is one in which the respondent must choose from a list of prescribed responses. An open question is one in which the respondent is free to choose his or her own response. Closed questions are easier to analyze, but limit the responses. Open questions allow respondents to state exactly how they feel, but are harder to analyze due to the variety of answers and possible misinterpretation of answers.

5. Trained interviewers generally obtain better survey results. For example, a talented interviewer will be able to elicit truthful responses even to sensitive questions.

7. A pro is that the interviewer is more likely to find the individual at home at this time. A con is that many individuals will be irritated at having their dinner interrupted and will refuse to respond.

9. Changing the order of questions and choices helps prevent bias due to previous question answers or situations where respondents are more likely to pick earlier choices.

11. Bias means that the results of the sample are not representative of the population. There are three types of bias: sampling bias, response bias, and nonresponse bias. Sampling bias is due to the use of a sample to describe a population. This includes bias due to convenience sampling. Response bias involves intentional or unintentional misinformation. This would include lying to a surveyor or entering responses incorrectly. Nonresponse bias results when individuals choose not to respond to questions or are unable to be reached. A census can suffer from response bias and nonresponse bias, but would not suffer from sampling bias.

13. (a) Sampling bias. The survey suffers from undercoverage because the first 60 customers are likely not representative of the entire customer population.

    (b) Since a complete frame is not possible, systematic random sampling could be used to make the sample more representative of the customer population.

15. (a) Response bias. The survey suffers from response bias because the question is poorly worded.

    (b) The survey should inform the respondent of the current penalty for selling a gun illegally and the question should be worded as: "Do you approve or disapprove of harsher penalties for individuals who sell guns illegally?" The order of "approve" and "disapprove" should be switched from one individual to the next.

17. (a) Nonresponse bias. Assuming the survey is written in English, non-English speaking homes will be unable to read the survey. This is likely the reason for the very low response rate.

    (b) The survey can be improved by using face-to-face or phone interviews, particularly if the interviewers are multi-lingual.

19. (a) The survey suffers from sampling bias due to undercoverage and interviewer error. The readers of the magazine may not be representative of all Australian women, and advertisements and images in the magazine could affect the women's view of themselves.

    (b) A well-designed sampling plan not in a magazine, such as a cluster sample, could make the sample more representative of the population.

21. (a) Response bias due to a poorly worded question.

    (b) The question should be reworded in a more neutral manner. One possible phrasing might be: "Do you believe that a marriage can be maintained after an extramarital relation?"

23. (a) Response bias. Students are unlikely to give honest answers if their teacher is administering the survey.

    (b) An impartial party should administer the survey in order to increase the rate of truthful responses.

25. No. The survey still suffers from sampling bias due to undercoverage, nonresponse bias, and potentially response bias.

27. It is very likely that the order of these two questions will affect the survey results. To alleviate the response bias, either question B could be asked first, or the order of the two questions could be rotated randomly.

29. The company is using a reward in the form of the $5.00 payment and an incentive by telling the reader that his or her input will make a difference.

31. For random digit dialing, the frame is anyone with a phone (whose number is not on a do-not-call registry). Even those with unlisted numbers can still be reached through this method.

Any household without a phone, households on the do-not-call registry, and homeless individuals are excluded. This could result in sampling bias due to undercoverage if the excluded individuals differ in some way than those included in the frame.

33. It is extremely likely, particularly if households on the do-not-call registry have a trait that is not part of those households that are not on the registry.

35. Some non-sampling errors presented in the article as leading to incorrect exit polls were poorly trained interviewers, interviewer bias, and over representation of female voters.

37. – 39. Answers will vary.

41. The *Literary Digest* made an incorrect prediction due to sampling bias (an incorrect frame led to undercoverage) and nonresponse bias (due to the low response rate).

43. (a) The population of interest is all vehicles that travel on the road (or portion of the road) in question.

(b) The variable of interest is the speed of the vehicles.

(c) The variable is quantitative.

(d) Because speed has a 'true zero', it is at the ratio level of measurement.

(e) A census is not feasible. It would be impossible to obtain a list of all the vehicles that travel on the road.

(f) A sample is feasible, but not a simple random sample (since a complete frame is impossible). A systematic random sample would be a feasible alternative.

(g) Answers will vary. One bias is sampling bias. If the city council wants to use the cars of residents who live in the neighborhood to gauge the prevailing speed, then individuals who are not part of the population were in the sample (likely a huge portion), so the sample is not representative of the intended population.

Section 1.6

1. (a) An experimental unit is a person, object, or some other well-defined item upon which a treatment is applied.

(b) A treatment is a condition applied to an experimental unit. It can be any combination of the levels of the explanatory variables.

(c) A response variable is a quantitative or qualitative variable that measures a response of interest to the experimenter.

(d) A factor is a variable whose effect on the response variable is of interest to the experimenter. Factors are also called explanatory variables.

(e) A placebo is an innocuous treatment, such as a sugar pill, administered to a subject in a manner indistinguishable from an actual treatment.

(f) Confounding occurs when the effect of two explanatory variables on a response variable cannot be distinguished.

3. In a single-blind experiment, subjects do not know which treatment they are receiving. In a double-blind experiment, neither the subject nor the researcher(s) in contact with the subjects knows which treatment is received.

5. completely randomized; matched-pair

7. True

9. Control groups are needed to serve as a baseline that other treatments can be compared against. This allows the researcher to take the 'placebo effect' into account when analyzing the results of the experiment.

11. (a) The researchers used an innocuous treatment to account for effects that would result from any treatment being given (i.e. the placebo effect). The placebo is the flavored water that looks and tastes like the sports drink. It serves as the baseline for which to compare the results when the noncaffeinated and caffeinated sports drinks are administered.

(b) Being double-blind means that neither the cyclists nor the researcher administering the treatments knew when the cyclists were given the caffeinated sports drink, the noncaffeinated sports drink, or the flavored-water placebo. This is necessary to avoid any intentional or unintentional bias due to knowing which treatment is being given.

(c) Randomization is used to determine the order of the treatments for each subject.

(d) The population of interest is all athletes or individuals involved in prolonged exercise. The sample consists of the 16 highly trained cyclists studied.

(e) There are three treatments in the study: caffeinated sports drink, noncaffeinated sports drink, and a flavored-water placebo.

(f) The response variable is total work completed.

(g) A repeated-measure design takes measurements on the same subject using multiple treatments. A matched-pairs design is a special case of the repeated-measures design that uses only two treatments.

13. **(a)** The response variable is the achievement test scores.

(b) Some factors are teaching methods, grade level, intelligence, school district, and teacher.
Fixed: grade level, school district, teacher.
Set at predetermined levels: teaching method.

(c) The treatments are the new teaching method and the traditional method. There are 2 levels of treatment.

(d) The factors that are not controlled are dealt with by random assignment into the two treatment groups.

(e) Group 2, using the traditional teaching method, serves as the control group.

(f) This experiment has a completely randomized design.

(g) The subjects are the 500 first-grade students from District 203 recruited for the study.

(h)

```
                        ┌────────────────┐      ┌────────────────┐
                   ┌──▶ │ Group 1:       │ ───▶ │ Treatment 1:   │ ───┐
                   │    │ 250 students   │      │ New method     │    │
┌──────────────┐   │    └────────────────┘      └────────────────┘    ▼
│ Random       │   │                                           ┌────────────────┐
│ assignment   │───┤                                           │ Compare        │
│ of students  │   │                                           │ achievement    │
│ to treatments│   │                                           │ test score     │
└──────────────┘   │    ┌────────────────┐      ┌────────────────┐  └──────────┘
                   └──▶ │ Group 2:       │ ───▶ │ Treatment 2:   │ ───┘
                        │ 250 students   │      │ Traditional method │
                        └────────────────┘      └────────────────┘
```

15. **(a)** This experiment has a matched-pairs design.

(b) The response variable is the level of whiteness.

(c) The explanatory variable is the whitening method. The treatments are Crest Whitestrips Premium in addition to brushing and flossing, and just brushing and flossing alone.

(d) Answers will vary. One other possible factor is diet. Certain foods and tobacco products are more likely to stain teeth. This could impact the level of whiteness.

(e) Answers will vary. One possibility is that using twins helps control for genetic factors such as weak teeth that may affect the results of the study.

17. **(a)** This experiment has a completely randomized design.

    (b) The population being studied is adults with insomnia.

    (c) The response variable is the terminal wake time after sleep onset (WASO).

    (d) The explanatory variable is the type of intervention. The treatments are cognitive behavioral therapy (CBT), muscle relaxation training (RT), and the placebo.

    (e) The experimental units are the 75 adults with insomnia.

    (f)

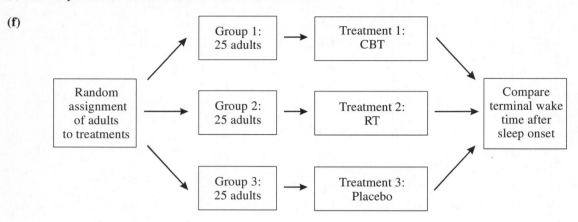

19. **(a)** This experiment has a completely randomized design.

    (b) The population being studied is adults over 60 years old and in good health.

    (c) The response variable is the standardized test of learning and memory.

    (d) The factor set to predetermined levels (explanatory variable) is the drug. The treatments are 40 milligrams of ginkgo 3 times per day and the matching placebo.

    (e) The experimental units are the 98 men and 132 women over 60 years old and in good health.

    (f) The control group is the placebo group.

    (g)

21. (a) This experiment has a matched-pairs design.

   (b) The response variable is the distance the yardstick falls.

   (c) The explanatory variable is hand dominance. The treatment is dominant versus non-dominant hand.

   (d) The experimental units are the 15 students.

   (e) Professor Neil used a coin flip to eliminate bias due to starting on the dominant or non-dominant hand first on each trial.

   (f)

23. (a) This experiment has a randomized block design.

   (b) The response variable is the score on the recall exam.

   (c) The explanatory variable is the type of advertising. The treatments are print, radio, and television.

   (d) Level of education is the variable that serves as the block.

   (e)

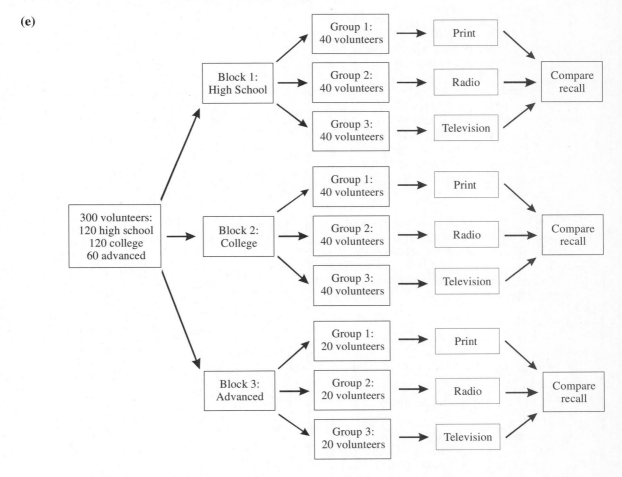

25. Answers will vary. Using a TI-84 Plus graphing calculator with a seed of 195, we would pick the volunteers numbered 8, 19, 10, 12, 13, 6, 17, 1, 4, and 7 to go into the experimental group. The rest would go into the control group. If the volunteers were numbered in the order listed, the experimental group would consist of Ann, Kevin, Christina, Eddie, Shannon, Randy, Tom, Wanda, Kim, and Colleen.

27. Answers will vary. A completely randomized design is likely the best.

29. Answers will vary. A matched-pairs design matched by type of exterior finish is likely the best.

31. Answers will vary. A randomized block design blocked by type of car is likely the best.

33. (a) The response variable is blood pressure.

    (b) Three factors that have been identified are daily consumption of salt, daily consumption of fruits and vegetables, and the body's ability to process salt.

    (c) The daily consumption of salt and the daily consumption of fruits and vegetables can be controlled. The body's ability to process salt cannot be controlled. To deal with variability of the body's ability to process salt, randomize experimental units to each treatment group.

    (d) Answers will vary. Three levels of treatment might be a good choice – one level below the recommended daily allowance, one equal to the recommended daily allowance, and one above the recommended daily allowance.

35. Answers will vary.

37. (a) The research objective is to determine if nymphs of *Brachytron pretense* will control mosquitoes.

    (b) The experiment is a completely randomized design.

    (c) The response variable is mosquito larvae density. This is a discrete (because the larvae are counted), quantitative variable.

(d) The researchers controlled the introduction of nymphs of *Brachytron pretense* into water tanks containing mosquito larvae by setting the factor at predetermined levels. The treatments are nymphs added or no nymphs added.

(e) Some other factors that may affect the larvae of mosquitoes are temperature, amount of rainfall, fish, presence of other larvae, and sunlight.

(f) The population is all mosquito larvae in all breeding places. The sample consists of the mosquito larvae in the ten 300-liter outdoor, open concrete water tanks.

(g) During the study period, mosquito larvae density changed from 7.34 to 0.83 to 6.83 larvae per dip in the treatment tanks. The mosquito larvae density changed from 7.12 to 6.83 to 6.79 larvae per dip in the control tanks.

(h) The researchers controlled the experiment by first clearing the ten tanks of all non-mosquito larvae, nymphs, and fish so that only mosquito larvae were left in the tanks. Five tanks were treated with 10 nymphs of *Brachyron pretense*, while five tanks were left untreated, serving as a control group for baseline comparison. The researchers attempted to make the tanks as identical as possible, except for the treatment.

(i)

(j) The researchers concluded that nymphs of *Brachytron pretense* can be used effectively as a strong, ecologically friendly means of controlling mosquitoes, and ultimately mosquito-borne diseases.

Chapter 1 Review Exercises

1. Statistics is the science of collecting, organizing, summarizing and analyzing information in order to draw conclusions.

2. The population is the group of individuals that is to be studied.

3. A sample is a subset of the population.

4. An observational study uses data obtained by studying individuals in a sample without trying to manipulate or influence the variable(s) of interest. Observational studies are often called *ex post facto* studies because the value of the response variable has already been determined.

5. In a designed experiment, a treatment is applied to the individuals in a sample in order to isolate the effects of the treatment on the response variable.

6. The three major types of observational studies are (1) Cross-sectional studies, (2) Case-control studies, and (3) Cohort studies.

   Cross-sectional studies collect data at a specific point in time or over a short period of time. Cohort studies are prospective and collect data over a period of time, sometimes over a long period of time. Case-controlled studies are retrospective, looking back in time to collect data either from historical records or from recollection by subjects in the study. Individuals possessing a certain characteristic are matched with those that do not.

7. The process of statistics refers to the approach used to collect, organize, analyze, and interpret data. The steps are
   (1) Identify the research objective
   (2) Collect the data needed to answer the research question.
   (3) Describe the data.
   (4) Perform inference.

8. The three types of bias are sampling bias, nonresponse bias, and response bias.
   Sampling bias occurs when the techniques used to select individuals to be in the sample favor one part of the population over another. This can be minimized by using chance to select the sample.
   Nonresponse bias occurs when the individuals selected to be in the sample that do not respond to the survey have different opinions from those that do respond. This can be minimized by using call-backs and follow-up visits to increase the response rate.
   Response bias occurs when the answers on a survey do not reflect the true feelings of the respondent. This can be minimized by using trained interviewers, using carefully worded questions, and rotating questions and answer selections.

9. Nonsampling errors are errors that result from undercoverage, nonresponse bias, response bias, and data-entry errors. These errors can occur even in a census. Sampling errors are errors that result from the use of a sample to estimate information about a population. These include random error and errors due to poor sampling plans, and result because samples contain incomplete information regarding a population.

10. The steps in conducting an experiment are:
    (1) *Identify the problem to be solved.*
        Gives direction and indicates the variables of interest (referred to as the claim).
    (2) *Determine the factors that affect the response variable.*
        List all variables that may affect the response, both controllable and uncontrollable.
    (3) *Determine the number of experimental units.*
        Determine the sample size. Use as many as time and money allow.
    (4) *Determine the level of each factor.*
        Factors can be controlled by fixing their level (e.g. only using men) or setting them at predetermined levels (e.g. different dosages of a new medicine). For factors that cannot be controlled, random assignment of units to treatments helps average out the effects of the uncontrolled factor over all treatments.
    (5) *Conduct the experiment.*
        Carry out the experiment using an equal number of units for each treatment. Collect and organize the data produced.
    (6) *Test the claim.*
        Analyze the collected data and draw conclusions.

11. 'Number of new automobiles sold at a dealership on a given day' is quantitative because its values are numerical measures on which addition and subtraction can be performed with meaningful results. The variable is discrete because its values result from a count.

12. 'Weight in carats of an uncut diamond' is quantitative because its values are numerical measures on which addition and subtraction can be performed with meaningful results. The variable is continuous because its values result from a measurement rather than a count.

13. 'Brand name of a pair of running shoes' is qualitative because its values serve only to classify individuals based on a certain characteristic.

14. 73% is a statistic because it describes a sample (the 1011 people age 50 or older who were surveyed).

15. 69% is a parameter because it describes a population (all the passes thrown by Chris Leak in the 2007 Championship Game).

16. Birth year has the *interval* level of measurement since differences between values have meaning, but it lacks a true zero.

17. Marital status has the *nominal* level of measurement since its values merely categorize individuals based on a certain characteristic.

18. Stock rating has the *ordinal* level of measurement because its values can be placed in rank order, but differences between values have no meaning.

19. Number of siblings has the *ratio* level of measurement because differences between values have meaning and there is a true zero.

20. This is an observational study because no attempt was made to influence the variable of interest. Sexual innuendos and curse words were merely observed.

21. This is an experiment because the researcher intentionally imposed treatments (experimental drug vs. placebo) on individuals in a controlled setting.

22. This was a cohort study because participants were identified to be included in the study and then followed over a period of time (over 26 years) with data being collected at regular intervals (every 2 years).

23. This is convenience sampling since the pollster simply asked the first 50 individuals she encountered.

24. This is a cluster sample since the ISP included all the households in the 15 randomly selected city blocks.

25. This is a stratified sample since individuals were randomly selected from each of the three grades.

26. This is a systematic sample since every 40th tractor trailer was tested.

27. (a) Sampling bias; undercoverage or nonrepresentative sample due to a poor sampling frame. Cluster sampling or stratified sampling are better alternatives.

    (b) Response bias due to interviewer error. A multi-lingual interviewer could reduce the bias.

    (c) Data-entry error due to the incorrect entries. Entries should be checked by a second reader.

28. Answers will vary. Using a TI-84 Plus graphing calculator with a seed of 1990, and numbering the individuals from 1 to 21, we would select individuals numbered 14, 6, 10, 17, and 11. If we numbered the businesses down each column, the businesses selected would be Jiffy Lube, Nancy's Flowers, Norm's Jewelry, Risky Business Security, and Solus, Maria, DDS.

29. Answers will vary. The first step is to select a random starting point among the first 9 bolts produced. Using row 9, column 17 from Table I in Appendix A, he will sample the 3rd bolt produced, then every 9th bolt after that until a sample size of 32 is obtained. In this case, he would sample bolts 3, 12, 21, 30, and so on until bolt 282.

30. Answers will vary. The goggles could be numbered 00 to 99, then a table of random digits could be used to select the numbers of the goggles to be inspected. Starting with row 12, column 1 of Table 1 in Appendix A and reading down, the selected labels would be 55, 96, 38, 85, 10, 67, 23, 39, 45, 57, 82, 90, and 76.

31. **(a)** This experiment has a completely randomized design.

(b) The response variable is the amount of energy required to light the bulb.

(c) The explanatory variable is the type of ballast. There are two treatments: old ballast and new ballast.

(d) The old ballast (group 2) serves as the control group (the baseline for comparison).

(e) The experimental units are the 200 fluorescent bulbs.

(f) By randomly assigning bulbs to the ballasts, the researchers are attempting to deal with factors that have not been controlled in the study. The idea is that the randomization will average out the effects of uncontrolled factors across all treatments.

(g)

32. **(a)** This experiment has a randomized block design.

(b) The response variable is the exam grade.

(c) The factor 'Notecard use' is set at predetermined levels. The treatments are 'with notecard' and 'without notecard'.

(d) The experimental units are the instructor's statistics students.

(e)

33. Answers will vary. Since there are ten digits (0 – 9), we will let a 0 or 1 indicate that (a) is to be the correct answer, 2 or 3 indicate that (b) is to be the correct answer, and so on. Beginning with row 1, column 8 of Table 1 in Appendix A, and reading downward, we obtain the following:
2, 6, 1, 4, 1, 4, 2, 9, 4, 3, 9, 0, 6, 4, 4, 8, 6, 5, 8, 5
Therefore, the sequence of correct answers would be:
b, d, a, c, a, c, b, e, c, b, e, a, d, c, c, e, d, c, e, c

34. (a) Answers will vary. One possible diagram is shown below.

(b) Answers will vary. One possible diagram is shown below.

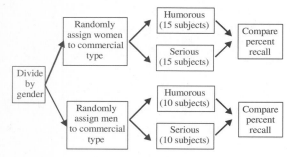

35. A matched-pairs design is an experimental design where experimental units are matched up so they are related in some way.

In a completely randomized design, the experimental units are randomly assigned to one of the treatments. The value of the response variable is compared for each treatment. In a matched-pairs design, experimental units are matched up on the basis of some common characteristic (such as husband-wife or twins). The differences between the matched units are analyzed.

Chapter 1 Test

1. Collect information, organize and summarize the information, analyze the information to draw conclusions, provide a measure of confidence in the conclusions drawn from the information collected.

2. The process of statistics refers to the approach used to collect, organize, analyze, and interpret data. The steps are
(1) Identify the research objective
(2) Collect the data needed to answer the research question.
(3) Describe the data.
(4) Perform inference.

3. Time to complete the 500-meter race in speed skating is at the *ratio* level of measurement because differences between values have meaning and there is a true zero. The variable is quantitative because its values are numerical measurements on which addition and subtraction have meaningful results. The variable is continuous because its values result from a measurement rather than a count.

4. Video game rating is at the *ordinal* level of measurement because its values can be placed in rank order, but differences between values have no meaning. The variable is qualitative because its values classify games based on certain characteristics but arithmetic operations have no meaningful results.

5. The number of surface imperfections is at the *ratio* level of measurement because differences between values have meaning and there is a true zero. The variable is quantitative because its values are numerical measurements on which addition and subtraction have meaningful results. The variable is discrete because its values result from a count.

6. This is an experiment because the researcher intentionally imposed treatments (brand-name battery versus plain-label battery) on individuals (cameras) in a controlled setting. The response variable is the battery life.

7. This is an observational study because no attempt was made to influence the variable of interest. Fan opinions about the asterisk were merely observed. The response variable is whether or not an asterisk should be placed on Barry Bonds' 756th homerun ball.

8. A *cross-sectional study* collects data at a specific point in time or over a short period of time; a *cohort study* collects data over a period of time, sometimes over a long period of time (prospective); a *case-controlled study* is retrospective, looking back in time to collect data.

9. An experiment involves the researcher actively imposing treatments on experimental units in order to observe any difference between the treatments in terms of effect on the response variable. In an observational study, the researcher observes the individuals in the study without attempting to influence the response variable in any way. Only an experiment will allow a researcher to establish causality.

10. A control group is necessary for a baseline comparison. This accounts for the placebo effect which says that some individuals will respond to any treatment. Comparing other treatments to the control group allows the researcher to identify which, if any, of the other treatments are superior to the current treatment (or no treatment at all). Blinding is important to eliminate bias due to the individual or experimenter knowing which treatment is being applied.

11. The steps in conducting an experiment are (1) Identify the problem to be solved, (2) Determine the factors that affect the response variable, (3) Determine the number of experimental units, (4) Determine the level of each factor, (5) Conduct the experiment, and (6) Test the claim.

12. Answers will vary. The franchise locations could be numbered 01 to 15 going across. Starting at row 7, column 14 of Table I in Appendix, and working downward, the selected numbers would be 08, 11, 03, and 02. The corresponding locations would be Ballwin, Chesterfield, Fenton, and O'Fallon.

13. Answers will vary. Using the available lists, obtain a simple random sample from each stratum and combine the results to form the stratified sample. Start at different points in Table I or use different seeds in a random number generator. Using a TI-84 Plus graphing calculator with a seed of 14 for Democrats, 28 for Republicans, and 42 for Independents, the selected numbers would be: Democrats: 3946, 8856, 1398, 5130, 5531,

1703, 1090, and 6369
Republicans: 7271, 8014, 2575, 1150, 1888, 3138, and 2008
Independents: 945, 2855, and 1401

14. Answers will vary. Number the blocks from 1 to 2500 and obtain a simple random sample of size 10. The blocks corresponding to these numbers represent the blocks analyzed. All trees in the selected blocks are included in the sample. Using a TI-84 Plus graphing calculator with a seed of 12, the selected blocks would be numbered 2367, 678, 1761, 1577, 601, 48, 2402, 1158, 1317, and 440.

15. Answers will vary. $\frac{600}{14} \approx 42.86$, so we let $k = 42$. Select a random number between 1 and 42 which represents the first slot machine inspected. Using a TI-84 Plus graphing calculator with a seed of 132, we select machine 18 as the first machine inspected. Starting with machine 18, every 42nd machine thereafter would also be inspected (60, 102, 144, 186, ..., 564).

16. In a completely randomized design, the experimental units are randomly assigned to one of the treatments. The value of the response variable is compared for each treatment. In a randomized block design, the experimental units are first divided according to some common characteristic (such as gender). Then each experimental unit within each block is randomly assigned to one treatment. Within each block the value of the response variable is compared for each treatment, but not between blocks. By blocking, we prevent the effect of the blocked variable from confounding with the treatment.

17. (a) Sampling bias due to voluntary response.

    (b) Nonresponse bias due to the low response rate.

    (c) Response bias due to poorly worded questions.

    (d) Sampling bias due to poor sampling plan (undercoverage).

18. (a) The response variable is the lymphocyte count.

    (b) The treatment is space flight.

(c) This experiment has a matched-pairs design.

(d) The experimental units are the 4 members of Skylab.

(e)

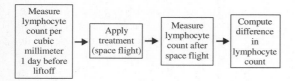

19. (a) This experiment has a completely randomized design.

(b) The response variable is the level of dermatitis improvement.

(c) The factor set to predetermined levels is the topical cream concentration. The treatments are 0.5% cream, 1.0% cream, and a placebo (0% cream).

(d) The study is double-blind if neither the subjects, nor the person administering the treatments, are aware of which topical cream is being applied.

(e) The control group is the placebo (0% topical cream).

(f) The experimental units are the 225 patients with skin irritations.

(g)

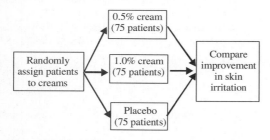

20. (a) This was a cohort study because participants were identified to be included in the study and then followed over a long period of time with data being collected at regular intervals (every 4 years).

(b) The response variable is bone mineral density. The explanatory variable is weekly cola consumption.

(c) The response variable is quantitative because its values are numerical measures on which addition and subtraction can be performed with meaningful results.

(d) The researchers observed values of variables that could potentially impact bone mineral density (besides cola consumption) so their effect could be isolated from the variable of interest.

(e) Answers will vary. Some possible lurking variables that should be accounted for are smoking status, alcohol consumption, physical activity, and calcium intake (form and quantity).

(f) The study concluded that women who consumed at least one cola per day (on average) had a bone mineral density that was significantly lower at the femoral neck than those who consumed less than one cola per day. The study cannot claim that increased cola consumption *causes* lower bone mineral density because it is only an observational study. The researchers can only say that increased cola consumption is *associated* with lower bone mineral density.

Chapter 2

Organizing and Summarizing Data

2.1 Organizing Qualitative Data

1. Raw data are the data as originally collected, before they have been organized or coded.

3. It is a good idea to add up the frequencies in a frequency distribution as a check to see if you missed data or possibly counted the same data more than once. If the total of the frequencies does not equal the total number of data values, the distribution should be done again.

5. A Pareto chart is a bar chart with bars drawn in order of decreasing frequency or relative frequency.

7. Arrange the ordinal data from least to greatest on either the horizontal or vertical axis. For example, if the possible data values are "poor", "fair", "good", or "excellent", then arrange the data in that order. The order of the data in a pie chart cannot be made apparent because pie charts don't have a natural starting point.

9. **(a)** The largest segment in the pie chart is for "Washing your hands" so the most commonly used approach to beat the flu bug is washing your hands. 61% of respondents selected this as their primary method for beating the flu.

   (b) The smallest segment in the pie chart is for "Drinking Orange Juice" so the least used method is drinking orange juice. 2% of respondents selected this as their primary method for beating the flu.

   (c) 25% of respondents felt that flu shots were the best way to beat the flu.

11. **(a)** The largest bar corresponds to the United States, so the U.S. had the most internet users in 2007.

   (b) The bar for Germany appears to reach the line for 50. Thus, we estimate that there were 50 million internet users in Germany in 2007.

   (c) The bar for China appears to reach 160 on the vertical axis. Since, $160 - 50 = 110$, we estimate that there were about 110 million more internet users in China than in Germany during 2007.

13. **(a)** 10.5% of game console owners were planning to buy a replacement device within the next 12 months.

   (b) $250(0.149) = 37.25$

   37.25 million cell phone owners planned to buy a replacement phone within the next 12 months.

   (c) No; the bars are not arranged in decreasing order.

   (d) No; in a relative frequency bar chart, the percents refer to the whole and sum to 1 (or 100%). In this chart, it is possible for someone to own several (if not all) of the listed devices so the percents refer to each category individually.

15. **(a)** In 2000, 55% of parents felt the Internet was a good thing. In 2006, 59% of parents felt this way.

   (b) Good Thing: $59 - 67 = -8$
   No Effect: $30 - 25 = 5$
   Bad Thing: $7 - 5 = 2$
   The opinion "No Effect Either Way" saw the greatest increase from 2004 to 2006.

   (c) No; the decrease of 8% for "Good Thing" corresponded to a 5% increase for "Bad Thing". The difference is accounted for by an increase in the "No Effect" category.

   (d) The percentages within each year may not add to 100% because some participants may not have answered the question (nonresponse). In addition, rounding error could lead to small discrepancies.

17. **(a)** Total students surveyed = 125 + 324 + 552 + 1257 + 2518 = 4776
Relative frequency of "Never"
= 125 / 4776 ≈ 0.0262 and so on.

Response	Relative Frequency
Never	0.0262
Rarely	0.0678
Sometimes	0.1156
Most of the time	0.2632
Always	0.5272

(b) 52.72%

(c) 0.0262 + 0.0678 = 0.0940 or 9.40%

(d)

"How Often Do You Wear Your Seat Belt?"

(e)

"How Often Do You Wear Your Seat Belt?"

(f)

"How Often Do You Wear Your Seat Belt?"

(g) The statement is inferential since it is inferring something about the entire population based on the results of a sample survey.

19. **(a)** Relative frequency of "More than 1 hour a day" = 377 / 1025 ≈ 0.3678 and so on.

Response	Relative Frequency
More than 1 hr a day	0.3678
Up to 1 hr a day	0.1873
A few times a week	0.1288
A few times a month or less	0.0790
Never	0.2371

(b) 0.2371 (about 24%)

(c)

Time Spent Online

(d)

Time Spent Online

(e)

Time Spent Online

(f) The statement provides an estimate, but no level of confidence is given.

21. **(a)** Total males = 92.2 (million)
Relative frequency for "Not HS graduate"

$= \dfrac{13.8}{92.2} \approx 0.1497$ and so on.

Educational Attainment	Males Rel. Freq.
Not a high school graduate	0.1497
High school graduate	0.3189
Some college, but no degree	0.1627
Associate's degree	0.0770
Bachelor's degree	0.1855
Advanced degree	0.1063

(b) Total females = 99.6 (million)
Relative frequency for "Not HS graduate"

$= \dfrac{14.1}{99.6} \approx 0.1416$ and so on.

Educational Attainment	Females Rel. Freq.
Not a high school graduate	0.1416
High school graduate	0.3163
Some college, but no degree	0.1767
Associate's degree	0.0964
Bachelor's degree	0.1817
Advanced degree	0.0873

(c)

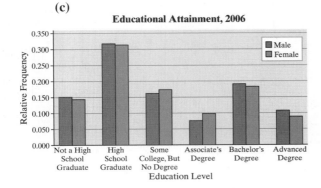

(d) Answers will vary. The percentages are fairly close for not graduating high school and just getting a high school diploma. It appears that men are more likely than women to complete bachelors degrees or higher while women are more likely than men to stop at an Associate's degree or after only a little college.

23. **(a)** Total male victims = 791 + 3762 + 3220 + 2977 + 860 = 11,610
Relative frequency for "Less than 17" =

$\dfrac{791}{11,610} \approx 0.0681$

Age	Relative Frequency
Less than 17	0.0681
17-24	0.3240
25-34	0.2773
35-54	0.2564
55 or older	0.0741

(b) Total female victims = 373 + 550 + 599 + 1102 + 465 = 3089
Relative frequency for "Less than 17" =

$\dfrac{373}{3089} \approx 0.1208$

Age	Relative Frequency
Less than 17	0.1208
17-24	0.1781
25-34	0.1939
35-54	0.3567
55 or older	0.1505

(c)

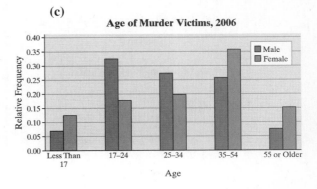

(d) Answers will vary. Children and older victims are more likely to be female. Young adult victims are more likely to be male.

25. **(a), (b)**

Total number of voters polled = 40
Relative frequency for Clinton voters =
$\frac{19}{40} = 0.475$ and so on.

Candidate	Frequency	Rel. Freq.
Clinton	19	0.475
Obama	8	0.2
Edwards	5	0.125
Kucinich	2	0.05
Biden	2	0.05
No Opinion	4	0.1

(c)

(d)

(e)

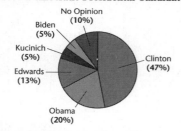

(f) The data indicates that Clinton would win the nomination. The conjecture would be inferential since a conclusion about the entire country is being made from sample data. Your confidence would increase with a larger sample since you have a larger portion of the voters in the country. Note that Obama ultimately won the nomination and went on to win the presidential election with Biden as his running mate.

27. **(a), (b)**

Total number of players surveyed = 25
Relative frequency for "First Base" =
$\frac{5}{25} = 0.20$ and so on.

Position	Freq.	Rel. Freq.
First Base	5	0.20
Second Base	0	0.00
Third Base	1	0.04
Shortstop	3	0.12
Pitcher	5	0.20
Catcher	0	0.00
Right Field	3	0.12
Center Field	2	0.08
Left Field	3	0.12
Designated Hitter	3	0.12

(c) Pitchers and first basemen appear to be more lucrative since they make up the highest percent of the top 25 highest paid players.

(d) Second base and catcher might be avoided since none of the top 25 highest paid players played either position.

(e)

(f)

(g)

Highest Paying Positions, 2007

(e)

**Foreign Languages
Studied by College Students**

29. (a), (b)

Total number of students = 30
Relative frequency for "Chinese"

$$= \frac{3}{30} = 0.100 \text{ and so on.}$$

Language	Freq.	Rel. Frequency
Chinese	3	0.100
French	3	0.100
German	3	0.100
Italian	2	0.067
Japanese	2	0.067
Latin	2	0.067
Russian	1	0.033
Spanish	14	0.467

(c)

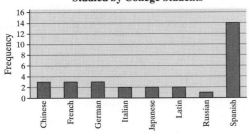

**Foreign Languages
Studied by College Students**

(d)

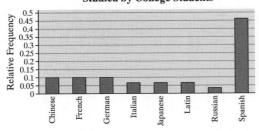

**Foreign Languages
Studied by College Students**

31. (a) It would make sense to draw a pie chart for land area since the 7 continents contain all the land area on Earth.
Total land area is 11,608,000 + 5,100,000 + … + 9,449,000 + 6,879,000 = 57,217,000 square miles
The relative frequency (percentage) for

Africa is $\frac{11,608,000}{57,217,000} = 0.2029$.

Continent	Land Area (mi^2)	Rel. Freq.
Africa	11,608,000	0.2029
Antarctica	5,100,000	0.0891
Asia	17,212,000	0.3008
Australia	3,132,000	0.0547
Europe	3,837,000	0.0671
North America	9,449,000	0.1651
South America	6,879,000	0.1202

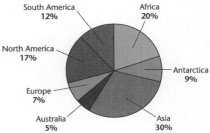

Land Area

(b) It would not make sense to draw a pie chart for the highest elevation because there is no whole to which to compare the parts.

2.2 Organizing Quantitative Data – The Popular Displays

1. Answers will vary. We have already seen that the pattern of a distribution is the same, whether we look at frequencies or relative frequencies. However, in statistics we will often use data from a sample to guide us to conclusions about the larger population from which the sample is drawn. (This is called inferential statistics.) The actual frequencies for a sample do not, by themselves, give much useful information about the population, but the relative frequencies for the sample data will usually be similar to the relative frequencies for the population.

3. A stem-and-leaf plot has the advantage that the raw data can be recovered from the plot whereas it cannot be from a histogram. On the other hand, stem-and-leaf plots are not well suited to large data sets and there is also more flexibility in the choice of classes for a histogram.

5. Classes

7. True; for example, if the class width is too large, the data may be lumped into only one or two classes making interpretation difficult.

9. False. The distribution shape shown has a longer tail to the right so it is skewed right.

11. **(a)** 8 **(b)** 2 **(c)** 15

   (d) $11 - 7 = 4$ **(e)** $\dfrac{15}{100} = 0.15$ or 15%

   (f) The distribution is roughly symmetric.

13. **(a)** Total frequency = 2 + 3 + 13 + 42 + 58 + 40 + 31 + 8 + 2 + 1 = 200

   (b) 10 (e.g. 70 – 60 = 10)

   (c)

IQ Score (class)	Frequency
60–69	2
70–79	3
80–89	13
90–99	42
100–109	58
110–119	40
120–129	31
130–139	8
140–149	2
150–159	1

   (d) The class '100 – 109' has the highest frequency.

   (e) The class '150 – 159' has the lowest frequency.

15. **(a)** Likely skewed right. Most household incomes will be to the left (perhaps in the $50,000 to $150,000 range), with fewer higher incomes to the right (in the millions).

   (b) Likely bell-shaped. Most scores will occur near the middle range, with scores tapering off equally in both directions.

   (c) Likely skewed right. Most households will have, say, 1 to 4 occupants, with fewer households having a higher number of occupants.

   (d) Likely skewed left. Most Alzheimer's patients will fall in older-aged categories, with fewer patients being younger.

17. **(a)** Total frequency $= 16 + 18 + 12 + 3 + 1 = 50$
   Relative frequency of 0 children = 16/50 = 0.32, and so on.

Number of Children Under Five	Relative Frequency
0	0.32
1	0.36
2	0.24
3	0.06
4	0.02

   (b) $\dfrac{12}{50} = 0.24$ or 24% of households have two children under the age of 5.

   (c) $\dfrac{18 + 12}{50} = \dfrac{30}{50} = 0.6$ or 60% of households have one or two children under the age of 5.

19. From the legend, 1|0 represents 10, so the original data set is:
   10, 11, 14, 21, 24, 24, 27, 29, 33, 35, 35, 35, 37, 37, 38, 40, 40, 41, 42, 46, 46, 48, 49, 49, 53, 53, 55, 58, 61, 62

21. From the legend, 1|2 represents 1.2, so the original data set is:
   1.2, 1.4, 1.6, 2.1, 2.4, 2.7, 2.7, 2.9, 3.3, 3.3, 3.3, 3.5, 3.7, 3.7, 3.8, 4.0, 4.1, 4.1, 4.3, 4.6, 4.6, 4.8, 4.8, 4.9, 5.3, 5.4, 5.5, 5.8, 6.2, 6.4

23. (a) 19 classes

(b) Lower class limits: 0, 2000, 4000, 6000, 8000, 10,000, 12,000, 14,000, 16,000, 18,000, 20,000, 22,000, 24,000, 26,000, 28,000, 30,000, 32,000, 34,000, 36,000
Upper class limits: 1999, 3999, 5999, 7999, 9999, 11,999, 13,999, 15,999, 17,999, 19,999, 21,999, 23,999, 25,999, 27,999, 29,999, 31,999, 33,999, 35,999, 37,999

(c) The class width is found by subtracting consecutive lower class limits. For example, $4,000 - 2,000 = 2,000$. Therefore, the class width is 2,000 (dollars).

25. (a) 6 classes

(b) Lower class limits: 15, 20, 25, 30, 35, 40; Upper class limits: 19, 24, 29, 34, 39, 44

(c) The class width is found by subtracting consecutive lower class limits. For example, $20 - 15 = 5$. Therefore, the class width is 5 (years).

27. (a) Total frequency $= 19 + 132 + 308 + \ldots + 46 + 32 + 6 = 1790$
Relative frequency for $0 - 1999 = 19/1790 = 0.0106$ and so on.

Tuition ($)	Rel. Freq.
0-1999	0.0106
2000-3999	0.0737
4000-5999	0.1721
6000-7999	0.1006
8000-9999	0.0592
10,000-11,999	0.0553
12,000-13,999	0.0497
14,000-15,999	0.0547
16,000-17,999	0.0698
18,000-19,999	0.0648
20,000-21,999	0.0654
22,000-23,999	0.0559
24,000-25,999	0.0464
26,000-27,999	0.0296
28,000-29,999	0.0251
30,000-31,999	0.0201
32,000-33,999	0.0257
34,000-35,999	0.0179
36,000-37,999	0.0034

(b)

Four-Year College Tuition

(c)

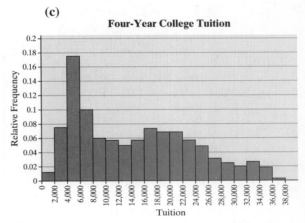

Four-Year College Tuition

Total number of colleges with tuition less than $4000 = 19 + 132 = 151$

$\dfrac{151}{1790} \cdot 100 \approx 8.44\%$ of colleges had tuition of less than $4000.

Total number of colleges with tuition of $30,000 or more $= 36 + 46 + 32 + 6 = 120$

$\dfrac{120}{1790} \cdot 100 \approx 6.70\%$ of colleges had tuition of $30,000 or more.

29. (a) Total births $= 414,406 + 1,040,399 + 1,132,293 + 952,013 + 483,401 + 104,644 = 4,127,156$
Relative frequency for $15\text{-}19 = 414,406/4,127,156 = 0.1004$ and so on.

Age	Relative Frequency
15-19	0.1004
20-24	0.2521
25-29	0.2744
30-34	0.2307
35-39	0.1171
40-44	0.0254

(b)

(c)

The relative frequency is 0.0254, so 2.54% of live births were to women 40-44 years of age.

0.1004 + 0.2521 = 0.3525 so 35.25% of live births were to women 24 years of age or younger.

31. (a) The data are discrete. The possible values for the number of color televisions in a household are countable.

(b), (c)

The relative frequency for 0 color televisions is 1/40 = 0.025, and so on.

Number of Color TVs	Frequency	Relative Frequency
0	1	0.025
1	14	0.350
2	14	0.350
3	8	0.200
4	2	0.050
5	1	0.025

(d) The relative frequency is 0.2 so 20% of the households surveyed had 3 color televisions.

(e) 0.05 + 0.025 = 0.075

7.5% of the households in the survey had 4 or more color televisions.

(f)

(g)

(h) The distribution is skewed right.

33. (a) and **(b)**

Relative frequency for 24,000-26,999 = 8/51 = 0.1569 and so on.

Disposable Income ($)	Freq.	Rel. Freq.
24,000 – 26,999	8	0.1569
27,000 – 29,999	15	0.2941
30,000 – 32,999	12	0.2353
33,000 – 35,999	10	0.1961
36,000 – 38,999	3	0.0588
39,000 – 41,999	2	0.0392
42,000 – 44,999	0	0.0000
45,000 – 47,999	1	0.0196

(c)

(d)

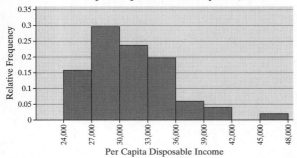

Per Capita Disposable Income by State, 2006

(e) The distribution appears to be skewed right.

(f) Relative frequency for 24,000-27,999 = 10/51 = 0.1961 and so on.

Disposable Income ($)	Freq.	Rel. Freq.
24,000 – 27,999	10	0.1961
28,000 – 31,999	22	0.4314
32,000 – 35,999	13	0.2549
36,000 – 39,999	4	0.0784
40,000 – 43,999	1	0.0196
44,000 – 47,999	1	0.0196

Per Capita Disposable Income by State, 2006

Per Capita Disposable Income by State, 2006

The distribution appears to be skewed right.

(g) Answers will vary. While both distributions indicate the data are skewed right, the first distribution provides a more detailed look at the data. The second distribution is a bit coarser.

35. (a) and **(b)**
Total number of data points = 40
Relative frequency of 20 – 29 = 1/40 = 0.025, and so on.

HDL Cholesterol	Frequency	Relative Frequency
20–29	1	0.025
30–39	6	0.150
40–49	10	0.250
50–59	14	0.350
60–69	6	0.150
70–79	3	0.075

(c)

Serum HDL of 20–29 Year Olds

(d)

Serum HDL of 20–29 Year Olds

(e) The distribution appears to be roughly bell-shaped.

(f) Relative frequency of 25 – 29 = 1/40
= 0.025, and so on.

HDL Cholesterol	Frequency	Relative Frequency
20–24	0	0.000
25–29	1	0.025
30–34	2	0.050
35–39	4	0.100
40–44	2	0.050
45–49	8	0.200
50–54	9	0.225
55–59	5	0.125
60–64	4	0.100
65–69	2	0.050
70–74	3	0.075

The distribution is roughly bell-shaped.

(g) Answers will vary. The first distribution gives a smoother pattern. The additional detail in the second case does not provide much more information.

37. Answers will vary. One possibility follows.

(a) We can determine a class width by subtracting the smallest value from the largest, dividing by the desired number of classes, then rounding up. For example,
$$\frac{23.59 - 6.37}{6} = 2.87 \to 3$$
Our first lower class limit should be a nice number below the smallest data value. In this case, 6 is a good first lower limit since it is the nearest whole number below the smallest data value. Thus, we will

have a class width of 3 and the first class will have a lower limit of 6.

(b), (c) Relative frequency for 6-8.99 = 15/35 = 0.4286 and so on.

Volume	Freq.	Rel. Freq.
6 – 8.99	15	0.4286
9 – 11.99	9	0.2571
12 – 14.99	4	0.1143
15 – 17.99	4	0.1143
18 – 20.99	2	0.0571
21 – 23.99	1	0.0286

(d)

(e)

(f) The distribution is skewed right.

39. (a) **President Ages at Inauguration**

```
4 | 23
4 | 667899
5 | 0011112244444
5 | 555566677778
6 | 0111244
6 | 589
```
Legend: 4 | 2 represents 42 years.

(b) The distribution appears to be roughly symmetric and bell shaped.

41. **(a)**   **Fat in McDonald's Breakfast**

```
0 | 39
1 | 1266
2 | 1224577
3 | 0012267
4 | 6
5 | 159
```
Legend: 5 | 1 represents 51 grams of fat.

(b) The distribution appears to be roughly symmetric and bell shaped.

43. (a) Rounded data:

7.1	12.8	8.2	7.0	12.8	7.6
14.8	10.1	11.1	10.5	7.6	20.7
4.9	7.1	6.5	7.0	6.9	5.4
8.3	11.8	10.0	15.5	8.1	7.0
8.3	6.3	6.9	6.1	9.6	13.8
11.9	7.4	15.3	7.5	6.2	7.7
7.3	6.5	8.7	14.0	7.0	6.7
7.0	10.3	6.0	11.4	6.9	6.1
5.0	8.1	5.3			

Average Electric Rates by State, 2006

```
 4 | 9
 5 | 034
 6 | 01123557999
 7 | 0000011345667
 8 | 112337
 9 | 6
10 | 0135
11 | 1489
12 | 88
13 | 8
14 | 08
15 | 35
16 |
17 |
18 |
19 |
20 | 7
```
Legend: 4 | 9 represents 4.9 cents/kWh.

(b) The distribution is skewed right.

(c) Hawaii's average retail price is 20.7 cents/kWh. Hawaii's rate is so much higher because it is an island far away from the mainland. Resources on the island are limited and importing resources increases the overall cost.

45. **(a)**   **Problems per 100 Vehicles, 2004**

```
16 | 2
17 |
18 | 79
19 | 46
20 | 9
21 | 26
22 | 4
23 |
24 | 0
25 |
26 | 22457
27 | 6
28 | 05589
29 | 578
30 |
31 | 044
32 | 77
33 |
34 | 6
35 |
36 | 5
37 | 5
38 | 6
39 | 3
40 |
41 | 1
42 |
43 | 2
44 |
45 |
46 |
47 | 2
```
Legend: 16 | 2 represents 162 problems.

(b) Number of problems per 100 vehicles, rounded to the nearest tens:

160	190	190	190	200	210
210	220	220	240	260	260
260	270	270	280	280	290
290	290	290	300	300	300
310	310	310	330	330	350
370	380	390	390	410	430
470					

(c)   **Problems per 100 Vehicles, 2004**

```
1 | 6999
2 | 01122466677889999
3 | 0001113357899
4 | 137
```
Legend: 1 | 6 represents 160 problems.

(d)   **Problems per 100 Vehicles, 2004**

```
1 | 6999
2 | 011224
2 | 66677889999
3 | 00011133
3 | 57899
4 | 13
4 | 7
```
Legend: 1 | 6 represents 160 problems.

(e) Answers will vary. The third display (with split stems) seems to be the best. The first display spreads the data too thin, while the second compresses the data too much. In both those cases, it is hard to get a good feel for what the data look like. The third display spreads the data enough to see the main features without diluting the details.

47. (a) **Ages of Academy Award Winners**

Best Actor		Best Actress
9	2	1566899
98877766220	3	0123333455689
76555332200	4	112599
4321	5	
20	6	11
6	7	4
	8	0

(b) Answers will vary. It appears that Academy Award winners for best actor tend to be older on the whole than winners for best actress.

49.

Televisions in Household

Number of Televisions

51. Because the data are quantitative, a stem-and-leaf plot or a histogram would be appropriate. One possible histogram is:

The data appear to be skewed right with a gap and one potential outlier. It seems as if the majority of surfers spent less than one minute viewing the page, while a few surfers spent several minutes viewing the page.

2.3 Additional Displays of Quantitative Data

1. Answers will vary. A stem-and-leaf plot is generally preferred for small data sets since the original data is retained, otherwise a histogram is preferred. A frequency polygon may be preferred if comparing more than one data set.

3. An ogive is a line graph of cumulative frequencies or cumulative relative frequencies against upper class limits.

5. True

7. (a) 10; The class width is the difference between successive class midpoints. e.g. $15 - 5 = 10$
There are 10 classes represented in the graph. The 12 plotted points are for the 10 class midpoints plus two additional points to connect the graph to the horizontal axis on the ends.

(b) The midpoint of the first class is 5. Assuming data values are given in whole numbers, the lower limit of the first class is 0 and the upper limit is 9.

(c) The midpoint of the last class is 95. Assuming data values are given in whole numbers, the lower limit of the last class is 90 and the upper limit is 99.

(d) The highest population occurs in the age group $40 - 49$; this population is about 45 million.

(e) The lowest population occurs in the age group $90 - 99$; this population is about 2 million.

(f) We estimate the total population by adding the estimates for the 10 classes.
$40 + 42 + 42 + 41 + 45 + 39 + 24 + 16 + 9 + 2 = 300$
The approximate total number of U.S. residents below 100 years of age on July 1, 2006 is 300 million.

(g) The population change is most extreme where the line segment is the longest. This occurs between the classes $50 - 59$ and $60 - 69$.
The population change is least extreme where the line segment is the shortest. This occurs between the classes $10 - 19$ and $20 - 29$.

9. (a) 4; The class width is the difference between successive class midpoints. e.g. $8 - 4 = 4$
    There are 9 classes represented in the graph. The 10 plotted points are for the 9 upper class limits plus one additional point to connect the graph to the horizontal axis on the left.

    (b) Since the points of an ogive are plotted at the upper class limits, the lower limit of the first class is 1 and the upper limit is 4.

    (c) Since the points of an ogive are plotted at the upper class limits, the lower limit of the last class is 33 and the upper limit is 36.

    (d) From the graph, it appears that about 76% of students had an ACT composite score of 24 or below.

    (e) From the graph, it appears that about 8% of students had an ACT composite score of 28 or above.

    (f) $(0.76)(1,300,599) \approx 988,455$ students had ACT composite scores of 24 or below.
    $(0.08)(1,300,599) \approx 104,048$ students had ACT composite scores of 28 or above.

11. (a) The closing price at the end of May 2006 was about $45.

    (b) The closing price at the end of December 2007 was about $215.

    (c) $\dfrac{215 - 45}{45} = \dfrac{170}{45} \approx 3.78$
    The price of the stock increased by about 378%.

    (d) $\dfrac{140 - 168}{168} = \dfrac{-28}{168} \approx -0.17$
    The price of the stock decreased by about 17%.

13. (a) For 1992, the unemployment rate was about 7.5% and the inflation rate was about 3.0%.

    (b) For 2006, the unemployment rate was about 4.6% and the inflation rate was about 3.4%.

    (c) $7.5\% + 3.0\% = 10.5\%$
    The misery index for 1992 was 10.5%.
    $4.6\% + 3.4\% = 8.0\%$
    The misery index for 2006 was 8.0%.

    (d) Answers may vary. One possibility: An increase in the inflation rate seems to be followed by an increase in the unemployment rate. Likewise, a decrease in the inflation rate seems to be followed by a decrease in the unemployment rate.

15. (a), (b)
    Second class cumulative frequency = 19 + 132 = 151 and so on.
    Total number of colleges = 19 + 132 + ... + 32 + 6 = 1790
    Second class cumulative relative
    frequency = $\dfrac{151}{1790} = 0.0844$ and so on.

Tuition ($)	Cumul. Freq.	Cumul. Rel. Freq.
0-1999	19	0.0106
2000-3999	151	0.0844
4000-5999	459	0.2564
6000-7999	639	0.3570
8000-9999	745	0.4162
10,000-11,999	844	0.4715
12,000-13,999	933	0.5212
14,000-15,999	1,031	0.5760
16,000-17,999	1,156	0.6458
18,000-19,999	1,272	0.7106
20,000-21,999	1,389	0.7760
22,000-23,999	1,489	0.8318
24,000-25,999	1,572	0.8782
26,000-27,999	1,625	0.9078
28,000-29,999	1,670	0.9330
30,000-31,999	1,706	0.9531
32,000-33,999	1,752	0.9788
34,000-35,999	1,784	0.9966
36,000-37,999	1,790	1

(c)

(d)

(e)

(c)

(d)

(e)

17. (a), (b)

Total births = 414,406 + 1,040,399 +
1,132,293 + 952,013 + 483,401 + 104,644
= 4,127,156
Second class cumulative frequency
= 414,406 + 1,040,399 = 1,454,805 and so
on.
Second class cumulative relative

frequency $= \dfrac{1,454,805}{4,127,156} \approx 0.3525$ and so

on.

Age	Cumul. Freq.	Cumul. Rel. Freq.
15-19	414,406	0.1004
20-24	1,454,805	0.3525
25-29	2,587,098	0.6268
30-34	3,539,111	0.8575
35-39	4,022,512	0.9746
40-44	4,127,156	1

19. (a) and **(b)**

Second class cumulative frequency = 8 + 15
= 23 and so on.
Second class cumulative relative frequency

$= \dfrac{23}{51} \approx 0.4510$ and so on.

Disposable Income ($)	Cumul. Freq.	Cumul. Rel. Freq.
$24,000 - 26,999$	8	0.1569
$27,000 - 29,999$	23	0.4510
$30,000 - 32,999$	35	0.6863
$33,000 - 35,999$	45	0.8824
$36,000 - 38,999$	48	0.9412
$39,000 - 41,999$	50	0.9804
$42,000 - 44,999$	50	0.9804
$45,000 - 47,999$	51	1

(c)

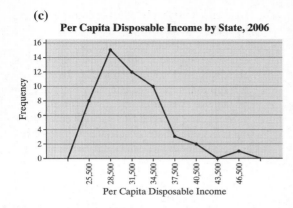

Per Capita Disposable Income by State, 2006

(d)

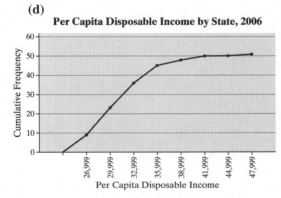

Per Capita Disposable Income by State, 2006

(e)

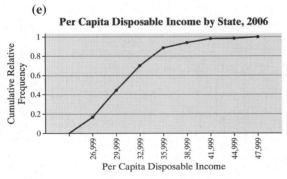

Per Capita Disposable Income by State, 2006

21. The price of Disney stock over the year seemed to fluctuate somewhat, with a general downward trend.

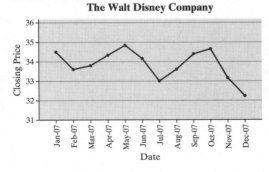

The Walt Disney Company

23. The percent of recent high school graduates enrolling in college seems to have increased slightly over the given time period amid a variety of fluctuations. The early 1990s showed no increase, but this was followed by an unusual jump and decline in the mid-to-late 1990s.

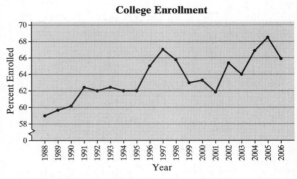

College Enrollment

25. (a) Total frequency for financial stocks = 35

Total frequency for energy stocks = 35

Rate of Return	Frequency (Financial)	Frequency (Retail)
−20 to −0.01	2	4
0 − 19.99	5	2
20 − 39.99	3	10
40 − 59.99	11	4
60 − 79.99	3	3
80 − 99.99	3	3
100 − 119.99	3	2
120 − 139.99	1	1
140 − 159.99	0	2
160 − 179.99	2	0
180 − 199.99	2	2
200 − 219.99	0	0
220 − 239.99	0	2

(b) Relative frequency of -20 to -0.01 for

financial stocks $= \dfrac{2}{35} \approx 0.0571$, and so

on.

Relative frequency of -20 to -0.01 for

retail stocks $= \dfrac{4}{35} \approx 0.1143$, and so on.

Rate of Return	Rel. Freq. (Financial)	Rel. Freq. (Retail)
-20 to -0.01	0.0571	0.1143
$0-19.99$	0.1429	0.0571
$20-39.99$	0.0857	0.2857
$40-59.99$	0.3143	0.1143
$60-79.99$	0.0857	0.0857
$80-99.99$	0.0857	0.0857
$100-119.99$	0.0857	0.0571
$120-139.99$	0.0286	0.0286
$140-159.99$	0	0.0571
$160-179.99$	0.0571	0
$180-199.99$	0.0571	0.0571
$200-219.99$	0	0
$220-239.99$	0	0.0571

(c)

Five-Year Rate of Return on Stocks

(d)

Five-Year Rate of Return on Stocks

(e) Answers will vary. While two of the retail stocks did exceptionally well, the financial stocks seem to have done slightly better overall since there are fewer low-return stocks.

27. Answers will vary. Reports should address the fact that the number of people going to the beach and participating in underwater activities (e.g. scuba diving, snorkeling) has also increased, so an increase in shark attacks is not unexpected. A better comparison would be the rate of attacks per 100,000 beach visitors. The number of fatalities could decrease due to better safety equipment (e.g. bite resistant suits) and better medical care.

2.4 Graphical Misrepresentations of Data

1. The lengths of the bars are not proportional. For example, the bar representing the cost of Clinton's inauguration should be slightly more than 9 times as long as the one for Carter's cost, and twice as long as the bar representing Reagan's cost.

3. (a) The vertical axis starts at 31.5 instead of 0. This tends to indicate that the median earnings for females decreased at a faster rate than actually occurred.

(b) This graph indicates that the median earnings for females has decreased slightly over the given time period.

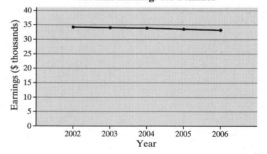

Median Earnings for Females

5. The bar for 12p-6p covers twice as many hours as the other bars. By combining two 3-hour periods, this bar looks larger compared to the others, making afternoon hours look more dangerous. If this bar were split into two periods, the graph may give a different impression. For example, the graph may show that daylight hours are safer.

7. (a) The vertical axis starts at 0.1 instead of 0. This might cause the reader to conclude, for example, that the proportion of people aged 25-34 years old who are not covered by any type of health insurance is more than 4 times the proportion for those aged 55-64 years old.

(b)

Proportion Not Covered by Health Insurance

9. (a) The vertical axis starts at 47 without including a gap.

(b) The graph may be trying to convey that median household income is on the rise after a period of decline.

(c)

U.S. Median Household Income

11. (a) The graphic is misleading because the bars are not proportional. The bar for housing should be a little more than twice the length of the bar for transportation, but it is not.

(b) The graphic could be improved by adjusting the bars so that their lengths are proportional.

13. (a)

ACT Composite Score

This graphic is misleading because the vertical scale starts at 20.7 instead of 0 without indicating a gap. This might cause the reader to think that ACT composite scores are increasing more quickly than they really are.

(b)

ACT Composite Score

(c) The graph in part (a) is preferred because the trend can be seen. ACT composite scores have been increasing, though not as sharply as indicated in the graph.

15. (a) The politician's view:

Health Care as a Percent of GDP

(b) The health care industry's view:

Health Care as a Percent of GDP

(c) A view that is not misleading:

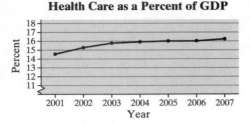

Health Care as a Percent of GDP

U.S. Corn Production

17. (a) A graph that is not misleading will use a vertical scale starting at $0 and bars of equal width. One example:

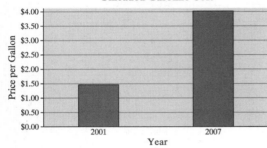

Unleaded Gasoline Cost

(b) A graph that is misleading might use bars of unequal width or will use a vertical scale that does not start at $0. One example, as follows, indicates that the average price for regular unleaded gasoline has increased ten-fold from 2001 to 2007.

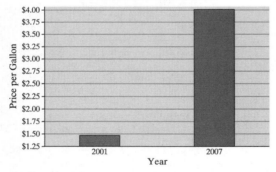

Unleaded Gasoline Cost

19. (a) The graph is a time series plot because the data are plotted in time order.

(b) The graph is too cluttered causing the grid and background to stand out more than the actual data; the axes are not labeled so it is not clear what production measure is being displayed.

(c) The following graph presents the same data, but makes the data stand out and clearly labels the axes.

Chapter 2 Review Exercises

1. (a) The bar for natural gas appears to be halfway between 20 and 25, so it appears that the U.S. consumed about 22.5 quadrillion Btu in energy from natural gas during 2006.

(b) The corresponding bar in the graph is more than halfway between 0 and 5, but not quite to 5. Therefore, we might estimate that the U.S. consumed about 3 quadrillion Btu in energy from biomass during 2006.

(c) $41 + 22.5 + 22.5 + 7.5 + 3 + 3 + 0.5 = 100$
Total energy consumption in the U.S. during 2006 was about 100 quadrillion Btu.

(d) "Other" (including geothermal, wind, and solar) has the lowest frequency.

(e) No; the data are qualitative, so order (and thus skewness) is irrelevant.

2. (a) Total homicides = 10,075 + 1,902 + 609 + 892 + 208 + 119 + 1,040 = 14,845
Relative frequency for Firearms =
$\dfrac{10,075}{14,845} \approx 0.6787$ and so on.

Type of Weapon	Relative Frequency
Firearms	0.6787
Knives or cutting intstruments	0.1281
Blunt objects (clubs, hammers, etc.)	0.0410
Personal weapons (hands, fists, etc.)	0.0601
Strangulation	0.0140
Fire	0.0080
Other weapon or not stated	0.0701

(b) The relative frequency is 0.0410, so 4.1% of the homicides were due to blunt objects.

(c)

(d)

(e)

3. (a), (b), and **(c)**
Total births (in thousands) = 6 + 435 + 1081 + 1182 + 950 + 499 + 105 = 4258
Relative frequency for 10-14 year old mothers = 6/4258 ≈ 0.0014 and so on.
Cumulative frequency for 15-19 year old mothers = 6 + 435 = 441 and so on.
Cumulative relative frequency for 15-19 year old mothers = 441/4258 ≈ 0.1036

and so on.

Age of Mother	Rel. Freq.	Cumul. Freq.	Cumul. Rel. Freq.
10 – 14	0.0014	6	0.0014
15 – 19	0.1022	441	0.1036
20 – 24	0.2539	1522	0.3574
25 – 29	0.2776	2704	0.6350
30 – 34	0.2231	3654	0.8582
35 – 39	0.1172	4153	0.9753
40 – 44	0.0247	4258	1

(d) The distribution is roughly symmetric and bell shaped.

(e)

(f)

(g)

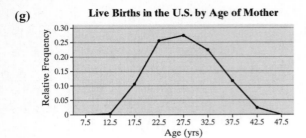

Live Births in the U.S. by Age of Mother

(h)

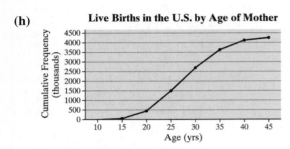

Live Births in the U.S. by Age of Mother

(i)

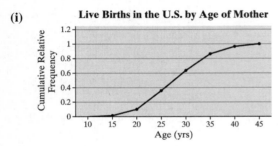

Live Births in the U.S. by Age of Mother

(j) From the relative frequency table, the relative frequency of 20-24 is 0.2539 and so the percentage is 25.39%.

(k) $\dfrac{950 + 499 + 105}{4258} = \dfrac{1554}{4258} \approx 0.3650$

36.5% of live births were to mothers aged 30 years or older.

4. (a) and **(b)**

Affiliation	Frequency	Relative Frequency
Democrat	46	0.46
Independent	16	0.16
Republican	38	0.38

(c)

Political Affiliation

(d)

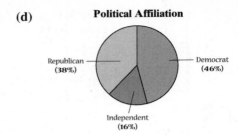

Political Affiliation

(e) Democrat appears to be the most common affiliation in Naperville.

5. (a), (b), (c), and **(d)**

Family Size	Freq.	Rel. Freq.	Cumul. Freq.	Cumul. Rel. Freq.
0	7	0.1167	7	0.1167
1	7	0.1167	14	0.2333
2	18	0.3000	32	0.5333
3	20	0.3333	52	0.8667
4	7	0.1167	59	0.9833
5	1	0.0167	60	1.0000

(e) The distribution is more or less symmetric.

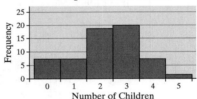

Number of Children for Couples Married 7 Years

(f)

Number of Children for Couples Married 7 Years

(g) From the relative frequency table, the relative frequency of two children is 0.3000 so 30% of the couples have two children.

(h) From the frequency table, the relative frequency of at least two children (i.e. two or more) is $0.3000 + 0.3333 + 0.1167 + 0.0167 = 0.7667$ or 76.67%. So, 76.67% of the couples have at least two children.

(i)

6. (a), (b), (c), and **(d)**

Class	Freq.	Rel. Freq.	Cumul. Freq.	Cumul. Rel. Freq.
1800 – 2199	3	0.0588	3	0.0588
2200 – 2599	3	0.0588	6	0.1176
2600 – 2999	10	0.1961	16	0.3137
3000 – 3399	3	0.0588	19	0.3725
3400 – 3799	7	0.1373	26	0.5098
3800 – 4199	4	0.0784	30	0.5882
4200 – 4599	9	0.1765	39	0.7647
4600 – 4999	6	0.1176	45	0.8824
5000 – 5399	5	0.0980	50	0.9804
5400 – 5799	0	0.0000	50	0.9804
5800 – 6199	0	0.0000	50	0.9804
6200 – 6599	1	0.0196	51	1

(e) The distribution is roughly symmetric.

(f)

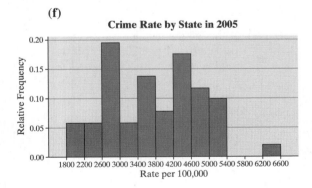

(g)

Class	Freq.	Rel. Freq.	Cumul. Freq.	Cumul. Rel. Freq.
1800 – 2799	8	0.1569	8	0.1569
2800 – 3799	18	0.3529	26	0.5098
3800 – 4799	16	0.3137	42	0.8235
4800 – 5799	8	0.1569	50	0.9804
5800 – 6799	1	0.0196	51	1

Answers will vary. Both class widths give a good overall picture of the distribution. The first class width provides a little more detail to the graph, but not necessarily enough to be worth the trouble. An intermediate value, say a width of 500, might be a reasonable compromise.

7. **(a), (b), (c),** and **(d)**

Answers will vary. Using 2.2000 as the lower class limit of the first class and 0.0200 as the class width, we obtain the following.

Class	Freq.	Rel. Freq.	Cumul. Freq.	Cumul. Rel.Freq.
2.2000 – 2.2199	2	0.0588	2	0.0588
2.2200 – 2.2399	3	0.0882	5	0.1470
2.2400 – 2.2599	5	0.1471	10	0.2941
2.2600 – 2.2799	6	0.1765	16	0.4706
2.2800 – 2.2999	4	0.1176	20	0.5882
2.3000 – 2.3199	7	0.2059	27	0.7941
2.3200 – 2.3399	5	0.1471	32	0.9412
2.3400 – 2.3599	1	0.0294	33	0.9706
2.3600 – 2.3799	1	0.0294	34	1

(e) **Diameter of Chocolate Chip Cookies**

The distribution is roughly symmetric.

(f) **Diameter of Chocolate Chip Cookies**

8. **Hours Spent Online**

```
12 |
13 | 467
14 | 05578
15 | 1236
16 | 456
17 | 113449
18 | 066889
19 | 2
20 | 168
21 | 119
22 | 29
23 | 48
24 | 4
25 | 7
26 |
```

Legend: 13 | 4 = average 13.4 hours per week.

The distribution is slightly skewed right.

9. **(a)**

The incident rate decreased dramatically over the given time period.

(b)

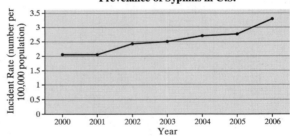

(c) Although the incident rate declined between 1990 and 2000, it increases gradually between 2002 and 2006. The plan needed to be adjusted to deal with this increase.

(d) No; a histogram will not allow the observer to see trends over time, only the distribution of yearly incident rates. To see a trend over time, a time series plot should be constructed.

10. The graph is misleading because there is no vertical scale.

11. (a) Graphs will vary. One way to mislead would be to start the vertical scale at a value other than 0. For example, starting the vertical scale at $20,000 might make the reader believe that college graduates earn more than three times what a high school graduate earns (on average).

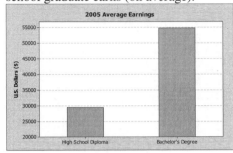

(b) A graph that does not mislead would use equal widths for the bars and would start the vertical scale at $0.

2005 Average Earnings

12. (a) Flats are preferred the most (40%) and extra-high heels are preferred the least (1%).

(b) The graph is misleading because the bar heights and areas for each category are not proportional.

Chapter 2 Test

1. (a) The United States won the most men's singles championships between 1968 and 2007 with 15 wins.

(b) $6 - 4 = 2$
Representatives from Australia have won 2 more championships than representatives from Germany.

(c) $15 + 7 + 6 + 5 + 4 + 1 + 1 + 1 = 40$ championships between 1968 and 2007.

$$\frac{7}{40} = 0.175$$

Representatives of Sweden won 17.5% of the championships between 1968 and 2007.

(d) No, it is not appropriate to describe the shape of the distribution as skewed right. The data represented by the graph are qualitative so the bars in the graph could be placed in any order.

2. (a) Total emissions = 5,934.4 + 605.1 + 378.6 + 157.6 = 7,075.7 million metric tons
Relative frequency of carbon dioxide = 5934.4/7075.7 ≈ 0.8387, and so on.

Gas	Relative Frequency
Carbon Dioxide	0.8387
Methane	0.0855
Nitrous Oxide	0.0535
Hydrofluorocarbons, perfluorocarbons, and sulfur hexafluoride	0.0223

(b) The relative frequency is 0.8387 so the percentage of emissions due to carbon dioxide is 83.87%.

(c)

(d)

(e)

U.S. Greenhouse Emissions

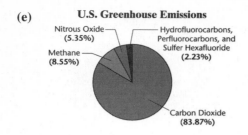

Nitrous Oxide (5.35%)
Hydrofluorocarbons, Perfluorocarbons, and Sulfer Hexafluoride (2.23%)
Methane (8.55%)
Carbon Dioxide (83.87%)

3. (a), (b)

Education	Freq.	Rel. Freq.
No high school diploma	9	0.18
High school graduate	16	0.32
Some college	9	0.18
Associate's degree	4	0.08
Bachelor's degree	8	0.16
Advanced degree	4	0.08

(c)

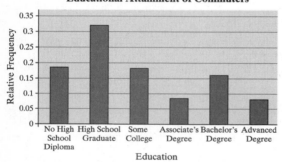

Educational Attainment of Commuters

(d) **Educational Attainment of Commuters**

Advanced Degree (8%)
Bachelor's Degree (16%)
Associate's Degree (8%)
Some College (18%)
No High School Degree (18%)
High School Graduate (32%)

(e) The largest bar (and largest pie segment) corresponds to 'High School Graduate', so high school graduate is the most common educational level of a commuter.

4. (a), (b), (c), and (d)

No. of Cars	Freq.	Rel. Freq.	Cumul. Freq.	Cumul. Rel. Freq.
1	5	0.10	5	0.10
2	7	0.14	12	0.24
3	12	0.24	24	0.48
4	6	0.12	30	0.60
5	8	0.16	38	0.76
6	5	0.10	43	0.86
7	2	0.04	45	0.90
8	4	0.08	49	0.98
9	1	0.02	50	1

(e) **Number of Cars Arriving at McDonald's**

The distribution is skewed right.

(f) **Number of Cars Arriving at McDonald's**

(g) The relative frequency of exactly 3 cars is 0.24. So, for 24% of the weeks, exactly three cars arrived between 11:50am and 12:00 noon.

(h) The relative frequency of 3 or more cars
= 1 − the relative frequency of 1 or 2 cars
= 1 − (0.10 + 0.14) = 0.76 . So, for 76% of the weeks, three or more cars arrived between 11:50am and 12:00 noon.

(i)

5. Answers may vary. One possibility follows:

  (a) We can determine a class width by subtracting the smallest value from the largest, dividing by the desired number of classes, then rounding up. For example,

$$\frac{459 - 100}{8} = \frac{359}{8} = 44.875 \to 50$$

Our first lower class limit should be a nice number below the smallest data value. In this case, 100 is a good first lower limit. Thus, we will have a class width of 50 grams and the first class will have a lower limit of 100 grams.

  (b), (c)
Relative frequency for 100-149 = 11/40 = 0.275 and so on.

Volume	Freq.	Rel. Freq.
100 – 149	11	0.275
150 – 199	7	0.175
200 – 249	9	0.225
250 – 299	8	0.200
300 – 349	1	0.025
350 – 399	2	0.050
400 – 449	1	0.025
450 – 499	1	0.025

  (d) The distribution is skewed right.

  (e)

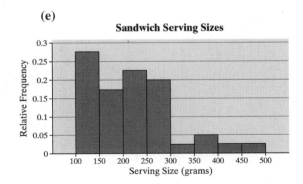

6. (a), (b)

Closing Price	Cumul. Freq.	Cumul. Rel. Freq.
50,000 – 99,999	4	0.08
100,000 – 149,999	17	0.34
150,000 – 199,999	36	0.72
200,000 – 249,999	43	0.86
250,000 – 299,999	46	0.92
300,000 – 349,999	48	0.96
350,000 – 399,999	49	0.98
400,000 – 449,999	49	0.98
450,000 – 499,999	50	1

  (c) The cumulative relative frequency for the class $150,000-$199,000 is 0.72. Therefore, 72% of the homes sold for less than $200,000.

  (d)

The distribution is skewed right.

  (e)

  (f)

7. (a) Rounded values:

67	67	68	57	78	65	68	64
77	59	69	62	53	75	79	76
82	67	73	69	64	66	94	
72	72	67	76	75	61	52	
72	74	71	68	61	57	66	
70	81	55	70	69	70	65	
59	67	64	75	69	78	59	

Fertility Rates by State, 2006

```
5 | 23
5 | 577999
6 | 112444
6 | 5566777778889999
7 | 000122234
7 | 555667889
8 | 12
8 |
9 | 4
```

Legend: 5 | 2 represents 52 births per 1000 women aged 15 to 44.

(b) The distribution is fairly symmetric, but does appear to be slightly skewed right.

8.

U.S. Average Milk Price

The price of milk decreased slightly from January 2006 to July 2006. It then increased slowly at first, then rapidly over the summer of 2007. Around late summer 2007 it appears that the price is leveled off.

9. Answers may vary. It is difficult to interpret this graph because it is not clear whether the scale is represented by the height of the steps, the width of the steps, or by the graphics above the steps. The graphics are misleading because they must be increased in size both vertically and horizontally to avoid distorting the image. Thus, the resulting areas are not proportionally correct. The graph could be redrawn using bars whose widths are the same and whose heights are proportional based on the given percentages. The use of graphics should be avoided, or a standard size graphic representing a fixed value could be used and repeated as necessary to illustrate the given percentages.

Chapter 3

Numerically Describing Data from One Variable

Section 3.1

1. A statistic is resistant if it is not sensitive to extreme data values. The median is resistant because it is a positional measure of central tendency and increasing the largest value or decreasing the smallest value does not affect the position of the center. The mean is not resistant because it is a function of the sum of the data values. Changing the magnitude of one value changes the sum of the values, and thus affects the mean.

3. HUD uses the median because the data are skewed. Explanations will vary. One possibility is that the price of homes has a distribution that is skewed to the right, so the median is more representative of the typical price of a home.

5. $\frac{10,000+1}{2} = 5000.5$. The median is between the 5000th and the 5001st ordered values.

7. $\bar{x} = \frac{20+13+4+8+10}{5} = \frac{55}{5} = 11$

9. $\mu = \frac{3+6+10+12+14}{5} = \frac{45}{5} = 9$

11. $\frac{162.5}{65} = 2.5$

    The mean price per ad slot was $2.5 million.

13. $\sum x_i = 976 + 2038 + 918 + 1899 = 5831$

    Mean = $\bar{x} = \frac{\sum x_i}{n} = \frac{5831}{4} = \$1457.75$

    Data in order: 918, 976, 1899, 2038

    Median = $\frac{976+1899}{2} = \frac{2875}{2} = \$1437.50$

    No data value occurs more than once, so there is no mode.

15. $\sum x_i = 3960 + 4090 + 3200 + 3100 + 2940 + 3830$
    $+ 4090 + 4040 + 3780$
    $= 33,030$ psi

    Mean = $\bar{x} = \frac{\sum x_i}{n} = \frac{33,030}{9} = 3670$ psi

Data in order: 2940, 3100, 3200, 3780, 3830, 3960, 4040, 4090, 4090
Median = the 5th ordered data value = 3830 psi
Mode = 4090 psi (because it is the only data value to occur twice)

17. (a) The histogram is skewed to the right, suggesting that the mean is greater than the median. That is, $\bar{x} > M$.
    (b) The histogram is symmetric, suggesting that the mean is approximately equal to the median. That is, $\bar{x} = M$.
    (c) The histogram is skewed to the left, suggesting that the mean is less than the median. That is, $\bar{x} < M$.

19. Answers will vary.

21. (a) Tap Water:
    $\sum x_i = 7.64 + 7.45 + 7.47 + 7.50 + 7.68 + 7.69$
    $+ 7.45 + 7.10 + 7.56 + 7.47 + 7.52 + 7.47$
    $= 90.00$

    Mean = $\bar{x} = \frac{\sum x_i}{n} = \frac{90.00}{12} = 7.50$

    Data in order: 7.10, 7.45, 7.45, 7.47, 7.47, 7.47, 7.50, 7.52, 7.56, 7.64, 7.68, 7.69

    Median = $\frac{7.47+7.50}{2} = \frac{14.97}{2} = 7.485$

    Mode = 7.47 (because it occurs three times, which is the most)

    Bottled Water:
    $\sum x_i = 5.15 + 5.09 + 5.26 + 5.20 + 5.02 + 5.23$
    $+ 5.28 + 5.26 + 5.13 + 5.26 + 5.21 + 5.24$
    $= 62.33$

    Mean = $\frac{\sum x_i}{n} = \frac{62.33}{12} \approx 5.194$

    Data in order: 5.02, 5.09, 5.13, 5.15, 5.20, 5.21, 5.23, 5.24, 5.26, 5.26, 5.26, 5.28

    Median = $\frac{5.21+5.23}{2} = \frac{10.44}{2} = 5.22$

    Mode = 5.26 (because it occurs three times, which is the most)

    The pH of the sample of tap water is substantially higher than the pH of the sample of bottled water.

(b) $\sum x_i = 7.64 + 7.45 + 7.47 + 7.50 + 7.68 + 7.69$
$\qquad + 7.45 + 1.70 + 7.56 + 7.47 + 7.52 + 7.47$
$\qquad = 84.60$

$\text{Mean} = \overline{x} = \dfrac{\sum x_i}{n} = \dfrac{84.60}{12} = 7.05$

Data in order: 1.70, 7.45, 7.45, 7.47, 7.47, 7.47, 7.50, 7.52, 7.56, 7.64, 7.68, 7.69

$\text{Median} = \dfrac{7.47 + 7.50}{2} = \dfrac{14.97}{2} = 7.485$

The incorrect entry causes the mean to decrease substantially while the median does not change. This illustrates that the median is resistant while the mean is not resistant.

23. (a) $\sum x_i = 76 + 60 + 60 + 81 + 72 + 80 + 80$
$\qquad + 68 + 73$
$\qquad = 650$

$\mu = \dfrac{\sum x_i}{N} = \dfrac{650}{9} \approx 72.2$ beats per minute

(b) Samples and sample means will vary.

(c) Answers will vary.

25. (a) $\sum x_i = 6{,}049{,}435 + 5{,}010{,}170 + 1{,}524{,}993$
$\qquad + 1{,}342{,}962 + 1{,}257{,}963 + 808{,}767$
$\qquad + 639{,}403 + 587{,}261 + 465{,}643$
$\qquad + 449{,}948$
$\qquad = 18{,}136{,}545$

$\mu = \dfrac{\sum x_i}{N} = \dfrac{18{,}136{,}545}{10} = 1{,}813{,}654.5$
thousand metric tons.

(b) Per capita is a better gauge because it adjusts for CO_2 emissions for population. After all, countries with more people will, in general, have higher CO_2 emissions.

(c) $\sum x_i = 5.61 + 1.05 + 2.89 + 0.34 + 2.69 + 2.67$
$\qquad + 5.46 + 2.67 + 2.64 + 2.12$
$\qquad = 28.14$

$\mu = \dfrac{28.14}{10} = 2.814$ thousand metric tons

Data in order: 0.34, 1.05, 2.12, 2.64, 2.67, 2.67, 2.69, 2.89, 5.46, 5.61

$\text{Median} = \dfrac{2.67 + 2.67}{2} = \dfrac{5.34}{2} = 2.67$

thousand metric tons.
The environmentalist will likely use the mean to support the position that CO_2 emissions are too high because the mean is higher.

27. The distribution is relatively symmetric as is evidenced by both the histogram and the fact that the mean and median are approximately equal. Therefore, the mean is the better measure of central tendency.

29. To create the histogram, we choose the lower class limit of the first class to be 0.78 and the class width to be 0.02. The resulting classes and frequencies follow:

Class	Freq.	Class	Freq.
0.78 – 0.79	1	0.88 – 0.89	10
0.80 – 0.81	1	0.90 – 0.91	9
0.82 – 0.83	4	0.92 – 0.93	4
0.84 – 0.85	8	0.94 – 0.95	2
0.86 – 0.87	11		

The histogram follows:

Weight of Plain M&Ms

To find the mean, we add all of the data values and divide by the sample size: $\sum x_i = 43.73$;

$\overline{x} = \dfrac{\sum x_i}{n} = \dfrac{43.73}{50} \approx 0.875$ grams

To find the median, we arrange the data in order. The median is the mean of the 25th and 26th data values: $M = \dfrac{0.87 + 0.88}{2} = 0.875$ grams

The mean is approximately equal to the median suggesting that the distribution is symmetric. This is confirmed by the histogram (though is does appear to be slightly skewed left). The mean is the better measure of central tendency.

31. To create the histogram, we choose the lower class limit of the first class to be 0 and the class width to be 5. The resulting classes and frequencies follow:

Class	Freq.	Class	Freq.
0 – 4	3	25 – 29	4
5 – 9	1	30 – 34	5
10 – 14	0	35 – 39	3
15 – 19	4	40 – 44	1
20 – 24	4		

The histogram follows:

To find the mean, we add all of the data values and divide by the sample size: $\sum x_i = 550$;

$$\bar{x} = \frac{\sum x_i}{n} = \frac{550}{25} = 22 \text{ hours}$$

To find the median, we arrange the data in order. The median is the 13th data value: $M = 25$ hours

The mean is smaller than the median suggesting that the distribution is skewed left. This is confirmed by the histogram. The median is the better measure of central tendency.

33. (a) The frequencies are:
Liberal = 10
Moderate = 12
Conservative = 8
The mode political view is Moderate.
(b) Yes. Rotating the choices will help to avoid response bias that might be caused by the wording of the question.

35. Sample size of 5:
All data recorded correctly: $\bar{x} = 99.8$; $M = 100$.
106 recorded at 160: $\bar{x} = 110.6$; $M = 100$.
Sample size of 12:
All data recorded correctly: $\bar{x} \approx 100.4$; $M = 101$.
106 recorded at 160: $\bar{x} \approx 104.9$; $M = 101$.
Sample size of 30:
All data recorded correctly: $\bar{x} = 100.6$; $M = 99$.
106 recorded at 160: $\bar{x} = 102.4$; $M = 99$.
For each sample size, the mean becomes larger while the median remains the same. As the sample size increases, the impact of the incorrectly recorded data value on the mean decreases.

37. No. Each state has a different population size of households who use natural gas. Each state's mean natural gas bill should be weighted by its population size in order to obtain an overall mean natural gas bill for the entire U.S.

39. NBA salaries are likely significantly skewed to the right. Therefore, since the median will be lower than the mean, the players would rather use the median salary to support the claim that the average player's salary need to be increases. The negotiator for the owners would rather use the mean salary.

41. The sum of the nineteen readable scores is $19 \cdot 84 = 1596$. The sum of all twenty scores is $20 \cdot 82 = 1640$. Therefore, the unreadable score is $1640 - 1596 = 44$.

43. (a) Mean:
$$\sum x_i = 30 + 30 + 45 + 50 + 50 + 50 + 55 + 55$$
$$+ 60 + 75$$
$$= 500$$
$$\mu = \frac{\sum x_i}{N} = \frac{500}{10} = 50$$
The mean is $50,000.

Median: The ten data values are in order. The median is the mean of the 5th and 6th data values: $M = \frac{50 + 50}{2} = \frac{100}{2} = 50$.
The median is $50,000.

Mode: The mode is $50,000 (the most frequent salary).

(b) Add $2500 ($2.5 thousand) to each salary to form the 2nd: New data set: 32.5, 32.5, 47.5, 52.5, 52.5, 52.5, 57.5, 57.5, 62.5, 77.5

2nd Mean:
$$\sum x_i = 32.5 + 32.5 + 47.5 + 52.5 + 52.5$$
$$+ 52.5 + 57.5 + 57.5 + 62.5 + 77.5$$
$$= 525$$
$$\mu_{2nd} = \frac{\sum x_i}{N} = \frac{525}{10} = 52.5 \quad \text{The mean for}$$
the 2nd data set is $52,500.

2nd Median: The ten data values are in order. The median is the mean of the 5th and 6th data values: $M_{2nd} = \frac{52.5 + 52.5}{2} = \frac{105}{2} = 52.5$.
The median for the 2nd data set is $52,500.

2nd Mode: The mode for the 2nd data set is $52,500 (the most frequent new salary).

All three measures of central tendency increased by $2500, which was the amount of the raises.

(c) Multiply each original data value by 1.05 to generate the 3rd data set: 31.5, 31.5, 47.25, 52.5, 52.5, 52.5, 57.75, 57.75, 63, 78.75

3rd Mean:
$$\sum x_i = 31.5 + 31.5 + 47.25 + 52.5 + 52.5$$
$$+ 52.5 + 57.75 + 57.75 + 63 + 78.75$$
$$= 525$$
$$\mu_{3rd} = \frac{\sum x_i}{N} = \frac{525}{10} = 52.5$$
The mean of the 3rd data set is $52,500.

3rd Median: The ten data values are in order. The median is the mean of the 5th and 6th data values:
$$M_{3rd} = \frac{52.5 + 52.5}{2} = \frac{105}{2} = 52.5 .$$
The median of the 3rd data set is $52,500.

3rd Mode: The mode of the 3rd data set is $52,500 (the most frequent new salary).

All three measures of central tendency increased by 5%, which was the amount of the raises.

(d) Add $25 thousand to the largest data value to form the new data set: 30, 30, 45, 50, 50, 50, 55, 55, 60, 100

4th Mean:
$$\sum x_i = 30 + 30 + 45 + 50 + 50 + 50 + 55 + 55$$
$$+ 60 + 100$$
$$= 525$$
$$\mu_{4th} = \frac{\sum x_i}{N} = \frac{525}{10} = 52.5$$
The mean of the 4th data set is $52,500.

4th Median: The ten data values are in order. The median is the mean of the 5th and 6th data values: $M_{4th} = \frac{50 + 50}{2} = \frac{100}{2} = 50 .$
The median of the 4th data set is $50,000.
4th Mode: The mode of the 4th data set is $50,000 (the most frequent salary).
The mean was increased by $2500, but the median and mode remained unchanged.

45. The largest value is 0.95 and the smallest is 0.79. After deleting these two values, we have:
$$\sum x_i = 41.99; \quad \overline{x} = \frac{\sum x_i}{n} = \frac{41.99}{48} \approx 0.875 \text{ grams} .$$
The mean after deleting these two data values is 0.875 grams. The trimmed mean is more resistant than the regular mean.

47. (a) The data are discrete.

(b) To construct the histogram, we first organize the data into a frequency table:

Number of Drinks	Frequency
0	23
1	17
2	4
3	3
4	2
5	1

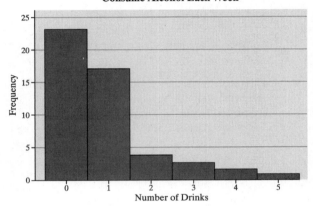

Number of Days High School Students Consume Alcohol Each Week

The distribution is skewed right.

(c) Since the distribution is skewed right, we would expect the mean to be greater than the median.

(d) To find the mean, we add all of the data values and divide by the sample size.
$$\sum x_i = 47; \quad \overline{x} = \frac{\sum x_i}{n} = \frac{47}{50} = 0.94$$

To find the median, we arrange the data in order. The median is the mean of the 25th and 26th data values. $M = \frac{1+1}{2} = 1 .$

This tells us that thee mean can be less than the median in skewed-right data. Therefore, the rule *mean greater than median implies the data are skewed right* is not always true.

(e) The mode is 0 (the most frequent value).

(f) Yes, Carlos' survey likely suffers from sampling bias. It is difficult to get truthful responses to this type of question. Carlos would need to ensure that the identity of the respondents is anonymous.

Section 3.2

1. No. In comparing two populations, the larger the standard deviation, the more dispersed the distribution, provided that the variable of interest in both populations has the same unit of measurement. Since 5 inches $\approx 5 \times 2.54 = 12.7$ centimeters, the distribution with a standard deviation of 5 inches is, in fact, more dispersed.

3. No, none of the measures of dispersion mentioned in this section – range, variance, and standard deviation – are resistant. The lowest and highest values are used to compute the range. All data values, including extreme values, are used in computing the variance and standard deviation. Since a statistic is resistant only if it is not influenced by extreme data values, they are not resistant.

5. A statistic is biased whenever it consistently overestimates or underestimates the population parameter.

7. Because male heights differ from female heights, this leads to a higher standard deviation when they are combined in a single group than when they segregated into separate groups.

9. True

11. From Section 3.1, Exercise 7, we know $\bar{x} = 11$.

x_i	$\bar{x}$	$x_i - \bar{x}$	$(x_i - \bar{x})^2$
20	11	$20 - 11 = 9$	$9^2 = 81$
13	11	$13 - 11 = 2$	$2^2 = 4$
4	11	$4 - 11 = -7$	$(-7)^2 = 49$
8	11	$8 - 11 = -3$	$(-3)^2 = 9$
10	11	$10 - 11 = -1$	$(-1)^2 = 1$
		$\sum(x_i - \bar{x}) = 0$	$\sum(x_i - \bar{x})^2 = 144$

$$s^2 = \frac{\sum(x_i - \bar{x})^2}{n-1} = \frac{144}{5-1} = \frac{144}{4} = 36$$

$$s = \sqrt{s^2} = \sqrt{\frac{\sum(x_i - \bar{x})^2}{n-1}} = \sqrt{\frac{144}{5-1}} = \sqrt{36} = 6$$

13. From Section 3.1, Exercise 9, we know $\mu = 9$.

x_i	μ	$x_i - \mu$	$(x_i - \mu)^2$
3	9	$3 - 9 = -6$	$(-6)^2 = 36$
6	9	$6 - 9 = -3$	$(-3)^2 = 9$
10	9	$10 - 9 = 1$	$1^2 = 1$
12	9	$12 - 9 = 3$	$3^2 = 9$
14	9	$14 - 9 = 5$	$5^2 = 25$
		$\sum(x_i - \mu) = 0$	$\sum(x_i - \mu)^2 = 80$

$$\sigma^2 = \frac{\sum(x_i - \mu)^2}{N} = \frac{80}{5} = 16$$

$$\sigma = \sqrt{\sigma^2} = \sqrt{\frac{\sum(x_i - \mu)^2}{N}} = \sqrt{\frac{80}{5}} = \sqrt{16} = 4$$

15. $\bar{x} = \dfrac{6 + 52 + 13 + 49 + 35 + 25 + 31 + 29 + 31 + 29}{10}$

$= \dfrac{300}{10} = 30$

x_i	$\bar{x}$	$x_i - \bar{x}$	$(x_i - \bar{x})^2$
6	30	$6 - 30 = -24$	$(-24)^2 = 576$
52	30	$52 - 30 = 22$	$22^2 = 484$
13	30	$13 - 30 = -17$	$(-17)^2 = 289$
49	30	$49 - 30 = 19$	$19^2 = 361$
35	30	$35 - 30 = 5$	$5^2 = 25$
25	30	$25 - 30 = -5$	$(-5)^2 = 25$
31	30	$31 - 30 = 1$	$1^2 = 1$
29	30	$29 - 30 = -1$	$(-1)^2 = 1$
31	30	$31 - 30 = 1$	$1^2 = 1$
29	30	$29 - 30 = -1$	$(-1)^2 = 1$
		$\sum(x_i - \bar{x}) = 0$	$\sum(x_i - \bar{x})^2 = 1764$

$$s^2 = \frac{\sum(x_i - \bar{x})^2}{n-1} = \frac{1764}{10-1} = \frac{1764}{9} = 196$$

$$s = \sqrt{s^2} = \sqrt{\frac{\sum(x_i - \bar{x})^2}{N}} = \sqrt{\frac{1764}{10-1}} = \sqrt{196} = 14$$

17. Range = Largest Value – Smallest Value
    $= 2038 - 918 = \$1120$.

    To calculate the sample variance and the sample standard deviation, we use the computational formulas:

x_i	x_i^2
976	952,576
2038	4,153,444
918	842,724
1899	3,606,201
$\sum x_i = 5831$	$\sum x_i^2 = 9,554,945$

$$s^2 = \frac{\sum x_i^2 - \frac{(\sum x_i)^2}{n}}{n-1}$$

$$= \frac{9{,}554{,}945 - \frac{(5831)^2}{4}}{4-1} \approx 351{,}601.6 \ (\$)^2$$

$$s = \sqrt{s^2} = \sqrt{\frac{9{,}554{,}945 - \frac{(5831)^2}{4}}{4-1}} \approx \$593.0$$

19. Range = Largest Value − Smallest Value
    = 4090 − 2940 = 1150 psi

To calculate the sample variance and the sample standard deviation, we use the computational formulas:

x_i	x_i^2
3960	15,681,600
4090	16,728,100
3200	10,240,000
3100	9,610,000
2940	8,643,600
3830	14,668,900
4090	16,728,100
4040	16,321,600
3780	14,288,400
$\sum x_i = 33{,}020$	$\sum x_i^2 = 122{,}910{,}300$

$$s^2 = \frac{\sum x_i^2 - \frac{(\sum x_i)^2}{n}}{n-1}$$

$$= \frac{122{,}910{,}300 - \frac{(33{,}030)^2}{9}}{9-1} \approx 211{,}275 \ (\text{psi})^2$$

$$s = \sqrt{\frac{122{,}910{,}300 - \frac{(33{,}030)^2}{9}}{9-1}} \approx 459.6 \ \text{psi}$$

21. Histogram (b) depicts a higher standard deviation because the data is more dispersed, with data values ranging from 30 to 75. In Histogram (a), the data values only range from 40 to 60.

23. (a) pH of tap water:
    Range = Largest Value − Smallest Value
        = 7.69 − 7.10 = 0.59

pH of bottled water:
    Range = Largest Value − Smallest Value
        = 5.28 − 5.02 = 0.26

Using range as the measure, the pH of tap water has more dispersion.

(b) To calculate the sample standard deviation, we use the computational formula:

pH of tap water:

x_i	x_i^2
7.64	58.3696
7.45	55.5025
7.47	55.8009
7.50	56.25
7.68	58.9824
7.69	59.1361
7.45	55.5025
7.10	50.41
7.56	57.1536
7.47	55.8009
7.52	56.5504
7.47	55.8009
$\sum x_i = 90$	$\sum x_i^2 = 675.2598$

$$s = \sqrt{\frac{\sum x_i^2 - \frac{(\sum x_i)^2}{n}}{n-1}}$$

$$= \sqrt{\frac{675.2598 - \frac{(90)^2}{12}}{12-1}} \approx 0.154$$

pH of bottled water:

x_i	x_i^2
5.15	26.5225
5.09	25.9081
5.26	27.6676
5.20	27.04
5.02	25.2004
5.23	27.3529
5.28	27.8784
5.26	27.6676
5.13	26.3169
5.26	27.6676
5.21	27.1441
5.24	27.4576
$\sum x_i = 62.33$	$\sum x_i^2 = 323.8237$

$$s = \sqrt{\frac{\sum x_i^2 - \frac{(\sum x_i)^2}{n}}{n-1}}$$

$$= \sqrt{\frac{323.8237 - \frac{(62.33)^2}{12}}{12-1}} \approx 0.081$$

Using standard deviation as the measure, the pH of tap water has more dispersion.

25. **(a)** We use the computational formula:

x_i	x_i^2
76	5776
60	3600
60	3600
81	6561
72	5184
80	6400
80	6400
68	4624
73	5329
$\sum x_i = 650$	$\sum x_i^2 = 47,474$

$$\sigma = \sqrt{\frac{\sum x_i^2 - \frac{(\sum x_i)^2}{N}}{N}}$$

$$= \sqrt{\frac{47,474 - \frac{(650)^2}{9}}{9}} \approx 7.7 \text{ beats/minute}$$

(b) Samples and sample standard deviations will vary.

(c) Answers will vary.

27. **(a)** Ethan:

$$\sum x_i = 9 + 24 + 8 + 9 + 5 + 8 + 9 + 10 + 8 + 10$$
$$= 100$$

$$\mu = \frac{\sum x_i}{N} = \frac{100}{10} = 10 \text{ fish}$$

Range = Largest Value – Smallest Value
$$= 24 - 5 = 19 \text{ fish}$$

Drew:

$$\sum x_i = 15 + 2 + 3 + 18 + 20 + 1 + 17 + 2 + 19 + 3$$
$$= 100$$

$$\mu = \frac{\sum x_i}{N} = \frac{100}{10} = 10 \text{ fish}$$

Range = Largest Value – Smallest Value
$$= 20 - 1 = 19 \text{ fish}$$

Both fishermen have the same mean and range, so these values do not indicate any differences between their catches per day.

(b) Ethan:

$$N = 10; \quad \sum x_i = 100$$

$$\sum x_i^2 = 9^2 + 24^2 + 8^2 + 9^2 + 5^2 + 8^2 + 9^2$$
$$+ 10^2 + 8^2 + 10^2$$
$$= 1236$$

$$\sigma = \sqrt{\frac{\sum x_i^2 - \frac{(\sum x_i)^2}{N}}{N}} = \sqrt{\frac{1236 - \frac{(100)^2}{10}}{10}}$$

$$\approx 4.9 \text{ fish}$$

Drew:

$$N = 10; \quad \sum x_i = 100$$

$$\sum x_i^2 = 15^2 + 2^2 + 3^2 + 18^2 + 20^2 + 1^2 + 17^2$$
$$+ 2^2 + 19^2 + 3^2$$
$$= 1626$$

$$\sigma = \sqrt{\frac{\sum x_i^2 - \frac{(\sum x_i)^2}{N}}{N}} = \sqrt{\frac{1626 - \frac{(100)^2}{10}}{10}}$$

$$\approx 7.9 \text{ fish}$$

Yes, now there appears to be a difference in the two fishermen's records. Ethan had a more consistent fishing record, which is indicated by the smaller standard deviation.

(c) Answers will vary. One possibility follows: The range is limited as a measure of dispersion because it does not take all of the data values into account. It is obtained by using only the two most extreme data values. Since the standard deviation utilizes all of the data values, it provides a better overall representation of dispersion.

29. **(a)** We use the computational formula:

$$\sum x_i = 43.73; \quad \sum x_i^2 = 38.3083; \quad n = 50;$$

$$s = \sqrt{\frac{\sum x_i^2 - \frac{(\sum x_i)^2}{n}}{n-1}}$$

$$= \sqrt{\frac{38.3083 - \frac{(43.73)^2}{50}}{50-1}} \approx 0.036 \text{ g}$$

(b) The histogram is approximately symmetric, so the Empirical Rule is applicable.

(c) Since 0.803 is two standard deviations below the mean [0.803 = 0.875 − 2(0.036)] and 0.943 is two standard deviations above the mean [0.947 = 0.875 + 2(0.036)], the Empirical Rule predicts that approximately 95% of the M&Ms will weigh between 0.803 and 0.947 grams.

(d) All except 2 of the M&Ms weigh between 0.803 and 0.947 grams. Thus, the actual percentage is 48/50 = 96%.

(e) Since 0.911 is one standard deviation above the mean [0.911 = 0.875 + 0.036], the Empirical Rule predicts that 13.5% + 2.35% + 0.15% = 16% of the M&Ms will weigh more than 0.911 grams.

(f) Six of the M&Ms weigh more than 0.911 grams. Thus, the actual percentage is 6/50 = 12%.

31. Car 1:

$\sum x_i = 3352; \ \sum x_i^2 = 755,712; \ n = 15$

Measures of Center:

$\overline{x} = \dfrac{\sum x_i}{n} = \dfrac{3352}{15} \approx 223.5$ miles

$M = 223$ miles (8th value in the ordered data)
Mode: none

Measures of Dispersion:
Range = Largest Value − Smallest Value
$= 271 - 178 = 93$ miles

$s^2 = \dfrac{\sum x_i^2 - \dfrac{(\sum x_i)^2}{n}}{n-1} = \dfrac{755,712 - \dfrac{(3352)^2}{15}}{15-1}$
≈ 475.1 (miles)^2

$s = \sqrt{s^2} = \sqrt{\dfrac{755,712 - \dfrac{(3352)^2}{15}}{15-1}} \approx 21.8$ miles

Car 2:

$\sum x_i = 3558; \ \sum x_i^2 = 877,654; \ n = 15$

Measures of Center:

$\overline{x} = \dfrac{\sum x_i}{n} = \dfrac{3558}{15} = 237.2$ miles

$M = 230$ miles (8th value in the ordered data)
Mode: none

Measures of Dispersion:
Range = Largest Value − Smallest Value
$= 326 - 160 = 166$ miles

$s^2 = \dfrac{\sum x_i^2 - \dfrac{(\sum x_i)^2}{n}}{n-1}$
$= \dfrac{877,654 - \dfrac{(3558)^2}{15}}{15-1} \approx 2406.9$ (miles)^2

$s = \sqrt{s^2} = \sqrt{\dfrac{877,654 - \dfrac{(3558)^2}{15}}{15-1}} \approx 49.1$ miles

We expect that the distribution for Car 1 is symmetric since the mean and median are approximately equal. We expect that the distribution for Car 2 is skewed right slightly since the mean is larger than the median. Both distributions have similar measures of center, but Car 2 has more dispersion which can be seen by its larger range, variance, and standard deviation. This means that the distance Car 1 can be driven on 10 gallons of gas is more consistent. Thus, Car 1 is probably the better car to buy.

33. (a) Financial Stocks:

$\sum x_i = 558.93; \ \sum x_i^2 = 21,101.5503; \ n = 25$

$\overline{x} = \dfrac{\sum x_i}{n} = \dfrac{558.93}{25} \approx 22.357\%$

$M = 16.01\%$ (13th value in the ordered data)

Energy Stocks:

$\sum x_i = 906.36; \ \sum x_i^2 = 42,536.5366; \ n = 25$

$\overline{x} = \dfrac{\sum x_i}{n} = \dfrac{906.36}{25} \approx 36.254\%$

$M = 34.18\%$ (13th value in the ordered data)
Energy Stocks have higher mean and median rates of return.

(b) Financial Stocks:

$s = \sqrt{\dfrac{\sum x_i^2 - \dfrac{(\sum x_i)^2}{n}}{n-1}}$

$= \sqrt{\dfrac{21,101.5503 - \dfrac{(558.93)^2}{25}}{25-1}} \approx 18.936\%$

Energy Stocks:

$$s = \sqrt{\dfrac{\sum x_i^2 - \dfrac{(\sum x_i)^2}{n}}{n-1}}$$

$$= \sqrt{\dfrac{42,536.5366 - \dfrac{(906.36)^2}{25}}{25-1}} \approx 20.080\%$$

Energy Stocks are riskier since they have a larger standard deviation. The investor is paying for the higher return. The higher returns are probably worth the cost.

35. (a) Since 70 is two standard deviations below the mean [70 = 100 − 2(15)] and 130 is two standard deviations above the mean [130 = 100 + 2(15)], the Empirical Rule predicts that approximately 95% of people has an IQ score between 70 and 130.

(b) Since about 95% of people has an IQ score between 70 and 30, then approximately 5% of people has an IQ score either less than 70 or greater than 130.

(c) Approximately 5% / 2 = 2.5% of people has an IQ score greater than 130.

37. (a) Approximately 95% of the data will be within two standard deviations of the mean. Now, 325 − 2(30) = 265 and 325 + 2(30) = 385. Thus, about 95% of pairs of kidneys will be between 265 and 385 grams.

(b) Since 235 is three standard deviations below the mean [235 = 325 − 3(30)] and 415 is three standard deviations above the mean [415 = 325 + 3(30)], the Empirical Rule predicts that about 99.7% of pairs of kidneys weighs between 235 and 415 grams.

(c) Since about 99.7% of pairs of kidneys weighs between 235 and 415 grams, then about 0.3% of pairs of kidneys weighs either less than 235 or more than 415 grams.

(d) Since 295 is one standard deviation below the mean [295 = 325 − 30] and 385 is two standard deviations above the mean [385 = 325 + 2(30)], the Empirical Rule predicts that approximately 34% + 34% + 13.5% = 81.5% of pairs of kidneys weighs between 295 and 385 grams.

39. (a) By Chebyshev's inequality, at least

$$\left(1 - \dfrac{1}{k^2}\right) \cdot 100\% = \left(1 - \dfrac{1}{3^2}\right) \cdot 100\% \approx 88.9\%$$

of gasoline prices have prices within three standard deviations of the mean.

(b) By Chebyshev's inequality, at least

$$\left(1 - \dfrac{1}{k^2}\right) \cdot 100\% = \left(1 - \dfrac{1}{2.5^2}\right) \cdot 100\% = 84\%$$

of gasoline prices has prices within $k = 2.5$ standard deviations of the mean. Now, $3.06 - 2.5(0.06) = 2.91$ and $3.06 + 2.5(0.06) = 3.21$. Thus, the gasoline prices that are within 2.5 standard deviations of the mean are from $2.91 to $3.21.

(c) Since 2.94 is $k = 2$ standard deviations below the mean [2.94 = 3.06 − 2(0.06)] and 3.18 is $k = 2$ standard deviations above the mean [3.18 = 3.06 + 2(0.06)], Chebyshev's theorem predicts that at least

$$\left(1 - \dfrac{1}{k^2}\right) \cdot 100\% = \left(1 - \dfrac{1}{2^2}\right) \cdot 100\% = 75\%$$

of gas stations has prices between $2.94 and $3.18 per gallon.

41. When calculating the variability in team batting averages, we are finding the variability of means. When calculating the variability of all players, we are finding the variability of individuals. Since there is more variability among individuals than among means, the teams will have less variability.

43. Sample size of 5:
All data recorded correctly: $s \approx 5.3$.
106 recorded incorrectly as 160: $s \approx 27.9$.

Sample size of 12:
All data recorded correctly: $s \approx 14.7$.
106 recorded incorrectly as 160: $s \approx 22.7$.

Sample size of 30:
All data recorded correctly: $s \approx 15.9$.
106 recorded incorrectly as 160: $s \approx 19.2$.

As the sample size increases, the impact of the misrecorded observation on the standard deviation decreases.

45. (a) We use the computational formula:

$\sum x_i = 1991.6$; $\sum x_i^2 = 199,033.1$; $n = 20$;

$$s = \sqrt{\frac{\sum x_i^2 - \frac{(\sum x_i)^2}{n}}{n-1}}$$

$$= \sqrt{\frac{199,033.1 - \frac{(1991.6)^2}{20}}{20-1}} \approx 6.11 \text{ cm}$$

(b) $\sum x_i = 969.4$; $\sum x_i^2 = 94,262.8$; $n = 10$;

$$s = \sqrt{\frac{\sum x_i^2 - \frac{(\sum x_i)^2}{n}}{n-1}}$$

$$= \sqrt{\frac{94,262.8 - \frac{(969.4)^2}{10}}{10-1}} \approx 5.67 \text{ cm}$$

(c) $\sum x_i = 1022.2$; $\sum x_i^2 = 104,770.3$; $n = 10$;

$$s = \sqrt{\frac{\sum x_i^2 - \frac{(\sum x_i)^2}{n}}{n-1}}$$

$$= \sqrt{\frac{104,770.3 - \frac{(1022.2)^2}{10}}{10-1}} \approx 5.59 \text{ cm}$$

(d) The standard deviation is lower for each block than it is for the combined group.

47. $\sum x_i = 5831$; $\bar{x} = \dfrac{\sum x_i}{n} = \dfrac{5831}{4} = \$1457.75$

x_i	$x_i - \bar{x}$	$|x_i - \bar{x}|$
976	−481.75	481.75
2038	580.25	580.25
918	−539.75	539.75
1899	441.25	441.25
$\sum(x_i - \bar{x}) = 0$		$\sum|x_i - \bar{x}| = 2043$

$\text{MAD} = \dfrac{\sum|x_i - \bar{x}|}{n} = \dfrac{\$2043}{4} \approx \$510.8$

This is somewhat less than the sample standard deviation of $s \approx \$593.0$ which we found in Problem 17.

49. (a) Reading from the graph, the average annual return for a portfolio that is 10% foreign is 14.9%. The level of risk is 14.7%.

(b) To best minimize risk, 30% should be invested in foreign stocks. According to the graph, a 30% investment in foreign stocks has the smallest standard deviation (level of risk) at about 14.3%.

(c) Answers will vary. One possibility follows: The risk decreases because a portfolio including foreign stocks is more diversified.

(d) According to Chebyshev's theorem, at least 75% of returns are within $k = 2$ standard deviations of the mean. Thus, at least 75% of returns are between $\bar{x} - ks = 15.8 - 2(14.3) = -12.8\%$ and

$\bar{x} + ks = 15.8 + 2(14.3) = 44.4\%$. By Chebyshev's theorem, at least 88.9% of returns are within $k = 3$ standard deviations of the mean, Thus, at least 88.9% of returns are between

$\bar{x} - ks = 15.8 - 3(14.3) = -27.1\%$ and

$\bar{x} + ks = 15.8 + 3(14.3) = 58.7\%$. An investor should not be surprised if she has a negative rate of return. Chebyshev's theorem indicates that a negative return is fairly common.

Section 3.3

1. When we approximate the mean and standard deviation from grouped data, we use the class midpoint to represent the average value of all the data values in the class.

3. To find the mean, we use the formula

$\mu = \dfrac{\sum x_i f_i}{\sum f_i}$. To find the standard deviation, we choose to use the computational formula

$$\sigma = \sqrt{\sigma^2} = \sqrt{\frac{\sum x_i^2 f_i - \frac{(\sum x_i f_i)^2}{\sum f_i}}{\sum f_i}} \text{ . We}$$

organize our computations of x_i, $\sum f_i$,

$\sum x_i f_i$, and $\sum x_i^2 f_i$ in the table that follows:

Class	Midpoint, x_i	Frequency, f_i	$x_i f_i$	x_i^2	$x_i^2 f_i$
0 – 999	$\dfrac{0+1,000}{2}=500$	30	15,000	250,000	7,500,000
1,000 – 1,999	$\dfrac{1,000+2,000}{2}=1,500$	97	145,500	2,250,000	218,250,000
2,000 – 2,999	2,500	935	2,337,500	6,250,000	5,843,750,000
3,000 – 3,999	3,500	2,698	9,443,000	12,250,000	33,050,500,000
4,000 – 4,999	4,500	344	1,548,000	20,250,000	6,966,000,000
5,000 – 5,999	5,500	5	27,500	30,250,000	151,250,000
		$\sum f_i = 4109$	$\sum x_i f_i = 13,516,500$		$\sum x_i^2 f_i = 46,237,250,000$

With the table complete, we compute the population mean and population standard deviation:

$$\mu = \frac{\sum x_i f_i}{\sum f_i} = \frac{13,516,500}{4109} \approx 3,289.5 \text{ grams}$$

$$\sigma = \sqrt{\sigma^2} = \sqrt{\frac{\sum x_i^2 f_i - \dfrac{\left(\sum x_i f_i\right)^2}{\sum f_i}}{\sum f_i}} = \sqrt{\frac{46,237,250,000 - \dfrac{(13,516,500)^2}{4109}}{4109}} \approx 657.2 \text{ grams}$$

5. To find the mean, we use the formula $\bar{x} = \dfrac{\sum x_i f_i}{\sum f_i}$. To find the standard deviation, we choose to use the

computational formula $s = \sqrt{s^2} = \sqrt{\dfrac{\sum x_i^2 f_i - \dfrac{\left(\sum x_i f_i\right)^2}{\sum f_i}}{\left(\sum f_i\right)-1}}$. We organize our computations of x_i, $\sum f_i$, $\sum x_i f_i$,

and $\sum x_i^2 f_i$ in the table that follows:

Class	Midpoint, x_i	Frequency, f_i	$x_i f_i$	x_i^2	$x_i^2 f_i$
61 – 64	$\dfrac{61+65}{2}=63$	31	1,953	3,969	123,039
65 – 67	$\dfrac{65+68}{2}=66.5$	67	4,455.5	4,422.25	296,290.75
68 – 69	69	198	13,662	4,761	942,678
70	70	195	13,650	4,900	955,500
71 – 72	72	120	8,640	5,184	622,080
73 – 76	75	89	6,675	5,625	500,625
77 – 80	79	50	3,950	6,241	312,050
		$\sum f_i = 750$	$\sum x_i f_i = 52,985.5$		$\sum x_i^2 f_i = 3,752,262.75$

With the table complete, we compute the sample mean and sample standard deviation:

$$\mu = \frac{\sum x_i f_i}{\sum f_i} = \frac{52,985.5}{750} \approx 70.6°\text{F} \; ; \quad s = \sqrt{s^2} = \sqrt{\frac{\sum x_i^2 f_i - \dfrac{\left(\sum x_i f_i\right)^2}{\sum f_i}}{\left(\sum f_i\right)-1}} = \sqrt{\frac{3,752,262.75 - \dfrac{(52,985.5)^2}{750}}{750-1}} \approx 3.5°\text{F}$$

7. **(a)** To find the mean, we use the formula $\mu = \dfrac{\sum x_i f_i}{\sum f_i}$. To find the standard deviation, we choose to use the

computational formula $\sigma = \sqrt{\sigma^2} = \sqrt{\dfrac{\sum x_i^2 f_i - \dfrac{\left(\sum x_i f_i\right)^2}{\sum f_i}}{\sum f_i}}$. We organize our computations of x_i, $\sum f_i$,

$\sum x_i f_i$, and $\sum x_i^2 f_i$ in the table that follows:

Class	Midpoint, x_i	Frequency, f_i	$x_i f_i$	x_i^2	$x_i^2 f_i$
$15-19$	$\dfrac{15+19}{2}=17.5$	84	1,470	306.25	25,725
$20-24$	$\dfrac{20+25}{2}=22.5$	431	9,697.5	506.25	218,193.75
$25-29$	27.5	1,753	48,207.5	756.25	1,325,706.25
$30-34$	32.5	2,752	89,440	1,056.25	2,906,800
$35-39$	37.5	1,785	66,937.5	1,406.25	2,510,156.25
$40-44$	42.5	378	16,065	1,806.25	682,762.5
$45-49$	47.5	80	3,800	2,256.25	180,500
$50-54$	52.5	9	472.5	2,756.25	24,806.25
		$\sum f_i = 7,272$	$\sum x_i f_i = 236,090$		$\sum x_i^2 f_i = 7,874,650$

With the table complete, we compute the population mean and population standard deviation:

$$\mu = \frac{\sum x_i f_i}{\sum f_i} = \frac{236,090}{7,272} \approx 32.5 \text{ years}$$

$$\sigma = \sqrt{\sigma^2} = \sqrt{\frac{\sum x_i^2 f_i - \frac{\left(\sum x_i f_i\right)^2}{\sum f_i}}{\sum f_i}} = \sqrt{\frac{7,874,650 - \frac{(236,090)^2}{7,272}}{7,272}} \approx 5.4 \text{ years}$$

(b)

Number of Multiple Births in 2004

(c) By the Empirical Rule, 95% of the observations will be within 2 standard deviations of the mean. Now, $\mu - 2\sigma = 32.5 - 2(5.4) = 21.7$ and $\mu + 2\sigma = 32.5 + 2(5.4) = 43.3$, so 95% of the mothers of multiple births will be between 21.7 and 43.3 years of age.

9. We organize our computations of x_i, $\sum f_i$, $\sum x_i f_i$, and $\sum x_i^2 f_i$ in the table that follows:

Class	Midpoint, x_i	Frequency, f_i	$x_i f_i$	x_i^2	$x_i^2 f_i$
$20-29$	$\dfrac{20+30}{2}=25$	1	25	625	625
$30-39$	35	6	210	$1,225$	$7,350$
$40-49$	45	10	450	$2,025$	$20,250$
$50-59$	55	14	770	$3,025$	$42,350$
$60-69$	65	6	390	$4,225$	$25,350$
$70-79$	75	3	225	$5,625$	$16,875$
		$\sum f_i = 40$	$\sum x_i f_i = 2,070$		$\sum x_i^2 f_i = 112,800$

With the table complete, we compute the sample mean and sample standard deviation:

$$\bar{x} = \frac{\sum x_i f_i}{\sum f_i} = \frac{2,070}{40} \approx 51.8 \; ; \; s = \sqrt{s^2} = \sqrt{\frac{\sum x_i^2 f_i - \frac{\left(\sum x_i f_i\right)^2}{\sum f_i}}{\left(\sum f_i\right)-1}} = \sqrt{\frac{112,800 - \frac{(2,070)^2}{40}}{40-1}} \approx 12.1$$

From the raw data, we find: $\sum x_i = 2,045$; $\sum x_i^2 = 109,151$; $n = 40$

$$\bar{x} = \frac{\sum x_i}{n} = \frac{2045}{40} \approx 51.1 \; ; \; s = \sqrt{s^2} = \sqrt{\frac{\sum x_i^2 - \frac{\left(\sum x_i\right)^2}{n}}{n-1}} = \sqrt{\frac{109,151 - \frac{(2,045)^2}{40}}{40-1}} \approx 10.9$$

The approximations from the grouped data are good estimates of the actual results from the raw data.

11. GPA $= \bar{x}_w = \dfrac{\sum w_i x_i}{\sum w_i} = \dfrac{5(3)+3(4)+4(4)+3(2)}{5+3+4+3} = \dfrac{49}{15} \approx 3.27$

13. Cost per pound $= \bar{x}_w = \dfrac{\sum w_i x_i}{\sum w_i} = \dfrac{4(\$3.50)+3(\$2.75)+2(\$2.25)}{4+3+2} = \dfrac{\$26.75}{9} \approx \$2.97 \,/\, \text{pound}$

15. (a) Male: We organize our computations of x_i, $\sum f_i$, $\sum x_i f_i$, and $\sum x_i^2 f_i$ in the following table.

Class	Midpoint, x_i	Frequency, f_i	$x_i f_i$	x_i^2	$x_i^2 f_i$
$0-9$	5	$20,518$	$102,590$	25	$512,950$
$10-19$	15	$22,496$	$337,440$	225	$5,061,600$
$20-29$	25	$21,494$	$537,350$	625	$13,433,750$
$30-39$	35	$20,629$	$722,015$	$1,225$	$25,270,525$
$40-49$	45	$22,461$	$1,010,745$	$2,025$	$45,483,525$
$50-59$	55	$18,872$	$1,037,960$	$3,025$	$57,087,800$
$60-69$	65	$11,216$	$729,040$	$4,225$	$47,387,600$
$70-79$	75	$6,950$	$521,250$	$5,625$	$39,093,750$
$80-89$	85	$3,327$	$282,795$	$7,225$	$24,037,575$
$90-99$	95	524	$49,780$	$9,025$	$4,729,100$
$100-109$	105	14	$1,470$	$11,025$	$154,350$
		$\sum f_i = 148,501$	$\sum x_i f_i = 5,332,435$		$\sum x_i^2 f_i = 262,252,525$

With the table complete, we compute the population mean and population standard deviation:

$$\mu = \frac{\sum x_i f_i}{\sum f_i} = \frac{5{,}332{,}435}{148{,}501} \approx 35.9 \text{ years}$$

$$\sigma = \sqrt{\sigma^2} = \sqrt{\frac{\sum x_i^2 f_i - \frac{(\sum x_i f_i)^2}{\sum f_i}}{\sum f_i}} = \sqrt{\frac{262{,}252{,}525 - \frac{(5{,}332{,}435)^2}{148{,}501}}{148{,}501}} \approx 21.8 \text{ years}$$

(b) Female: We organize our computations of x_i, $\sum f_i$, $\sum x_i f_i$, and $\sum x_i^2 f_i$ in the following table.

Class	Midpoint, x_i	Frequency, f_i	$x_i f_i$	x_i^2	$x_i^2 f_i$
0-9	5	19,607	98,035	25	490,175
10-19	15	20,453	306,795	225	4,601,925
20-29	25	20,326	508,150	625	12,703,750
30-39	35	20,261	709,135	1,225	24,819,725
40-49	45	22,815	1,026,675	2,025	46,200,375
50-59	55	19,831	1,090,705	3,025	59,988,775
60-69	65	12,519	813,735	4,225	52,892,775
70-79	75	8,970	672,750	5,625	50,456,250
80-89	85	5,678	482,630	7,225	41,023,550
90-99	95	1,356	128,820	9,025	12,237,900
100-109	105	59	6,195	11,025	650,475
		$\sum f_i = 151{,}875$	$\sum x_i f_i = 5{,}843{,}625$		$\sum x_i^2 f_i = 306{,}065{,}675$

With the table complete, we compute the population mean and population standard deviation:

$$\mu = \frac{\sum x_i f_i}{\sum f_i} = \frac{5{,}843{,}625}{151{,}875} \approx 38.5 \text{ years}$$

$$\sigma = \sqrt{\sigma^2} = \sqrt{\frac{\sum x_i^2 f_i - \frac{(\sum x_i f_i)^2}{\sum f_i}}{\sum f_i}} = \sqrt{\frac{306{,}065{,}675 - \frac{(5{,}843{,}625)^2}{151{,}875}}{151{,}875}} \approx 23.1 \text{ years}$$

(c) Females have a higher mean age.

(d) Females also have more dispersion in age, which is indicated by the larger standard deviation.

17.

Class	f_i	CF
0-999	30	30
1,000-1,999	97	127
2,000-2,999	935	1,062
3,000-3,999	2,698	3,760
4,000-4,999	344	4,104
5,000-5,999	5	4,109

The distribution contains $n = 4109$ data values. The position of the median is $\frac{n+1}{2} = \frac{4109+1}{2} = 2055$, which is in the fourth class, $3{,}000 - 3{,}999$. Then,

$$M = L + \frac{\frac{n}{2} - CF}{f} \cdot i$$

$$= 3{,}000 + \frac{\frac{4109}{2} - 1{,}062}{2{,}698} (4{,}000 - 3{,}000)$$

$$\approx 3{,}367.9 \text{ grams}$$

19.

Class	f_i	CF
15 – 19	84	84
20 – 24	431	515
25 – 29	1,753	2,268
30 – 34	2,752	5,020
35 – 39	1,785	6,805
40 – 44	378	7,183
45 – 49	80	7,263
50 – 54	9	7,272

The distribution contains $n = 7{,}272$ data values. The position of the median is

$$\frac{n+1}{2} = \frac{7{,}272+1}{2} = 3{,}636.5 \text{, which is in the}$$

fourth class, 30 – 34. Then,

$$M = L + \frac{\frac{n}{2} - CF}{f} \cdot i$$

$$= 30 + \frac{\frac{7{,}272}{2} - 2{,}268}{2{,}752}(35 - 30)$$

$$\approx 32.5 \text{ years}$$

21. From the table in Problem 3, the highest frequency is 2,698. So, the modal class is 3,000 – 3,999 grams.

Section 3.4

1. Answers will vary. The kth percentile separates the lower k percent of the data from the upper $(100–k)$ percent of the data. For example, if a data value lies at the 60th percentile, then approximately 60% of the data is below it and approximately 40% is above this value.

3. A four-star mutual fund is in the 4th *quintile* of the funds. That is, it is above the bottom 60%, but below the top 20% of the ranked funds.

5. To qualify for Mensa, one needs to have an IQ that is in the top 2% of people.

7. The interquartile range is the preferred measure of dispersion when the data are skewed or have outliers. An advantage of the standard deviation is that it uses all of the observations in the computation.

9. 34-week gestation:

$$z = \frac{x - \mu}{\sigma} = \frac{2{,}400 - 2{,}600}{660} \approx -0.30$$

40-week gestation:

$$z = \frac{x - \mu}{\sigma} = \frac{3{,}300 - 3{,}500}{470} \approx -0.43$$

The weight of the 34-week gestation baby is 0.30 standard deviations below the mean, while the weight of the 40-week gestation baby is 0.43 standard deviations below the mean. Thus, the 40-week gestation baby weighs less relative to the gestation period.

11. 75-inch man:

$$z = \frac{x - \mu}{\sigma} = \frac{75 - 69.6}{3.0} = 1.8$$

70-inch woman:

$$z = \frac{x - \mu}{\sigma} = \frac{70 - 64.1}{3.8} \approx 1.55$$

The height of the 75-inch man is 1.8 standard deviations above the mean, while the height of a 70-inch woman is 1.55 standard deviations above the mean. Thus, the 75-inch man is relatively taller than the 70-inch woman.

13. Jake Peavy:

$$z = \frac{x - \mu}{\sigma} = \frac{2.54 - 4.119}{0.697} \approx -2.27$$

John Lackey:

$$z = \frac{x - \mu}{\sigma} = \frac{3.01 - 4.017}{0.678} \approx -1.49$$

Jake Peavy's 2007 ERA was 2.27 standard deviations below the mean, while John Lackey's 2007 ERA was 1.49 standard deviations below the mean. Thus, Peavy had the better year relative to his peers.

15. $z = \dfrac{x - \mu}{\sigma}$

$$\frac{x - 200}{26} = 1.5$$

$$x - 200 = 1.5(26)$$

$$x - 200 = 39$$

$$x = 239$$

An applicant must make a minimum score of 239 to be accepted into the school.

17. (a) 15% of 3- to 5-month-old males have a head circumference that is 41.0 cm or less, and $(100 - 15)\% = 85\%$ of 3- to 5-month-old males have a head circumference that is greater than 41.0 cm.

(b) 90% of 2-year-old females have a waist circumference that is 52.7 cm or less, and $(100 - 90)\% = 10\%$ of 2-year-old females have a waist circumference that is more than 52.7 cm.

(c) The heights at each percentile decrease (except for the 40-49 age group) as the age increases. This implies that adults males are getting taller.

19. (a) 25% of the states have a violent crime rate that is 272.8 crimes per 100,000 population or less, and $(100 - 25)\% = 75\%$ of the states have a violent crime rate more than 272.8. 50% of the states have a violent crime rate that is 387.4 crimes per 100,000 population or less, while $(100 - 50)\% = 50\%$ of the states have a violent crime rate more than 387.4. 75% of the states have a violent crime rate that is 529.7 crimes per 100,000 population or less, and $(100 - 75)\% = 25\%$ of the states have a violent crime rate more than 529.7.

(b) $IQR = Q_3 - Q_1 = 529.7 - 272.8 = 256.9$ crimes per 100,000 population. This means that the middle 50% of all observations have a range of 256.9 crimes per 100,000 population.

(c) $LF = Q_1 - 1.5(IQR)$
$= 272.8 - 1.5(256.9) = -112.55$
$UF = Q_3 + 1.5(IQR)$
$= 529.7 + 1.5(256.9) = 915.05$
Since 1,459 is above the upper fence, the Washington, D.C. crime rate is an outlier.

(d) Skewed right. The difference between Q_1 and Q_2 (114.6) is quite a bit less than the difference between Q_2 and Q_3 (142.3), and the outlier is in the right tail of the distribution, which implies that the distribution is skewed right.

21. (a) Computing the sample mean ($\bar{x}$) and sample standard deviation (s) for the data yields $\bar{x} = 3.9935$ inches and $s \approx 1.7790$ inches. Using these values as estimates for the μ and σ, the z-score for $x = 0.97$ inches $z = \dfrac{x - \mu}{\sigma} \approx \dfrac{0.97 - 3.9935}{1.7790} \approx -1.70$.
The 1971 rainfall amount is 1.70 standard deviations below the mean.

(b) Notice that the $n = 20$ data values are already arranged in order (moving down the columns). The second quartile (median) is the mean of the values that lie in positions $\dfrac{n}{2} = \dfrac{20}{2} = 10$ and $\dfrac{n}{2} + 1 = \dfrac{20}{2} + 1 = 11$, which are 3.97 and 4.00, respectively. So,
$Q_2 = M = \dfrac{3.97 + 4.00}{2} = 3.985$ inches.
The first quartile is the median of the bottom 10 data values, which is the mean of the data values that lie in the 5th and 6th positions. These values are 2.47 and 2.78, so $Q_1 = \dfrac{2.47 + 2.78}{2} = 2.625$ inches.
The third quartile is the median of the top 10 data values, which is the mean of the data values that lie in the 15th and 16th positions. These values are 5.22 and 5.50, so $Q_3 = \dfrac{5.22 + 5.50}{2} = 5.36$ inches.

Note: Results from MINITAB differ:
$Q_1 = 2.548$ inches, $Q_2 = 3.985$ inches, and $Q_3 = 5.430$ inches.

(c) $IQR = Q_3 - Q_1 = 5.36 - 2.625 = 2.735$ inches. The range of the middle 50% of the observations of Chicago rainfall in April is 2.734 inches.
Note: If using MINITAB, the result will be: $IQR = 5.430 - 2.548 = 2.882$ inches.

(d) $LF = Q_1 - 1.5(IQR)$
$= 2.625 - 1.5(2.735) = -1.478$ inches
$UF = Q_3 + 1.5(IQR)$
$= 5.36 + 1.5(2.735) = 9.463$ inches

Note: If using MINITAB, the result will be:
$LF = 2.548 - 1.5(2.882) = -1.775$ inches
$UF = 5.430 + 1.5(2.882) = 9.753$ inches

There are no outliers, according to this criteria.

23. (a) There are $n = 40$ data values, and we put them in ascending order:

–0.16	–0.02	0.02	0.06	0.13
–0.11	–0.02	0.02	0.06	0.18
–0.10	–0.02	0.03	0.07	0.19
–0.08	–0.02	0.03	0.08	0.22
–0.05	–0.01	0.04	0.09	0.25
–0.05	0.01	0.05	0.09	0.26
–0.04	0.01	0.06	0.10	0.26
–0.04	0.02	0.06	0.11	0.47

The second quartile (median) is the mean of the values that lie the 20th and 21th positions, which are 0.03 and 0.04, respectively. So,

$$Q_2 = M = \frac{0.03 + 0.04}{2} = 0.035 .$$

The first quartile is the median of the bottom 20 data values, which is the mean of the data values that lie in the 10th and 11th positions. These values are –0.02 and –0.02, so $Q_1 = \frac{-0.02 + (-0.02)}{2} = -0.02$.

The third quartile is the median of the top 20 data values, which is the mean of the data values that lie in the 30th and 31st positions. These values are 0.09 and 0.10, so $Q_3 = \frac{0.09 + 0.10}{2} = 0.095$.

Note: Results from MINITAB differ: $Q_1 = -0.02$, $Q_2 = 0.035$, and $Q_3 = 0.0975$.

Interpretation: Using the by-hand results, 25% of the monthly returns are less than or equal to the first quartile, –0.02, and about 75% of the monthly returns are greater than –0.02; 50% of the monthly returns are less than or equal to the second quartile, 0.035, and about 50% of the monthly returns are greater than 0.035; about 75% of the monthly returns are less than or equal to the third quartile, 0.095, and about 25% of the monthly returns are greater than 0.095.

(b) $IQR = Q_3 - Q_1 = 0.095 - (-0.02) = 0.115$

$$LF = Q_1 - 1.5(IQR)$$
$$= -0.02 - 1.5(0.115) = -0.1925$$
$$UF = Q_3 + 1.5(IQR)$$
$$= 0.095 + 1.5(0.115) = 0.2675$$

Note: If using MINITAB, the result will be:
$IQR = Q_3 - Q_1 = 0.0975 - (-0.02) = 0.1175$
$LF = -0.02 - 1.5(0.1175) = -0.19675$
$UF = 0.0975 + 1.5(0.1175) = 0.27375$
The return 0.47 is an outlier because it is greater than the upper fence.

25. To find the upper fence, we must find the third quartile and the interquartile range. There are $n = 20$ data values, and we put them in ascending order:

345	429	461	471	505
346	437	466	480	515
358	442	466	489	516
372	442	470	490	549

The first quartile is the median of the bottom 10 data values, which is the mean of the data values that lie in the 5th and 6th positions. These values are 429 and 437, so $Q_1 = \frac{429 + 437}{2} = 433$ min.

The third quartile is the median of the top 10 data values, which is the mean of the data values that lie in the 15th and 16st positions. These values are 489 and 490, so

$$Q_3 = \frac{489 + 490}{2} = 489.5 \text{ min.}$$

$IQR = 489.5 - 433 = 56.5$ min.
$UF = Q_3 + 1.5(IQR)$
$= 489.5 + 1.5(56.5) = 574.25$ min.

The customer is contacted if more than 574 minutes are used.

Note: Results from MINITAB differ:
$Q_1 = 431$ minutes and $Q_3 = 489.8$ minutes
$IQR = 489.8 - 431 = 58.8$ minutes
$UF = 489.8 + 1.5(58.8) = 578$ min.

Using MINITAB, the customer is contacted if more than 578 minutes are used.

27. a. To find outliers, we must find the first and third quartiles and the interquartile range. There are $n = 50$ data values, and we put them in ascending order:

0	0	188	347	547
0	0	203	367	567
0	67	244	375	579
0	82	262	389	628
0	83	281	403	635
0	95	289	454	650
0	100	300	476	671
0	149	310	479	719
0	159	316	521	736
0	181	331	527	12,777

The first quartile is the median of the bottom 25 data values, which is the value that lies in the 13th position. So, $Q_1 = \$67$.

The third quartile is the median of the top 25 data values, which is value that lies in the 38th position. So, $Q_3 = \$479$.

$$IQR = 479 - 67 = \$412$$
$$LF = Q_1 - 1.5(IQR)$$
$$= 67 - 1.5(412) = -\$551$$
$$UF = Q_3 + 1.5(IQR)$$
$$= 479 + 1.5(412) = \$1,097$$

Note: Results from MINITAB differ:
$Q_1 = \$50$ and $Q_3 = \$490$
$$IQR = 490 - 50 = \$440$$
$$LF = 50 - 1.5(440) = -\$610$$
$$UF = 490 + 1.5(440) = \$1,150$$

So, the only outlier is $12,777 because it is greater than the upper fence.

(b) To create the histogram, we choose the lower class limit of the first class to be 0 and the class width to be 100. The resulting classes and frequencies follow:

Class	Freq.	Class	Freq.
0 – 99	16	500 – 599	5
100 – 199	5	600 – 699	4
200 – 299	5	700 – 799	2
300 – 3999	8	⋮	
400 – 499	4	12,700 – 12,799	1

The histogram follows:

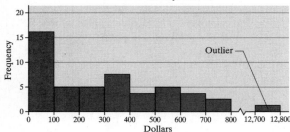

(c) Answers will vary. One possibility is that a student may have provided his or her annual income instead of his or her weekly income.

29. From Problem 23 in Section 3.1 and Problem 25 in Section 3.2, we have $\mu \approx 72.2$ beats per minute and $\sigma \approx 7.7$ beats per minute.

Student	Pulse, x_i	z-score, z_i
P. Bernpah	76	$\dfrac{76 - 72.2}{7.7} = 0.49$
M. Brooks	60	$\dfrac{60 - 72.2}{7.7} = -1.58$
J. Honeycutt	60	$\dfrac{60 - 72.2}{7.7} = -1.58$
C. Jefferson	81	$\dfrac{81 - 72.2}{7.7} = 1.14$
C. Kurtenbach	72	$\dfrac{72 - 72.2}{7.7} = -0.03$
J. Laotka	80	$\dfrac{80 - 72.2}{7.7} = 1.01$
K. McCarthy	80	$\dfrac{80 - 72.2}{7.7} = 1.01$
T. Ohm	68	$\dfrac{68 - 72.2}{7.7} = -0.55$
K. Wojdyla	73	$\dfrac{73 - 72.2}{7.7} = 0.10$

The mean of the z-scores is 0.001 and the standard deviation is 1.054. These are off slightly from the true mean of 0 and the true standard deviation of 1 because of rounding.

31. (a) To find the standard deviation, we use the computational formula:

$$\sum x_i = 9,049; \ \sum x_i^2 = 4,158,129; \ n = 20$$

$$s = \sqrt{\dfrac{\sum x_i^2 - \dfrac{(\sum x_i)^2}{n}}{n-1}}$$

$$= \sqrt{\dfrac{4,158,129 - \dfrac{(9,049)^2}{20}}{20 - 1}} \approx 58.0 \text{ minutes}$$

To find the interquartile range, we look back to the solution of Problem 25. There we found $Q_1 = 433$ minutes, . $Q_3 = 489.5$ minutes, and $IQR = 489.5 - 433 = 56.5$ minutes.

Note: Results from MINITAB differ:
$Q_1 = 431$ minutes, $Q_3 = 489.8$ minutes, and $IQR = 489.8 - 431 = 58.8$ minutes.

(b) Again, to find the standard deviation, we use the computational formula:

$$\sum x_i = 8,703 \; ; \; \sum x_i^2 = 4,038,413 \; ; \; n = 20$$

$$s = \sqrt{\frac{\sum x_i^2 - \frac{(\sum x_i)^2}{n}}{n-1}}$$

$$= \sqrt{\frac{4,038,413 - \frac{(8,703)^2}{20}}{20-1}} \approx 115.0 \text{ minutes}$$

To find the interquartile range, we first put the $n = 20$ data values in ascending order:

0	429	461	471	505
345	437	466	480	515
358	442	466	489	516
372	442	470	490	549

The first quartile is the median of the bottom 10 data values, which is the mean of the data values that lie in the 5th and 6th positions. These values are 429 and 437,

so $Q_1 = \dfrac{429 + 437}{2} = 433$ minutes.

The third quartile is the median of the top 10 data values, which is the mean of the data values that lie in the 15th and 16st positions. These values are 489 and 490,

so $Q_3 = \dfrac{489 + 490}{2} = 489.5$ minutes.

So, $IQR = 489.5 - 433 = 56.5$ minutes.
Note: Results from MINITAB differ:
$Q_1 = 431$ minutes and $Q_3 = 489.8$ minutes
$IQR = 489.8 - 431 = 58.8$ minutes
$UF = 489.8 + 1.5(58.8) = 578$ min.

Changing the value 346 to 0 causes the standard deviation to nearly double in size, while the interquartile range does not change at all. This illustrates the property of resistance. The standard deviation is not resistant, but the interquartile range is resistant.

Section 3.5

1. **Using the boxplot:** If the median is left of center in the box, and the right whisker is longer than the left whisker, the distribution is skewed right. If the median is in the center of the box, and the left and right whiskers are roughly the same length, the distribution is symmetric. If the median is right of center in the box, and the left whisker is longer than the right whisker, the distribution is skewed left.

Using the quartiles: If the distance from the median to the first quartile is less than the distance from the median to the third quartile, or the distance from the median to the minimum value in the data set is less than the distance from the median to the maximum value in the data set, then the distribution is skewed right. If the distance from the median to the first quartile is the same as the distance from the median to the third quartile, or the distance from the median to the minimum value in the data set is the same as the distance from the median to the maximum value in the data set, the distribution is symmetric. If the distance from the median to the first quartile is more than the distance from the median to the third quartile, or the distance from the median to the minimum value in the data set is more than the distance from the median to the maximum value in the data set, the distribution is skewed left.

3. **(a)** The median is to the left of the center of the box and the right line is substantially longer than the left line, so the distribution is skewed right.

    (b) Reading the boxplot, the five-number summary is: 0, 1, 3, 6, 16.

5. **(a)** For the variable x: $M = 40$

    (b) For the variable y: $Q_3 = 52$

    (c) The variable y has more dispersion. This can be seen by the much broader range (span of the lines) and the much broader interquartile range (span of the box).

    (d) The distribution of the variable x is symmetric. This can be seen because the median is near the center of the box and the horizontal lines are approximately the same in length.

    (e) The distribution of the variable y is skewed right. This can be seen because the median is to the left of the center of the box and the right line is substantially longer than the left line.

7.

Statistics Exams Scores

Percent

9. (a) Notice that the $n = 43$ data values are already arranged in order (moving down the columns). The smallest value (youngest president) in the data set is 42. The largest value (oldest president) in the data set is 69. The second quartile (median) is data value that lies in the $\frac{43+1}{2} = 22^{nd}$ position. So, $Q_2 = M = 55$.

The first quartile is the median of the bottom 21 data values, which is the data value that lies in the 11th position. So, $Q_1 = 51$.

The third quartile is the median of the top 21 data values, which is the data value that lies in the 33rd position. So, $Q_3 = 58$.

So, the five-number summary is:
42, 51, 55, 58, 69

(b) $IQR = 58 - 51 = 7$

$LF = Q_1 - 1.5(IQR) = 51 - 1.5(7) = 40.5$

$UF = Q_3 + 1.5(IQR) = 58 + 1.5(7) = 68.5$

Thus, 69 is an outlier.

Age of Presidents at Inauguration

(c) The median is near the center of the box and the horizontal lines are approximately the same in length, so the distribution is symmetric, with an outlier.

11. (a) We arrange the $n = 25$ data values into ascending order.

0.598	0.603	0.607	0.608	0.610
0.600	0.605	0.607	0.608	0.610
0.600	0.605	0.608	0.609	0.611
0.601	0.605	0.608	0.610	0.611
0.602	0.606	0.608	0.610	0.612

The smallest data value is 0.598. The largest data value is 0.612.

The second quartile (median) is data value that lies in the $\frac{25+1}{2} = 13^{th}$ position. So, $Q_2 = M = 0.608$.

The first quartile is the median of the bottom 12 data values, which is the mean of the data values that lie in the 6th and 7th

positions, which are 0.603 and 0.605. So, $Q_1 = \frac{0.603+0.605}{2} = 0.604$.

The third quartile is the median of the top 12 data values, which is the mean of the data values that lies in the 19th and 20th positions, which are 0.610 and 0.610. So, $Q_3 = \frac{0.610+0.610}{2} = 0.610$.

So, the five-number summary is:
0.598, 0.604, 0.608, 0.610, 0.612

$IQR = 0.610 - 0.604 = 0.006$

$LF = Q_1 - 1.5(IQR)$
$= 0.604 - 1.5(0.006) = 0.595$

$UF = Q_3 + 1.5(IQR)$
$= 0.610 + 1.5(0.006) = 0.619$

Thus, there are no outliers.

Weight of Tylenol Tablets

(b) The median is to the right of the center of the box and the left line is longer than the right line, so the distribution is skewed left.

13. To find the five-number summary, we arrange the $n = 50$ data values in ascending order:

0.79	0.84	0.87	0.88	0.91
0.81	0.84	0.87	0.88	0.91
0.82	0.85	0.87	0.89	0.91
0.82	0.85	0.87	0.89	0.91
0.83	0.86	0.87	0.89	0.92
0.83	0.86	0.88	0.90	0.93
0.84	0.86	0.88	0.90	0.93
0.84	0.86	0.88	0.90	0.93
0.84	0.86	0.88	0.90	0.94
0.84	0.86	0.88	0.91	0.95

The smallest data value is 0.79 grams, and the largest data value is 0.95 grams.

The second quartile (median) is the mean of the data values that lie in the 25th and 26th positions, which are 0.87 and 0.88. So, $Q_2 = M = \frac{0.87+0.88}{2} = 0.875$ grams.

The first quartile is the median of the bottom 25 data values, which is the value that lies in the 13th position. So, $Q_1 = 0.85$ grams.

The third quartile is the median of the top 25 data values, which is value that lies in the 38th position. So, $Q_3 = 0.90$ grams.

So, the five-number summary is:
0.79, 0.85, 0.875, 0.90, 0.95
$IQR = 0.90 - 0.85 = 0.05$ grams

$LF = Q_1 - 1.5(\text{IQR})$
$= 0.85 - 1.5(0.05) = 0.775$ grams

$UF = Q_3 + 1.5(\text{IQR})$
$= 0.90 + 1.5(0.05) = 0.975$ grams

So, there are no outliers.

Note: Results from MINITAB differ:
0.79, 0.8475, 0.875, 0.90, 0.95
$IQR = 0.0525$, $LF = 0.76875$, $UF = 0.97875$

Using the by-hand computations for the five-number summary, the boxplot follows:

Weight of M&Ms

Weight (in grams)

Since the range of the data between the minimum value and the median is roughly the same as the range between the median and the maximum value, and because the range of the data between the first quartile and median is the same as the range of the data between the median and third quartile, the distribution is symmetric.

15. (a) To find the five-number summary for each vitamin type, we arrange each data set in ascending order:

Centrum

2.15	2.57	2.80	3.12	3.85
2.15	2.60	2.95	3.25	3.92
2.23	2.63	3.02	3.30	4.00
2.25	2.67	3.02	3.35	4.02
2.30	2.73	3.03	3.53	4.17
2.38	2.73	3.07	3.63	4.33

Generic Brand

4.97	5.55	6.17	6.50	7.17
5.03	5.57	6.23	6.50	7.18
5.25	5.77	6.30	6.57	7.25
5.35	5.78	6.33	6.60	7.42
5.38	5.92	6.35	6.73	7.42
5.50	5.98	6.47	7.13	7.58

For Centrum, the smallest value is 2.15, and the largest value is 4.33. For the generic brand, the smallest value is 4.97, and the largest value is 7.58.

Since both sets of data contain $n = 30$ data points, the quartiles are in the same positions for both sets.

The second quartile (median) is the mean of the values that lie in the 15th and 16th positions. For Centrum, these values are both 3.02. So, $Q_2 = M = \frac{3.02 + 3.02}{2} = 3.02$.

For the generic brand, these values are 6.30 and 6.33, so $Q_2 = M = \frac{6.30 + 6.33}{2} = 6.315$.

The first quartile is the median of the bottom 15 data values. This is the value that lies in the 8th position. So, for Centrum, $Q_1 = 2.60$. For the generic brand, $Q_1 = 5.57$.

The third quartile is the median of the top 15 data values. This is the value that lies in the 23rd position. So, for Centrum, $Q_3 = 3.53$. For the generic brand, $Q_3 = 6.73$.

So, the five-number summaries are:
Centrum: 2.15, 2.60, 3.02, 3.53, 4.33
Generic: 4.97, 5.57, 6.315, 6.73, 7.58

The fences for Centrum are:
$LF = Q_1 - 1.5(\text{IQR})$
$= 2.60 - 1.5(3.53 - 2.60) = 1.205$

$UF = Q_3 + 1.5(\text{IQR})$
$= 3.53 + 1.5(3.53 - 2.60) = 4.925$

The fences for the generic brand are:
$LF = Q_1 - 1.5(\text{IQR})$
$= 5.57 - 1.5(6.73 - 5.57) = 3.83$

$UF = Q_3 + 1.5(\text{IQR})$
$= 6.73 + 1.5(6.73 - 5.57) = 8.47$

So, neither data set has any outliers.

Note: Results from MINITAB differ:
Centrum: 2.15, 2.593, 3.02, 3.555, 4.33
$LF = 1.15$ and $UF = 4.998$
Generic: 4.97, 5.565, 6.315, 6.83, 7.58
$LF = 3.6675$ and $UF = 8.7275$

Using the by-hand computations for the five-number summaries, the side-by-side boxplots follow:

Dissolving Time of Vitamins

(b) From the boxplots, we can see that the generic brand has both a larger range and a larger interquartile range. Therefore, the generic brand has more dispersion.

(c) From the boxplots, we can see that the Centrum vitamins dissolve in less time than the generic vitamins. That is, Centrum vitamins dissolve faster.

17. (a) This is an observational study. The researchers did not attempt to manipulate the outcomes, but merely observed them.

(b) The explanatory variable is whether or not the father smoked. The response variable is the birth weight.

(c) Answers will vary. Some possible lurking variables are eating habits, exercise habits, and whether the mother received prenatal care.

(d) This means that the researchers attempted to adjust their results for any variable that may also be related to birth weight.

(e) Nonsmokers:
Adding up the $n = 30$ data values, we obtain: $\sum x_i = 109,965$. So,

$$\bar{x} = \frac{\sum x_i}{n} = \frac{109,965}{30} = 3,665.5 \text{ grams}$$

To find the standard deviation, we use the computational formula. To do so, we square each data value and add them to obtain: $\sum x_i^2 = 406,751,613$. So,

$$s = \sqrt{s^2} = \sqrt{\frac{\sum x_i^2 - \frac{(\sum x_i)^2}{n}}{n-1}}$$

$$= \sqrt{\frac{406,751,613 - \frac{(109,965)^2}{30}}{30-1}}$$

$$\approx 356.0 \text{ grams}$$

To find the median and quartiles, we arrange the data in order:

Nonsmokers

2976	3423	3544	3771	4019
3062	3436	3544	3783	4054
3128	3454	3668	3823	4067
3263	3471	3719	3884	4194
3290	3518	3732	3976	4248
3302	3522	3746	3994	4354

The median (second quartile) is the mean of the data values that lie in the 15th and 16th positions, which are 3668 and 3719. So,

$$M = Q_2 = \frac{3668 + 3719}{2} = 3693.5 \text{ grams.}$$

The first quartile is the median of the bottom 15 data values, which is the value in the 8th position. So, $Q_1 = 3436$ grams.

The third quartile is the median of the top 15 data values, which is the value in the 23rd position. So, $Q_3 = 3976$ grams.

Note: Results from MINITAB for the quartiles differ: $Q_1 = 3432.8$ grams and $Q_3 = 3980.5$ grams.

Smokers:
Adding up the $n = 30$ data values, we obtain: $\sum x_i = 103,812$. So,

$$\bar{x} = \frac{\sum x_i}{n} = \frac{103,812}{30} = 3,460.4 \text{ grams}$$

To find the standard deviation, we use the computational formula. To do so, we square each data value and add them to obtain: $\sum x_i^2 = 365,172,886$. So,

$$s = \sqrt{s^2} = \sqrt{\frac{\sum x_i^2 - \frac{(\sum x_i)^2}{n}}{n-1}}$$

$$= \sqrt{\frac{365,172,886 - \frac{(103,812)^2}{30}}{30-1}}$$

$$\approx 452.6 \text{ grams}$$

To find the median and quartiles, we arrange the data in order:

Smokers

2746	3066	3282	3548	3963
2768	3129	3455	3629	3998
2851	3145	3457	3686	4104
2860	3150	3493	3769	4131
2918	3234	3502	3807	4216
2986	3255	3509	3892	4263

The median (second quartile) is the mean of the data values that lie in the 15th and 16th positions, which are 3457 and 3493. So,

$$M = Q_2 = \frac{3457 + 3129}{2} = 3475 \text{ grams.}$$

The first quartile is the median of the bottom 15 data values, which is the value in the 8th position. So, $Q_1 = 3129$ grams.

The third quartile is the median of the top 15 data values, which is the value in the 23rd position. So, $Q_3 = 3807$ grams.

Note: Results from MINITAB for the quartiles differ: $Q_1 = 3113.3$ grams and $Q_3 = 3828.3$ grams.

(f) For nonsmoking fathers, 25% of infants have a birth weight that is more than 3,436 grams (or 3,432.8 grams, if using MINITAB). For smoking fathers, 25% of infants have a birth weight that is more than 3,129 grams (or 3,113.3 grams, if using MINITAB).

(g) Using the by-hand results from part (e), the five number summaries are:
Nonsmokers: 2976, 3436, 3693.5, 3976, 4354
Smokers: 2746, 3129, 3475, 3807, 4263
The side-by-side boxplots follow:

Birthweight

Chapter 3 Review Exercises

1. (a) $\sum x_i = 793.8 + 793.1 + 792.4 + 794.0 + 791.4$
$\qquad\qquad + 792.4 + 791.7 + 792.3 + 789.6 + 794.4$
$\qquad = 7,925.1$

$$\text{Mean} = \overline{x} = \frac{\sum x_i}{n} = \frac{7,925.1}{10} = 792.51 \text{ m/sec.}$$

Data in order:
789.6, 791.4, 791.7, 792.3, 792.4,
792.4, 793.1, 793.8, 794.0, 794.4

$$\text{Median} = \frac{792.4 + 792.4}{2} = 792.4 \text{ m/sec.}$$

(b) Range = Largest Value – Smallest Value
$\qquad = 794.4 - 789.6 = 4.8$ m/sec

To calculate the sample variance and the sample standard deviation, we use the computational formulas:

x_i	x_i^2
793.8	630,118.44
793.1	629,007.61
792.4	627,897.76
794.0	630,436
791.4	626,313.96
792.4	627,897.76
791.7	626,788.89
792.3	627,739.29
789.6	623,468.16
794.4	631,071.36
$\sum x_i = 7925.1$	$\sum x_i^2 = 6,280,739.23$

$$s^2 = \frac{\sum x_i^2 - \frac{\left(\sum x_i\right)^2}{n}}{n-1}$$

$$= \frac{6,280,739.23 - \frac{\left(7,925.1\right)^2}{10}}{10-1}$$

$$\approx 2.03 \text{ (m/sec)}^2$$

$$s = \sqrt{\frac{6,280,739.23 - \frac{\left(7,925.1\right)^2}{10}}{10-1}}$$

$$\approx 1.42 \text{ m/sec}$$

2. (a) Add up the 9 data values: $\sum x = 91,610$

$$\overline{x} = \frac{\sum x}{n} = \frac{91,610}{9} \approx \$10,178.9$$

Data in order:
5500, 7200, 7889, 8998, 9980, 10995, 12999, 13999, 14050
The median is in the 5th position, so
$M = \$9,980$.

(b) Range = Largest Value – Smallest Value
$\qquad = 14,050 - 5,500 = \$8,550$

To find the interquartile range, we must find the first and third quartiles. The first quartile is the median of the bottom four data values, which is the mean of the values in the 2nd and 3rd positions. So,

$$Q_1 = \frac{7,200 + 7,889}{2} = 7,544.5.$$

The third quartile is the median of the top

four data values, which is the mean of the values in the 7th and 8th positions. So,

$$Q_3 = \frac{12,999 + 13,999}{2} = 13,499.$$

Finally, the interquartile range is:
IQR = 13,499 – 7,544.5 = \$5,954.5.

Note: Results from MINITAB differ:
$Q_1 = \$7,545$ and $Q_3 = \$13,499$, so
IQR = \$5,954.

To calculate the sample standard deviation, we use the computational formulas:

Data, x_i	x_i^2
14,050	197,402,500
13,999	195,972,001
12,999	168,974,001
10,995	120,890,025
9,980	99,600,400
8,998	80,964,004
7,889	62,236,321
7,200	51,840,000
5,550	30,250,000
$\sum x_i = 91,610$	$\sum x_i^2 = 1,008,129,252$

$$s = \sqrt{\frac{\sum x_i^2 - \frac{(\sum x_i)^2}{n}}{n-1}}$$

$$= \sqrt{\frac{1,008,129,252 - \frac{(91,610)^2}{9}}{9-1}} \approx \$3,074.9$$

(c) Add up the 9 data values: $\sum x = 118,610$

$$\bar{x} = \frac{\sum x}{n} = \frac{118,610}{9} \approx \$13,178.9$$

Data in order:
5500, 7200, 7889, 8998, 9980, 10995, 12999, 13999, 41050

The median is in the 5th position, so
$M = \$9,980$.

Range = 41,050 – 5,500 = \$35,550.

The first quartile is the mean of the values in the 2nd and 3rd positions. So,

$$Q_1 = \frac{7,200 + 7,889}{2} = 7,544.5.$$

The third quartile is the mean of the values in the 7th and 8th positions. So,

$$Q_3 = \frac{12,999 + 13,999}{2} = 13,499.$$

Finally, the interquartile range is:
IQR = 13,499 – 7,544.5 = \$5,954.5
(or \$5,954 if using MINITAB)

To calculate the sample standard deviation, we again use the computational formulas:

Data, x_i	x_i^2
41,050	1,685,102,500
13,999	195,972,001
12,999	168,974,001
10,995	120,890,025
9,980	99,600,400
8,998	80,964,004
7,889	62,236,321
7,200	51,840,000
5,550	30,250,000
$\sum x_i = 118,610$	$\sum x_i^2 = 2,495,829,252$

$$s = \sqrt{\frac{\sum x_i^2 - \frac{(\sum x_i)^2}{n}}{n-1}}$$

$$= \sqrt{\frac{2,495,829,252 - \frac{(118,610)^2}{9}}{9-1}}$$

$$\approx \$10,797.5$$

The mean, range, and standard deviation are all changed considerably by the incorrectly entered data value. The median and interquartile range did not change. The median and interquartile range are resistant, while the mean, range and standard deviation are not resistant.

3. (a) $\mu = \dfrac{\sum x}{N} = \dfrac{983}{17} \approx 57.8$ years

Data in order:
44, 46, 50, 51, 55, 56, 56, 56, 58, 59, 62, 62, 62, 64, 65, 68, 69

The median is the data value in the
$\dfrac{17+1}{2} = 9^{th}$ position. So, $M = 58$ years.

The data is bimodal: 56 years and 62 years. Both have frequencies of 3.

(b) Range = 69 – 44 = 25 years
To calculate the population standard deviation, we use the computational formula:

Data, x_i	x_i^2
44	1936
56	3136
51	2601
46	2116
59	3481
56	3136
58	3364
55	3025
65	4225
64	4096
68	4624
69	4761
56	3136
62	3844
62	3844
62	3844
50	2500
$\sum x_i = 983$	$\sum x_i^2 = 57{,}669$

$$\sigma = \sqrt{\frac{\sum x_i^2 - \frac{\left(\sum x_i\right)^2}{N}}{N}} = \sqrt{\frac{57{,}669 - \frac{(983)^2}{17}}{17}}$$

$$\approx 7.0 \text{ years}$$

(c) Answers will vary depending on samples selected.

4. (a) To construct the histogram, we first organize the data into a frequency table:

Tickets Issued	Frequency
0	18
1	11
2	1

Number of Tickets Issued in a Month

The distribution is skewed right.

(b) Since the distribution is skewed right, we would expect the mean to be greater than the median.

(c) To find the mean, we add all of the data values and divide by the sample size.

$$\sum x_i = 13 \ ; \ \overline{x} = \frac{\sum x_i}{n} = \frac{13}{30} = 0.4$$

To find the median, we arrange the data in order. The median is the mean of the mean of the 15th and 16th data values.

$$M = \frac{0+0}{2} = 0 \ .$$

(d) The mode is 0 (the most frequent value).

5. (a) By the Empirical Rule, approximately 99.7% of the data will be within 3 standard deviations of the mean. Now, $600 - 3(53) = 441$ and $600 + 3(53) = 759$. Thus, about 99.7% of light bulbs have lifetimes between 441 and 759 hours.

(b) Since 494 is exactly 2 standard deviations below the mean $[494 = 600 - 2(53)]$ and 706 is exactly 2 standard deviations above the mean $[706 = 600 + 2(53)]$, the Empirical Rule predicts that about 95% of the light bulbs will have lifetimes between 494 and 706 hours.

(c) Since 547 is exactly 1 standard deviations below the mean $[547 = 600 - 1(53)]$ and 706 is exactly 2 standard deviations above the mean $[706 = 600 + 2(53)]$, the Empirical Rule predicts that about $34 + 47.5 = 81.5\%$ of the light bulbs will have lifetimes between 547 and 706 hours.

(d) Since 441 hours is 3 standard deviations below the mean $[441 = 600 - 3(53)]$, the Empirical Rule predicts that 0.15% of light bulbs will last less than 441 hours. Thus, the company should expect to replace about 0.15% of the light bulbs.

(e) By Chebyshev's theorem, at least

$$\left(1 - \frac{1}{k^2}\right) \cdot 100\% = \left(1 - \frac{1}{2.5^2}\right) \cdot 100\% = 84\%$$

of all the light bulbs are within $k = 2.5$ standard deviations of the mean.

(f) Since 494 is exactly $k = 2$ standard deviations below the mean $[494 = 600 - 2(53)]$ and 706 is exactly 2 standard deviations above the mean $[706 = 600 + 2(53)]$, Chebyshev's inequality indicates that at least

$$\left(1 - \frac{1}{k^2}\right) \cdot 100\% = \left(1 - \frac{1}{2^2}\right) \cdot 100\% = 75\%$$

of the light bulbs will have lifetimes between 494 and 706 hours.

6. (a) To find the mean, we use the formula $\bar{x} = \dfrac{\sum x_i f_i}{\sum f_i}$. To find the standard deviation in part (b), we will

choose to use the computational formula $s = \sqrt{s^2} = \sqrt{\dfrac{\sum x_i^2 f_i - \dfrac{(\sum x_i f_i)^2}{\sum f_i}}{(\sum f_i) - 1}}$. We organize our

computations of x_i, $\sum f_i$, $\sum x_i f_i$, and $\sum x_i^2 f_i$ in the table that follows:

Class	Midpoint, x_i	Frequency, f_i	$x_i f_i$	x_i^2	$x_i^2 f_i$
$0-9$	$\dfrac{0+10}{2} = 5$	28	140	25	700
$10-19$	$\dfrac{10+20}{2} = 15$	151	2,265	225	33,975
$20-29$	25	177	4,425	625	110,625
$30-39$	35	206	7,210	1,225	252,350
$40-49$	45	94	4,230	2,025	190,350
$50-59$	55	93	5,115	3,025	281,325
$60-69$	65	88	5,720	4,225	371,800
$70-79$	75	45	3,375	5,625	253,125
$80-89$	85	13	1,105	7,225	93,925
		$\sum f_i = 895$	$\sum x_i f_i = 33,585$		$\sum x_i^2 f_i = 1,588,175$

With the table complete, we compute the sample mean:

$\mu = \dfrac{\sum x_i f_i}{\sum f_i} = \dfrac{33,585}{895} = 37.5$ minutes

(b) $s = \sqrt{s^2} = \sqrt{\dfrac{\sum x_i^2 f_i - \dfrac{(\sum x_i f_i)^2}{\sum f_i}}{(\sum f_i) - 1}} = \sqrt{\dfrac{1,588,175 - \dfrac{(33,585)^2}{895}}{895 - 1}} \approx 19.2$ minutes

---

7. $\text{GPA} = \bar{x}_w = \dfrac{\sum w_i x_i}{\sum w_i}$

$= \dfrac{5(4) + 4(3) + 3(4) + 3(2)}{5 + 4 + 3 + 3} = \dfrac{50}{15} \approx 3.33$

8. Female: $z = \dfrac{x - \mu}{\sigma} = \dfrac{160 - 156.5}{51.2} \approx 0.07$

Male: $z = \dfrac{x - \mu}{\sigma} = \dfrac{185 - 183.4}{40.0} = 0.04$

The weight of the 160-pound female is 0.07 standard deviations above the mean, while the weight of the 185-pound male is 0.04 standard deviations above the mean. Thus, the 160-pound female is relatively heavier.

9. (a) The European Union is more densely populated than the United States. From the side-by-side boxplots, we can see that all three quartiles are larger for the European Union.

(b) If we use the range as the measure, then it appears that the dispersion in population density is about the same for the United States and the European Union. However, if we use the interquartile range as the measure, then the European Union has more dispersion in population density than the United States.

(c) There is one outlier for the density of the European Union. It is approximately 490 people per square kilometer.

(d) Both distributions are skewed right.

10. (a) Add up the 54 data values: $\sum x_i = 127{,}611$

$$\mu = \frac{\sum x_i}{n} = \frac{127{,}611}{54} \approx 2{,}363 \text{ words}$$

To find the median and quartiles, we must arrange the data in order:

135	1,337	1,802	2,446	3,634
559	1,340	1,807	2,449	3,801
698	1,355	1,883	2,463	3,838
985	1,425	2,015	2,480	3,967
996	1,437	2,130	2,546	4,059
1,087	1,507	2,158	2,821	4,388
1,125	1,526	2,170	2,906	4,467
1,128	1,571	2,217	2,978	4,776
1,172	1,668	2,242	3,217	5,433
1,175	1,681	2,283	3,318	8,445
1,209	1,729	2,308	3,319	

The median is the mean of the data values that lie in the 27th and 28th positions, which are 2,130 and 2,168. So,

$$M = \frac{2{,}130 + 2{,}158}{2} = 2{,}144 \text{ words.}$$

(b) The first quartile is the median of the bottom 27 data values, which is the value in the 14th position. So, $Q_1 = 1{,}355$ words.

The second quartile is the median, $Q_2 = M = 2{,}144$ words.

The third quartile is the median of the top 27 data values, which is the value in the 41st position. So, $Q_3 = 2{,}978$ words.

Note: Results from MINITAB differ: $Q_1 = 1{,}351$, $Q_2 = 2{,}144$, and $Q_3 = 3{,}038$.

We use the by-hand quartiles for the interpretations: 25% of the inaugural addresses had 1,355 words or less, 75% of the inaugural addresses had more than 1,355 words; 50% of the inaugural addresses had 2,144 words or less, 50% of the inaugural addresses had more than 2,144 words; 75% of the inaugural addresses had 2,978 words or less, 25% of the inaugural addresses had more than 2,978 words.

(c) The smallest value is 135 and the largest is 8,445. Combining these with the quartiles, we obtain the five-number summary:
135, 1,355, 2,144, 2,978, 8,445

Note: results from MINITAB differ:
135, 1,351, 2,144, 3,038, 8,445

(d) To calculate the sample standard deviation, we use the computational formulas:
We add up the 54 data values:
$\sum x_i = 127{,}611$.

We square each of the 54 data values and add up the results: $\sum x_i^2 = 410{,}172{,}767$

$$\sigma = \sqrt{\frac{\sum x_i^2 - \dfrac{\left(\sum x_i\right)^2}{N}}{N}}$$

$$= \sqrt{\frac{410{,}172{,}767 - \dfrac{(127{,}611)^2}{54}}{54}}$$

$$\approx 1418 \text{ words}$$

The interquartile range is:
$$IQR = Q_3 - Q_1$$
$$= 2{,}978 - 1{,}355 = 1{,}623 \text{ words}$$

Note: results from MINITAB differ:
IQR = 3,038 − 1,351 = 1,687 words

(e) $LF = Q_1 - 1.5(IQR)$
$$= 1{,}355 - 1.5(1{,}623) = -1{,}079.5 \text{ words}$$

$UF = Q_3 + 1.5(IQR)$
$$= 2{,}978 + 1.5(1{,}623) = 5{,}412.5 \text{ words}$$

So, there are two outliers: 5,433 and 8,445.

Note: results from MINITAB differ:
$LF = 1{,}351 - 1.5(1687) = -1{,}179.5$ words
$UF = 3{,}038 + 1.5(1687) = 5{,}568.5$ words
Using MINITAB only 8,445 is an outlier.

(f) We use the by-hand results to construct the boxplot.

Number of Words in Inaugural Addresses

Number of Words

(g) The distribution is skewed right. We can tell because the median is slightly left of center in the box and the right whisker is longer than the left whisker (even without considering the outlier)

(h) The median is the better measure because the distribution is skewed right and the outliers inflate the value of the mean.

(i) The interquartile range is the better measure because the outliers inflate the value of the standard deviation.

11. This means that 85% of 19-year-old females have a height that is 67.1 inches or less, and $(100 - 85)\% = 15\%$ of 19-year-old females have a height that is more than 67.1 inches.

12. The median is used for three measures since it is likely the case that one of the three measures is extreme relative to the other two, thus substantially affecting the value of the mean. Since the median is resistant to extreme values, it is the better measure of central tendency.

Chapter 3 Test

1. (a) $\sum x_i = 399 + 542 + 347 + 381 + 400 + 289$
$+ 358 + 439 + 602 + 661$
$= 4,418$

Mean $= \bar{x} = \dfrac{\sum x_i}{n} = \dfrac{4,418}{10} = 441.8$ inches

(b) Data in order: 289, 347, 358, 381, 399, 400, 439, 542, 602, 661
The median is the mean of the values in the 5th and 6th positions:
Median $= \dfrac{399 + 400}{2} = 399.5$ inches

(c) $\sum x_i = 399 + 542 + 347 + 381 + 400 + 289$
$+ 358 + 439 + 602 + 1,661$
$= 5,418$

Mean $= \bar{x} = \dfrac{\sum x_i}{n} = \dfrac{5,418}{10} = 541.8$ inches

Data in order: 289, 347, 358, 381, 399, 400, 439, 542, 602, 1661
The median is the mean of the values in the 5th and 6th positions:
Median $= \dfrac{399 + 400}{2} = 399.5$ inches

The mean was changed significantly by the incorrectly entered data value. The median did not change. The median is resistant, while the mean is not resistant.

2. The mode type of larceny is "From motor vehicles." It has the highest frequency.

3. Range = Largest Value – Smallest Value
$= 661 - 289 = 372$ inches.

4. (a) To calculate the sample standard deviation, we use the computational formula:

x_i	x_i^2
399	159,201
542	293,764
347	120,409
381	145,161
400	160,000
289	83,521
358	128,164
439	192,721
602	362,404
661	436,921
$\sum x_i = 4418$	$\sum x_i^2 = 2,082,266$

$$s = \sqrt{\dfrac{\sum x_i^2 - \dfrac{(\sum x_i)^2}{n}}{n-1}}$$

$$= \sqrt{\dfrac{2,082,266 - \dfrac{(4,418)^2}{10}}{10-1}} \approx 120.4 \text{ inches}$$

(b) To find the interquartile range, we must find the first and third quartiles. The first quartile is the median of the bottom five data values, which is the value in the 3rd position. So, $Q_1 = 358$ inches.
The third quartile is the median of the top five data values, which is the value in the 8th position. So, $Q_3 = 542$ inches.
Finally, the interquartile range is:
IQR $= 542 - 358 = 184$ inches.
Interpretation: The middle 50% of all snowfalls have a range of 184 inches.

(c) The interquartile range is resistant; the standard deviation is not resistant.

5. (a) By the Empirical Rule, approximately 99.7% of the data will be within 3 standard deviations of the mean. Now, $4302 - 3(340) = 3282$ and $4302 + 3(340) = 5322$. So, about 99.7% of toner cartridges will print between 3282 and 5322 page.

(b) Since 3622 is =2 standard deviations below the mean $[3622 = 4302 - 2(340)]$ and 4982 is 2 standard deviations above the mean $[4982 = 4302 + 2(340)]$, the Empirical Rule predicts that about 95% of the toner cartridges will print between 3622 and 4982 hours.

(c) Since 3622 is 2 standard deviations below the mean [3622 = 4302 − 2(340)], the Empirical Rule predicts that 0.15 + 2.35 = 2.5% of the toner cartridges will last less than 3622 pages. So, the company can expect to replace about 2.5% of the toner cartridges.

(d) By Chebyshev's theorem, at least

$$\left(1-\frac{1}{k^2}\right)\cdot100\% = \left(1-\frac{1}{1.5^2}\right)\cdot100\% \approx 55.6\%$$

of all the toner cartridges are within $k = 1.5$ standard deviations of the mean.

(e) Since 3282 is $k = 3$ standard deviations below the mean [3282 = 4302 − 3(340)] and 5322 is 3 standard deviations above the mean [5322 = 4302 + 3(340)], so by Chebyshev's inequality at least

$$\left(1-\frac{1}{k^2}\right)\cdot100\% = \left(1-\frac{1}{3^2}\right)\cdot100\% \approx 88.9\%$$

of the toner cartridges will print between 3282 and 5322 pages.

---

6. (a) To find the mean, we use the formula $\bar{x} = \dfrac{\sum x_i f_i}{\sum f_i}$. To find the standard deviation in part (b), we will use

the computational formula $s = \sqrt{s^2} = \sqrt{\dfrac{\sum x_i^2 f_i - \dfrac{(\sum x_i f_i)^2}{\sum f_i}}{(\sum f_i)-1}}$. We organize our computations of x_i, $\sum f_i$,

$\sum x_i f_i$, and $\sum x_i^2 f_i$ in the table that follows:

Class	Midpoint, x_i	Frequency, f_i	$x_i f_i$	x_i^2	$x_i^2 f_i$
40 – 49	$\frac{40+50}{2}=45$	8	360	2,025	16,200
50 – 59	$\frac{50+60}{2}=55$	44	2,420	3,025	133,100
60 – 69	65	23	1,495	4,225	97,175
70 – 79	75	6	450	5,625	33,750
80 – 89	85	107	9,095	7,225	773,075
90 – 99	95	11	1,045	9,025	99,275
100 – 109	105	1	105	11,025	11,025
		$\sum f_i = 200$	$\sum x_i f_i = 14{,}970$		$\sum x_i^2 f_i = 1{,}163{,}600$

With the table complete, we compute the sample mean: $\mu = \dfrac{\sum x_i f_i}{\sum f_i} = \dfrac{14{,}970}{200} \approx 74.9$ minutes

(b) $s = \sqrt{s^2} = \sqrt{\dfrac{\sum x_i^2 f_i - \dfrac{(\sum x_i f_i)^2}{\sum f_i}}{(\sum f_i)-1}} = \sqrt{\dfrac{1{,}163{,}600 - \dfrac{(14{,}970)^2}{200}}{200-1}} \approx 14.7$ minutes

7. Cost per pound $= \bar{x}_w = \dfrac{\sum w_i x_i}{\sum w_i} = \dfrac{2(\$2.70)+1(\$1.30)+\frac{1}{2}(\$1.80)}{2+1+\frac{1}{2}} \approx \$2.17/\text{lb}$

8. (a) Material A: $\sum x_i = 64.04$ and $n = 10$, so

$$\bar{x}_A = \frac{64.04}{10} = 6.404 \text{ million cycles}.$$

Material B: $\sum x_i = 113.32$ and $n = 10$,

so $\bar{x}_B = \frac{113.32}{10} = 11.332 \text{ million cycles}.$

(b) Notice that each set of $n = 10$ data values is already arranged in order. For each set, the median is the mean of the values in the 5th and 6th positions.

$$M_A = \frac{5.69 + 5.88}{2} = 5.785 \text{ million cycles}$$

$$M_B = \frac{8.20 + 9.65}{2} = 8.925 \text{ million cycles}$$

(c) To find the sample standard deviation, we choose to use the computational formula:
Material A:

$\sum x_i = 64.04$ and $\sum x_i^2 = 472.177$, so

$$s_A = \sqrt{\frac{\sum x_i^2 \sum x_i^2 - \frac{(\sum x_i)^2}{n}}{n-1}}$$

$$= \sqrt{\frac{472.177 - \frac{(64.04)^2}{10}}{10-1}}$$

$$\approx 2.626 \text{ million cycles}$$

Material B:

$\sum x_i = 113.32$ and $\sum x_i^2 = 1{,}597.4002$, so

$$s_B = \sqrt{\frac{1597.4002 - \frac{(113.32)^2}{10}}{10-1}}$$

$$\approx 5.900 \text{ million cycles}$$

Material B has more dispersed failure times because it has a much larger standard deviation.

(d) For each set, the first quartile is the data value in the 3rd position and the third quartile is the data value in the 8th position.

Material A:
3.17; 4.52; 5.785; 8.01; 11.92

Material B:
5.78; 6.84; 8.925; 14.71; 24.37

(e) Before drawing the side-by-side boxplots, we check each data set for outliers.

Fences for Material A:
$$LF = Q_1 - 1.5(IQR)$$
$$= 4.52 - 1.5(8.01 - 4.52)$$
$$= -0.715 \text{ million cycles}$$
$$UF = Q_3 + 1.5(IQR)$$
$$= 8.01 + 1.5(8.01 - 4.52)$$
$$= 13.245 \text{ million cycles}$$
Material A has no outliers.

Fences for Material B:
$$LF = 6.84 - 1.5(14.71 - 6.84)$$
$$= -4.965 \text{ million cycles}$$
$$UF = 14.71 + 1.5(14.71 - 6.84)$$
$$= 26.515 \text{ million cycles}$$
Material B has no outliers

Bearing Failures

Material A

Material B

Failure Time (million cycles)

Annotated remarks will vary. Possible remarks. Material A generally has lower failure times. Material B has more dispersed failure times.

(f) In both boxplots, the median is to the left of the center of the box and the right line is substantially longer than the left line, so both distributions are skewed right.

(h) For both distributions, the median is the better measure of central tendency since the distributions are skewed.

9. Notice that the data set is already arranged in order. The second quartile (median) is the mean of the values in the 25th and 26th positions, which are both 5.60. So,

$$Q_2 = \frac{5.60 + 5.60}{2} = 5.60 \text{ grams}.$$

The first quartile is the median of the bottom 25 data values, which is the value in the 13th position. So, $Q_1 = 5.58$ grams.

The third quartile is the median of the top 25 data values, which is the value in the 38th position. So, $Q_3 = 5.66$ grams.

$LF = Q_1 - 1.5(IQR)$

$\quad = 5.58 - 1.5(5.66 - 5.58) = 5.46$ grams

$UF = Q_3 + 1.5(IQR)$

$\quad = 5.66 + 1.5(5.66 - 5.58) = 5.78$ grams

So, the quarter that weighs 5.84 grams is an outlier.

Note: Results from MINITAB differ: $Q_1 = 5.58$, $Q_2 = 5.60$, $Q_3 = 5.6625$, $LF = 5.45625$, and $UF = 5.78625$.

For the interpretations, we use the by-hand quartiles: 25% of quarters have a weight that is 5.58 grams or less, 75% of the quarters have a weight more than 5.58 grams; 50% of the quarters have a weight that is 5.60 grams or less, 50% of the quarters have a weight more than 5.60 grams; 75% of the quarters have a weight that is 5.66 grams or less, 25% of the quarters have a weight that is more than 5.66 grams.

10. SAT: $z = \dfrac{x - \mu}{\sigma} = \dfrac{610 - 515}{114} \approx 0.83$

    ACT: $z = \dfrac{x - \mu}{\sigma} = \dfrac{27 - 21.0}{5.1} \approx 1.18$

    Armando's SAT score is 0.83 standard deviation above the mean, his ACT score is 1.18 standard deviations above the mean. So, Armando should report his ACT score since it is more standard deviations above the mean.

11. This means that 15% of 10-year-old males have a height that is 53.5 inches or less, and $(100 - 15)\% = 85\%$ of 10-year-old males have a height that is more than 53.5 inches.

12. You should report the median. Income data will be skewed right which means the median will be less than the mean.

13. (a) Report the mean since the distribution is symmetric.
    (b) Histogram I has more dispersion. The range of classes is larger.

Chapter 4

Describing the Relation between Two Variables

Section 4.1

1. Univariate data measures the value of a single variable for each individual in the study. Bivariate data measures values of two variables for each individual.

3. If $r = 1$, a perfect positive linear relation exits between the variables, and the points of the scatter diagram will lie exactly on a straight line with positive slope.

5. The linear correlation coefficient can only be calculated from bivariate *quantitative* data. The gender of a driver is a qualitative variable.

7. A lurking variable is a variable that is related either to both the explanatory variable and the response variable or to the explanatory variable, but has not been considered in the study. Examples will vary. One possibility: The number of firemen responding to a fire can be used to predict the amount of damage done. Both variables are related to the seriousness of the fire.

9. Nonlinear

11. Linear; positive

13. (a) III   (b) IV   (c) II   (d) I

15. (a) Looking at the scatter diagram, the points tend to increase at a relatively consistent rate from left to right. So, there appears to be a positive, linear association between level of education and median income.

   (b) The point with roughly the coordinates (48.5, 46000) appears to stick out. Reasons may vary. One possibility: A high concentration of government jobs that require a bachelor's degree but pay less than the private sector in a region with a high cost of living.

   (c) This illustrates that the correlation coefficient is not resistant.

17. (a)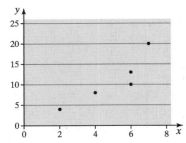

   (b) We compute the mean and standard deviation for both variables: $\bar{x} = 5$, $s_x = 2$, $\bar{y} = 11$, and $s_y = 6$. We determine $\dfrac{x_i - \bar{x}}{s_x}$ and $\dfrac{y_i - \bar{y}}{s_y}$ in columns 3 and 4. We multiply these entries to determine to obtain the entries in column 5.

x_i	y_i	$\dfrac{x_i - \bar{x}}{s_x}$	$\dfrac{y_i - \bar{y}}{s_y}$	$\left(\dfrac{x_i - \bar{x}}{s_x}\right)\left(\dfrac{y_i - \bar{y}}{s_y}\right)$
2	4	−1.5	−1.16667	1.75
4	8	−0.5	−0.5	0.25
6	10	0.5	−0.16667	−0.08334
6	13	0.5	0.33333	0.16667
7	20	1	1.5	1.5

   We add the entries in column 5 to obtain
   $$\sum\left(\frac{x_i - \bar{x}}{s_x}\right)\left(\frac{y_i - \bar{y}}{s_y}\right) = 3.58333 .$$

   Finally, we use this result to compute r:
   $$r = \frac{\sum\left(\dfrac{x_i - \bar{x}}{s_x}\right)\left(\dfrac{y_i - \bar{y}}{s_y}\right)}{n - 1} = \frac{3.58333}{5 - 1} \approx 0.896$$

   (c) A linear relation exists between x and y.

19. (a)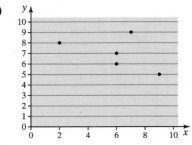

(b) We compute the mean and standard deviation for both variables: $\bar{x} = 6$, $s_x = 2.54951$, $\bar{y} = 7$, and $s_y = 1.58114$.

We determine $\dfrac{x_i - \bar{x}}{s_x}$ and $\dfrac{y_i - \bar{y}}{s_y}$ in columns 3 and 4. We multiply these entries to determine to obtain the entries in column 5.

x_i	y_i	$\dfrac{x_i - \bar{x}}{s_x}$	$\dfrac{y_i - \bar{y}}{s_y}$	$\left(\dfrac{x_i - \bar{x}}{s_x}\right)\left(\dfrac{y_i - \bar{y}}{s_y}\right)$
2	8	−1.56893	0.63246	−0.99229
6	7	0	0	0
6	6	0	−0.63246	0
7	9	0.39223	1.26491	0.49614
9	5	1.17670	−1.26491	−1.48842

We add the entries in column 5 to obtain

$$\sum\left(\frac{x_i - \bar{x}}{s_x}\right)\left(\frac{y_i - \bar{y}}{s_y}\right) = -1.98457.$$

Finally, we use this result to compute r:

$$r = \frac{\sum\left(\dfrac{x_i - \bar{x}}{s_x}\right)\left(\dfrac{y_i - \bar{y}}{s_y}\right)}{n - 1} = \frac{-1.98457}{5 - 1}$$
$$\approx -0.496$$

(c) No linear relation exists between x and y.

21. (a) Positive correlation. The more infants the more diapers will be needed.

(b) Negative correlation. The lower the interest rates the more people can afford to buy a car.

(c) Negative correlation. More exercise can lower cholesterol.

(d) Negative correlation. The higher the price of a Big Mac, the fewer Big Macs and French fries will be sold.

(e) No correlation. There is no correlation between shoe size and intelligence.

23. (a) Explanatory variable: height; Response variable: head circumference

(b)

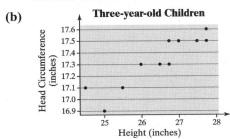

(c) To calculate the correlation coefficient, we use the computational formula:

x_i	y_i	x_i^2	y_i^2	$x_i y_i$
27.75	17.5	770.0625	306.25	485.625
24.5	17.1	600.25	292.41	418.95
25.5	17.1	650.25	292.41	436.05
26	17.3	676	299.29	449.8
25	16.9	625	285.61	422.5
27.75	17.6	770.0625	309.76	488.4
26.5	17.3	702.25	299.29	458.45
27	17.5	729	306.25	472.5
26.75	17.3	715.5625	299.29	462.775
26.75	17.5	715.5625	306.25	468.125
27.5	17.5	756.25	306.25	481.25
291	190.6	7710.25	3303.06	5044.425

From the table, we have $n = 11$, $\sum x_i = 291$, $\sum y_i = 190.6$, $\sum x_i^2 = 7710.25$, $\sum y_i^2 = 3303.06$, and $\sum x_i y_i = 5044.425$.
So, the correlation coefficient is

$$r = \frac{\sum x_i y_i - \dfrac{\sum x_i \sum y_i}{n}}{\sqrt{\left(\sum x_i^2 - \dfrac{(\sum x_i)^2}{n}\right)\left(\sum y_i^2 - \dfrac{(\sum y_i)^2}{n}\right)}}$$

$$= \frac{5044.425 - \dfrac{(291)(190.6)}{11}}{\sqrt{\left(7710.25 - \dfrac{(291)^2}{11}\right)\left(3303.06 - \dfrac{(190.6)^2}{11}\right)}}$$

$$\approx 0.911$$

(d) There is a strong positive linear association between the height and head circumference of a child.

25. (a) Explanatory variable: weight; Response variable: miles per gallon

(b)

(c) To calculate the correlation coefficient, we use the computational formula.

From the table that follows, we have $n = 11$, $\sum x_i = 36{,}194$, $\sum y_i = 239$, $\sum x_i^2 = 121{,}648{,}000$, $\sum y_i^2 = 5329$, and $\sum x_i y_i = 768{,}408$.

x_i	y_i	x_i^2	y_i^2	$x_i y_i$
3765	19	14,175,225	361	71,535
3984	18	15,872,256	324	71,712
3530	21	12,460,900	441	74,130
3175	22	10,080,625	484	69,850
2580	27	6,656,400	729	69,660
3730	18	13,912,900	324	67,140
2605	26	6,786,025	676	67,730
3772	17	14,227,984	289	64,124
3310	20	10,956,100	400	66,200
2991	25	8,946,081	625	74,775
2752	26	7,573,504	676	71,552
36,194	239	121,648,000	5329	768,408

So, the correlation coefficient is

$$r = \frac{\sum x_i y_i - \frac{\sum x_i \sum y_i}{n}}{\sqrt{\left(\sum x_i^2 - \frac{(\sum x_i)^2}{n}\right)\left(\sum y_i^2 - \frac{(\sum y_i)^2}{n}\right)}}$$

$$= \frac{768,408 - \frac{(36,194)(239)}{11}}{\sqrt{\left(121,648,000 - \frac{(36,194)^2}{11}\right)\left(5329 - \frac{(239)^2}{11}\right)}}$$

$$\approx -0.964$$

(d) Yes, there is a strong negative linear association between the weight of a car and its miles per gallon in the city.

27. (a)

HDL Cholesterol

(b) To calculate the correlation coefficient, we use the computational formula. From the table that follows, we have $n = 17$,

$\sum x_i = 765$, $\sum y_i = 764$, $\sum x_i^2 = 36,851$,

$\sum y_i^2 = 35,862$, and $\sum x_i y_i = 34,065$. So, the correlation coefficient is

$$r = \frac{\sum x_i y_i - \frac{\sum x_i \sum y_i}{n}}{\sqrt{\left(\sum x_i^2 - \frac{(\sum x_i)^2}{n}\right)\left(\sum y_i^2 - \frac{(\sum y_i)^2}{n}\right)}}$$

$$= \frac{34,065 - \frac{(765)(764)}{17}}{\sqrt{\left(36,851 - \frac{(765)^2}{17}\right)\left(35,862 - \frac{(764)^2}{17}\right)}}$$

$$\approx -0.164$$

x_i	y_i	x_i^2	y_i^2	$x_i y_i$
38	57	1444	3249	2166
42	54	1764	2916	2268
46	34	2116	1156	1564
32	56	1024	3136	1792
55	35	3025	1225	1925
52	40	2704	1600	2080
61	42	3721	1764	2562
61	38	3721	1444	2318
26	47	676	2209	1222
38	44	1444	1936	1672
66	62	4356	3844	4092
30	53	900	2809	1590
51	36	2601	1296	1836
27	45	729	2025	1215
52	38	2704	1444	1976
49	55	2401	3025	2695
39	28	1521	784	1092
765	764	36851	35862	34065

(c) No linear relation exists between age and HDL cholesterol.

29. (a)

Brain Size vs. IQ

(b) To calculate the correlation coefficient, we use the computational formula:

x_i	y_i	x_i^2	y_i^2	$x_i y_i$
816,932	133	667,377,892,624	17,689	108,651,956
951,545	137	905,437,887,025	18,769	130,361,665
991,305	138	982,685,603,025	19,044	136,800,090
833,868	132	695,335,841,424	17,424	110,070,576
856,472	140	733,544,286,784	19,600	119,906,080
852,244	132	726,319,835,536	17,424	112,496,208
790,619	135	625,078,403,161	18,225	106,733,565
866,662	130	751,103,022,244	16,900	112,666,060
857,782	133	735,789,959,524	17,689	114,085,006
948,066	133	898,829,140,356	17,689	126,092,778
949,395	140	901,350,866,025	19,600	132,915,300
1,001,121	140	1,002,243,256,641	19,600	140,156,940
1,038,437	139	1,078,351,402,969	19,321	144,342,743
965,353	133	931,906,414,609	17,689	128,391,949
955,466	133	912,915,277,156	17,689	127,076,978
1,079,549	141	1,165,426,043,401	19,881	152,216,409
924,059	135	853,885,035,481	18,225	124,747,965
955,003	139	912,030,730,009	19,321	132,745,417
935,494	141	875,149,024,036	19,881	131,904,654
949,589	144	901,719,268,921	20,736	136,740,816
18,518,961	2,728	17,256,479,190,951	372,396	2,529,103,155

From the table, we have $n = 20$,
$\sum x_i = 18{,}518{,}961$, $\sum y_i = 2{,}728$, $\sum x_i^2 = 17{,}256{,}479{,}190{,}951$, $\sum y_i^2 = 373{,}396$,
and $\sum x_i y_i = 2{,}529{,}103{,}155$. So, the correlation coefficient is

$$r = \frac{\sum x_i y_i - \dfrac{\sum x_i \sum y_i}{n}}{\sqrt{\left(\sum x_i^2 - \dfrac{(\sum x_i)^2}{n}\right)\left(\sum y_i^2 - \dfrac{(\sum y_i)^2}{n}\right)}}$$

$$= \frac{2{,}529{,}103{,}155 - \dfrac{(18{,}518{,}961)(2{,}728)}{20}}{\sqrt{\left(17{,}256{,}479{,}190{,}951 - \dfrac{(18{,}518{,}961)^2}{20}\right)\left(372{,}396 - \dfrac{(2{,}728)^2}{20}\right)}}$$

$$\approx 0.548$$

A weak positive linear association exists between MRI count and IQ.

(c)

Brain Size vs. IQ

Looking at the scatter diagram, we can see that females tend to have lower MRI counts. When separating the two groups, even the weak linear relationship seems to disappear. Neither group presents any clear relationship between IQ and MRI counts.

(d) We first use the computational formula on the data from the female participants:

x_i	y_i	x_i^2	y_i^2	$x_i y_i$
816,932	133	667,377,892,624	17,689	108,651,956
951,545	137	905,437,887,025	18,769	130,361,665
991,305	138	982,685,603,025	19,044	136,800,090
833,868	132	695,335,841,424	17,424	110,070,576
856,472	140	733,544,286,784	19,600	119,906,080
852,244	132	726,319,835,536	17,424	112,496,208
790,619	135	625,078,403,161	18,225	106,733,565
866,662	130	751,103,022,244	16,900	112,666,060
857,782	133	735,789,959,524	17,689	114,085,006
948,066	133	898,829,140,356	17,689	126,092,778
8,765,495	1,343	7,721,501,871,703	180,453	1,177,863,984

From the table, we have $n = 10$,
$\sum x_i = 8{,}765{,}495$, $\sum y_i = 1{,}343$, $\sum x_i^2 = 7{,}721{,}501{,}871{,}703$, $\sum y_i^2 = 180{,}453$, and
$\sum x_i y_i = 1{,}177{,}863{,}984$. So, the correlation coefficient for the female data is

$$r = \frac{\sum x_i y_i - \dfrac{\sum x_i \sum y_i}{n}}{\sqrt{\left(\sum x_i^2 - \dfrac{(\sum x_i)^2}{n}\right)\left(\sum y_i^2 - \dfrac{(\sum y_i)^2}{n}\right)}}$$

$$= \frac{1{,}177{,}863{,}984 - \dfrac{(8{,}765{,}495)(1{,}343)}{10}}{\sqrt{\left(7{,}721{,}501{,}871{,}703 - \dfrac{(8{,}765{,}495)^2}{10}\right)\left(180{,}453 - \dfrac{(1{,}343)^2}{10}\right)}}$$

$$\approx 0.359$$

We next use the computational formula on the data from the male participants:

x_i	y_i	x_i^2	y_i^2	$x_i y_i$
949,395	140	901,350,866,025	19,600	132,915,300
1,001,121	140	1,002,243,256,641	19,600	140,156,940
1,038,437	139	1,078,351,402,969	19,321	144,342,743
965,353	133	931,906,414,609	17,689	128,391,949
955,466	133	912,915,277,156	17,689	127,076,978
1,079,549	141	1,165,426,043,401	19,881	152,216,409
924,059	135	853,885,035,481	18,225	124,747,965
955,003	139	912,030,730,009	19,321	132,745,417
935,494	141	875,149,024,036	19,881	131,904,654
949,589	144	901,719,268,921	20,736	136,740,816
9,753,466	1,385	9,534,977,319,248	191,943	1,351,239,171

From the table, we have $n = 10$,
$\sum x_i = 9{,}753{,}466$, $\sum y_i = 1{,}385$,
$\sum x_i^2 = 9{,}534{,}977{,}319{,}248$,
$\sum y_i^2 = 191{,}943$, and
$\sum x_i y_i = 1{,}351{,}239{,}171$. So, the correlation coefficient for the male data is

$$r = \frac{\sum x_i y_i - \dfrac{\sum x_i \sum y_i}{n}}{\sqrt{\left(\sum x_i^2 - \dfrac{(\sum x_i)^2}{n}\right)\left(\sum y_i^2 - \dfrac{(\sum y_i)^2}{n}\right)}}$$

$$= \frac{1{,}351{,}239{,}171 - \dfrac{(9{,}753{,}466)(1{,}385)}{10}}{\sqrt{\left(9{,}534{,}977{,}319{,}248 - \dfrac{(9{,}753{,}466)^2}{10}\right)\left(191{,}943 - \dfrac{(1{,}385)^2}{10}\right)}}$$

$$\approx 0.236$$

There is no linear association between MRI count and IQ. The moral of the story is to beware of lurking variables. Mixing distinct populations can produce misleading results that result in incorrect conclusions.

31. (a)

Weight vs. Mileage

(Miles per Gallon vs. Weight (lb) scatter plot with y-axis "Miles per Gallon" from 16 to 28, x-axis "Weight (lb)" from 2,500 to 4,000)

(d)

Miles per Gallon vs Weight (lb)

(b) To calculate the correlation coefficient, we use the computational formula:

x_i	y_i	x_i^2	y_i^2	$x_i y_i$
3305	19	10,923,025	361	62,795
3765	19	14,175,225	361	71,535
3984	18	15,872,256	324	71,712
3530	21	12,460,900	441	74,130
3175	22	10,080,625	484	69,850
2580	27	6,656,400	729	69,660
3730	18	13,912,900	324	67,140
2605	26	6,786,025	676	67,730
3772	17	14,227,984	289	64,124
3310	20	10,956,100	400	66,200
2991	25	8,946,081	625	74,775
2752	26	7,573,504	676	71,552
39,499	258	132,571,025	5690	831,203

From the table, we have $n = 12$, $\sum x_i = 39,499$, $\sum y_i = 258$, $\sum x_i^2 = 132,571,025$, $\sum y_i^2 = 5690$, and $\sum x_i y_i = 831,203$. So, the correlation coefficient is

$$r = \frac{\sum x_i y_i - \frac{\sum x_i \sum y_i}{n}}{\sqrt{\left(\sum x_i^2 - \frac{(\sum x_i)^2}{n}\right)\left(\sum y_i^2 - \frac{(\sum y_i)^2}{n}\right)}}$$

$$= \frac{831,203 - \frac{(39,499)(258)}{12}}{\sqrt{\left(132,571,025 - \frac{(39,499)^2}{12}\right)\left(5690 - \frac{(258)^2}{12}\right)}}$$

$$\approx -0.943$$

(c) The scatter diagram in part (a) looks very similar to the one from problem 25, and the correlation coefficient from part (b) is similar to the one computed in problem 25 (-0.964). These results are reasonable because the Taurus follows the overall pattern of the data.

(e) To recalculate the correlation coefficient using the computational formula:

x_i	y_i	x_i^2	y_i^2	$x_i y_i$
2890	60	8,352,100	3600	173,400
3765	19	14,175,225	361	71,535
3984	18	15,872,256	324	71,712
3530	21	12,460,900	441	74,130
3175	22	10,080,625	484	69,850
2580	27	6,656,400	729	69,660
3730	18	13,912,900	324	67,140
2605	26	6,786,025	676	67,730
3772	17	14,227,984	289	64,124
3310	20	10,956,100	400	66,200
2991	25	8,946,081	625	74,775
2752	26	7,573,504	676	71,552
39,084	299	130,000,100	8929	941,808

From the table, we have $n = 12$, $\sum x_i = 39,084$, $\sum y_i = 299$, $\sum x_i^2 = 130,000,100$, $\sum y_i^2 = 8929$, and $\sum x_i y_i = 941,808$. So, the correlation coefficient is

$$r = \frac{\sum x_i y_i - \frac{\sum x_i \sum y_i}{n}}{\sqrt{\left(\sum x_i^2 - \frac{(\sum x_i)^2}{n}\right)\left(\sum y_i^2 - \frac{(\sum y_i)^2}{n}\right)}}$$

$$= \frac{941,808 - \frac{(39,084)(299)}{12}}{\sqrt{\left(130,000,100 - \frac{(39,084)^2}{12}\right)\left(8929 - \frac{(299)^2}{12}\right)}}$$

$$\approx -0.507$$

(f) The Prius is a hybrid car, so it gets much better gas mileage than the other cars which are not hybrids.

33. (a) <u>Data Set 1:</u>

x_i	y_i	x_i^2	y_i^2	x_iy_i
10	8.04	100	64.6416	80.40
8	6.95	64	48.3025	55.60
13	7.58	169	57.4564	98.54
9	8.81	81	77.6161	79.29
11	8.33	121	69.3889	91.63
14	9.96	196	99.2016	139.44
6	7.24	36	52.4176	43.44
4	4.26	16	18.1476	17.04
12	10.84	144	117.5056	130.08
7	4.82	49	23.2324	33.74
5	5.68	25	32.2624	28.40
99	82.51	1001	660.1727	797.60

From the table, we have $n = 11$,

$\sum x_i = 99$, $\sum y_i = 82.51$, $\sum x_i^2 = 1001$,

$\sum y_i^2 = 660.1727$, and $\sum x_iy_i = 797.60$.

So, the correlation coefficient is

$$r = \frac{\sum x_iy_i - \dfrac{\sum x_i \sum y_i}{n}}{\sqrt{\left(\sum x_i^2 - \dfrac{(\sum x_i)^2}{n}\right)\left(\sum y_i^2 - \dfrac{(\sum y_i)^2}{n}\right)}}$$

$$= \frac{797.60 - \dfrac{(99)(82.51)}{11}}{\sqrt{\left(1001 - \dfrac{(99)^2}{11}\right)\left(660.1727 - \dfrac{(82.51)^2}{11}\right)}}$$

≈ 0.816

<u>Data Set 2:</u>

x_i	y_i	x_i^2	y_i^2	x_iy_i
10	9.14	100	83.5396	91.40
8	8.14	64	66.2596	65.12
13	8.74	169	76.3876	113.62
9	8.77	81	76.9129	78.93
11	9.26	121	85.7476	101.86
14	8.10	196	65.6100	113.40
6	6.13	36	37.5769	36.78
4	3.10	16	9.6100	12.40
12	9.13	144	83.3569	109.56
7	7.26	49	52.7076	50.82
5	4.47	25	19.9809	22.35
99	82.24	1001	657.6896	796.24

From the table, we have $n = 11$,

$\sum x_i = 99$, $\sum y_i = 82.24$, $\sum x_i^2 = 1001$,

$\sum y_i^2 = 657.6896$, and $\sum x_iy_i = 796.24$.

So, the correlation coefficient is

$$r = \frac{\sum x_iy_i - \dfrac{\sum x_i \sum y_i}{n}}{\sqrt{\left(\sum x_i^2 - \dfrac{(\sum x_i)^2}{n}\right)\left(\sum y_i^2 - \dfrac{(\sum y_i)^2}{n}\right)}}$$

$$= \frac{796.24 - \dfrac{(99)(82.24)}{11}}{\sqrt{\left(1001 - \dfrac{(99)^2}{11}\right)\left(657.6896 - \dfrac{(82.24)^2}{11}\right)}}$$

≈ 0.817

<u>Data Set 3:</u>

x_i	y_i	x_i^2	y_i^2	x_iy_i
10	7.46	100	55.6516	74.60
8	6.77	64	45.8329	54.16
13	12.74	169	162.3076	165.62
9	7.11	81	50.5521	63.99
11	7.81	121	60.9961	85.91
14	8.84	196	78.1456	123.76
6	6.08	36	36.9664	36.48
4	5.39	16	29.0521	21.56
12	8.15	144	66.4225	97.80
7	6.42	49	41.2164	44.94
5	5.73	25	32.8329	28.65
99	82.50	1001	659.9762	797.47

From the table, we have $n = 11$,

$\sum x_i = 99$, $\sum y_i = 82.50$, $\sum x_i^2 = 1001$,

$\sum y_i^2 = 659.9762$, and $\sum x_iy_i = 797.47$.

So, the correlation coefficient is

$$r = \frac{\sum x_iy_i - \dfrac{\sum x_i \sum y_i}{n}}{\sqrt{\left(\sum x_i^2 - \dfrac{(\sum x_i)^2}{n}\right)\left(\sum y_i^2 - \dfrac{(\sum y_i)^2}{n}\right)}}$$

$$= \frac{797.47 - \dfrac{(99)(82.50)}{11}}{\sqrt{\left(1001 - \dfrac{(99)^2}{11}\right)\left(659.9762 - \dfrac{(82.50)^2}{11}\right)}}$$

≈ 0.816

<u>Data Set 4:</u>

x_i	y_i	x_i^2	y_i^2	x_iy_i
8	6.58	64	43.2964	52.64
8	5.76	64	33.1776	46.08
8	7.71	64	59.4441	61.68
8	8.84	64	78.1456	70.72
8	8.47	64	71.7409	67.76
8	7.04	64	49.5616	56.32
8	5.25	64	27.5625	42.00
8	5.56	64	30.9136	44.48
8	7.91	64	62.5681	63.28
8	6.89	64	47.4721	55.12
19	12.5	361	156.2500	237.50
99	82.51	1001	660.1325	797.58

From the table, we have $n = 11$,
$\sum x_i = 99$, $\sum y_i = 82.51$, $\sum x_i^2 = 1001$,
$\sum y_i^2 = 660.1325$, and $\sum x_i y_i = 797.58$.
So, the correlation coefficient is

$$r = \dfrac{\sum x_i y_i - \dfrac{\sum x_i \sum y_i}{n}}{\sqrt{\left(\sum x_i^2 - \dfrac{(\sum x_i)^2}{n}\right)\left(\sum y_i^2 - \dfrac{(\sum y_i)^2}{n}\right)}}$$

$$= \dfrac{797.58 - \dfrac{(99)(82.51)}{11}}{\sqrt{\left(1001 - \dfrac{(99)^2}{11}\right)\left(660.1325 - \dfrac{(82.51)^2}{11}\right)}}$$

≈ 0.817

(b)

Even though the correlation coefficients for the four data sets are roughly equal, the scatter plots are clearly different. Thus, linear correlation coefficients and scatter diagrams must be used together in a statistical analysis of bivariate data to determine whether a linear relations exists.

35. Begin by finding the linear correlation coefficient between each pair of variables. The following correlation matrix summarizes the correlation coefficients.

	Cisco	Disney	GE	Exxon
Disney	0.296			
GE	0.766	0.535		
Exxon	0.464	0.772	0.726	
TECO	−0.246	0.205	−0.009	0.265

If the goal is to have the lowest correlation between two stocks (i.e., a correlation close to zero), then invest in General Electric and TECO Energy since their correlation coefficient is only −0.009.

If the goal is to have one stock go up when the other goes down (i.e., a negative association), then invest in Cisco Systems and TECO Energy since they have the strongest negative association, −0.246.

37. $r = 0.599$ implies that a positive linear relation exists between the number of television stations and life expectancy. However, this is correlation, not causation. The more television stations a country has, the more affluent it is. The more affluent, the better the healthcare is, which in turn helps increase the life expectancy. So, wealth is a likely lurking variable.

39. No, increasing the percentage of the population that has a cell phone will not decrease the violent crime rate. A likely lurking variable is the economy. In a strong economy, crime rates tend to decrease, and consumers are better able to afford cell phones.

41. **(a)**

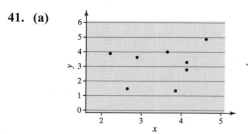

To calculate the correlation coefficient, we use the computational formula:

x_i	y_i	x_i^2	y_i^2	$x_i y_i$
2.2	3.9	4.84	15.21	8.58
3.7	4.0	13.69	16.00	14.80
3.9	1.4	15.21	1.96	5.46
4.1	2.8	16.81	7.84	11.48
2.6	1.5	6.76	2.25	3.90
4.1	3.3	16.81	10.89	13.53
2.9	3.6	8.41	12.96	10.44
4.7	4.9	22.09	24.01	23.03
28.2	25.4	104.62	91.12	91.22

From the table, we have $n = 8$, $\sum x_i = 28.2$, $\sum y_i = 25.4$, $\sum x_i^2 = 104.62$, $\sum y_i^2 = 91.12$, and $\sum x_i y_i = 91.22$. So, the correlation coefficient is

$$r = \frac{\sum x_i y_i - \dfrac{\sum x_i \sum y_i}{n}}{\sqrt{\left(\sum x_i^2 - \dfrac{(\sum x_i)^2}{n}\right)\left(\sum y_i^2 - \dfrac{(\sum y_i)^2}{n}\right)}}$$

$$= \frac{91.22 - \dfrac{(28.2)(25.4)}{8}}{\sqrt{\left(104.62 - \dfrac{(28.2)^2}{8}\right)\left(91.12 - \dfrac{(25.4)^2}{8}\right)}}$$

$$\approx 0.228$$

(b)

To calculate the correlation coefficient, we use the computational formula.

x_i	y_i	x_i^2	y_i^2	$x_i y_i$
10.4	9.3	108.16	86.49	96.72
2.2	3.9	4.84	15.21	8.58
3.7	4.0	13.69	16.00	14.80
3.9	1.4	15.21	1.96	5.46
4.1	2.8	16.81	7.84	11.48
2.6	1.5	6.76	2.25	3.90
4.1	3.3	16.81	10.89	13.53
2.9	3.6	8.41	12.96	10.44
4.7	4.9	22.09	24.01	23.03
38.6	34.7	212.78	177.61	187.94

From the table, we have $n = 9$,
$\sum x_i = 38.6$, $\sum y_i = 34.7$, $\sum x_i^2 = 212.78$,
$\sum y_i^2 = 177.61$, and $\sum x_i y_i = 187.94$.

So, the correlation coefficient is

$$r = \frac{\sum x_i y_i - \dfrac{\sum x_i \sum y_i}{n}}{\sqrt{\left(\sum x_i^2 - \dfrac{(\sum x_i)^2}{n}\right)\left(\sum y_i^2 - \dfrac{(\sum y_i)^2}{n}\right)}}$$

$$= \frac{187.94 - \dfrac{(38.6)(34.7)}{9}}{\sqrt{\left(212.78 - \dfrac{(38.6)^2}{9}\right)\left(177.61 - \dfrac{(34.7)^2}{9}\right)}}$$

$$\approx 0.860$$

The additional data point increases r from a value that suggests no linear correlation to one that suggests a fairly strong linear correlation. However, the second scatter diagram shows that the new data point is located very far away from the rest of the data, so this new data point has a strong influence on the value of r, even though there is no apparent correlation between the variables. Therefore, correlations should always be reported along with scatter diagrams in order to check for potentially influential observations.

43. Answers may vary. For those who feel "in-class" evaluations are meaningful, a correlation of 0.68 would lend some validity, because it indicates that students respond in the same general way on RateMyProfessors.com (high with high, low with low). The correlations between quality and easiness or hotness tend to indicate that evaluations at RateMyProfessors.com are based more on likability rather than actual quality of teaching.

45. Individual results will vary.

Section 4.2

1. The least-squares regression line is the line that minimizes the sum of the squared residuals. That is, it minimizes the sum of the squared vertical deviations from the line.

3. Values of the explanatory (or predictor) variable that are much larger or much smaller than those observed are considered to be "outside the scope of the model". That is, they lie outside the data values used to construct the model. It is dangerous to make such predictions because we do not know the behavior of the data for which we have no observations.

5. Answers will vary.

7. Each point's y-coordinate on the least-squares regression line represents the mean value of the response variable for the corresponding value of the explanatory variable.

9. (a)

There appears to be a negative linear relationship between the x and y.

(b) $b_1 = r \cdot \dfrac{s_y}{s_x} = -0.9477 \left(\dfrac{1.8239}{2.4221} \right) \approx -0.7136$

$b_0 = \overline{y} - b_1 \overline{x}$

$\quad = 3.9333 - (-0.7136)(3.6667)$

$\quad \approx 6.5499$

So, the least-squares regression line is
$\hat{y} = -0.7136x + 6.5499$

(c)

11. (a)

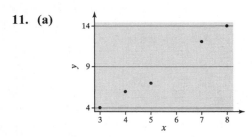

(b) Using the points $(3, 4)$ and $(8, 14)$:

$m = \dfrac{14 - 4}{8 - 3} = 2$

$y - y_1 = m(x - x_1)$

$y - 4 = 2(x - 3)$

$y - 4 = 2x - 6$

$\quad\; y = 2x - 2$

(c)

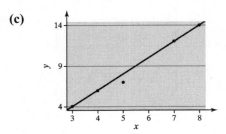

(d) We compute the mean and standard deviation for each variable and the correlation coefficient to be $\overline{x} = 5.4$, $\overline{y} = 8.6$, $s_x \approx 2.07364$, $s_y \approx 4.21901$, and $r \approx 0.99443$. Then the slope and intercept for the least-squares regression line are:

$b_1 = r \cdot \dfrac{s_y}{s_x} = 0.99443 \left(\dfrac{4.21901}{2.07364} \right)$

$\quad \approx 2.02326$

$b_0 = \overline{y} - b_1 \overline{x} = 8.6 - (2.02326)(5.4)$

$\quad \approx -2.3256$

Rounding to four decimal places, the least-squares regression line is
$\hat{y} = 2.0233x - 2.3256$

(e)

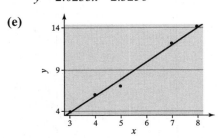

(f)

x	y	$\hat{y} = 2x - 2$	$y - \hat{y}$	$(y - \hat{y})^2$
3	4	4	0	0
4	6	6	0	0
5	7	8	−1	1
7	12	12	0	0
8	14	14	0	0

Total $= 1$

Sum of squared residuals (computed line): 1.0

(g)

x	y	$\hat{y}$	$y - \hat{y}$	$(y - \hat{y})^2$
3	4	3.7443	0.2557	0.0654
4	6	5.7676	0.2324	0.0540
5	7	7.7909	−0.7909	0.6255
7	12	11.8375	0.1625	0.0264
8	14	13.8608	0.1392	0.0194

Total $= 0.7907$

Sum of squared residuals (regression line): 0.7907

(h) Answers will vary. The regression line gives a smaller sum of squared residuals, so it has a better fit.

13. (a)

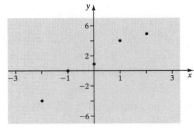

(b) Using the points $(-2, -4)$ and $(2, 5)$:

$$m = \frac{5 - (-4)}{2 - (-2)} = \frac{9}{4}$$

$$y - y_1 = m(x - x_1)$$

$$y - 5 = \frac{9}{4}(x - 2)$$

$$y - 5 = \frac{9}{4}x - \frac{9}{2}$$

$$y = \frac{9}{4}x + \frac{1}{2} \quad \text{or} \quad y = 2.25x + 0.5$$

(c)

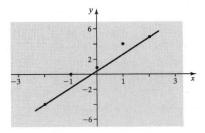

(d) We compute the mean and standard deviation for each variable and the correlation coefficient to be $\bar{x} = 0$, $\bar{y} = 1.2$, $s_x \approx 1.58114$, $s_y \approx 3.56371$, and $r \approx 0.97609$. Then the slope and intercept for the least-squares regression line are:

$$b_1 = r \cdot \frac{s_y}{s_x} = 0.97609\left(\frac{3.56371}{1.58114}\right) \approx 2.20000$$

$$b_0 = \bar{y} - b_1\bar{x} = 1.2 + (2.20000)(0) = 1.2$$

So, the least-squares regression line is $\hat{y} = 2.2x + 1.2$

(e)

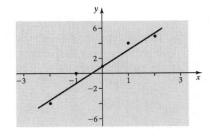

(f)

x	y	$y = \frac{9}{4}x + \frac{1}{2}$	$y - \hat{y}$	$(y - \hat{y})^2$
-2	-4	-4.00	0.00	0.0000
-1	0	-1.75	1.75	3.0625
0	1	0.50	0.50	0.2500
1	4	2.75	1.25	1.5625
2	5	5.00	0.00	0.0000

Total = 4.8750

Sum of squared residuals (calculated line): 4.8750

(g)

x	y	$\hat{y} = 2.2x + 1.2$	$y - \hat{y}$	$(y - \hat{y})^2$
-2	-4	-3.2	-0.8	0.64
-1	0	-1.0	1.0	1.00
0	1	1.2	-0.2	0.04
1	4	3.4	0.6	0.36
2	5	5.6	-0.6	0.36

Total = 2.40

Sum of squared residuals (regression line): 2.40

(h) Answers will vary. The regression line gives a smaller sum of squared residuals, so it has a better fit.

15. (a)

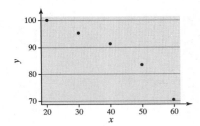

(b) Using the points $(30, 95)$ and $(60, 70)$:

$$m = \frac{70 - 95}{60 - 30} = \frac{-25}{30} = -\frac{5}{6}$$

$$y - y_1 = m(x - x_1)$$

$$y - 95 = -\frac{5}{6}(x - 30)$$

$$y - 95 = -\frac{5}{6}x + 25$$

$$y = -\frac{5}{6}x + 120$$

(c)

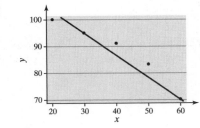

(d) We compute the mean and standard deviation for each variable and the correlation coefficient to be $\bar{x} = 40$, $\bar{y} = 87.8$, $s_x \approx 15.81139$, $s_y \approx 11.73456$, and $r \approx -0.97014$. Then the slope and intercept for the least-squares regression line are:

$$b_1 = r \cdot \frac{s_y}{s_x} = -0.97014\left(\frac{11.73456}{15.81139}\right)$$
$$\approx -0.72000$$
$$b_0 = \bar{y} - b_1\bar{x} = 87.8 - (-0.72000)(40)$$
$$\approx 116.6$$

So, the least-squares regression line is
$$\hat{y} = -0.72x + 116.6$$

(e)

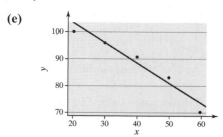

(f)

x	y	$\hat{y} = -\frac{5}{6}x + 120$	$y - \hat{y}$	$(y - \hat{y})^2$
20	100	103.333	−3.333	11.1111
30	95	95.000	0.000	0.0000
40	91	86.667	4.333	18.7778
50	83	78.330	4.667	21.7778
60	70	70.000	0.000	0.0000

Total = 51.6667

Sum of squared residuals (calculated line): 51.6667.

(g)

x	y	$\hat{y}$	$y - \hat{y}$	$(y - \hat{y})^2$
20	100	102.2	−2.2	4.84
30	95	95.0	0.0	0.00
40	91	87.8	3.2	10.24
50	83	80.6	2.4	5.76
60	70	73.4	−3.4	11.56

Total = 32.40

Sum of squared residuals (regression line): 32.4

(h) Answers will vary. The regression line gives a smaller sum of squared residuals, so it has a better fit.

17. (a) Let $x = 8$ in the regression equation.
$$\hat{y} = -0.0526(8) + 2.9342 \approx 2.51$$
So, we predict that the GPA of a student who plays video games 8 hours per week will be 2.51.

(b) The slope is -0.0526. For each additional hour a student spends playing video games in a week, the GPA will decrease by 0.0526 points, on average.

(c) The y-intercept is 2.9342. The grade-point average for a student who does not play video games is 2.9342.

(d) Let $x = 7$ in the regression equation.
$$\hat{y} = -0.0526(7) + 2.9342 \approx 2.57$$
So, the average GPA among all students who play video games 7 hours per week is 2.57. Therefore, a student with a grade-point average of 2.68 is above average for those who play video games 7 hours per week.

19. (a) The slope is 2.94. For each percentage point increase in on-base percentage, the winning percentage will increase by 2.94 percentage points, on average.

(b) The y-intercept is -0.4871. This is outside the scope of the model. An on-base percentage of 0 is not reasonable for this model, since a team cannot win if players do not reach base.

(c) No, it would not be a good idea to use this model to predict the winning percentage of a team whose on-base percentage was 0.250 because 0.250 is outside the scope of the model.

(d) Let $x = 0.322$ in the regression equation.
$$\hat{y} = 2.94(0.322) - 0.4871 \approx 0.4596$$
So, the residual is
$$y - \hat{y} = 0.546 - 0.4596 = 0.0864$$
This residual indicates that San Diego's winning percentage is above average for teams with an on-base percentage of 0.322.

21. (a) In Problem 23, Section 4.1, we computed the correlation coefficient. Rounded to six decimal places, it is $r = 0.911073$. We compute the mean and standard deviation for each variable to be $\bar{x} \approx 26.454545$, $s_x \approx 1.094407$, $\bar{y} \approx 17.327273$, and $s_y \approx 0.219504$.

$$b_1 = r \cdot \frac{s_y}{s_x} = 0.911073 \left(\frac{0.219504}{1.094407} \right)$$

$$\approx 0.182733$$

$$b_0 = \bar{y} - b_1 \bar{x}$$

$$= 17.327273 - (0.182733)(26.454545)$$

$$\approx 12.4932$$

Rounding to four decimal places, the least-squares regression line is

$$\hat{y} = 0.1827x + 12.4932$$

(b) If height increases by 1 inch, head circumference increases by about 0.1827 inches, on average. It is not appropriate to interpret the *y*-intercept since it is outside the scope of the model. In addition, it makes no sense to consider the head circumference of a child with a height of 0 inches.

(c) Let $x = 25$ in the regression equation.

$$\hat{y} = 0.1827(25) + 12.4932 \approx 17.06$$

We predict the head circumference of a child who is 25 inches tall is 17.06 inches.

(d) $y - \hat{y} = 16.9 - 17.06 = -0.16$ inches

This indicates that the head circumference for this child is below average.

(e)

Head Circumference vs. Height

(f) The head circumferences of children who are 26.75 inches tall naturally vary.

(g) No, a height of 32 inches is well outside the scope of the model.

23. (a) In Problem 25, Section 4.1, we computed the correlation coefficient. Rounded to seven decimal places, it is $r = -0.964086$. We compute the mean and standard deviation for each variable to be $\bar{x} \approx 3290.363636$, $s_x \approx 505.626200$, $\bar{y} \approx 21.727273$, and $s_y \approx 3.690282$.

$$b_1 = r \cdot \frac{s_y}{s_x} = -0.9640858 \left(\frac{3.6902821}{505.6262004} \right)$$

$$\approx -0.0070363$$

$$b_0 = \bar{y} - b_1 \bar{x}$$

$$= 21.7272727 - (-0.0070363)(3290.3636364)$$

$$\approx 44.8793$$

Rounding to four decimal places, the least-squares regression line is

$$\hat{y} = -0.0070x + 44.8793$$

(b) For every pound added to the weight of the car, gas mileage in the city will decrease by about 0.0070 mile per gallon, on average. It is not appropriate to interpret the *y*-intercept because it is well beyond the scope of the model. Also, a weight of 0 pounds for a car does not make sense.

(c) Let $x = 2780$ in the regression equation.

$$\hat{y} = -0.0070(2780) + 44.8793 \approx 25.4$$

We predict that a car weighing 2780 pounds will get 25.4 miles per gallon. So, the Chevy Cobalt getting 22 miles per gallon is below average.

(d) No, it is not reasonable to use this least-squares regression line to predict the miles per gallon of a Toyota Prius. The data given are for domestic cars with traditional internal combustion engines. The Toyota Prius is a foreign-made hybrid car.

25. (a) We compute the correlation coefficient, rounded to seven decimal places, to be $r = -0.8056468$. We compute the mean and standard deviation for each variable to be $\bar{x} = 3.6$, $s_x \approx 2.6403463$, $\bar{y} \approx 0.8756667$, and $s_y \approx 0.0094692$.

$$b_1 = r \cdot \frac{s_y}{s_x} = -0.8056468 \left(\frac{0.0094692}{2.6403463} \right)$$

$$\approx -0.0028893$$

$$b_0 = \bar{y} - b_1 \bar{x}$$

$$= 0.8756667 - (-0.0028893)(3.6)$$

$$\approx 0.8861$$

Rounding to four decimal places, the least-squares regression line is

$$\hat{y} = -0.0029x + 0.8861$$

(b) For each additional cola consumed per week, bone mineral density will decrease by 0.0029 g/cm^2, on average.

(c) For a woman who does not drink cola, bone mineral density will be 0.8861 g/cm^2.

(d) Let $x = 4$ in the regression equation.

$\hat{y} = -0.0029(4) + 0.8861 = 0.8745$

We predict the bone mineral density of the femoral neck of a woman who consumes four colas per week is 0.8745 g/cm^2.

(e) Since 0.873 is smaller than the result in part (d), this woman's bone mineral density is below average among women who consume four colas per week.

(f) No. Two cans of soda per day equates to 14 cans of soda per week, which is outside the scope of the model.

27. (a) In Problem 29, Section 4.1, we computed the correlation coefficient. Unrounded, it is $r \approx 0.3590979$. We compute the mean and standard deviation for each variable to be $\bar{x} = 876549.5$, $s_x \approx 65073.9854153$, $\bar{y} = 134.3$, and $s_y \approx 3.1287200$.

$$b_1 = r \cdot \frac{s_y}{s_x} = 0.3590979 \left(\frac{3.1287200}{65073.9854153} \right)$$

$$\approx 0.00001726522$$

$$b_0 = \bar{y} - b_1 \bar{x}$$

$$= 134.3 - (0.00001726522)(876549.5)$$

$$\approx 119.1662$$

Rounding to four decimal places, the least-squares regression line is

$$\hat{y} = 0.00002x + 119.1662$$

(b) The slope is close to 0. While this may partially be due to the weak linear relationship that is present, it could also be due to the magnitude of the values for the explanatory variable. The values for MRI counts are in the hundreds of thousands. If we coded our data so that x was the MRI count in hundreds of thousands, then the slope would be around 3. Even if the data were all in a straight line, the slope would be very small because of the size of the values for x. Therefore, before assuming the slope is 0, first check the magnitude of the data.

(c) We previously found that there was no apparent relation between the MRI count and IQ. Therefore, in both cases we would use $\bar{y}$ as our estimate. That is, for both cases, our estimate is $\hat{y} = \bar{y} = 134.3$.

29. Answers will vary. Mark Twain was poking fun at those who carelessly go outside the scope of their data (model) by providing an extreme example of what can happen.

31. (a) Square footage is the explanatory variable.

(b)

(c) To calculate the correlation coefficient, we use the computational formula.

x_i	y_i	x_i^2	y_i^2	$x_i y_i$
1,148	154.0	1,317,904	23,716.00	176,792.0
1,096	159.9	1,201,216	25,568.01	175,250.4
1,142	169.0	1,304,164	28,561.00	192,998.0
1,288	169.9	1,658,944	28,866.01	218,831.2
1,322	170.0	1,747,684	28,900.00	224,740.0
1,466	179.9	2,149,156	32,364.01	263,733.4
1,344	180.0	1,806,336	32,400.00	241,920.0
1,544	189.0	2,383,936	35,721.00	291,816.0
1,494	189.9	2,232,036	36,062.01	283,710.6
11,844	1561.6	15,801,376	272,158.04	2,069,791.6

From the table, we have $n = 9$, $\sum x_i = 11,844$, $\sum y_i = 1,561.6$, $\sum x_i^2 = 15,801,376$, $\sum y_i^2 = 272,158.04$, and $\sum x_i y_i = 2,069,791.6$. So, the correlation coefficient is

$$r = \frac{\sum x_i y_i - \dfrac{\sum x_i \sum y_i}{n}}{\sqrt{\left(\sum x_i^2 - \dfrac{(\sum x_i)^2}{n} \right) \left(\sum y_i^2 - \dfrac{(\sum y_i)^2}{n} \right)}}$$

$$= \frac{2,069,791.6 - \dfrac{(11,844)(1,561.6)}{9}}{\sqrt{\left(15,801,376 - \dfrac{(11,844)^2}{9} \right) \left(272,158.04 - \dfrac{(1,561.6)^2}{9} \right)}}$$

$$\approx 0.916$$

(d) Yes, a linear relation exists between square footage and asking price.

(e) We found the correlation coefficient in part (b). Unrounded, it is $r \approx 0.91632226$. We compute the mean and standard deviation for each variable to be $\bar{x} = 1316$, $s_x \approx 163.81086655$, $\bar{y} = 173.51111111$, and $s_y \approx 12.26320150$.

$$b_1 = r \cdot \frac{s_y}{s_x} = 0.91632226 \left(\frac{12.26320150}{163.81086655} \right)$$
$$\approx 0.068597675$$
$$b_0 = \bar{y} - b_1 \bar{x}$$
$$= 173.51111111 - (0.068597675)(1316)$$
$$\approx 83.2366$$

Rounding to four decimal places, the least-squares regression line is
$$\hat{y} = 0.0686x + 83.2366$$

(f) For each square foot added to the area, the asking price of the house will increase by $68.60 (that is 0.0686 thousand dollars), on average.

(g) It is not reasonable to interpret the y-intercept because a house with an area of 0 square feet makes no sense.

(h) Let $x = 1092$ in the regression equation:
$$\hat{y} = 0.0686(1092) + 83.2366 \approx 158.1$$

The average price of a home with 1,092 square feet is about $158,100. Since the actual list price of this particular home is $189,900, it is above average.

Reasons provided may vary. Some factors that could affect the price include location and updates such as thermal windows or new siding.

Section 4.3

1. 75% of the total variation in the response variable is explained by the regression line.

3. A pattern in the residual plot indicates that the response and explanatory variables may not be linearly related.

5. An influential observation is one that significantly affects the correlation coefficient, the slope, or the y-intercept of the least-squares regression line. If an influential observation is observed, it should be removed only if there is justification to do so. If the observation cannot be removed, one can collect more data so that

additional points near the influential observation are obtained, or apply other techniques (beyond the scope of this text) that reduce the influence of the observation.

7. An influential observation can change either the slope or y-intercept of the least-squares regression line, or both.

9. Knowing R^2 alone is not sufficient to determine the value of r. In this case, if $R^2 = 0.81$, then we have either $r = 0.9$ or $r = -0.9$ since both $(0.9)^2 = 0.81$ and $(-0.9)^2 = 0.81$. To determine which is the correct value for r, we need to know whether the relation has positive or negative association. We can determine this by looking at a scatter diagram or the slope of the regression line.

11. The variance of the residuals is not constant. The variance of the residuals increases as the value of x increases.

13. There is an outlier in the residual plot (a residual of 4).

15. **(a)** III   **(b)** II
    (c) IV   **(d)** I

17. Yes, the point is influential because the two lines have substantially different slopes and y-intercepts.

19. No, the point is not influential because there is very little change in either the slope or the y-intercept. The lines are almost identical.

21. **(a)** From the scatter diagram, a positive linear association appears to exist between time between eruptions and length of eruption.

    (b) Yes, the residual plot does not show any obvious departures from the requirements of the model.

    (c) 83.0% of the variation in the length of eruption is explained by the least-squares regression equation.

23. (a)

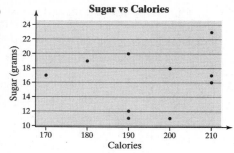

Sugar vs Calories

No relation appears to exist between calories and sugar content.

(b) To calculate the correlation coefficient, we use the computational formula.

x_i	y_i	x_i^2	y_i^2	x_iy_i
200	18	40,000	324	3600
210	23	44,100	529	4830
170	17	28,900	289	2890
190	20	36,100	400	3800
200	18	40,000	324	3600
180	19	32,400	361	3420
210	23	44,100	529	4830
210	16	44,100	256	3360
210	17	44,100	289	3570
190	12	36,100	144	2280
190	11	36,100	121	2090
200	11	40,000	121	2200
2360	205	466,000	3687	40,470

From the table, we have $n = 12$,

$\sum x_i = 2306$, $\sum y_i = 205$, $\sum x_i^2 = 466,000$,

$\sum y_i^2 = 3687$, and $\sum x_iy_i = 40,470$. So, the correlation coefficient is

$$r = \frac{\sum x_iy_i - \frac{\sum x_i \sum y_i}{n}}{\sqrt{\left(\sum x_i^2 - \frac{(\sum x_i)^2}{n}\right)\left(\sum y_i^2 - \frac{(\sum y_i)^2}{n}\right)}}$$

$$= \frac{40,470 - \frac{(2360)(205)}{12}}{\sqrt{\left(466,000 - \frac{(2360)^2}{12}\right)\left(3687 - \frac{(205)^2}{12}\right)}}$$

$$\approx 0.261$$

No, there is no significant linear relation between calories and sugar content.

(c)

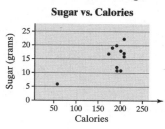

Sugar vs. Calories

We add the following values to the table in part (a): $x_{13} = 80$, $y_{13} = 6$, $x_{13}^2 = 6400$, $y_{13}^2 = 36$, and $x_{13}y_{13} = 480$. We now have $n = 13$, $\sum x_i = 2440$, $\sum y_i = 211$, $\sum x_i^2 = 472,400$, $\sum y_i^2 = 3723$, and $\sum x_iy_i = 40,950$. So, the new correlation coefficient is

$$r = \frac{40,950 - \frac{(2440)(211)}{13}}{\sqrt{\left(472,400 - \frac{(2440)^2}{13}\right)\left(3723 - \frac{(211)^2}{13}\right)}}$$

$$\approx 0.649$$

Yes, according to the new correlation coefficient, there is a significant linear relation between calories and sugar content. All-Bran is an influential observation.

(d) It is necessary to show a scatter diagram with the correlation coefficient when claiming that a linear relation exists between two variables because influential observations can cause the correlation coefficient to increase substantially, thereby increasing the apparent strength of the linear relation between the two variables.

25. (a) From Problem 23(c) in Section 4.1, we have $r = 0.911073$ (unrounded). So, $R^2 = (0.911073)^2 \approx 0.830 = 83.0\%$.

(b) From Problem 21(a) in Section 4.2, we have $\hat{y} = 0.1827x + 12.4932$. We compute the residuals and construct the residual plot.

x_i	y_i	$\hat{y} = 0.1827x + 12.4932$	$y - \hat{y}$
27.75	17.5	17.56	−0.06
24.50	17.1	16.97	0.13
25.50	17.1	17.15	−0.05
26.00	17.3	17.24	0.06
25.00	16.9	17.06	−0.16
27.75	17.6	17.56	0.04
26.50	17.3	17.33	−0.03
27.00	17.5	17.43	0.07
26.75	17.3	17.38	−0.08
26.75	17.5	17.38	0.12
27.50	17.5	17.52	−0.02

Residuals vs. Height

(c) 83.0% of the variation in head circumference is explained by the least-squares regression line. The residual plot does not reveal any problems, so the linear model appears to be appropriate.

27. (a) From Problem 25(c) in Section 4.1, we have $r = -0.964086$ (unrounded). So,

$$R^2 = (-0.964086)^2 \approx 0.929 = 92.9\%.$$

(b) From Problem 23(a) in Section 4.2, we have $\hat{y} = -0.0070x + 44.8793$. We compute the residuals and construct the residual plot.

x_i	y_i	$\hat{y} = -0.0070x + 44.8793$	$y - \hat{y}$
3765	19	18.5	0.5
3984	18	17.0	1.0
3530	21	20.2	0.8
3175	22	22.7	−0.7
2580	27	26.8	0.2
3730	18	18.8	−0.8
2605	26	26.6	−0.6
3772	17	18.5	−1.5
3310	20	21.7	−1.7
2991	25	23.9	1.1
2752	26	25.6	0.4

Residuals vs. Weight

(c) 92.9% of the variation in gas mileage is explained by the least-squares regression line. The residual plot does not reveal any problems, so the linear model appears to be appropriate.

29. (a)

Planetary Motion

(b) To calculate the correlation coefficient, we use the computational formula.

x_i	y_i	x_i^2	y_i^2	$x_i y_i$
36	0.24	1,296	0.0576	8.64
67	0.62	4,489	0.3844	41.54
93	1.00	8,649	1	93
142	1.88	20,164	3.5344	266.96
483	11.9	233,289	141.61	5,747.7
887	29.5	786,769	870.25	26,166.5
1,785	84	3,186,225	7,056	14,9940
2,797	165	7,823,209	27,225	46,1505
3,675	248	13,505,625	61,504	91,1400
9,965	542.14	25,569,715	96,801.8364	1,555,169.34

From the table, we have $n = 9$,
$\sum x_i = 9,965$, $\sum y_i = 542.14$,
$\sum x_i^2 = 25,569,715$, $\sum y_i^2 = 96,801.8364$,
and $\sum x_i y_i = 1,555,169.34$. So, the correlation coefficient is

$$r = \dfrac{\sum x_i y_i - \dfrac{\sum x_i \sum y_i}{n}}{\sqrt{\left(\sum x_i^2 - \dfrac{(\sum x_i)^2}{n}\right)\left(\sum y_i^2 - \dfrac{(\sum y_i)^2}{n}\right)}}$$

$$= \dfrac{1,555,169.346 - \dfrac{(9965)(542.14)}{9}}{\sqrt{\left(25,569,715 - \dfrac{(9965)^2}{9}\right)\left(96,801.8364 - \dfrac{(542.14)^2}{9}\right)}}$$

$$= 0.9888989$$

$$\approx 0.989$$

Yes, there appears to be a linear relation between distance and sidereal year.

(c) We compute the mean and standard deviation for each variable to be
$\bar{x} \approx 1107.2222222$, $s_x \approx 1347.972809$,
$\bar{y} \approx 60.237778$, and $s_y \approx 89.543654$.

$$b_1 = r \cdot \dfrac{s_y}{s_x} = 0.9888989\left(\dfrac{89.543654}{1347.972809}\right)$$

$$\approx 0.06569095$$

$$b_0 = \bar{y} - b_1 \bar{x}$$

$$= 60.237778 - (0.06569095)(1107.222222)$$

$$\approx -12.4967$$

Rounding to four decimal places, the least-squares regression line is
$$\hat{y} = 0.0657x - 12.4967$$

(d) We compute the residuals and construct the residual plot.

x_i	y_i	$\hat{y} = 0.0657x - 12.4967$	$y - \hat{y}$
36	0.24	−10.132	10.372
67	0.62	−8.095	8.715
93	1.00	−6.387	7.387
142	1.88	−3.167	5.047
483	11.9	19.236	−7.336
887	29.5	45.779	−16.279
1,785	84	104.778	−20.778
2,797	165	171.266	−6.266
3,675	248	228.951	19.049

Residuals vs. Distance

(e) No; the U-shaped pattern in the residual plot indicates that a linear model is not appropriate.

31. (a) From Problem 25(c) in Section 4.1, we found $r = -0.964086$ (unrounded). In Problem 27 of Section 4.3, we found $R^2 = 92.9\%$.

Including the Dodge Viper data ($x_{12} = 3425$, $y_{12} = 11$), we obtain
$n = 12$, $\sum x_i = 39,619$, $\sum y_i = 250$,
$\sum x_i^2 = 133,378,625$, $\sum y_i^2 = 5450$, and
$\sum x_i y_i = 806,083$. So, the new correlation coefficient is

$$r = \frac{\sum x_i y_i - \dfrac{\sum x_i \sum y_i}{n}}{\sqrt{\left(\sum x_i^2 - \dfrac{(\sum x_i)^2}{n}\right)\left(\sum y_i^2 - \dfrac{(\sum y_i)^2}{n}\right)}}$$

$$= \frac{806,083 - \dfrac{(39,619)(250)}{12}}{\sqrt{\left(133,378,625 - \dfrac{(39,619)^2}{12}\right)\left(5450 - \dfrac{(250)^2}{12}\right)}}$$

$$\approx -0.774464$$

Therefore, the coefficient of determination with the Dodge Viper included is
$$R^2 = (-0.774464)^2 \approx 0.600 = 60.0\%.$$

Adding the Viper reduces the amount of variability explained by the model by approximately 33%.

(b) With the Viper included, we recomputed the means and standard deviations for each variable to be $\bar{x} = 3301.583333$, $s_x \approx 483.659809$, $\bar{y} = 20.833333$, and $s_y \approx 4.687184$.

$$b_1 = r \cdot \frac{s_y}{s_x} = -0.774464\left(\frac{4.687184}{483.659809}\right)$$
$$\approx -0.00750539$$
$$b_0 = \bar{y} - b_1\bar{x}$$
$$= 20.83333 - (-0.00750539)(3301.58333)$$
$$\approx 45.6130$$

Rounding to four decimal places, the new least-squares regression line with the Viper included is $\hat{y} = -0.0075x + 45.6130$

Since neither the slope or y-intercept are altered significantly from the original equation found in Problem 23(a) of Section 4.2, the Viper is not influential. We illustrate this further with the following scatter plot showing the original regression line (solid) and the new regression line that includes the Viper (dashed).

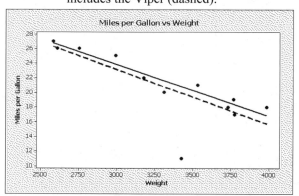

Looking at the scatter plot above, since the point (3425, 11) that corresponds to the Dodge Viper, lies far below the rest of the data, it is an outlier.

33. Using all 14 data points, we compute the regression line for carbon dioxide emissions, y, versus energy generated, x, to be $\hat{y} = 0.5824x + 828.9540$. The correlation coefficient is $r = 0.836$ and the coefficient of determination is $R^2 = 69.9\%$.

Excluding the Laramie River Power Plant, the regression line for the remaining 13 data points becomes $\hat{y} = 1.1452x - 241.3009$. The correlation coefficient is $r = 0.994$ and the coefficient of determination is $R^2 = 98.8\%$.

By removing the Laramie River Power Plant, the slope and y-intercept both change significantly. Likewise, removing this plant increases the amount of variability in carbon emissions that is explained by the model by approximately 29%.

The Laramie River Power Plant is unusual in that it emits less carbon that expected from the overall trend. The observation is an outlier and it is influential. We illustrate this further with the following scatter plot showing the original regression line (solid) and the new regression line with Laramie River Power Plant removed (dashed).

Excluding the Jim Bridger power plant (and putting back in the Laramie River Power Plant), the regression line for the remaining 13 data points becomes $\hat{y} = 0.4340x + 784.1952$. The correlation coefficient is $r = 0.878$ and the coefficient of determination is $R^2 = 77.1\%$.

By removing the Jim Bridger power plant, neither the slope nor the y-intercept change significantly. Likewise, removing this plant does not significantly decrease the amount of variability in carbon missions that is explained by the model.

If we exclude both the Jim Bridger power plant and the Laramie River Power Plant, the regression line for the remaining 12 data points becomes $\hat{y} = 1.1720x - 276.4573$. The correlation coefficient is $r = 0.975$ and the coefficient of determination is $R^2 = 95.1\%$. This result is not significantly different from the regression line obtained when only the Laramie River Power Plant was excluded.

In short, the Jim Bridger power plant is not unusual. It follows the overall trend of the data. It in not influential either. We illustrate this further with the following scatter plot showing the original regression line (dashed) and the third regression line with Jim Bridger power plant removed with the Laramie River plant included (solid).

For completion, we also show a comparison of the regression line with only the Laramie River plant removed (solid) and the regression line with both the Laramie River plant and the Jim Bridger plant removed (dashed).

35. (a)

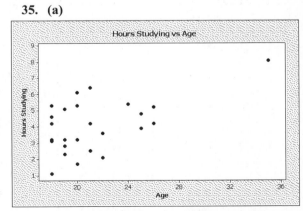

The 35 year old student who studied 8.1 hours is potentially influential.

93

(b) Using all 26 observations, we compute the regression line to be $\hat{y} = 0.2320x - 0.9204$.

(c) Excluding the point (35, 8.1), we compute the regression line for the remaining 25 observations to be $\hat{y} = 0.1549x + 0.6346$.

(d) We graph the original regression line (solid) and the new regression line with (35, 8.1) excluded (dashed).

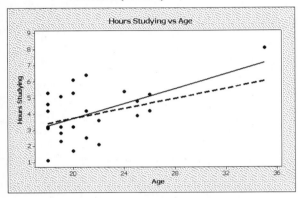

(e) Since the point (35, 8.1) significantly affects both the slope and the y-intercept of the least-squares regression line, it is an influential observation.

37. (a)

(b) To calculate the correlation coefficient, we use the computational formula.

x_i	y_i	x_i^2	y_i^2	$x_i y_i$
0.203	0.402	0.041209	0.161604	0.081606
0.310	0.510	0.096100	0.260100	0.158100
0.267	0.410	0.071289	0.168100	0.109470
0.195	0.536	0.038025	0.287296	0.104520
−0.101	−0.060	0.010201	0.003600	0.006060
−0.130	−0.151	0.016900	0.022801	0.019630
−0.234	−0.377	0.054756	0.142129	0.088218
0.264	0.308	0.069696	0.094864	0.081312
0.090	0.207	0.008100	0.042849	0.018630
0.030	−0.014	0.000900	0.000196	−0.000420
0.128	0.093	0.016384	0.008649	0.011904
−0.035	0.027	0.001225	0.000729	−0.000945
0.987	1.891	0.424785	1.192917	0.678085

From the table, we have $n = 12$,

$\sum x_i = 0.987$, $\sum y_i = 1.891$,

$\sum x_i^2 = 0.424785$, $\sum y_i^2 = 1.192917$, and

$\sum x_i y_i = 0.678085$. So, the correlation coefficient is

$$r = \frac{\sum x_i y_i - \dfrac{\sum x_i \sum y_i}{n}}{\sqrt{\left(\sum x_i^2 - \dfrac{(\sum x_i)^2}{n}\right)\left(\sum y_i^2 - \dfrac{(\sum y_i)^2}{n}\right)}}$$

$$= \frac{0.678085 - \dfrac{(0.987)(1.891)}{12}}{\sqrt{\left(0.424785 - \dfrac{(0.987)^2}{12}\right)\left(1.192917 - \dfrac{(1.891)^2}{12}\right)}}$$

$= 0.94233433$

≈ 0.942

(c) Yes, a linear relation exists between the rates of return of the S&P 500 and GE.

(d) We compute the mean and standard deviation for each variable to be
$\bar{x} = 0.08225$, $s_x \approx 0.1767392$,
$\bar{y} \approx 0.1575833$, and $s_y \approx 0.2852315$.

$$b_1 = r \cdot \frac{s_y}{s_x} = 0.94233433\left(\frac{0.2852315}{0.1767392}\right)$$

≈ 1.5207913

$b_0 = \bar{y} - b_1\bar{x}$

$= 0.1575833 - (1.5207913)(0.08225)$

≈ 0.0325

Rounding to four decimal places, the least-squares regression line is
$\hat{y} = 1.5208x + 0.0325$.

(e) Let $x = 0.10$ in the regression equation:
$\hat{y} = 1.5208(0.10) + 0.0325 \approx 0.1846$

We predict the rate of return of GE stock will be 0.1846 (or 18.46%) if the rate of return of the S&P 500 is 0.10 (or 10%).

(f) Since the actual rate of return of 13.2% (or 0.132) is below the predicted rate of return from part (e), GE's performance was below average among all years for which the S&P 500's returns were 10%.

(g) The slope indicates that, for each percentage point increase in the rate of return for the S&P 500, the rate of return of GE stock will increase by about 1.52 percentage points, on average.

(h) The *y*-intercept indicates that the rate of return for GE stock will be 0.0325 (or 3.25%) when there is no change to the S&P 500 (i.e., when there is a 0% rate of return for the S&P 500).

(i) $R^2 = (0.94233433)^2 \approx 0.888 = 88.8\%$

(j) We compute the residuals and construct the residual plot.

x_i	y_i	$\hat{y} = 1.5208x + 0.0325$	$y - \hat{y}$
0.203	0.402	0.3412	0.0608
0.310	0.510	0.5039	0.0061
0.267	0.410	0.4386	−0.0286
0.195	0.536	0.3291	0.2069
−0.101	−0.060	−0.1211	0.0611
−0.130	−0.151	−0.1652	0.0142
−0.234	−0.377	−0.3234	−0.0536
0.264	0.308	0.4340	−0.1260
0.090	0.207	0.1694	0.0376
0.030	−0.014	0.0781	−0.0921
0.128	0.093	0.2272	−0.1342
−0.035	0.027	−0.0207	0.0477

Residuals Versus S&P 500

Overall, the residual plot supports the linear relation. However, there may be a slight concern of a nonconstant variance.

(k) The five-number summary for the residuals is −0.1342, −0.0729, 0.0102, 0.0543, and 0.2069. We compute the fences to look for outliers:

Lower Fence
$= -0.0729 - 1.5(0.0543 - (-0.0729))$
$= -0.2637$

Upper Fence
$= 0.0543 + 1.5(0.0543 - (-0.0729))$
$= 0.2451$

None of the residuals fall outside the fences, so the residuals do not show any unusual observations.

Section 4.4

1. A *marginal distribution* is a frequency or relative frequency distribution of either the row or column variable in a contingency table. A *conditional distribution* is the relative frequency of each category of one variable, given a specific value of the other variable in a contingency table.

3. The term correlation is used with quantitative variables. In this section, we are considering the association between qualitative variables (variables which take non-numeric values).

5. (a) We find the frequency marginal distribution for the row variable, *y*, by finding the total of each row. We find the frequency marginal distribution for the column variable, *x*, by finding the total for each column.

	x_1 x_2 x_3	Frequency Marg. Dist.
y_1	20 25 30	75
y_2	30 25 50	105
Frequency Marg. Dist.	50 50 80	180

(b) We find the relative frequency marginal distribution for the row variable, *y*, by dividing the row total for each y_i by the total table, 180. For example, for the relative frequency for y_1 is $\frac{75}{180} \approx 0.417$.

We find the relative frequency marginal distribution for the column variable, *x*, by dividing the column total for each x_i by the table total. For example, for the relative frequency for x_1 is $\frac{50}{180} \approx 0.278$.

	x_1	x_2	x_3	Rel. Freq. Marg. Dist.
y_1	20	25	30	0.417
y_2	30	25	50	0.583
Rel. Freq. Marg. Dist.	0.278	0.278	0.444	1

(c) Beginning with x_1, we determine the relative frequency of each y_i given x_1 by dividing each frequency by the column total for x_1. For example, the relative frequency of y_1 given x_1 is $\frac{20}{50} = 0.4$. Next, we compute each relative frequency of each y_i given x_2. For example, the relative frequency of y_1 given x_2 is $\frac{25}{50} = 0.5$. Finally, we compute the relative frequency of each y_i given x_3.

	x_1	x_2	x_3
y_1	$\frac{20}{50} = 0.4$	$\frac{25}{50} = 0.5$	$\frac{30}{80} = 0.375$
y_2	$\frac{30}{50} = 0.6$	$\frac{25}{50} = 0.5$	$\frac{50}{80} = 0.625$
Total	1	1	1

(d) We draw two bars, side by side, for each y_i. The horizontal axis represents x_i and the vertical axis represents the relative frequency.

Conditional Distribution

7. **(a)** 780 Americans were surveyed.
195 Hispanics were surveyed.

(b) The relative frequency marginal distribution for the row variable, response to immigration question, is found by dividing the row total for each answer choice by the table total, 780. For example, the relative frequency for the response "Good Thing" is $\frac{478}{780} \approx 0.613$. The relative frequency marginal distribution for the column variable, ethnicity, is found by dividing the column total for each ethnicity by the table total. For example, the relative frequency for Non-Hispanic Whites is $\frac{305}{780} \approx 0.391$.

	Non-Hispanic Whites	Blacks	Hispanics	Rel. Freq. Marg. Dist.
Good Thing	180	154	144	0.613
Bad Thing	107	101	33	0.309
Good & bad	9	14	12	0.045
No Opinion	9	11	6	0.033
Rel. Freq. Marg. Dist.	0.391	0.359	0.250	1

(c) The proportion of Americans who feel that immigration is a good thing for this country is 0.613.

(d) Beginning with Non-Hispanic Whites, we determine the relative frequency for each response to the immigration question by dividing each frequency by the column total for Non-Hispanic Whites. For example, the relative frequency for the response "Good Thing" given that the respondent is Non-Hispanic White is $\frac{180}{305} \approx 0.590$. Next, we compute each relative frequency for each response given the respondent is Black. Finally, we compute the relative frequency for each response given the respondent is Hispanic.

	Non-Hispanic Whites	Blacks	Hispanics
Good Thing	$\frac{180}{305} \approx 0.590$	$\frac{154}{280} = 0.550$	$\frac{144}{195} \approx 0.738$
Bad Thing	$\frac{107}{305} \approx 0.351$	$\frac{101}{280} \approx 0.361$	$\frac{33}{195} \approx 0.169$
Good & bad	$\frac{9}{305} \approx 0.030$	$\frac{14}{280} = 0.50$	$\frac{12}{195} \approx 0.062$
No Opinion	$\frac{9}{305} \approx 0.030$	$\frac{11}{280} \approx 0.039$	$\frac{6}{195} \approx 0.031$
Total	1	1	1

(e) We draw four bars, side by side, for each response to the immigration question. The horizontal axis represents ethnicity and the vertical axis represents the relative frequency.

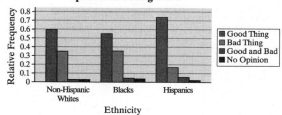

Opinion on Immigration

(f) Yes, ethnicity is associated with opinion regarding immigration. Hispanics are more likely to feel that immigration is a good thing for the country and much less likely to feel it is a bad thing.

9. (a) We find the frequency marginal distribution for the row variable, party affiliation, by finding the total of each row. We find the frequency marginal distribution for the column variable, gender, by finding the total for each column.

	Female	Male	Frequency Marg. Dist.
Republican	105	115	220
Democrat	150	103	253
Independent	150	179	329
Frequency Marg. Dist.	405	397	802

(b) The relative frequency marginal distribution for the row variable, party affiliation, is found by dividing the row total for each party by the table total, 802. For example, the relative frequency for Republican is $\frac{220}{802} \approx 0.274$. The relative frequency marginal distribution for the column variable, gender, is found by dividing the column total for each gender by the table total. For example, the relative frequency for female is $\frac{405}{802} \approx 0.505$.

	Female	Male	Rel. Freq. Marg. Dist.
Republican	105	115	0.274
Democrat	150	103	0.315
Independent	150	179	0.410
Rel. Freq. Marg. Dist.	0.505	0.495	1

(c) The proportion of registered voters who consider themselves Independent is 0.410.

(d) Beginning with females, we determine the relative frequency for each political party by dividing each frequency by the column total for females. For example, the relative frequency for Republicans given that the voter is female is $\frac{105}{405} \approx 0.259$. We then compute each relative frequency for each political party given that the voter is male.

	Female	Male
Republican	$\frac{105}{405} \approx 0.259$	$\frac{115}{397} \approx 0.290$
Democrat	$\frac{150}{405} \approx 0.370$	$\frac{103}{397} \approx 0.259$
Independent	$\frac{150}{405} \approx 0.370$	$\frac{179}{397} \approx 0.451$
Total	1	1

(e) We draw three bars, side by side, for each political affiliation. The horizontal axis represents gender and the vertical axis represents the relative frequency.

(f) Yes, gender is associated with party affiliation. Male voters are more likely to be Independents and less likely to be Democrats.

11. To determine whether healthier people tend to also be happier, we construct a conditional distribution of people's happiness by health and draw a bar graph of the conditional distribution.

	Poor Health	Fair Health	Good Health	Excellent Health
Not too happy	$\frac{696}{1,996} \approx 0.349$	$\frac{1,386}{6,585} \approx 0.210$	$\frac{1,629}{15,791} \approx 0.103$	$\frac{732}{11,022} \approx 0.066$
Pretty happy	$\frac{950}{1,996} \approx 0.476$	$\frac{3,817}{6,585} \approx 0.580$	$\frac{9,642}{15,791} \approx 0.611$	$\frac{5,195}{11,022} \approx 0.471$
Very happy	$\frac{350}{1,996} \approx 0.175$	$\frac{1,382}{6,585} \approx 0.210$	$\frac{4,520}{15,791} \approx 0.286$	$\frac{5,095}{11,022} \approx 0.462$
Total	1	1	1	1

We draw three bars, side by side, for each level of happiness. The horizontal axis represents level of health and the vertical axis represents the relative frequency.

There is an association between health and happiness. Based on the conditional distribution by health, we can see that healthier people tend to be happier. As health increases, the proportion who are very happy increases, while the proportion who are not happy decreases.

13. (a) Of the 582 smokers in the study, 139 were dead after 20 years, so the proportion of smokers who were dead after 20 years was $\frac{139}{582} \approx 0.239$. Of the 732 nonsmokers in the study, 230 were dead after 20 years, so the proportion of nonsmokers who were dead after 20 years was $\frac{230}{732} \approx 0.314$. This result implies that it is healthier to smoke.

   (b) Of the $2 + 53 = 55$ smokers in the 18- to 24-year-old category, 2 were dead after 20 years, so the proportion who were dead was $\frac{2}{55} \approx 0.036$. Of the $1 + 61 = 62$ nonsmokers in the 18-to 24-year old category, 1 was dead after 20 years, so the proportion was $\frac{1}{62} \approx 0.016$.

   (c) We repeat the procedure from part (b) for each age category and organize the results in the table that follows:

Age	Smoker	Nonsmoker
18-24	$\frac{2}{2+53} \approx 0.036$	$\frac{1}{1+61} \approx 0.016$
25-34	$\frac{3}{3+121} \approx 0.024$	$\frac{5}{5+152} \approx 0.032$
35-44	$\frac{14}{14+95} \approx 0.128$	$\frac{7}{7+114} \approx 0.058$
45-54	$\frac{27}{27+103} \approx 0.208$	$\frac{12}{12+66} \approx 0.154$
55-64	$\frac{51}{51+64} \approx 0.443$	$\frac{40}{40+81} \approx 0.331$
65-74	$\frac{29}{29+7} \approx 0.806$	$\frac{101}{101+28} \approx 0.783$
75+	$\frac{13}{13+0} \approx 1$	$\frac{64}{64+0} \approx 1$

   (d) We draw two bars, side by side, for smokers and nonsmokers. The horizontal axis represents age and the vertical axis represents the relative frequency of deaths.

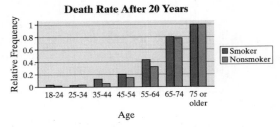

Death Rate After 20 Years

   (e) Reports will vary. When taking age into account, the direction of association changed. In almost all age groups, smokers had a higher death rate than nonsmokers. The most notable exception is for the 25 to 34 age group, the largest age group for the nonsmokers. A possible explanation could be rigorous physical activity (e.g., rock climbing) that nonsmokers are more likely to participate in than smokers.

Section 4.5 (on CD)

1. The logarithm to the base a, where $a > 0$, $a \neq 1$, is $y = \log_a x$ if and only if $x = a^y$. The logarithm is only defined for $x > 0$.

3. $\log y = \log(ab^x)$
   $= \log a + \log b^x$
   $= \log a + x \log b$
   The linear form of $y = ab^x$
   is: $\log y = \log a + x \log b$.
   If we let $Y = \log y$, $B = \log b$ and $A = \log a$, then the linear form is $Y = A + Bx$.

5. $4^2 = 16 \iff 2 = \log_4 16$

7. $x^5 = 18 \iff 5 = \log_x 18$

9. $7^x = 15 \iff x = \log_7 15$

11. $5^3 = y \iff 3 = \log_5 y$

13. $\log_3 81 = 4 \iff 3^4 = 81$

15. $\log_a 14 = 3 \iff a^3 = 14$

17. $\log_4 c = 3 \iff 4^3 = c$

19. $\log_2 5 = k \iff 2^k = 5$

21. $\log_3(uv) = \log_3 u + \log_3 v$ (Property 1)

23. $\log_5 x^4 = 4\log_5 x$ (Property 2)

25. $\log_a(uv^2) = \log_a u + \log_a v^2$ (Property 1)
$\qquad\quad = \log_a u + 2\log_a v$ (Property 2)

27. $\log(x^4 y^3) = \log x^4 + \log y^3$ (Property 1)
$\qquad\quad = 4\log x + 3\log y$ (Property 2)

29. $\log 34 \approx 1.531$

```
log(34)
        1.531478917
■
```

31. $\log 13.2 \approx 1.121$

```
log(13.2)
        1.120573931
■
```

33. $10^{1.6} \approx 39.811$

```
10^1.6
        39.81071706
■
```

35. $10^{0.36} \approx 2.291$

```
10^0.36
        2.290867653
■
```

37. (a)

E-coli Population vs. Time

(graph with axes Population, y (0.1 to 0.5) and Time (hours), x (0 to 6))

(b)

x	y	$Y = \log y$
0	0.09	$\log 0.09 \approx -1.0458$
2.5	0.18	$\log 0.18 \approx -0.7447$
3.5	0.26	$\log 0.26 \approx -0.5850$
4.5	0.35	$\log 0.35 \approx -0.4559$
6	0.50	$\log 0.50 \approx -0.3010$

Logarithm of E-coli Population vs. Time

(c) Using technology we find the least-squares regression line of the transformed data to be: $\hat{Y} = 0.1266x - 1.0443$

(d) $\hat{y} = 10^{\hat{Y}}$
$\quad = 10^{0.1266x - 1.0443}$
$\quad = 10^{-1.0443} \cdot (10^{0.1266})^x$
$\quad = 0.0903(1.3384)^x$

(e) $y = 0.0903(1.3384)^7 = 0.69$
The predicted population after 7 hours is 0.69.

39. (a)

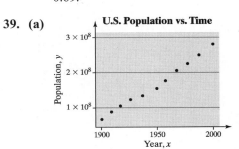

U.S. Population vs. Time

(b)

x	y	$Y = \log y$
1900	76,212,168	7.8820
1910	92,228,496	7.9649
1920	106,021,537	8.0254
1930	123,202,624	8.0906
1940	132,164,569	8.1211
1950	151,325,798	8.1799
1960	179,323,175	8.2536
1970	203,302,031	8.3081
1980	226,542,203	8.3551
1990	248,709,873	8.3957
2000	281,421,906	8.4494

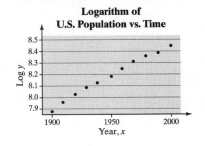

Logarithm of U.S. Population vs. Time

(c) Using technology we find the least-squares regression line of the transformed data to be: $\hat{Y} = 0.0056x - 2.6593$

(d) $\hat{y} = 10^{\hat{Y}}$

$= 10^{0.0056x - 2.6593}$

$= 10^{-2.6593} \cdot \left(10^{0.0056}\right)^{x}$

$= 0.0022(1.013)^{x}$

(e) $\hat{y} = 0.0022(1.013)^{2010} = 414,388,653.$

The predicted population for the U.S. in 2010 is 414,388,653, or about 414 million.

(f) Answers may vary.

(g) Using the U.S. Census Bureau website, the predicted U.S. population in 2010 (as of August 14, 2008) was 310,233,000. This is substantially lower than the prediction from our regression equation.

41. (a)

Planetary Motion

(b)

x	y	$X = \log x$	$Y = \log y$
36	0.24	1.5563	−0.6198
67	0.62	1.8261	−0.2076
93	1.00	1.9685	0
142	1.88	2.1523	0.2742
483	11.9	2.6839	1.0755
887	29.5	2.9479	1.4698
1785	84.0	3.2516	1.9243
2797	165.0	3.4467	2.2175
3675	248.0	3.5653	2.3945

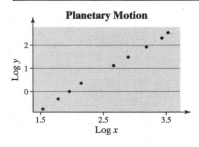

Planetary Motion

(c) Using technology we find the least-squares regression line of the transformed data to be: $\hat{Y} = 1.4994X - 2.9506$

(d) $\hat{y} = 10^{\hat{Y}}$

$= 10^{1.4994X - 2.9506}$

$= 10^{-2.9506} \cdot 10^{1.4994\log x}$

$= 10^{-2.9506} \cdot \left(10^{\log x}\right)^{1.4994}$

$= 0.0011x^{1.4994}$

43. (a)

Intensity vs. Distance

(b)

x	y	$X = \log x$	$Y = \log y$
1.0	0.29645	0	−0.5280
1.1	0.25215	0.0414	−0.5983
1.2	0.20547	0.0792	−0.6873
1.3	0.17462	0.1139	−0.7579
1.4	0.15342	0.1461	−0.8141
1.5	0.13521	0.1761	−0.8690
1.6	0.11450	0.2041	−0.9412
1.7	0.10243	0.2304	−0.9896
1.8	0.09231	0.2553	−1.0348
1.9	0.08321	0.2788	−1.0798
2.0	0.07342	0.3010	−1.1342

Logarithm of Intensity vs. Logarithm of Distance

(c) Using technology we find the least-squares regression line of the transformed data to be: $\hat{Y} = -2.012X - 0.5236$

(d) $\hat{y} = 10^{\hat{Y}}$

$= 10^{-2.012X - 0.5236}$

$= 10^{-0.5236} \cdot 10^{-2.012\log x}$

$= 10^{-0.5236} \cdot \left(10^{\log x}\right)^{-2.012}$

$= 0.2995 x^{-2.012}$

(e) $\hat{y} = 0.2995(2.3)^{-2.012} = 0.05605$

The predicted intensity of a 100-watt light bulb that is 2.3 meters away is 0.05605.

45. (a) Using technology we find the least-squares regression line to be:
$\hat{y} = -3.711.8x + 30,984.3$

We compute the residuals and construct the residual plot.

x	y	$\hat{y} = -3.711.8x + 30,984.3$	$y - \hat{y}$
1	27,417	27272.5	145
1	28,199	27272.5	927
2	22,995	23560.7	−566
2	23,195	23560.7	−366
3	18,499	19848.9	−1,350
4	16,995	16137.1	858
4	16,490	16137.1	353

The residual plot shows a clear U-shaped pattern, so the relation between asking price and age is not linear.

(b) To transform the data, we determine the logarithm of the y-values.

x	y	$Y = \log y$
1	27,417	4.43802
1	28,199	4.45023
2	22,995	4.36163
2	23,195	4.36539
3	18,499	4.26715
4	16,995	4.23032
4	16,490	4.21722

Using technology we find the least-squares regression line of the transformed data to be:
$\hat{Y} = -0.074589216x + 4.513998311$.

So, the exponential equation of best fit is:

$y = 10^{4.513998311} \cdot \left(10^{-0.074589216}\right)^x$

$= 32,658.6562(0.8422)^x$

We compute the residuals and construct the residual plot.

x	$Y = \log y$	$\hat{Y} = -0.074589x + 4.513998$	$Y - \hat{Y}$
1	4.43802	4.43941	−0.00139
1	4.45023	4.43941	0.01082
2	4.36163	4.36482	−0.00319
2	4.36539	4.36482	0.00057
3	4.26715	4.29023	−0.02308
4	4.23032	4.21564	0.01468
4	4.21722	4.21564	0.00158

The relation between asking price and age could be exponential since the residual plot does not show a discernable pattern.

(c) The coefficient of determination for the linear model found in part (a) is $R^2 \approx 0.971$. The coefficient of determination for the exponential model found in part (b) is $R^2 \approx 0.984$. Thus, the exponential model does a better job of explaining the variation in asking price. In fact, 98.4% of the variation in asking price is explained by the age of the car in the exponential model.

Chapter 4 Review Exercises

1. (a) The explanatory variable is fat content.

(b)

(c) To calculate the correlation coefficient, we use the computational formula.

x_i	y_i	x_i^2	y_i^2	$x_i y_i$
20	430	400	184,900	8,600
39	750	1521	562,500	29,250
27	480	729	230,400	12,960
29	540	841	291,600	15,660
26	510	676	260,100	13,260
47	760	2209	577,600	35,720
35	690	1225	476,100	24,150
38	632	1444	399,424	24,016
261	4792	9045	2,982,624	163,616

From the table, we have $n = 8$, $\sum x_i = 261$, $\sum y_i = 4792$, $\sum x_i^2 = 9045$, $\sum y_i^2 = 2{,}982{,}624$, and $\sum x_i y_i = 163{,}616$. So, the correlation coefficient is

$$r = \frac{\sum x_i y_i - \frac{\sum x_i \sum y_i}{n}}{\sqrt{\left(\sum x_i^2 - \frac{(\sum x_i)^2}{n}\right)\left(\sum y_i^2 - \frac{(\sum y_i)^2}{n}\right)}}$$

$$= \frac{163{,}616 - \frac{(261)(4792)}{8}}{\sqrt{\left(9045 - \frac{(261)^2}{8}\right)\left(2{,}982{,}624 - \frac{(4792)^2}{8}\right)}}$$

$$\approx 0.944$$

(d) Yes, a strong positive linear relation exists between fat content and calories in fast-food restaurant sandwiches.

2. (a)

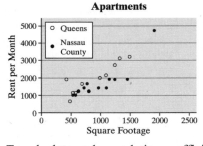

(b) To calculate each correlation coefficient, we use the computational formula.

From the table for the Queens (New York City) apartments at the top of the next column, we have $n = 12$, $\sum x_i = 10{,}360$, $\sum y_i = 22{,}475$, $\sum x_i^2 = 10{,}245{,}152$, $\sum y_i^2 = 49{,}773{,}275$, and $\sum x_i y_i = 22{,}277{,}560$.

Queens (New York City)

x_i	y_i	x_i^2	y_i^2	$x_i y_i$
500	650	250,000	422,500	325,000
588	1215	345,744	1,476,225	714,420
1000	2000	1,000,000	4,000,000	2,000,000
688	1655	473,344	2,739,025	1,138,640
825	1250	680,625	1,562,500	1,031,250
460	1805	211,600	3,258,025	830,300
1259	2700	1,585,081	7,290,000	3,399,300
650	1200	422,500	1,440,000	780,000
560	1250	313,600	1,562,500	700,000
1073	2350	1,151,329	5,522,500	2,521,550
1452	3300	2,108,304	10,890,000	4,791,600
1305	3100	1,703,025	9,610,000	4,045,500
10,360	22,475	10,245,152	49,773,275	22,277,560

So, the correlation coefficient for Queens is:

$$r = \frac{\sum x_i y_i - \frac{\sum x_i \sum y_i}{n}}{\sqrt{\left(\sum x_i^2 - \frac{(\sum x_i)^2}{n}\right)\left(\sum y_i^2 - \frac{(\sum y_i)^2}{n}\right)}}$$

$$= \frac{22{,}277{,}560 - \frac{(10{,}360)(22{,}475)}{12}}{\sqrt{\left(10{,}245{,}152 - \frac{(10{,}360)^2}{12}\right)\left(49{,}773{,}275 - \frac{(22{,}475)^2}{12}\right)}}$$

$$\approx 0.909$$

Nassau County (Long Island)

x_i	y_i	x_i^2	y_i^2	$x_i y_i$
1100	1875	1,210,000	3,515,625	2,062,500
588	1075	345,744	1,155,625	632,100
1250	1775	1,562,500	3,150,625	2,218,750
556	1050	309,136	1,102,500	583,800
825	1300	680,625	1,690,000	1,072,500
743	1475	552,049	2,175,625	1,095,925
660	1315	435,600	1,729,225	867,900
975	1400	950,625	1,960,000	1,365,000
1429	1900	2,042,041	3,610,000	2,715,100
800	1650	640,000	2,722,500	1,320,000
1906	4625	3,632,836	21,390,625	8,815,250
1077	1395	1,159,929	1,946,025	1,502,415
11,909	20,835	13,521,085	46,148,375	24,251,240

From the table for the Nassau County (Long Island) apartments, we have $n = 12$, $\sum x_i = 11{,}909$, $\sum y_i = 20{,}835$, $\sum x_i^2 = 13{,}521{,}085$, $\sum y_i^2 = 46{,}148{,}375$, and $\sum x_i y_i = 24{,}251{,}240$. So, the correlation coefficient for Nassau County is:

$$r = \frac{24{,}251{,}240 - \frac{(11{,}909)(20{,}835)}{12}}{\sqrt{\left(13{,}521{,}085 - \frac{(11{,}909)^2}{12}\right)\left(46{,}148{,}375 - \frac{(20{,}835)^2}{12}\right)}}$$

$$\approx 0.867$$

(c) Yes, both locations have a positive linear association between square footage and monthly rent.

(d) For small apartments (those less than 1000 square feet in area), there seems to be no difference in rent between Queens and Nassau County. In larger apartments, however, Queens seems to have higher rents than Nassau County.

3. (a) In Problem 1, we computed the correlation coefficient. Rounded to eight decimal places, it is $r = 0.94370937$. We compute the mean and standard deviation for each variable to be $\bar{x} = 32.625$, $s_x \approx 8.7003695$, $\bar{y} = 599$, and $s_y \approx 126.6130212$.

$$b_1 = r \cdot \frac{s_y}{s_x} = 0.94370937 \left(\frac{126.6130212}{8.7003695} \right)$$

$$\approx 13.7334276$$

$$b_0 = \bar{y} - b_1 \bar{x}$$

$$= 599 - (13.7334276)(32.625)$$

$$\approx 150.9469$$

Rounding to four decimal places, the least-squares regression line is

$$\hat{y} = 13.7334x + 150.9469$$

(b)

Calories vs Fat Content

(c) The slope indicates that each additional gram of fat in a sandwich adds 13.73 calories, on average. The y-intercept indicates that a sandwich with no fat will contain about 151 calories.

(d) Let $x = 30$ in the regression equation.
$$\hat{y} = 13.7334(30) + 150.9469 \approx 562.9$$
We predict that a sandwich with 30 grams of fat will have 562.9 calories.

(e) Let $x = 42$ in the regression equation.
$$\hat{y} = 13.7334(42) + 150.9469 \approx 727.7$$
Sandwiches with 42 grams of fat have an average of 727.7 calories. So, the number of calories for this cheeseburger from Sonic is below average.

4. (a) In Problem 2, we computed the correlation coefficient. Rounded to eight decimal places, it is $r = 0.90928694$. We compute the mean and standard deviation for each variable to be $\bar{x} \approx 863.333333$, $s_x \approx 343.9104887$, $\bar{y} \approx 1872.916667$, and $s_y \approx 835.5440751$.

$$b_1 = r \cdot \frac{s_y}{s_x} = 0.90928694 \left(\frac{835.5440751}{343.9104887} \right)$$

$$\approx 2.20914843$$

$$b_0 = \bar{y} - b_1 \bar{x}$$

$$= 1872.916667 - (2.2091484)(863.333333)$$

$$\approx -34.3148$$

Rounding to four decimal places, the least-squares regression line is
$$\hat{y} = 2.2091x - 34.3148 .$$

(b) The slope of the least-squares regression equation indicates that, for each additional square foot of floor area, the rent increases by \$2.21, on average. It is not appropriate to interpret the y-intercept since it is not possible to have an apartment with an area of 0 square feet.

(c) Let $x = 825$ in the regression equation.
$$\hat{y} = 2.2091(825) - 34.3148 \approx 1788$$
Apartments in Queens with 825 square feet have an average rent of \$1788. Since the rent of the apartment in the data set with 825 square feet is \$1250, it is below average.

5. (a)

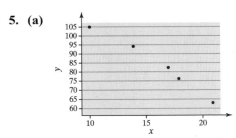

(b) Using the two points $(10, 105)$ and $(18, 76)$ gives:

$$m = \frac{76 - 105}{18 - 10} = \frac{-29}{8} = -\frac{29}{8}$$

$$y - 105 = -\frac{29}{8}(x - 10)$$

$$y - 105 = -\frac{29}{8}x + \frac{145}{4}$$

$$y = -\frac{29}{8}x + \frac{565}{4}$$

(c)

(d) We compute the mean and standard deviation for each variable and the correlation coefficient to be $\bar{x} = 16$, $\bar{y} = 84$, $s_x \approx 4.18330$, $s_y \approx 16.20185$, and $r \approx -0.99222$. Then the slope and intercept for the least-squares regression line are:

$$b_1 = r \cdot \frac{s_y}{s_x} = -0.992221\left(\frac{16.20185}{4.18330}\right)$$
$$\approx -3.842855$$
$$b_0 = \bar{y} - b_1\bar{x} = 84 - (-3.842855)(16)$$
$$\approx 145.4857$$

So, rounding to four decimal places, the least-squares regression line is $\hat{y} = -3.8429x + 145.4857$.

(e)

(f)

x	y	$\hat{y} = -\frac{29}{8}x + \frac{565}{4}$	$y - \hat{y}$	$(y - \hat{y})^2$
10	105	105	0	0
14	94	90.5	3.5	12.25
17	82	79.625	2.375	5.6406
18	76	76	0	0
21	63	65.125	-2.125	4.5156

Total = 22.4062

(g)

x	y	$\hat{y}$	$y - \hat{y}$	$(y - \hat{y})^2$
10	105	107.0567	-2.0567	4.2300
14	94	91.6851	2.3149	5.3588
17	82	80.1564	1.8436	3.3989
18	76	76.3135	-0.3135	0.0983
21	63	64.7848	-1.7848	3.1855

Total = 16.2715

(h) The regression line gives a smaller sum of squared residuals, so it is a better fit.

6. A linear model is appropriate. There is no pattern to the residuals and the variability stays the same.

7. A linear model is not appropriate because a discernible pattern is present in the plot.

8. A linear model is not appropriate because the amount of scatter in the residuals is clearly decreasing, indicating that the error variance is not constant.

9. (a) In Problem 1(c), we found $r \approx 0.944$. So, $R^2 = (0.944)^2 \approx 0.891 = 89.1\%$. This means that 89.1% of the variation in calories is explained by the least-squares regression line.

(b) From Problem 3(a), we have $\hat{y} = 13.7334x + 150.9469$. We compute the residuals and construct the residual plot.

x_i	y_i	$\hat{y} = 13.7334x + 150.9469$	$y - \hat{y}$
20	430	425.6	4.4
39	750	686.5	63.5
27	480	521.7	-41.7
29	540	549.2	-9.2
26	510	508.0	2.0
47	760	796.4	-36.4
35	690	631.6	58.4
38	632	672.8	-40.8

(c) Yes, based on the residual plot, a linear model is appropriate.

(d) There are no outliers or influential observations present.

(e) The observation (9, 80) is an outlier since it is far away from the rest of the data. Adding it to the data set, the regression equation becomes $\hat{y} = 17.7154x + 9.8713$. Since there is a significant change in both the slope and y-intercept, the observation is influential.

10. (a) From Problem 2(b), the correlation coefficient for the Queens apartments is $r = 0.90928694$ (unrounded). Therefore, $R^2 = (0.90928694)^2 \approx 0.827 = 82.7\%$. This means that 82.7% of the variation in monthly rent is explained by the least-squares regression equation.

(b) From Problem 4(a), we have $\hat{y} = 2.2091x - 34.3148$. We compute the residuals and construct the residual plot.

x_i	y_i	$\hat{y} = 2.2091x - 34.3148$	$y - \hat{y}$
500	650	1070.2	-420.2
588	1215	1264.6	-49.6
1000	2000	2174.8	-174.8
688	1655	1485.5	169.5
825	1250	1788.2	-538.2
460	1805	981.9	823.1
1259	2700	2746.9	-46.9
650	1200	1401.6	-201.6
560	1250	1202.8	47.2
1073	2350	2336.0	14.0
1452	3300	3173.3	126.7
1305	3100	2848.6	251.4

Apartments

Square Footage

(c) A linear model is appropriate. There is no pattern to the residuals and the variability stays the same.

(d) The observation (460, 1805) is an outlier because it lies far away from the rest of the data. It is also an influential point because, if it is removed from the data set, the linear regression equation becomes $\hat{y} = 2.5315x - 399.2496$ in which the y-intercept is significantly different.

11. (a) In Problem 2(b), we computed the correlation coefficient. Rounded to eight decimal places, it is $r = 0.86741573$. We compute the mean and standard deviation for each variable to be $\bar{x} \approx 992.416667$, $s_x \approx 393.399510$, $\bar{y} = 1736.25$, and $s_y \approx 952.203485$.

$$b_1 = r \cdot \frac{s_y}{s_x} = 0.86741573 \left(\frac{952.203485}{393.399510} \right)$$
$$\approx 2.0995356$$

$$b_0 = \bar{y} - b_1\bar{x}$$
$$= 1736.25 - (2.0995356)(992.416667)$$
$$\approx -347.3641$$

Rounding to four decimal places, the least-squares regression line is $\hat{y} = 2.0995x - 347.3641$.

(b) The observation (1906, 4625) is an outlier because it lies far away from the rest of the data. It is also influential point because, if it is removed from the data set, the linear regression equation becomes $\hat{y} = 0.8779x + 675.2945$ in which both the slope and y-intercept are significantly different.

12. (a)

Camaro Depreciation

```
Price (dollars)
20,000 ┤ •
       │ •
       │   •
15,000 ┤   •
       │          •
       │              •     •
10,000 ┤               •     •
       └─┬──┬──┬──┬──┬──┬──
         1  2  3  4  5  6
           Age (years)
```

(b) To calculate the correlation coefficient, we use the computational formula.

x_i	y_i	x_i^2	y_i^2	$x_i y_i$
2	15,900	4	252,810,000	31,800
5	10,988	25	120,736,144	54,940
2	16,980	4	288,320,400	33,960
5	9,995	25	99,900,025	49,975
4	11,995	16	143,880,025	47,980
5	10,995	25	120,890,025	54,975
1	20,365	1	414,733,225	20,365
2	16,463	4	271,030,369	32,926
6	10,824	36	117,158,976	64,944
1	19,995	1	399,800,025	19,995
1	18,650	1	347,822,500	18,650
4	10,488	16	109,998,144	41,952
38	173,638	158	2,687,079,858	472,462

From the table, we have $n = 12$, $\sum x_i = 38$, $\sum y_i = 173,638$, $\sum x_i^2 = 158$, $\sum y_i^2 = 2,687,079,858$, and $\sum x_i y_i = 472,462$. So, the correlation coefficient is

$$r = \frac{472,462 - \frac{(38)(173,638)}{12}}{\sqrt{\left(158 - \frac{(38)^2}{12}\right)\left(2,687,079,858 - \frac{(173,638)^2}{12}\right)}}$$
$$\approx -0.95440918 \approx -0.954$$

Yes, there is strong negative linear relation between age and price.

(c) We compute the mean and standard deviation for each variable: $\bar{x} \approx 3.1666667$, $\bar{y} \approx 14{,}469.8333333$, $s_x \approx 1.85047087$, and $s_y \approx 3983.68191577$. Then the slope and intercept for the least-squares regression line are:

$$b_1 = r \cdot \frac{s_y}{s_x}$$

$$= -0.95440918\left(\frac{3983.68191577}{1.85047087}\right)$$

$$\approx -2054.646011$$

$$b_0 = \bar{y} - b_1\bar{x}$$

$$= 14{,}469.8333333 - (-2054.646011)(3.1666667)$$

$$\approx 20{,}976.2124$$

So, rounding to four decimal places, the least-squares regression line is $\hat{y} = -2054.6460x + 20{,}976.2124$.

(d) We compute the residuals and construct the residual plot.

x_i	y_i	$\hat{y} = -2054.6460x + 20{,}976.2124$	$y - \hat{y}$
2	15,900	16,866.9	−966.9
5	10,988	10,703.0	285.0
2	16,980	16,866.9	113.1
5	9,995	10,703.0	−708.0
4	11,995	12,757.6	−762.6
5	10,995	10,703.0	292.0
1	20,365	18,921.6	1,443.4
2	16,463	16,866.9	−403.9
6	10,824	8,648.3	2,175.7
1	19,995	18,921.6	1,073.4
1	18,650	18,921.6	−271.6
4	10,488	12,757.6	−2,269.6

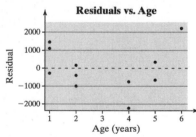

Residuals vs. Age

(e) No, the residuals follow a U-shaped pattern indicating that the constant variance assumption is not met.

13. No. Correlation does not imply causation. Florida has a large number of tourists in warmer months, times when more people will be in the water to cool off. The larger number of people in the water splashing around lends to a larger number of shark attacks.

14. (a) 203 buyers were extremely satisfied with their automobile purchase.

(b) The relative frequency marginal distribution for the row variable, level of satisfaction, is found by dividing the row total for each level of satisfaction by the table total, 396. For example, the relative frequency for "Not too satisfied" is $\frac{36}{396} \approx 0.091$. The relative frequency marginal distribution for the column variable, purchase type (new versus used), is found by dividing the column total for each purchase type by the table total. For example, the relative frequency for "New" is $\frac{207}{396} \approx 0.523$.

	New	Used	Rel. Freq. Marg. Dist.
Not Too Satisfied	11	25	$\frac{36}{396} \approx 0.091$
Pretty Satisfied	78	79	$\frac{157}{396} \approx 0.396$
Extremely Satisfied	118	85	$\frac{203}{396} \approx 0.513$
Rel. Freq. Marg. Dist.	$\frac{207}{396} \approx 0.523$	$\frac{189}{396} \approx 0.477$	1

(c) The proportion of consumers who were extremely satisfied with their automobile purchase was 0.513.

(d) Beginning with new automobiles, we determine the relative frequency for level of satisfaction by dividing each frequency by the column total for new automobiles. For example, the relative frequency for "Not Too Satisfied" given that the consumer purchased a new automobile is $\frac{11}{207} \approx 0.053$. We then compute each relative frequency for the used automobiles.

	New	Used
Not Too Satisfied	$\frac{11}{207} \approx 0.053$	$\frac{25}{189} \approx 0.132$
Pretty Satisfied	$\frac{78}{207} \approx 0.377$	$\frac{79}{189} \approx 0.418$
Extremely Satisfied	$\frac{118}{207} \approx 0.570$	$\frac{85}{189} \approx 0.450$
Total	1	1

(e) We draw three bars, side by side, for each level of satisfaction. The horizontal axis represents purchase type (new versus used) and the vertical axis represents the relative frequency.

Customer Satisfaction

(f) Yes, there appears to be some association between purchase type (new versus used) and level of satisfaction. Buyers of used cars are more likely to be dissatisfied and are less likely to be extremely satisfied than buyers of new cars.

15. (a) For each hospital, we determine the survival status proportions by dividing the number of patients in each survival status by the total number of patients admitted for surgery in that hospital:

	PCH	GH
Survived	$\dfrac{150}{150+20} \approx 0.882$	$\dfrac{230}{230+55} \approx 0.807$
Died	$\dfrac{20}{150+20} \approx 0.118$	$\dfrac{55}{230+55} \approx 0.193$

(b) The proportion of surgeries at PCH that resulted in death is 0.118. The proportion of surgeries at GH that resulted in death was 0.193.

(c) A higher proportion of surgeries at GH result in death than at PCH.

(d) The proportion of planned surgeries at PCH having patients that survived was:
$$\frac{138}{138+12} = 0.92$$
The proportion of planned surgeries at GH having patients that survived was:
$$\frac{160}{160+13} \approx 0.925$$

(e) The proportion of emergency surgeries at PCH having patients that survived was:
$$\frac{12}{12+8} = 0.6$$
The proportion of emergency surgeries at GH having patients that survived was:
$$\frac{70}{70+42} = 0.625$$

(f) Answers will vary. The direction of the association reversed after taking the type of surgery (planned versus emergency) into account. For each type of surgery, GH had higher survival rates than PCH.

16. (a) A positive linear relation appears to exist between number of marriages and number unemployed.

(b) Population is highly correlated with both the number of marriages and the number unemployed. The size of the population affects both variables.

(c) It appears that no association exists between the two variables.

(d) Answers may vary. A strong correlation between two variables may be due to a third variable that is highly correlated with the two original variables.

17. The eight properties of a linear correlation coefficient are:
1. The linear correlation coefficient is always between −1 and 1, inclusive. That is, $-1 \le r \le 1$.
2. If $r = +1$, there is a perfect positive linear relation between the two variables.
3. If $r = -1$, there is a perfect negative linear relation between the two variables.
4. The closer r is to $+1$, the stronger is the evidence of positive association between the two variables.
5. The closer r is to -1, the stronger is the evidence of negative association between the two variables.
6. If r is close to 0, there is no evidence of a *linear* relation between the two variables. Because the linear correlation coefficient is a measure of the strength of the linear relation; an r close to 0 does not imply no relation, just no linear relation.
7. The linear correlation coefficient is a unitless measure of association. So, the unit of measure for x and y plays no role in the interpretation of r.
8. The correlation coefficient is not resistant.

18. (a) Answers will vary.

(b) The slope can be interpreted as "the school day decreases by about 0.01 hours for every 1 percent increase in the percentage of the district with low income", on average. The y-intercept can be interpreted as the length of the school day for a district with no low income families.

(c) $\hat{y} = -0.0102(20) + 7.11 \approx 6.91$

The school day is predicted to be about 6.91 hours long if 20% of the district is low income.

(d) Answers will vary. Because so many of the data values are clumped together, it is difficult to see any patterns. As shown, it does not appear that any of the model requirements are badly violated.

(e) Answers will vary. The likely candidates will be the three points with the highest percentage of low income population.

(f) Answers will vary.

(g) Answers will vary. There is some indication that there is a positive association between length of school day and PSAE score.

(h) Answers will vary. The scatter diagram indicates that there is some negative association between PSAE score and percentage of population as low income. However, it is likely not as strong as the correlation coefficient indicates. A few influential observations could be inflating the correlation coefficient.

(i) Answers will vary.

Chapter 4 Test

1. (a) The likely explanatory variable is temperature because crickets would likely chirp more frequently in warmer temperatures and less frequently in colder temperatures, so the temperature could be seen as explaining the number of chirps.

(b)

Cricket Chirp Rates

(c) To calculate the correlation coefficient, we use the computational formula.

x_i	y_i	x_i^2	y_i^2	$x_i y_i$
88.6	20.0	7849.96	400.00	1772.00
93.3	19.8	8704.89	392.04	1847.34
80.6	17.1	6496.36	292.41	1378.26
69.7	14.7	4858.09	216.09	1024.59
69.4	15.4	4816.36	237.16	1068.76
79.6	15.0	6336.16	225.00	1194.00
80.6	16.0	6496.36	256.00	1289.60
76.3	14.4	5821.69	207.36	1098.72
71.6	16.0	5126.56	256.00	1145.60
84.3	18.4	7106.49	338.56	1551.12
75.2	15.5	5655.04	240.25	1165.60
82.0	17.1	6724.00	292.41	1402.20
83.3	16.2	6938.89	262.44	1349.46
82.6	17.2	6822.76	295.84	1420.72
83.5	17.0	6972.25	289.00	1419.50
1200.6	249.8	96,725.86	4200.56	20,127.47

From the table, we have $n = 15$, $\sum x_i = 1200.6$, $\sum y_i = 249.8$, $\sum x_i^2 = 96,725.86$, $\sum y_i^2 = 4200.56$, and $\sum x_i y_i = 20,127.47$. So, the correlation coefficient is

$$r = \frac{\sum x_i y_i - \frac{\sum x_i \sum y_i}{n}}{\sqrt{\left(\sum x_i^2 - \frac{(\sum x_i)^2}{n}\right)\left(\sum y_i^2 - \frac{(\sum y_i)^2}{n}\right)}}$$

$$= \frac{20,127.47 - \frac{(1200.6)(249.8)}{15}}{\sqrt{\left(96,725.86 - \frac{(1200.6)^2}{15}\right)\left(4200.56 - \frac{(249.8)^2}{15}\right)}}$$

$$\approx 0.8351437868$$

$$\approx 0.835$$

(d) Based on the scatter diagram and the linear correlation coefficient, a positive linear relation between temperature and chirps per second.

(e) We compute the mean and standard deviation for each variable: $\bar{x} = 80.04$, $\bar{y} \approx 16.6533333$, $s_x \approx 6.70733074$, and $s_y \approx 1.70204359$. The slope and intercept for the least-squares regression line are:

$$b_1 = r \cdot \frac{s_y}{s_x} = 0.8351437868\left(\frac{1.70204359}{6.70733074}\right)$$

$$\approx 0.21192501$$

$$b_0 = \bar{y} - b_1\bar{x}$$

$$= 16.6533333 - (0.21192501)(80.04)$$

$$\approx -0.3091$$

So, rounding to four decimal places, the least-squares regression line is $\hat{y} = 0.2119x - 0.3091$.

(f) If the temperature increases $1°F$, the number of chirps per second increases by 0.2119, on average. Since there are no observations near $0°F$, it does not make sense to interpret the y-intercept.

(g) Let $x = 83.3$ in the regression equation.
$\hat{y} = 0.2119(83.3) - 0.3091 \approx 17.3$
We predict that, if the temperature is $83.3°F$, then there will be 17.3 chirps per second.

(h) Let $x = 82$ in the regression equation.
$\hat{y} = 0.2119(82) - 0.3091 \approx 17.1$
There will be an average of 17.1 chirps per second when the temperature is $82°F$. Therefore, 15 chirps per second is below average.

(i) No, we should not use this model to predict the number of chirps when the temperature is $55°F$, because $55°F$ is outside the scope of the model.

(j) $R^2 = (0.8351437868)^2 \approx 0.697 = 69.7\%$
69.7% of the variation in number of chirps per second is explained by the least-squares regression line.

(k) Yes, a linear model is appropriate. There is no pattern to the residuals and the variability stays the same.

2. If we add the point (92, 14.5) to the data set, then the linear regression equation becomes $\hat{y} = 0.1431x + 4.9565$. Since both the slope and y-intercept change substantially, the observation is influential.

3. We compute the correlation coefficient to be $r \approx 0.992$, so from this it appears that there is a strong positive linear relation between speed and braking distance. However, we must analyze this further by computing the least-squares regression equation and construct a residual plot. The least-squares regression line is $\hat{y} = 5.3943x - 131.3524$. We compute the residuals and construct the residual plot.

x_i	y_i	$\hat{y} = 5.3943x - 131.3524$	$y - \hat{y}$
30	48	30.5	17.5
40	79	84.4	−5.4
50	123	138.4	−15.4
60	184	192.3	−8.3
70	243	246.2	−3.2
80	315	300.2	14.8

The residual plot has a clear U-shaped pattern, so a linear model is not appropriate despite the correlation coefficient.

4. Correlation does not imply causation. It is possible that a lurking variable, such as income level or educational level, is affecting both the explanatory and response variables.

5. (a) The relative frequency marginal distribution for the row variable, level of education, is found by dividing the row total for each level of satisfaction by the table total, 2375. For example, the relative frequency for "Less than high school" is $\frac{412}{2375} \approx 0.173$. The relative frequency marginal distribution for the column variable, response concerning belief in Heaven, is found by dividing the column total for each response by the table total. For example, the relative frequency for "Yes, Definitely" is $\frac{1539}{2375} = 0.648$.

	Yes, Definitely	Yes, Probably	No, Probably Not	No, Definitely Not	Rel. Freq. Marg. Dist.
Less than High School	316	66	21	9	0.173
High School	956	296	122	65	0.606
Bachelor's	267	131	62	64	0.221
Rel. Freq. Marg. Dist.	0.648	0.208	0.086	0.058	1

(b) The proportion of adult Americans in the survey who definitely believe in Heaven is 0.648.

(c) Beginning with "Less than high school", we determine the relative frequency for responses concerning belief in Heaven by dividing each frequency by the row total for "Less than high school." For example, the relative frequency for "Yes, Definitely" given that the respondent has less than a high school education is $\frac{316}{412} \approx 0.767$.

We then compute each relative frequency for the other levels of education.

	Yes, Definitely	Yes, Probably	No, Probably Not	No, Definitely Not	Total
< HS	$\frac{316}{412} \approx 0.767$	$\frac{66}{412} \approx 0.160$	$\frac{21}{412} \approx 0.051$	$\frac{9}{412} \approx 0.022$	1
HS	$\frac{956}{1439} \approx 0.664$	$\frac{296}{1439} \approx 0.206$	$\frac{122}{1439} \approx 0.085$	$\frac{65}{1439} \approx 0.045$	1
BS	$\frac{267}{524} \approx 0.510$	$\frac{131}{524} \approx 0.250$	$\frac{62}{524} \approx 0.118$	$\frac{64}{524} \approx 0.122$	1

(e) We draw four bars, side by side, for each response concerning belief in Heaven. The horizontal axis represents level of education and the vertical axis represents the relative frequency.

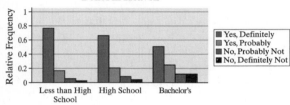

Belief in Heaven

(f) Yes, as education level increases, the proportion who definitely believe in Heaven decreases (i.e., doubt in Heaven increases).

6. (a) For each gender, we determine the proportions for acceptance status by dividing the number of applicants in each acceptance status by the total number of applicants of that gender:

	Accepted	Denied
Male	$\frac{98}{98+522} \approx 0.158$	$\frac{522}{98+522} \approx 0.842$
Female	$\frac{90}{90+200} \approx 0.310$	$\frac{200}{90+200} \approx 0.690$

(b) The proportion of males that was accepted is 0.158. The proportion of females that was accepted is 0.310.

(c) The college accepted a higher proportion of females than males.

(d) The proportion of male applicants accepted into the business school was:
$$\frac{90}{90+510} = 0.15$$

The proportion of female applicants accepted into the business school was:
$$\frac{10}{10+60} \approx 0.143$$

(e) The proportion of male applicants accepted into the social work school was:
$$\frac{8}{8+12} = 0.4$$

The proportion of male applicants accepted into the social work school was:
$$\frac{80}{80+140} \approx 0.364$$

(f) Answers will vary. A larger number of males applied to the business school, which has an overall lower acceptance rate than the social work school, so more male applicants were declined.

7. A set of quantitative bivariate data whose linear coefficient is −1 would have a perfect negative linear relation. A scatter diagram would show all the observations being collinear (falling on the same line) with a negative slope.

8. If the slope of the least-squares regression line is negative, then the correlation between the explanatory and response variables is also negative.

9. If a linear correlation is close to zero, then this means that there is no linear relation between the explanatory and response variables. This does not necessarily mean that there is no relation at all, however, just no *linear* relation.

Chapter 5
Probability

5.1 Probability Rules

1. Empirical probability is based on the outcome of a probability experiment and is approximately equal to the relative frequency of the event. Classical probability is based on counting techniques and is equal to the ratio of the number of ways an event can occur to the number of possible outcomes in the experiment.

3. Outcomes are equally likely when each outcome has the same probability of occurring.

5. True

7. experiment

9. From the Law of Large Numbers, as the number of repetitions of a probability experiment increases (long term), the proportion with which a certain outcome is observed (relative frequency) gets closer to the probability of the outcome.

11. Rule 1 is satisfied since all of the probabilities in the model are greater than or equal to zero and less than or equal to one. Rule 2 is satisfied since the sum of the probabilities in the model is one: $0.3 + 0.15 + 0 + 0.15 + 0.2 + 0.2 = 1$. In this model, the outcome "blue" is an impossible event.

13. This cannot be a probability model because $P(\text{green}) < 0$.

15. Probabilities must be between 0 and 1, inclusive, so the only values which could be probabilities are: 0, 0.01, 0.35, and 1.

17. The probability of 0.42 means that approximately 42 out of every 100 hands will contain two cards of the same value and three cards of different value. No, probability deals with long-term behavior. While we would expect to see about 42 such hands out of every 100, on average, this does not mean we will always get a pair exactly 42 out of every 100 hands.

19. The empirical probability that the next flip would result in a head is $\frac{95}{100} = 0.95$.

21. $P(2) \neq \frac{1}{11}$ because the 11 possible outcomes are not equally likely.

23. The sample space is $S = \{$1H, 2H, 3H, 4H, 5H, 6H, 1T, 2T, 3T, 4T, 5T, 6T$\}$

25. The probability that a randomly selected three-year-old is enrolled in daycare is $P = 0.428$.

27. Event E contains 3 of the 10 equally likely outcomes, so $P(E) = \frac{3}{10} = 0.3$.

29. There are 10 equally likely outcomes and 4 are even numbers less than 9, so
$$P(E) = \frac{4}{10} = \frac{2}{5} = 0.4.$$

31. (a) $P(\text{plays organized sports}) = \frac{288}{500} = 0.576$

    (b) If we sampled a large number of high school students, we would expect that about 57.6% of the students play organized sports.

33. (a) Since 40 of the 100 equally likely tulip bulbs are red, $P(\text{red}) = \frac{40}{100} = 0.4$.

    (b) Since 25 of the 100 equally likely tulip bulbs are purple,
    $$P(\text{purple}) = \frac{25}{100} = 0.25.$$

    (c) If we sampled a large number of tulip bulbs from the bag (replacing the bulb each time), we would expect about 40% of the bulbs to be red and about 25% of the bulbs to be purple.

35. (a) The sample space is $S = \{$0, 00, 1, 2, 3, 4,…, 36$\}$.

(b) Since the slot marked 8 is one of the 38 equally likely outcomes,

$P(8) = \dfrac{1}{38} \approx 0.0263$. This means that, in many spins of such a roulette wheel, the long-run relative frequency of the ball landing on "8" will be close to

$\dfrac{1}{38} \approx 0.0263 = 2.63\%$. That is, if we spun the wheel 1000 times, we would expect about 26 of those times to result in the ball landing in slot 8.

(c) Since there are 18 odd slots (1, 3, 5, 7, 9, 11, 13, 15, 17, 19, 21, 23, 25, 27, 29, 31, 33, 35) in the 38 equally likely outcomes,

$P(\text{odd}) = \dfrac{18}{38} = \dfrac{9}{19} \approx 0.4737$. This means that, in many spins of such a roulette wheel, the long-run relative frequency of the ball landing in an odd slot will be

close to $\dfrac{9}{19} \approx 0.4737 = 47.37\%$. That is,

if we spun the wheel 100 times, we would expect about 47 of those times to result in an odd number.

37. (a) The sample space of possible genotypes is {*SS, Ss, sS, ss*}.

(b) Only one of the four equally likely genotypes gives rise to sickle-cell anemia, namely *ss*. Thus, the probability

is $P(ss) = \dfrac{1}{4} = 0.25$. This means that of the many children who are offspring of two *Ss* parents, approximately 25% will have sickle-cell anemia.

(c) Two of the four equally likely genotypes result in a carrier, namely *Ss* and *sS*. Thus, the probability of this is

$P(Ss \text{ or } sS) = \dfrac{2}{4} = \dfrac{1}{2} = 0.5$. This means that of the many children who are offspring of two *Ss* parents, approximately 50% will not themselves have sickle-cell anemia, but will be carriers of sickle-cell anemia.

39. (a) There are

$125 + 324 + 552 + 1257 + 2518 = 4776$ college students in the survey. The individuals can be thought of as the trials of the probability experiment. The relative frequency of "Never" is

$\dfrac{125}{4776} \approx 0.026$. We compute the relative frequencies of the other outcomes similarly and obtain the probability model below.

Response	Probability
Never	0.026
Rarely	0.068
Sometimes	0.116
Most of the time	0.263
Always	0.527

(b) Yes, it is unusual to find a college student who never wears a seatbelt when riding in a car driven by someone else. The approximate probability of this is only 0.026, which is less than 0.05.

41. (a) There are

$4 + 6 + 133 + 219 + 90 + 42 + 143 + 5 = 642$ police records included in the survey. The individuals can be thought of as the trials of the probability experiment. The relative frequency of "Pocket picking" is

$\dfrac{4}{642} \approx 0.006$. We compute the relative frequencies of the other outcomes similarly and obtain the probability model below.

Type of Larceny Theft	Probability
Pocket picking	0.006
Purse snatching	0.009
Shoplifting	0.207
From motor vehicles	0.341
Motor vehicle accessories	0.140
Bicycles	0.065
From buildings	0.223
From coin-operated machines	0.008

(b) Yes, purse-snatching larcenies are unusual since the probability is only $0.009 < 0.05$.

(c) No, bicycle larcenies are not unusual since the probability is 0.065 > 0.05.

43. Assignments A, B, C, and F are consistent with the definition of a probability model. Assignment D cannot be a probability model because it contains a negative probability, and Assignment E cannot be a probability model because is does not add up to 1.

45. Assignment B should be used if the coin is known to always come up tails.

47. (a) The sample space is S = {John-Roberto; John-Clarice; John-Dominique; John-Marco, Roberto-Clarice; Roberto-Dominique; Roberto-Marco; Clarice-Dominique; Clarice-Marco; Dominique-Marco}.

(b) Clarice-Dominique is one of the ten possible samples from part (a). Thus,
$$P(\text{Clarice and Dominique}) = \frac{1}{10} = 0.1.$$

(c) Four of the ten samples from part (a) include Clarice. Thus,
$$P(\text{Clarice attends}) = \frac{4}{10} = \frac{2}{5} = 0.4.$$

(d) Six of the ten samples from part (a) do not include John. Thus,
$$P(\text{John stays home}) = \frac{6}{10} = \frac{3}{5} = 0.6.$$

49. (a) Since 24 of the 73 homeruns went to right field, $P(\text{right field}) = \frac{24}{73} \approx 0.329$.

(b) Since 2 of the 73 homeruns went to left field, $P(\text{right field}) = \frac{2}{73} \approx 0.027$.

(c) Yes, it was unusual for Barry Bonds to his a homerun to left field. The probability is below 0.05.

51. (a)-(d) Answers will vary depending on the results from the simulation.

53. If the dice were fair, then each outcome should occur approximately $\frac{400}{6} \approx 67$ times. Since 1 and 6 occurred with considerably higher frequency, the dice appear to be loaded.

55. Half of all families are above the median and half are below, so
$$P(\text{Income greater than } \$58,500) = \frac{1}{2} = 0.5.$$

57. Answers will vary.

5.2 The Addition Rule and Complements

1. Two events are disjoint (mutually exclusive) if they have no outcomes in common.

3. $P(E) + P(F) - P(E \text{ and } F)$

5. E and F = {5, 6, 7}. No, E and F are not mutually exclusive because they have simple events in common.

7. F or G = {5, 6, 7, 8, 9, 10, 11, 12}.
$P(F \text{ or } G) = P(F) + P(G) - P(F \text{ and } G)$
$= \frac{5}{12} + \frac{4}{12} - \frac{1}{12} = \frac{8}{12} = \frac{2}{3}.$

9. E and G = { }. Yes, E and G are mutually exclusive because they have no simple events in common.

11. E^c = {1, 8, 9, 10, 11, 12}.
$$P(E^c) = 1 - P(E) = 1 - \frac{6}{12} = \frac{1}{2}$$

13. $P(E \text{ or } F) = P(E) + P(F) - P(E \text{ and } F)$
$= 0.25 + 0.45 - 0.15 = 0.55$

15. $P(E \text{ or } F) = P(E) + P(F) = 0.25 + 0.45 = 0.7$

17. $P(E^c) = 1 - P(E) = 1 - 0.25 = 0.75$

19. $P(E \text{ or } F) = P(E) + P(F) - P(E \text{ and } F)$
$0.85 = 0.60 + P(F) - 0.05$
$P(F) = 0.85 - 0.60 + 0.05 = 0.30$

21. $P(\text{Titleist or Maxfli}) = \frac{9+8}{20} = \frac{17}{20} = 0.85$

23. $P(\text{not Titleist}) = 1 - P(\text{Titleist})$
$= 1 - \frac{9}{20} = \frac{11}{20} = 0.55$

25. (a) Rule 1 is satisfied since all of the probabilities in the model are between 0 and 1.
Rule 2 is satisfied since the sum of the probabilities in the model is one: 0.671 + 0.126 + 0.044 + 0.063 + 0.010 + 0.009 + 0.077 = 1.

(b) $P(\text{gun or knife}) = 0.671 + 0.126 = 0.797$.
This means that there is a 79.7% probability of randomly selecting a murder committed with a gun or a knife.

(c) $P(\text{knife, blunt object, or strangulation})$.
$$= 0.126 + 0.044 + 0.010$$
$$= 0.180$$
This means that there is a 18.0% probability of randomly selecting a murder committed with a knife, with a blunt object, or by strangulation.

(d) $P(\text{not a gun}) = 1 - 0.671 = 0.329$.
This means that there is a 32.9% probability of randomly selecting a murder that was not committed with a gun.

(e) Yes, murders by strangulation are unusual since the probability is 0.010 < 0.05.

27. No; for example, on one draw of a card from a standard deck, let $E = \text{diamond}$, $F = \text{club}$, and $G = \text{red card}$. Here, E and F are disjoint, as are F and G. However, E and G are *not* disjoint since diamond cards are red.

29. The total number of multiple births was
$83 + 465 + 1635 + 2443 + 1604 + 344 + 120 = 6694$

(a) $P(30-39) = P(30-34) + P(35-39)$
$$= \frac{2443 + 1604}{6694} \approx 0.605$$
This means that there is a 60.5% probability that a mother involved in a multiple birth was between 30 and 39 years old.

(b) $P(\text{not } 30-39) = 1 - P(30-39)$
$$\approx 1 - 0.605 = 0.395$$
This means that there is a 39.5% probability that a mother involved in a multiple birth was not between 30 and 39. That is, the mother was younger than 30 or older than 39.

(c) $P(\text{younger than } 45) = 1 - \frac{120}{6694} \approx 0.982$
This means that there is a 98.2% probability that a randomly selected mother involved in a multiple birth was younger than 45.

(d) $P(\text{at least } 20) = 1 - \frac{83}{6694} \approx 0.988$
This means that there is a 98.8% probability that a randomly selected mother involved in a multiple birth was at least 20 years old.

31. (a) $P(\text{Heart or Club}) = P(\text{Heart}) + P(\text{Club})$
$$= \frac{13}{52} + \frac{13}{52}$$
$$= \frac{1}{2} = 0.5$$

(b) $P(\text{Heart, Club, or Diamond})$
$= P(\text{Heart}) + P(\text{Club}) + P(\text{Diamond})$
$$= \frac{13}{52} + \frac{13}{52} + \frac{13}{52}$$
$$= \frac{3}{4} = 0.75$$

(c) $P(\text{Ace or Heart})$
$= P(\text{Ace}) + P(\text{Heart}) - P(\text{Ace of Hearts})$
$$= \frac{4}{52} + \frac{13}{52} - \frac{1}{52}$$
$$= \frac{4}{13} \approx 0.308$$

33. (a) $P(\text{not on Nov. 8}) = 1 - P(\text{on Nov. 8})$
$$= 1 - \frac{1}{365}$$
$$= \frac{364}{365} \approx 0.997$$

(b) $P\left(\text{not on the } 1^{st}\right) = 1 - P\left(\text{on the } 1^{st}\right)$
$$= 1 - \frac{12}{365}$$
$$= \frac{353}{365} \approx 0.967$$

(c) $P\left(\text{not on the } 31^{st}\right) = 1 - \left(\text{on the } 31^{st}\right)$
$$= 1 - \frac{7}{365}$$
$$= \frac{358}{365} \approx 0.981$$

(d) $P(\text{not in Dec.}) = 1 - P(\text{in Dec.})$

$$= 1 - \frac{31}{365}$$

$$= \frac{334}{365} \approx 0.915$$

35. No, we cannot compute the probability of randomly selecting a citizen of the U.S. who has hearing problems or vision problems by adding the given probabilities because the events "hearing problems" and "vision problems" are not disjoint. That is, some people have both vision and hearing problems, but we do not know the proportion.

37. (a) $P(\text{between 6 and 10 yrs old}) = \frac{5}{24} \approx 0.208$

This is not unusual because $0.208 > 0.05$.

(b) $P(\text{more than 5 yrs old}) = \frac{5+7+5}{24}$

$$= \frac{17}{24} \approx 0.708$$

(c) $P(\text{less than 1 yr old}) = \frac{1}{24} \approx 0.042$

This is unusual because $0.042 < 0.05$.

39. (a) P(drives or takes public transportation)
= P(drives) + P(takes public transportation)
= 0.867 + 0.048 = 0.915

(b) P(neither drives nor takes pub. trans.)
= 1 − P(drives or takes pub. trans.)
= 1 − 0.915 = 0.085

(c) P(does not drive) = 1 − P(drives)
= 1 − 0.867 = 0.133

(d) No, the probability that a randomly selected worker walks to work cannot equal 0.15 because the sum of the probabilities would be more than 1 and there would be no probability model.

41. (a) Of the 137,243 men included in the study, 782 + 91 + 141 = 1014 died from cancer. Thus,

$$P(\text{died from cancer}) = \frac{1014}{137,243} \approx 0.007 .$$

(b) Of the 137,243 men included in the study, 141 + 7725 = 7866 were current cigar smokers. Thus,

$$P(\text{current cigar smoker}) = \frac{7866}{137,243}$$

$$\approx 0.057 .$$

(c) Of the 137,243 men included in the study, 141 were current cigar smokers who died from cancer. Thus,

$$P(\text{died from cancer and current smoker})$$

$$= \frac{141}{137,243} \approx 0.001 .$$

(d) Of the 137,243 men included in the study, 1014 died from cancer, 7866 were current cigar smokers, and 141 were current cigar smokers who died from cancer. Thus,

$$P(\text{died from cancer or current smoker})$$

$$= \frac{1014}{137,243} + \frac{7866}{137,243} - \frac{141}{137,243}$$

$$= \frac{8739}{137,243} \approx 0.064 .$$

43. (a) Of the 375 students surveyed, 231 were satisfied with student government. Thus,

$$P(\text{satisfied}) = \frac{231}{375} = \frac{77}{125} = 0.616 .$$

(b) Of the 375 students surveyed, 94 were juniors. Thus, $P(\text{junior}) = \frac{94}{375} \approx 0.251 .$

(c) Of the 375 students surveyed, 64 were juniors who were satisfied with student government. Thus,

$$P(\text{satisfied and junior}) = \frac{64}{375} \approx 0.171 .$$

(d) Of the 375 students surveyed, 231 were satisfied with student government, 94 were juniors, and 64 were juniors who were satisfied with student government. Thus,

$$P(\text{satisfied or junior}) = \frac{231}{375} + \frac{94}{375} - \frac{64}{375}$$

$$= \frac{261}{375} = \frac{87}{125}$$

$$= 0.696$$

45. (a) Of the 1,355,137 military personnel, 220,506 are officers. Thus,
$$P(\text{officer}) = \frac{220,506}{1,355,137} \approx 0.163 \,.$$

(b) Of the 1,355,137 military personnel, 329,360 are in the Navy. Thus,
$$P(\text{Navy}) = \frac{329,360}{1,355,137} \approx 0.243 \,.$$

(c) Of the 1,355,137 military personnel, 51,167 are Navy officers. Thus,
$$P(\text{Navy officer}) = \frac{51,167}{1,355,137} \approx 0.038 \,.$$

(d) Of the 1,355,137 military personnel, 220,506 are officers, 329,360 are in the Navy, and 51,167 are Navy officers. Thus,

$P(\text{officer or Navy})$
$$= \frac{220,506}{1,355,137} + \frac{329,360}{1,355,137} - \frac{51,167}{1,355,137}$$
$$= \frac{498,699}{1,355,137} \approx 0.368 \,.$$

47. (a) The variables presented in the table are Crash Type, System, and Projected Number of Crashes.

(b) Crash Type is qualitative because it describes an attribute or characteristic.

System is qualitative because it describes an attribute or characteristic.

Projected Number of Crashes is quantitative because it is a numerical measure. It is discrete because the numerical measurement is the result of a count.

(c) Of the 1,121 projected crashes under the current system, 289 were projected to have reported injuries. Of the 922 projected crashes under the new system, 221 were projected to have reported injuries. The relative frequency for reported injury crashes would be
$\frac{289}{1,121} \approx 0.26$ under the current system
and $\frac{221}{922} \approx 0.24$ under the new system.
Similar computations can be made for the remaining crash types for both systems.

Crash Type	Current System	w/Red-Light Cameras
Reported Injury	0.26	0.24
Reported Property Damage Only	0.35	0.36
Unreported Injury	0.07	0.07
Unreported Property Damage Only	0.32	0.33
Total	1	1

(d)

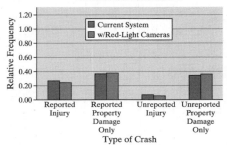

Crashes at Intersections

(e) Mean for Current System:
$\frac{1121}{13} \approx 86.2$ crashes per intersection
Mean for Red-light Camera System:
$\frac{922}{13} \approx 70.9$ crashes per intersection

(f) It is not possible to compute the standard deviation because we do not know the number of crashes at each intersection.

(g) Since the mean number of crashes is less with the cameras, it appears that the program will be beneficial.

(h) There are 1,121 crashes under the current system and 289 with reported injuries. Thus,
$$P(\text{reported injuries}) = \frac{289}{1121} \approx 0.258 \,.$$

(i) There are 922 crashes under the camera system. Of these, 333 had reported property damage only and 308 had unreported property damage only. Thus,
$$P\left(\begin{array}{c}\text{property damage}\\\text{only}\end{array}\right) = \frac{333+308}{922}$$
$$= \frac{641}{922} \approx 0.695$$

(j) Simpson's Paradox refers to situations when conclusions reverse or change direction when accounting for specific variable. When accounting for the cause of the crash (rear-end vs. red-light running), the camera system does not reduce all types of accidents. Under the camera system, red-light running crashes decreased, but rear-end crashes increased.

(k) Recommendations may vary. The benefits of the decrease in red-light running crashes must be weighed against the negative of increased rear-end crashes. Seriousness of injuries and amount of property damage may need to be considered.

5.3 Independence and the Multiplication Rule

1. independent

3. Addition

5. $P(E) \cdot P(F)$

7. (a) Dependent. Speeding on the interstate increases the probability of being pulled over by a police officer.

(b) Dependent: Eating fast food affects the probability of gaining weight.

(c) Independent: Your score on a statistics exam does not affect the probability that the Boston Red Sox win a baseball game.

9. Since E and F are independent.,
$$P(E \text{ and } F) = P(E) \cdot P(F)$$
$$= (0.3)(0.6)$$
$$= 0.18$$

11. $P(5 \text{ heads in a row}) = \left(\dfrac{1}{2}\right)\left(\dfrac{1}{2}\right)\left(\dfrac{1}{2}\right)\left(\dfrac{1}{2}\right)\left(\dfrac{1}{2}\right)$
$$= \left(\dfrac{1}{2}\right)^5 = \dfrac{1}{32} = 0.03125$$

If we flipped a coin five times, 100 different times, we would expect to observe 5 heads in a row about 3 times.

13. $P(2 \text{ left-handed people}) = (0.13)(0.13)$
$$= 0.0169$$
$$P\left(\begin{matrix}\text{At least 1 is}\\\text{right-handed}\end{matrix}\right) = 1 - P\left(\begin{matrix}2 \text{ left-handed}\\\text{people}\end{matrix}\right)$$
$$= 1 - 0.0169 = 0.9831$$

15. (a) $P(\text{all 5 negative})$
$$= (0.995)(0.995)(0.995)(0.995)(0.995)$$
$$= (0.995)^5 \approx 0.9752$$

(b) $P(\text{at least one positive})$
$$= 1 - P(\text{all 5 negative})$$
$$= 1 - 0.9752$$
$$= 0.0248$$

17. (a) $P(\text{Both live to be 41}) = (0.99757)(0.99757)$
$$\approx 0.99515$$

(b) $P(\text{All 5 live to be 41}) = (0.99757)^5$
$$\approx 0.98791$$

(c) This is the complement of the event in (b), so the probability is $1 - 0.98791 = 0.01209$ which is unusual since $0.01209 < 0.05$.

19. (a) $P(\text{Both have Rh}^+ \text{ blood}) = (0.99)(0.99)$
$$= 0.9801$$

(b) $P(\text{All 6 have Rh}^+ \text{ blood}) = (0.99)^6$
$$\approx 0.9415$$

(c) This is the complement of the event in (b), so the probability is $1 - 0.9415 = 0.0585$ which is not unusual since $0.0585 > 0.05$.

21. (a) $P(\text{one failure}) = 0.15$; this is not unusual because $0.15 > 0.05$.
Since components fail independent of each other, we get
$P(\text{two failures}) = (0.15)(0.15) = 0.0225$;
this is unusual because $0.0225 < 0.05$.

(b) This is the complement of both components failing, so
$$P(\text{system succeeds}) = 1 - P(\text{both fail})$$
$$= 1 - 0.0225$$
$$= 0.9775$$

(c) From part (b) we know that two components are not enough, so we increase the number.

3 components:

$$P(\text{system succeeds}) = 1 - (0.15)^3$$
$$\approx 0.99663$$

4 components:

$$P(\text{system succeeds}) = 1 - (0.15)^4$$
$$\approx 0.99949$$

5 components:

$$P(\text{system succeeds}) = 1 - (0.15)^5$$
$$\approx 0.99992$$

Therefore, 5 components would be needed to make the probability of the system succeeding greater than 0.9999.

23. (a) $P\left(\begin{array}{c}\text{batter makes 10}\\\text{consecutive outs}\end{array}\right) = (0.70)^{10} \approx 0.02825$

(b) Yes, cold streaks are unusual since the probability is $0.02825 < 0.05$.

(c) In repeated sets of 10 consecutive at-bats, we expect the hitter to make an out in all 10 at-bats about 28 times out of 1,000.

25. (a) $P\left(\begin{array}{c}\text{two strikes}\\\text{in a row}\end{array}\right) = (0.3)(0.3) = 0.09$

(b) $P(\text{turkey}) = (0.3)^3 = 0.027$

(c) $P\left(\begin{array}{c}\text{strike}\\\text{followed}\\\text{by a}\\\text{non-strike}\end{array}\right) = P(\text{strike}) \cdot P(\text{non-strike})$

$$= (0.3)(0.7) = 0.21$$

27. (a) $P\left(\begin{array}{c}\text{all 3 have}\\\text{driven under}\\\text{the influence}\\\text{of alcohol}\end{array}\right) = (0.29)^3 \approx 0.0244$

(b) $P\left(\begin{array}{c}\text{at least one}\\\text{has not driven}\\\text{under the}\\\text{influence}\\\text{of alcohol}\end{array}\right) = 1 - 0.0244 = 0.9756$

(c) The probability that an individual 21- to 25-year-old has not driven while under the influence of alcohol is $1 - 0.29 = 0.71$, so

$$P\left(\begin{array}{c}\text{none of the}\\\text{3 have driven}\\\text{under the}\\\text{influence of}\\\text{alcohol}\end{array}\right) = (0.71)^3 \approx 0.3579$$

(d) $P\left(\begin{array}{c}\text{at least one}\\\text{of the 3 has}\\\text{driven under}\\\text{the influence}\\\text{of alcohol}\end{array}\right) = 1 - 0.3579 = 0.6421$

29. (a) $P(\text{audit}) = 0.0177$; this is unusual since $0.0177 < 0.05$.

(b) $P(\text{both audited}) = (0.0177)^2 \approx 0.000313$

(c) The probability that a single return with income of $100,000 or more will be not be audited is $1 - 0.0177 = 0.9823$. Therefore,

$$P(\text{neither audited}) = (0.9823)^2 \approx 0.9649.$$

(d) $P\left(\begin{array}{c}\text{at least}\\\text{one audited}\end{array}\right) = 1 - P(\text{neither audited})$

$$= 1 - 0.9649 = 0.0351$$

31. (a) No, the events "male" and "bet on professional sports" are not mutually exclusive. The two events can happen at the same time. That is, an adult male can bet on professional sports.

(b) $P\left(\begin{array}{c}\text{male and bets on}\\\text{professional sports}\end{array}\right)$

$$= P(\text{male}) \cdot P\left(\begin{array}{c}\text{bets on}\\\text{professional}\\\text{sports}\end{array}\right)$$

$$= (0.484)(0.170)$$

$$\approx 0.0823$$

(c) $P\left(\begin{array}{c}\text{male or}\\\text{bets on}\\\text{professional}\\\text{sports}\end{array}\right) = 0.17 + 0.484 - 0.0823$

$$= 0.5717$$

(d) Since $P\left(\dfrac{\text{male and bets on}}{\text{professional sports}}\right) = 0.106,$

but we computed it as 0.0823 assuming independence, it appears that the independence assumption is not correct.

(e) $P\left(\begin{array}{c}\text{male or}\\\text{bets on}\\\text{professional}\\\text{sports}\end{array}\right) = 0.17 + 0.484 - 0.106$

$= 0.548$

The actual probability is lower than we computed assuming independence.

5.4 Conditional Probability and the General Multiplication Rule

1. $F; E$

3. $P(F\mid E) = \dfrac{P(E \text{ and } F)}{P(E)} = \dfrac{0.6}{0.8} = 0.75$

5. $P(F\mid E) = \dfrac{N(E \text{ and } F)}{N(E)} = \dfrac{420}{740} = 0.568$

7. $P(E \text{ and } F) = P(E) \cdot P(F\mid E)$
$= (0.8)(0.4)$
$= 0.32$

9. No, the events "earn more than \$75,000 per year" and "earned a bachelor's degree" are not independent because P("earn more than \$75,000 per year" | "earned a bachelor's degree") $\neq$ P("earn more than \$75,000 per year").

11. $P(\text{club}) = \dfrac{13}{52} = \dfrac{1}{4};$

$P(\text{club} \mid \text{black card}) = \dfrac{13}{26} = \dfrac{1}{2}$

13. $P(\text{rainy} \mid \text{cloudy}) = \dfrac{P(\text{rainy and cloudy})}{P(\text{cloudy})}$
$= \dfrac{0.21}{0.37} \approx 0.568$

15. $P(\text{white} \mid \text{16-17 year-old dropout})$
$= \dfrac{P(\text{white and 16-17 year-old dropout})}{P(\text{16-17 year-old dropout})}$
$= \dfrac{0.062}{0.084} = \dfrac{31}{42} \approx 0.738$

17. (a) $P(\text{no health insurance} \mid \text{under 18})$
$= \dfrac{N(\text{no health insurance and under 18})}{N(\text{under 18})}$
$= \dfrac{8,661}{47,906 + 22,109 + 8,661}$
$= \dfrac{8,661}{78,676} \approx 0.110$

(b) $P(\text{under 18} \mid \text{no health insurance})$
$= \dfrac{N(\text{under 18 and no health insurance})}{N(\text{no health insurance})}$
$= \dfrac{8,661}{8,661 + 27,054 + 10,737 + 541}$
$= \dfrac{8,661}{46,993} \approx 0.184$

19. (a) $P(\text{passenger} \mid \text{female})$
$= \dfrac{N(\text{passenger and female})}{N(\text{female})}$
$= \dfrac{4,896}{6,598 + 4,896} = \dfrac{4,896}{11,494} \approx 0.426$

(b) $P(\text{female} \mid \text{passenger})$
$= \dfrac{N(\text{female and passenger})}{N(\text{passengers})}$
$= \dfrac{4,896}{5,190 + 4,896} = \dfrac{4,896}{10,086} \approx 0.485$

(c) $P(\text{male} \mid \text{driving})$
$= \dfrac{N(\text{male and driving})}{N(\text{driving})}$
$= \dfrac{20,795}{20,795 + 6,598} = \dfrac{20,795}{27,393} \approx 0.759$

$P(\text{female} \mid \text{driving})$
$= \dfrac{N(\text{female and driving})}{N(\text{driving})}$
$= \dfrac{6,598}{20,795 + 6,598} = \dfrac{6,598}{27,393} \approx 0.241$

The victim is more likely to be male. Knowing that the victim was driving, the victim is roughly three times as likely to be male as female.

21. $P(\text{both televisions work})$

$= P(\text{1st works}) \cdot P(\text{2nd works} \mid \text{1st works})$

$= \dfrac{4}{6} \cdot \dfrac{3}{5} = 0.4$

$P(\text{at least one television does not work})$

$= 1 - P(\text{both televisions work})$

$= 1 - 0.4 = 0.6$

23. (a) $P(\text{both kings})$

$= P(\text{first king}) \cdot P(\text{2nd king} \mid \text{first king})$

$= \dfrac{4}{52} \cdot \dfrac{3}{51} = \dfrac{1}{221} \approx 0.005$

(b) $P(\text{both kings})$

$= P(\text{first king}) \cdot P(\text{2nd king} \mid \text{first king})$

$= \dfrac{4}{52} \cdot \dfrac{4}{52} = \dfrac{1}{169} \approx 0.006$

25. $P(\text{Dave 1st and Neta 2nd})$

$= P(\text{Dave 1st}) \cdot P(\text{Neta 2nd} \mid \text{Dave 1st})$

$= \dfrac{1}{5} \cdot \dfrac{1}{4} = \dfrac{1}{20} = 0.05$

27. (a) $P(\text{like both songs})$

$= P(\text{like 1st}) \cdot P(\text{like 2nd} \mid \text{like 1st})$

$= \dfrac{5}{13} \cdot \dfrac{4}{12} = \dfrac{5}{39} \approx 0.128$

This is not a small enough probability to be considered unusual.

(b) $P(\text{dislike both songs})$

$= P(\text{dislike 1st}) \cdot P(\text{dislike 2nd} \mid \text{dislike 1st})$

$= \dfrac{8}{13} \cdot \dfrac{7}{12} = \dfrac{14}{39} \approx 0.359$

(c) Since you either like both or neither or exactly one (and these are disjoint) then the probability that you like exactly one is given by

$P(\text{like exactly one song})$

$= 1 - \big(P(\text{like both}) + P(\text{dislike both})\big)$

$= 1 - \left(\dfrac{5}{39} + \dfrac{14}{39}\right) = \dfrac{20}{39} \approx 0.513$

(d) $P(\text{like both songs})$

$= P(\text{like 1st}) \cdot P(\text{like 2nd} \mid \text{like 1st})$

$= \dfrac{5}{13} \cdot \dfrac{5}{13} = \dfrac{25}{169} \approx 0.148$

$P(\text{dislike both songs})$

$= P(\text{dislike 1st}) \cdot P(\text{dislike 2nd} \mid \text{dislike 1st})$

$= \dfrac{8}{13} \cdot \dfrac{8}{13} = \dfrac{64}{169} \approx 0.379$

$P(\text{like exactly one song})$

$= 1 - \big(P(\text{like both}) + P(\text{dislike both})\big)$

$= 1 - \left(\dfrac{25}{169} + \dfrac{64}{169}\right) = \dfrac{80}{169} \approx 0.473$

29.

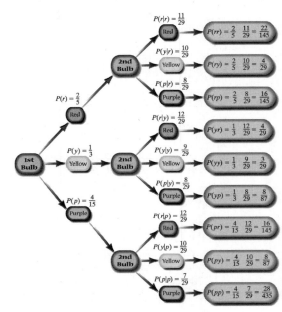

(a) $P(\text{both red}) = \dfrac{22}{145} \approx 0.152$

(b) $P(\text{1st red and 2nd yellow}) = \dfrac{4}{29} \approx 0.138$

(c) $P(\text{1st yellow and 2nd red}) = \dfrac{4}{29} \approx 0.138$

(d) Since one each of red and yellow must be either 1st red, 2nd yellow or vice versa, by the addition rule this probability is

$P(\text{one red and one yellow}) = \dfrac{4}{29} + \dfrac{4}{29}$

$= \dfrac{8}{29} \approx 0.276$

31. $P(\text{female and smoker})$

$= P(\text{female} \mid \text{smoker}) \cdot P(\text{smoker})$

$= (0.334)(0.209) \approx 0.070$

It is not unusual to select a female who smokes.

33. **(a)** $P(10 \text{ different birthdays})$

$= \dfrac{365}{365} \cdot \dfrac{364}{365} \cdot \dfrac{363}{365} \cdots \dfrac{358}{365} \cdot \dfrac{357}{365} \cdot \dfrac{356}{365}$

≈ 0.883

(b) $P(\text{At least 2 of the 10 share same birthday})$

$\approx 1 - 0.883$

$= 0.117$

35. **(a)** $P(\text{male}) = \dfrac{200}{400} = \dfrac{1}{2}$

$P(\text{male} \mid 0 \text{ activ.}) = \dfrac{21}{42} = \dfrac{1}{2}$

$P(\text{male}) = P(\text{male} \mid 0 \text{ activities})$, so the events "male" and "0 activities" are independent.

(b) $P(\text{female}) = \dfrac{200}{400} = \dfrac{1}{2} = 0.5$

$P(\text{female} \mid 5\text{-}7 \text{ activ.}) = \dfrac{71}{109} \approx 0.651$

$P(\text{female}) \neq P(\text{female} \mid 5\text{-}7 \text{ activ.})$, so the events "female" and "5-7 activities" are not independent.

(c) Yes, the events "1-2 activities" and "3-4 activities" are mutually exclusive because the two events cannot occur at the same time. $P(1\text{-}2 \text{ activ. and } 3\text{-}4 \text{ activ.}) = 0$

(d) No, the events "male" and "1-2 activities" are not mutually exclusive because the two events can happen at the same time.

$P(\text{male and } 1\text{-}2 \text{ activ.}) = \dfrac{81}{400}$

$= 0.2025 \neq 0$

37. **(a)** $P(\text{being dealt 5 clubs})$

$= \dfrac{13}{52} \cdot \dfrac{12}{51} \cdot \dfrac{11}{50} \cdot \dfrac{10}{49} \cdot \dfrac{9}{48}$

$= \dfrac{33}{66,640} \approx 0.000495$

(b) $P(\text{bing dealt a flush}) = 4\left(\dfrac{33}{66,640}\right)$

$= \dfrac{33}{16,660} \approx 0.002$

39. **(a)** $P(\text{selecting two defective chips})$

$= \dfrac{50}{10,000} \cdot \dfrac{49}{9,999}$

≈ 0.0000245

(b) Assuming independence,

$P(\text{selecting two defective chips})$

$\approx (0.005)^2$

$= 0.000025$

The difference in the results of parts (a) and (b) is only 0.0000005, so the assumption of independence did not significantly affect the probability.

41. $P(< 18 \text{ yrs old}) = \dfrac{78,676}{328,953} \approx 0.239$

$P(\text{under } 18 \mid \text{no health insurance}) = \dfrac{8,661}{46,993}$

≈ 0.184

No, the events "<18 years old" and "no health insurance" are not independent since the preceding probabilities are not equal.

43. $P(\text{female}) = P(\text{female}) = \dfrac{11,494}{37,479} \approx 0.307$

$P(\text{female} \mid \text{driver}) = \dfrac{6,598}{27,393} \approx 0.241$

No, the events "female" and "driver" are not independent since the preceding probabilities are not equal.

45. **(a)-(d)** Answers will vary.

For part (d), of the 18 boxes in the outermost ring, 12 indicate you win if you switch while 6 indicate you lose if you switch. Assuming random selection so each box is equally likely, we get $P(\text{win if switch}) = \dfrac{12}{18} = \dfrac{2}{3} \approx 0.667$.

5.5 Counting Techniques

1. permutation

3. True

5. $5! = 5 \cdot 4 \cdot 3 \cdot 2 \cdot 1 = 120$

7. $10! = 10 \cdot 9 \cdot 8 \cdot 7 \cdot 6 \cdot 5 \cdot 4 \cdot 3 \cdot 2 \cdot 1 = 3,628,800$

9. $0! = 1$

11. $_6P_2 = \dfrac{6!}{(6-2)!} = \dfrac{6!}{4!} = 6 \cdot 5 = 30$

13. $_4P_4 = \dfrac{4!}{(4-4)!} = \dfrac{4!}{0!} = \dfrac{24}{1} = 24$

15. $_5P_0 = \dfrac{5!}{(5-0)!} = \dfrac{5!}{5!} = 1$

17. $_8P_3 = \dfrac{8!}{(8-3)!} = \dfrac{8!}{5!} = 8 \cdot 7 \cdot 6 = 336$

19. $_8C_3 = \dfrac{8!}{3!(8-3)!} = \dfrac{8!}{3!5!} = \dfrac{8 \cdot 7 \cdot 6}{3 \cdot 2 \cdot 1} = 56$

21. $_{10}C_2 = \dfrac{10!}{2!(10-2)!} = \dfrac{10!}{2!8!} = \dfrac{10 \cdot 9}{2 \cdot 1} = 45$

23. $_{52}C_1 = \dfrac{52!}{1!(52-1)!} = \dfrac{52!}{1!51!} = \dfrac{52}{1} = 52$

25. $_{48}C_3 = \dfrac{48!}{3!(48-3)!} = \dfrac{48!}{3!45!} = \dfrac{48 \cdot 47 \cdot 46}{3 \cdot 2 \cdot 1}$
$= 17,296$

27. *ab, ac, ad, ae, ba, bc, bd, be, ca, cb, cd, ce, da, db, dc, de, ea, eb, ec, ed*
Since there are 20 permutations, $_5P_2 = 20$.

29. *ab, ac, ad, ae, bc, bd, be, cd, ce, de*
Since there are 10 combinations, $_5C_2 = 10$.

31. Here we use the Multiplication Rule of Counting. There are six shirts and four ties, so there are $6 \cdot 4 = 24$ different shirt-and-tie combinations the man can wear.

33. There are 12 ways Dan can select the first song, 11 ways to select the second song, etc. From the Multiplication Rule of Counting, there are $12 \cdot 11 \cdot ... \cdot 2 \cdot 1 = 12! = 479,001,600$ ways that Dan can arrange the 12 songs.

35. There are 8 ways to pick the first city, 7 ways to pick the second, etc. From the Multiplication Rule of Counting, there are $8 \cdot 7 \cdot ... \cdot 2 \cdot 1 = 8! = 40,320$ different routes possible for the salesperson.

37. Since the company name can be represented by 1, 2, or 3 letters, we find the total number of 1 letter abbreviations, 2 letter abbreviations, and 3 letter abbreviations, then we sum the results to obtain the total number of abbreviations possible. There are 26 letters that can be used for abbreviations, so there are 26 one-letter abbreviations. Since repetitions are allowed, there are $26 \cdot 26 = 26^2$ different two-letter abbreviations, and $26 \cdot 26 \cdot 26 = 26^3$ different three-letter abbreviations. Therefore, the maximum number of companies that can be listed on the New York Stock Exchange is 26 (one letter) $+ 26^2$ (two letters) $+ 26^3$ (three letters) $= 18,278$ companies.

39. **(a)** $10 \cdot 10 \cdot 10 \cdot 10 = 10^4 = 10,000$ different codes are possible.

(b) $P(\text{guessing the correct code}) = \dfrac{1}{10,000}$
$= 0.0001$

41. Since lower and uppercase letters are considered the same, each of the 8 letters to be selected has 26 possibilities. Therefore, there are $26^8 \approx 2.08827 \times 10^{11}$ different usernames possible for the local area network.

43. **(a)** $50 \cdot 50 \cdot 50 = 50^3 = 125,000$ different combinations are possible.

(b) $P(\text{guessing combination}) = \dfrac{1}{50^3} = \dfrac{1}{125,000}$
$= 0.000008$

45. Since order matters, we use the permutation formula $_nP_r$.

$$_{40}P_3 = \frac{40!}{(40-3)!} = \frac{40!}{37!} = 40 \cdot 39 \cdot 38 = 59,280$$

There are 59,280 ways in which the top three cars can result.

47. Since the order of selection determines the office of the member, order matters. Therefore, we use the permutation formula $_nP_r$.

$$_{20}P_4 = \frac{20!}{(20-4)!} = \frac{20!}{16!} = 20 \cdot 19 \cdot 18 \cdot 17$$
$$= 116,280$$

There are 116,280 different leadership structures possible.

49. Since the problem states the numbers must be matched in order, this is a permutation problem.

$$_{25}P_4 = \frac{25!}{(25-4)!} = \frac{25!}{21!} = 25 \cdot 24 \cdot 23 \cdot 22$$
$$= 303,600$$

There are 303,600 different outcomes possible for this game.

51. Since order of selection does not matter, this is a combination problem.

$$_{50}C_5 = \frac{50 \cdot 49 \cdot 48 \cdot 47 \cdot 46}{5 \cdot 4 \cdot 3 \cdot 2 \cdot 1} = 2,118,760$$

There are 2,118,760 different simple random samples of size 5 possible.

53. There are 6 children and we need to determine the number of ways two can be boys. The order of the two boys does not matter, so this is a combination problem. There are

$$_6C_2 = \frac{6!}{2!(6-2)!} = \frac{6!}{2! \, 4!} = \frac{6 \cdot 5}{2 \cdot 1} = 15 \text{ ways to}$$

select 2 of the 6 children to be boys (the rest are girls), so there are 15 different birth and gender orders possible.

55. Since there are three A's, two C's, two G's, and three T's from which to form a DNA sequence, we can make $\frac{10!}{3! \, 2! \, 2! \, 3!} = 25,200$ distinguishable DNA sequences.

57. Arranging the trees involves permutations with repetitions. Using Formula (3), we find that there are $\frac{11!}{4! \, 5! \, 2!} = 6930$ different ways the landscaper can arrange the trees.

59. Since the order of the balls does not matter, this is a combination problem. Using Formula (2), we find there are $_{39}C_5 = 575,757$ possible choices (without regard to order), so

$$P(\text{winning}) = \frac{1}{575,757}.$$

61. (a) $P(\text{all students}) = \frac{_8C_5}{_{18}C_5}$

$$= \frac{8 \cdot 7 \cdot 6 \cdot 5 \cdot 4}{5 \cdot 4 \cdot 3 \cdot 2 \cdot 1} \cdot \frac{5 \cdot 4 \cdot 3 \cdot 2 \cdot 1}{18 \cdot 17 \cdot 16 \cdot 15 \cdot 14}$$
$$= \frac{1}{153} \approx 0.0065$$

(b) $P(\text{all faculty}) = \frac{_{10}C_5}{_{18}C_5}$

$$= \frac{10 \cdot 9 \cdot 8 \cdot 7 \cdot 6}{5 \cdot 4 \cdot 3 \cdot 2 \cdot 1} \cdot \frac{5 \cdot 4 \cdot 3 \cdot 2 \cdot 1}{18 \cdot 17 \cdot 16 \cdot 15 \cdot 14}$$
$$= \frac{1}{34} \approx 0.0294$$

(c) $P(\text{2 students and 3 faculty}) = \frac{_8C_2 \cdot _{10}C_3}{_{18}C_5}$

$$= \frac{8 \cdot 7}{2 \cdot 1} \cdot \frac{10 \cdot 9 \cdot 8}{3 \cdot 2 \cdot 1} \cdot \frac{5 \cdot 4 \cdot 3 \cdot 2 \cdot 1}{18 \cdot 17 \cdot 16 \cdot 15 \cdot 14}$$
$$= \frac{20}{51} \approx 0.3922$$

63. $P(\text{one or more defective})$
$= 1 - P(\text{none defective})$
$$= 1 - \frac{_{116}C_4}{_{120}C_4}$$
$$= 1 - \frac{116}{120} \cdot \frac{115}{119} \cdot \frac{114}{118} \cdot \frac{113}{117} \approx 0.1283$$

There is a 12.83% probability that the shipment is rejected.

65. (a) $P(\text{you like 2 of the 4 songs}) = \frac{_5C_2 \cdot _8C_2}{_{13}C_4}$
$$\approx 0.3916$$

There is a 39.16% probability that you will like 2 of the 4 songs played.

(b) $P(\text{you like 3 of the 4 songs}) = \dfrac{_5C_3 \cdot _8C_1}{_{13}C_4}$

≈ 0.1119

There is a 11.19% probability that you will like 3 of the 4 songs played.

(c) $P(\text{you like all 4 songs}) = \dfrac{_5C_4 \cdot _8C_0}{_{13}C_4}$

≈ 0.0070

There is a 0.7% probability that you will like all 4 songs played.

67. (a) Five cards can be selected from a deck in $_{52}C_5 = 2{,}598{,}960$ ways.

(b) There are $_4C_3 = 4$ ways of choosing 3 two's, and so on for each denomination. Hence, there are $13 \cdot 4 = 52$ ways of choosing three of a kind.

(c) There are $_{12}C_2 = 66$ choices of two additional denominations (different from that of the three of a kind) and 4 choices of suit for the first remaining card and then, for each choice of suit for the first remaining card, there are 4 choices of suit for the last card. This gives a total of $66 \cdot 4 \cdot 4 = 1056$ ways of choosing the last two cards.

(d) $P(\text{three of a kind}) = \dfrac{52 \cdot 1056}{2{,}598{,}960} \approx 0.0211$

69. $P(\text{all 4 modems work}) = \dfrac{17}{20} \cdot \dfrac{16}{19} \cdot \dfrac{15}{18} \cdot \dfrac{14}{17}$

≈ 0.4912

There is a 49.12% probability that the shipment will be accepted.

71. (a) Using the Multiplication Rule of Counting and the given password format, there are

$21 \cdot 5 \cdot 21 \cdot 21 \cdot 5 \cdot 21 \cdot 10 \cdot 10 = 486{,}202{,}500$ different passwords that are possible.

(b) If letters are case sensitive, there are

$42 \cdot 10 \cdot 42 \cdot 42 \cdot 10 \cdot 42 \cdot 10 \cdot 10 = 420^4 = 31{,}116{,}960{,}000$ different passwords that are possible.

5.6 Putting it Together: Which Method Do I Use?

1. In a permutation, the order in which the objects are chosen matters; in a combination, the order in which the objects are chosen is unimportant.

3. 'AND' is generally associated with multiplication, while 'OR' is generally associated with addition.

5. $P(E) = \dfrac{4}{10} = \dfrac{2}{5} = 0.4$

7. *abc, acb, abd, adb, abe, aeb, acd, adc, ace, aec, ade, aed, bac, bca, bad, bda, bae, bea, bcd, bdc, bce, bec, bde, bed, cab, cba, cad, cda, cae, cea, cbd, cdb, cbe, ceb, cde, ced, dab, dba, dac, dca, dae, dea, dbc, dcb, dbe, deb, dce, dec, eab, eba, eac, eca, ead, eda, ebc, ecb, ebd, edb, ecd, edc*

9. $P(E \text{ or } F) = P(E) + P(F) - P(E \text{ and } F)$

$= 0.7 + 0.2 - 0.15$

$= 0.75$

11. $_7P_3 = \dfrac{7!}{(7-3)!} = \dfrac{7!}{4!} = 7 \cdot 6 \cdot 5 = 210$

13. $P(E \text{ and } F) = P(E) \cdot P(F)$

$= (0.8)(0.5)$

$= 0.4$

15. $P(E \text{ and } F) = P(E) \cdot P(F \mid E)$

$= (0.9)(0.3)$

$= 0.27$

17. $P(\text{soccer}) \approx \dfrac{22}{500} = \dfrac{11}{250} = 0.044$

19. Since the actual order of the men in their three seats matters, there are $_3P_3 = 6$ ways to arrange the men. Similarly, there are $_3P_3 = 6$ ways to arrange the women among their seats. Therefore, there are $6 \cdot 6 = 36$ ways to arrange the men and women.

Using the Multiplication Rule of Counting, and starting with a female, we get:

$\underset{F}{\dfrac{3}{}} \cdot \underset{M}{\dfrac{3}{}} \cdot \underset{F}{\dfrac{2}{}} \cdot \underset{M}{\dfrac{2}{}} \cdot \underset{F}{\dfrac{1}{}} \cdot \underset{M}{\dfrac{1}{}} = 3 \cdot 3 \cdot 2 \cdot 2 \cdot 1 \cdot 1 = 36$

Again, there are 36 ways to arrange the men and women.

21. (a) $P(\text{survived}) = \dfrac{711}{2224} \approx 0.320$

(b) $P(\text{female}) = \dfrac{425}{2224} \approx 0.191$

(c) $P(\text{female or child}) = \dfrac{425 + 109}{2224} = \dfrac{534}{2224}$

$= \dfrac{267}{1112} \approx 0.240$

(d) $P(\text{female and survived}) = \dfrac{316}{2224}$

$= \dfrac{79}{556} \approx 0.142$

(e) $P(\text{female or survived})$

$= P(\text{female}) + P(\text{survived})$

$- P(\text{female and survived})$

$= \dfrac{425}{2224} + \dfrac{711}{2224} - \dfrac{316}{2224}$

$= \dfrac{820}{2224} = \dfrac{205}{556} \approx 0.369$

(f) $P(\text{survived} \mid \text{female}) = \dfrac{316}{425} \approx 0.744$

(g) $P(\text{survived} \mid \text{child}) = \dfrac{57}{109} \approx 0.523$

(h) $P(\text{survived} \mid \text{male}) = \dfrac{338}{1690} = \dfrac{1}{5} = 0.2$

(i) Yes; the survival rate was much higher for women and children than for men.

(j) $P(\text{both females survived})$

$= P(\text{1st surv.}) \cdot P(\text{2nd surv.} \mid \text{1st surv.})$

$= \dfrac{316}{425} \cdot \dfrac{315}{424} \approx 0.552$

23. Since the order in which the colleges are selected does not matter, there are $_{12}C_3 = 220$ different ways to select 3 colleges from the 12 he is interested in.

25. (a) Since the games are independent of each other, we get

$P(\text{win both}) = \dfrac{1}{5,200,000} \cdot \dfrac{1}{705,600}$

≈ 0.00000000000027

(b) Since the games are independent of each other, we get

$P\left(\begin{array}{c}\text{win Jubilee} \\ \text{twice}\end{array}\right) = \left(\dfrac{1}{705,600}\right)^2$.

≈ 0.000000000002

27. $P(\text{parent owns} \mid \text{teenager owns})$

$= \dfrac{P(\text{both own})}{P(\text{teenager owns})}$

$= \dfrac{0.220}{0.51} \approx 0.431$

29. The order in which the questions are answered does not matter so there are $_{12}C_8 = 495$ different sets of questions that could be answered.

31. Using the Multiplication Rule of Counting, there are $2 \cdot 3 \cdot 2 \cdot 8 \cdot 2 = 192$ different Elantras that are possible.

5.7 Bayes's Rule (on CD)

1.

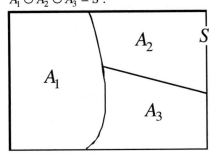

3. To partition a sample space S into three subsets A_1, A_2, and A_3 means that S is separated into three nonempty subsets where the subsets are all pairwise-disjoint. That is, $A_1 \cup A_2 \cup A_3 = S$.

5. Answers will vary. The Rule of Total Probability expresses a probability as a weighted average of conditional probabilities.

7. From the figure, $P(E \mid A) = 0.6$.

9. From the figure, $P(E^c \mid A) = 0.4$.

11. From the figure, $P(E \mid C) = 0.4$.

13. $P(E) = P(A) \cdot P(E \mid A)$
$\qquad + P(B) \cdot P(E \mid B)$
$\qquad + P(B) \cdot P(E \mid B)$
$\qquad = 0.2(0.6) + 0.55(0.7) + 0.25(0.4)$
$\qquad = 0.12 + 0.385 + 0.1$
$\qquad = 0.605$

15. $P(A \mid E) = \dfrac{P(A) \cdot P(E \mid A)}{P(E)}$

$\qquad = \dfrac{0.2(0.6)}{0.2(0.6) + 0.55(0.7) + 0.25(0.4)}$

$\qquad = \dfrac{0.12}{0.605} \approx 0.198$

17. $P(C \mid E) = \dfrac{P(C) \cdot P(E \mid C)}{P(E)}$

$\qquad = \dfrac{0.25(0.4)}{0.2(0.6) + 0.55(0.7) + 0.25(0.4)}$

$\qquad = \dfrac{0.1}{0.605} \approx 0.165$

19. $P(B \mid E) = \dfrac{P(B) \cdot P(E \mid B)}{P(E)}$

$\qquad = \dfrac{0.55(0.7)}{0.2(0.6) + 0.55(0.7) + 0.25(0.4)}$

$\qquad = \dfrac{0.385}{0.605} \approx 0.636$

21. $P(E) = P(A_1) \cdot P(E \mid A_1) + P(A_2) \cdot P(E \mid A_2)$
$\qquad = 0.55(0.06) + 0.45(0.08)$
$\qquad = 0.069$

23. $P(E) = P(A_1) \cdot P(E \mid A_1) + P(A_2) \cdot P(E \mid A_2)$
$\qquad + P(A_3) \cdot P(E \mid A_3)$
$\qquad = 0.35(0.25) + 0.45(0.18) + 0.2(0.14)$
$\qquad = 0.1965$

25. **(a)** $P(A_1 \mid E) = \dfrac{P(A_1) \cdot P(E \mid A_1)}{P(E)}$

$\qquad = \dfrac{0.55(0.06)}{0.069} \approx 0.4783$

$\quad$ **(b)** $P(A_2 \mid E) = \dfrac{P(A_2) \cdot P(E \mid A_2)}{P(E)}$

$\qquad = \dfrac{0.45(0.08)}{0.069} \approx 0.5217$

27. **(a)** $P(A_1 \mid E) = \dfrac{P(A_1) \cdot P(E \mid A_1)}{P(E)}$

$\qquad = \dfrac{0.35(0.25)}{0.1965} \approx 0.4453$

$\quad$ **(b)** $P(A_2 \mid E) = \dfrac{P(A_2) \cdot P(E \mid A_2)}{P(E)}$

$\qquad = \dfrac{0.45(0.18)}{0.1965} \approx 0.4122$

$\quad$ **(c)** $P(A_3 \mid E) = \dfrac{P(A_3) \cdot P(E \mid A_3)}{P(E)}$

$\qquad = \dfrac{0.2(0.14)}{0.1965} \approx 0.1425$

29. **(a)**

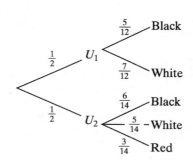

$P(B) = B \cap U_1 + B \cap U_2$

$\qquad = \dfrac{1}{2}\left(\dfrac{5}{12}\right) + \dfrac{1}{2}\left(\dfrac{6}{14}\right)$

$\qquad = \dfrac{71}{168} \approx 0.4226$

(b) $P(B) = P(U_1) \cdot P(B|U_1)$
$\qquad\qquad + P(U_2) \cdot P(B|U_2)$
$\qquad = \dfrac{1}{2}\left(\dfrac{5}{12}\right) + \dfrac{1}{2}\left(\dfrac{6}{14}\right)$
$\qquad = \dfrac{71}{168} \approx 0.4226$

(c) $P(U_1|B) = \dfrac{P(U_1) \cdot P(B|U_1)}{P(B)}$
$\qquad = \dfrac{(1/2)(5/12)}{71/168}$
$\qquad = \dfrac{35}{71} \approx 0.4930$

(d) $P(U_1|R) = 0$ since there are no red balls in urn I.
$\quad P(U_2|R) = 1$ since the only red balls are in urn II.

31. (a)

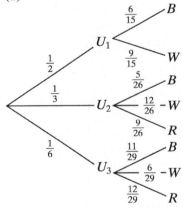

$P(B) = B \cap U_1 + B \cap U_2 + B \cap U_3$
$\qquad = \dfrac{1}{2}\left(\dfrac{6}{15}\right) + \dfrac{1}{3}\left(\dfrac{5}{26}\right) + \dfrac{1}{6}\left(\dfrac{11}{29}\right)$
$\qquad = \dfrac{617}{1885} \approx 0.3273$

(b) $P(B) = P(U_1) \cdot P(B|U_1)$
$\qquad\qquad + P(U_2) \cdot P(B|U_2)$
$\qquad\qquad + P(U_3) \cdot P(B|U_3)$
$\qquad = \dfrac{1}{2}\left(\dfrac{6}{15}\right) + \dfrac{1}{3}\left(\dfrac{5}{26}\right) + \dfrac{1}{6}\left(\dfrac{11}{29}\right)$
$\qquad = \dfrac{617}{1885} \approx 0.3273$

(c) $P(U_1|B) = \dfrac{P(U_1) \cdot P(B|U_1)}{P(B)}$
$\qquad = \dfrac{(1/2)(6/15)}{617/1885}$
$\qquad = \dfrac{377}{617} \approx 0.6110$

(d) $P(R) = P(U_1) \cdot P(R|U_1)$
$\qquad\qquad + P(U_2) \cdot P(R|U_2)$
$\qquad\qquad + P(U_3) \cdot P(R|U_3)$
$\qquad = \dfrac{1}{2}(0) + \dfrac{1}{3}\left(\dfrac{9}{26}\right) + \dfrac{1}{6}\left(\dfrac{12}{29}\right)$
$\qquad = \dfrac{139}{754} \approx 0.1844;$
$P(U_1|R) = \dfrac{P(U_1) \cdot P(R|U_1)}{P(R)}$
$\qquad = \dfrac{(1/2)(0)}{139/754} = 0$
(note that no red balls are in urn I);
$P(U_2|R) = \dfrac{P(U_2) \cdot P(R|U_2)}{P(R)}$;
$\qquad = \dfrac{(1/3)(9/26)}{139/754} = \dfrac{87}{139} \approx 0.6259$
$P(U_3|R) = \dfrac{P(U_3) \cdot P(R|U_3)}{P(R)}$
$\qquad = \dfrac{(1/6)(12/29)}{139/754} = \dfrac{52}{139} \approx 0.3741$

33. (a) $P(\text{colorblind})$
$\quad = P(\text{male}) \cdot P(\text{colorblind}|\text{male})$
$\qquad + P(\text{female}) \cdot P(\text{colorblind}|\text{female})$
$\quad = 0.491(0.08) + 0.509(0.0064)$
$\quad = 0.0425376 \approx 0.0425$

(b) $P(\text{female}|\text{colorblind})$
$\quad = \dfrac{P(\text{female}) \cdot P(\text{colorblind}|\text{female})}{P(\text{colorblind})}$
$\quad = \dfrac{0.509(0.0064)}{0.0425376} \approx 0.0766$

35. (a) $P(M) = P(A_1) \cdot P(M \mid A_1)$
$\qquad +P(A_2) \cdot P(M \mid A_2)$
$\qquad +P(A_3) \cdot P(M \mid A_3)$
$\qquad +P(A_4) \cdot P(M \mid A_4)$
$\qquad +P(A_5) \cdot P(M \mid A_5)$
$\qquad +P(A_6) \cdot P(M \mid A_6)$
$\qquad = 0.145(0.496) + 0.317(0.482)$
$\qquad + 0.170(0.460) + 0.087(0.424)$
$\qquad + 0.183(0.486) + 0.098(0.530)$
$\qquad = 0.48068 \approx 0.4807$

(b) $P(A_6 \mid M) = \dfrac{P(A_6) \cdot P(M \mid A_6)}{P(M)}$

$\qquad = \dfrac{0.098(0.530)}{0.48068} \approx 0.1081$

(c) $P(A_1 \mid M) = \dfrac{P(A_1) \cdot P(M \mid A_1)}{P(M)}$

$\qquad = \dfrac{0.145(0.496)}{0.48068} \approx 0.1496$

37. (a) $P(D) = P(A_1) \cdot P(D \mid A_1)$
$\qquad +P(A_2) \cdot P(D \mid A_2)$
$\qquad +P(A_3) \cdot P(D \mid A_3)$
$\qquad = 0.076(0.690) + 0.591(0.465)$
$\qquad +0.333(0.500)$
$\qquad = 0.493755 \approx 0.4938$

(b) $P(A_3 \mid D) = \dfrac{P(A_3) \cdot P(D \mid A_3)}{P(D)}$

$\qquad = \dfrac{0.333(0.500)}{0.493755} \approx 0.3372$

(c) $P(A_1 \mid D) = \dfrac{P(A_1) \cdot P(D \mid A_1)}{P(D)}$

$\qquad = \dfrac{0.076(0.690)}{0.493755} \approx 0.1062$

39. $P(\text{found } G) = P(G) \cdot P(\text{found } G \mid G)$
$\qquad + P(I) \cdot P(\text{found } G \mid I)$
$\qquad = 0.95(0.90) + 0.05(0.01)$
$\qquad = 0.8555$

$P(I \mid \text{found } G) = \dfrac{P(I) \cdot P(\text{found } G \mid I)}{P(\text{found } G)}$

$\qquad = \dfrac{0.05(0.01)}{0.8555} \approx 0.00058$

Chapter 5 Review Exercises

1. (a) Probabilities must be between 0 and 1, so the possible probabilities are: 0, 0.75, 0.41.

(b) Probabilities must be between 0 and 1, so the possible probabilities are: $\dfrac{2}{5}$, $\dfrac{1}{3}$, $\dfrac{6}{7}$.

2. Event E contains 1 of the 5 equally likely outcomes, so $P(E) = \dfrac{1}{5} = 0.2$.

3. Event F contains 2 of the 5 equally likely outcomes, so $P(F) = \dfrac{2}{5} = 0.4$.

4. Event E contains 3 of the 5 equally likely outcomes, so $P(E) = \dfrac{3}{5} = 0.6$.

5. Since $P(E) = \dfrac{1}{5} = 0.2$, we have

$P(E^c) = 1 - \dfrac{1}{5} = \dfrac{4}{5} = 0.8$.

6. $P(E \text{ or } F) = P(E) + P(F) - P(E \text{ and } F)$
$\qquad = 0.76 + 0.45 - 0.32 = 0.89$

7. Since events E and F are mutually exclusive,
$P(E \text{ or } F) = P(E) + P(F)$
$\qquad = 0.36 + 0.12$
$\qquad = 0.48$

8. Since events E and F are independent,
$P(E \text{ and } F) = P(E) \cdot P(F) = 0.45 \cdot 0.2 = 0.09$.

9. No, events E and F are not independent because
$$P(E)\cdot P(F) = 0.8\cdot 0.5 = 0.40 \neq P(E \text{ and } F).$$

10. $P(E \text{ and } F) = P(E)\cdot P(F \mid E)$
$$= 0.59\cdot 0.45 = 0.2655$$

11. $P(E \mid F) = \dfrac{P(E \text{ and } F)}{P(F)} = \dfrac{0.35}{0.7} = 0.5$

12. **(a)** $7! = 7\cdot 6\cdot 5\cdot 4\cdot 3\cdot 2\cdot 1 = 5040$

(b) $0! = 1$

(c) $_9C_4 = \dfrac{9!}{4!(9-4)!} = \dfrac{9!}{4!5!} = \dfrac{9\cdot 8\cdot 7\cdot 6}{4\cdot 3\cdot 2\cdot 1} = 126$

(d) $_{10}C_3 = \dfrac{10!}{3!(10-3)!} = \dfrac{10!}{3!7!} = \dfrac{10\cdot 9\cdot 8}{3\cdot 2\cdot 1} = 120.$

(e) $_9P_2 = \dfrac{9!}{(9-2)!} = \dfrac{9!}{7!} = 9\cdot 8 = 72$

(f) $_{12}P_4 = \dfrac{12!}{(12-4)!} = \dfrac{12!}{8!}$
$$= 12\cdot 11\cdot 10\cdot 9 = 11{,}880.$$

13. **(a)** $P(\text{green}) = \dfrac{2}{38} = \dfrac{1}{19} \approx 0.0526$. This means that, in many games of roulette, the metal ball lands on green approximately 5.26% of the time.

(b) $P(\text{green or red}) = \dfrac{2+18}{38} = \dfrac{20}{38}$
$$= \dfrac{10}{19} \approx 0.5263$$
This means that, in many games of roulette, the metal ball lands on green or red approximately 52.63% of the time.

(c) $P(00 \text{ or red}) = \dfrac{1+18}{38} = \dfrac{19}{38} = \dfrac{1}{2} = 0.5$
This means that, in many games of roulette, the metal ball lands on 00 or red approximately 50% of the time.

(d) Since 31 is an odd number and the odd slots are colored red,
$P(31 \text{ and black}) = 0$. This is called an impossible event.

14. **(a)** Of the 454 accidents, 197 were alcohol related, so
$$P(\text{alcohol related}) = \dfrac{193}{454} \approx 0.425.$$

(b) Of the 454 accidents, $454 - 193 = 261$ were not alcohol related, so
$$P(\text{not alcohol related}) = \dfrac{261}{454} \approx 0.575.$$

(c) $P(\text{both were alcohol related}) = \dfrac{193}{454}\cdot \dfrac{193}{454}$
$$\approx 0.181$$

(d) $P(\text{neither was alcohol related}) = \dfrac{261}{454}\cdot \dfrac{261}{454}$
$$\approx 0.330$$

(e) $P(\text{at least one of the two was alcohol related})$
$$= 1 - \dfrac{261}{454}\cdot \dfrac{261}{454} \approx 0.670$$

15. **(a)** There are $219{,}852 + 39{,}351 + 17{,}060 + 498 = 276{,}761$ workers included in the table. The individuals can be thought of as the trials of the probability experiment. The relative frequency of "Private wage and salary worker" is $\dfrac{219{,}852}{276{,}761} \approx 0.7944$.

We compute the relative frequencies of the other outcomes similarly and obtain the probability model below.

Type of Worker	Probability
Private wage and salary worker	0.7944
Government worker	0.1422
Self-employed worker	0.0616
Unpaid family worker	0.0018

(b) Yes, it is unusual for a Memphis worker to be an unpaid family worker since the probability is only 0.0018.

(c) No, it is not unusual for a Memphis worker to be self-employed since the probability is $0.0616 > 0.05$.

129

16. (a) Of the 4,105,371 births included in the table, 239,706 are postterm. Thus,

$$P(\text{postterm}) = \frac{239,706}{4,105,371} \approx 0.058 .$$

(b) Of the 4,105,371 births included in the table, 2,693,809 weighed 3000 to 3999 grams. Thus, P(3000 to 3999 grams)

$$= \frac{2,693,809}{4,105,371} \approx 0.656 .$$

(c) Of the 4,105,371 births included in the table, 172,957 both weighed 3000 to 3999 grams and were postterm. Thus, P(3000 to 3999 grams and postterm)

$$= \frac{172,957}{4,105,371} \approx 0.042 .$$

(d) P(3000 to 3999 grams or postterm)
= P(3000 to 3999 grams) + P(postterm)
– P(3000 to 3999 grams and postterm)

$$= \frac{2,693,809}{4,105,371} + \frac{239,706}{4,105,371} - \frac{172,957}{4,105,371}$$

$$= \frac{2,760,558}{4,105,371} \approx 0.672$$

(e) Of the 4,105,371 births included in the table, 15 both weighed less than 1000 grams and were postterm. Thus, P(less than 1000 grams and postterm)

$$= \frac{15}{4,105,371} \approx 0.000004 .$$

This event is highly unlikely, but not impossible.

(f) P(3000 to 3999 grams | postterm)

$$= \frac{N(\text{3000 to 3999 grams and postterm})}{N(\text{postterm})}$$

$$= \frac{172,957}{239,706} \approx 0.722$$

(g) No, the events "postterm baby" and "weighs 3000 to 3999 grams" are not independent since
P(3000 to 3999 grams) · P(postterm)
$\approx 0.656(0.058) = 0.038$
$\neq$ P(3000 to 3999 grams and postterm)
$\approx 0.042 .$

17. (a) P(complaint settled) = 0.722

(b) P(complaint not settled) = 1 – 0.722
= 0.278

(c) P(all 5 settled) = $(0.722)^5 \approx 0.196$

(d) P(at least one of the five not settled)
= 1 – P(all settled)
= 1 – 0.196 = 0.804

(e) P(none of the 5 settled) = $(0.278)^5 \approx 0.002$

(f) P(at least one of the five settled)
= 1 – P(none settled)
= 1 – 0.002 = 0.998

18. $P\big(\text{matching the winning PICK 3 numbers}\big)$

$$= \frac{1}{10} \cdot \frac{1}{10} \cdot \frac{1}{10}$$

$$= \frac{1}{1000} = 0.001$$

19. $P\big(\text{matching the winning PICK 4 numbers}\big)$

$$= \frac{1}{10} \cdot \frac{1}{10} \cdot \frac{1}{10} \cdot \frac{1}{10}$$

$$= \frac{1}{10,000} = 0.0001$$

20. $P\big(\text{three aces}\big) = \dfrac{4}{52} \cdot \dfrac{3}{51} \cdot \dfrac{2}{50} \approx 0.00018$

21. $26 \cdot 26 \cdot 10^4 = 6,760,000$ license plates are possible.

22. $_{10}P_4 = 10 \cdot 9 \cdot 8 \cdot 7 = 5040$ seating arrangements are possible.

23. $\dfrac{10!}{4!3!2!} = 12,600$ different vertical arrangements of flags are possible.

24. $_{55}C_8 = 1,217,566,350$ simple random samples are possible.

25. $P\left(\begin{array}{c}\text{winning Arizona's}\\ \text{Pick 5}\end{array}\right) = \dfrac{1}{_{35}C_5}$

$$= \frac{1}{324,632} \approx 0.000003$$

26. (a) $P\big(\text{all three are Merlot}\big) = \dfrac{5}{12} \cdot \dfrac{4}{11} \cdot \dfrac{3}{10}$

$$= \frac{1}{22} \approx 0.0455$$

(b) $P\left(\begin{array}{c}\text{exactly two}\\\text{are Merlot}\end{array}\right)=\dfrac{_5C_2\cdot{_7}C_1}{_{12}C_3}$

$=\dfrac{\frac{5\cdot4}{2\cdot1}\cdot\frac{7}{1}}{\frac{12\cdot11\cdot10}{3\cdot2\cdot1}}$

$=\dfrac{7}{22}\approx0.3182$

(c) $P(\text{none are Merlot})=\dfrac{7}{12}\cdot\dfrac{6}{11}\cdot\dfrac{5}{10}$

$=\dfrac{7}{44}\approx0.1591$

27. **(a), (b)** Answers will vary depending on the results of the simulation. Results should be reasonably close to $\frac{1}{38}$ for part (a) and $\frac{1}{19}$ for part (b).

28. Subjective probabilities are probabilities based on personal experience or intuition. Examples will vary. Some examples are the likelihood of life on other planets or the chance that the Packers will make it to the 2012 NFL playoffs.

29. **(a)** There are 13 clubs in the deck.

(b) There are 37 cards remaining in the deck. There are also 4 cards unknown by you. So, there are $37 + 4 = 41$ cards not known to you. Of the unknown cards, $13-(3+2)=8$ are clubs.

(c) $P(\text{next card dealt is a club})=\dfrac{8}{41}$

≈0.1951

(d) $P(\text{two clubs in a row})=\dfrac{8}{41}\cdot\dfrac{7}{40}\approx0.0341$

(e) Answers will vary. One possibility follows: No, you should not stay in the game because the probability of completing the flush is low (0.0341 is less than 0.05).

30. **(a)** Since 20 of the 70 homeruns went to left field, $P(\text{left field})=\dfrac{34}{70}=\dfrac{17}{35}\approx0.486$. Mark McGwire hit 48.6% of his home runs to left field that year.

(b) Since none of the 70 homeruns went to right field, $P(\text{right field})=\dfrac{0}{70}=0$.

(c) No. While Mark McGwire did not hit any home runs to right field in 1998, this does not imply that it is impossible for him to hit a right-field homerun. He just never did it.

31. Someone winning a lottery twice is not that unlikely considering millions of people play lotteries (sometimes more than one) each week, and many lotteries have multiple drawings each week. So, while it is highly unlikely that a specific person will win the lottery twice, it is not that unlikely that *someone* will win twice.

32. **(a)** $P(\text{Bryce})=\dfrac{119}{1009}\approx0.118$

This is not unusual since the probability is greater than 0.05.

(b) $P(\text{Gourmet})=\dfrac{264}{1009}\approx0.262$

(c) $P(\text{Mallory}\mid\text{Single Cup})=\dfrac{75}{625}=0.12$

(d) $P(\text{Bryce}\mid\text{Gourmet})=\dfrac{9}{264}=\dfrac{3}{88}\approx0.034$

This is unusual since the probability is less than 0.05.

(e) While it is not unusual for Bryce to sell a case, it is unlikely that he will sell a Gourmet case.

(f) $P(\text{Mallory})=\dfrac{186}{1009}\approx0.184$

$P(\text{Mallory}\mid\text{Filters})=\dfrac{40}{120}=\dfrac{1}{3}\approx0.333$

No, the events 'Mallory' and 'Filters' are not independent since $P(\text{Mallory})\ne P(\text{Mallory}\mid\text{Filters})$.

(g) No, the events 'Paige' and 'Gourmet' are not mutually exclusive because the two events can happen at the same time.

$P(\text{Paige and Gourmet})=\dfrac{42}{1009}\approx0.042\ne0$

Chapter 5 Test

1. Probabilities must be between 0 and 1, so the possible probabilities are: 0.23, 0, and $\frac{3}{4}$.

2. Event E contains 1 of the 5 equally likely outcomes, so $P(E) = \frac{1}{5} = 0.2$.

3. Event F contains 2 of the 5 equally likely outcomes, so $P(F) = \frac{2}{5} = 0.4$.

4. Since $P(E) = \frac{1}{5} = 0.2$, we have

   $P(E^c) = 1 - \frac{1}{5} = \frac{4}{5} = 0.8$.

5. **(a)** Since events E and F are mutually exclusive, $P(E \text{ or } F) = P(E) + P(F)$
   $$= 0.37 + 0.22$$
   $$= 0.59$$

   (b) Since events E and F are independent,
   $P(E \text{ and } F) = P(E) \cdot P(F)$
   $$= (0.37)(0.22)$$
   $$= 0.0814.$$

6. **(a)** $P(E \text{ and } F) = P(E) \cdot P(F \mid E)$
   $$= 0.15 \cdot 0.70$$
   $$= 0.105$$

   (b) $P(E \text{ or } F) = P(E) + P(F) - P(E \text{ and } F)$
   $$= 0.15 + 0.45 - 0.105$$
   $$= 0.495$$

   (c) $P(E \mid F) = \dfrac{P(E \text{ and } F)}{P(F)} = \dfrac{0.105}{0.45}$
   $$= \frac{7}{30} \approx 0.233$$

   (d) No, the events E and F are not independent because $P(F \mid E) \neq F(F)$.

7. **(a)** $8! = 8 \cdot 7 \cdot 6 \cdot 5 \cdot 4 \cdot 3 \cdot 2 \cdot 1 = 40{,}320$

(b) $_{12}C_6 = \dfrac{12!}{6!(12-6)!} = \dfrac{12!}{6!6!}$
$$= \frac{12 \cdot 11 \cdot 10 \cdot 9 \cdot 8 \cdot 7}{6 \cdot 5 \cdot 4 \cdot 3 \cdot 2 \cdot 1}$$
$$= 924$$

(c) $_{14}P_8 = \dfrac{14!}{(14-8)!} = \dfrac{14!}{6!}$
$$= 14 \cdot 13 \cdot 12 \cdot 11 \cdot 10 \cdot 9 \cdot 8 \cdot 7$$
$$= 121{,}080{,}960$$

8. **(a)** A "7" can occur in six ways: (1, 6), (2, 5), (3, 4), (4,3), (5, 2), or (6, 1). An "11" can occur two ways: (6, 5) or (5, 6). Thus, 8 of the 36 possible outcomes of the first roll results in a win, so

   $P(\text{wins on first roll}) = \dfrac{8}{36} = \dfrac{2}{9} \approx 0.2222$.

   This means that there is a 22.2% probability that a player will win on the first roll.

   (b) A "2" can occur in only one way: (1, 1). A "3" can occur in two ways: (1, 2) or (2, 1). A "12" can occur in only one way: (6, 6). Thus, 4 of the 36 possible outcomes of the first roll results in a loss, so

   $P(\text{loses on first roll}) = \dfrac{4}{36} = \dfrac{1}{9} \approx 0.1111$.

   This means that there is an 11.1% probability that a player will lose on the first roll.

9. $P(\text{structure}) = 0.32$;
   $P(\text{not in a structure}) = 1 - P(\text{structure})$
   $$= 1 - 0.32$$
   $$= 0.68$$

10. **(a)** Rule 1 is satisfied since all of the probabilities in the model are between 0 and 1.
    Rule 2 is satisfied since the sum of the probabilities in the model is one:
    $0.25 + 0.19 + 0.13 + 0.11 + 0.09 + 0.23 = 1$.

    (b) $P\left(\begin{array}{l}\text{PB Patties/Tagalongs or}\\ \text{PB Sandwich/Do-si-dos}\end{array}\right) = 0.13 + 0.11$
    $$= 0.24$$

(c) $P\left(\begin{array}{c}\text{Thin Mints, Samoas,}\\ \text{Shortbread}\end{array}\right) = 0.25 + 0.19 + 0.09$

$$= 0.53$$

(d) $P(\text{not Thin Mints}) = 1 - 0.25 = 0.75$

11. (a) Of the 162 medals won by the top eight countries, 62 were gold. Thus,

$$P(\text{gold}) = \frac{62}{162} = \frac{31}{81} \approx 0.383.$$

(b) Of the 162 medals won by the top eight countries, 22 were won by Russia. Thus,

$$P(\text{won by Russia}) = \frac{22}{162} = \frac{11}{81} \approx 0.136.$$

(c) Of the 162 medals won by the top eight countries, 8 were gold and won by Russia. Thus, $P(\text{gold and won by Russia})$

$$= \frac{8}{162} = \frac{4}{81} \approx 0.049.$$

(d) $P\left(\begin{array}{c}\text{gold or}\\ \text{won by}\\ \text{Russia}\end{array}\right) = P(\text{gold}) + P\left(\begin{array}{c}\text{won by}\\ \text{Russia}\end{array}\right)$

$$- P(\text{gold and won by Russia})$$

$$= \frac{62}{162} + \frac{22}{162} - \frac{8}{162} = \frac{76}{162}$$

$$= \frac{38}{81} \approx 0.469$$

(e) Of the 47 bronze medals won by the top eight countries, 5 were won by Sweden. Thus,

$$P(\text{won by Sweden} \mid \text{bronze}) = \frac{5}{47} \approx 0.106.$$

(f) Of the 14 medals won by Sweden, 5 were bronze. Thus,

$$P(\text{bronze} \mid \text{won by Sweden}) = \frac{5}{14} \approx 0.357.$$

12. (a) Assuming the outcomes are independent, we get $P(\text{win two games in a row})$

$$= (0.593)^2 \approx 0.352.$$

(b) Assuming the outcomes are independent, we get $P(\text{win seven games in a row})$

$$= (0.593)^7 \approx 0.026.$$

(c) Losing at least 1 of the next 7 is the complement of winning all 7. Thus, $P(\text{lose at least one of next seven games})$
$\approx 1 - 0.026 = 0.974.$

13. $P(\text{accept}) = P(\text{1st works}) \cdot P(\text{2nd works} \mid \text{1st works})$

$$= \frac{9}{10} \cdot \frac{8}{9} = 0.8$$

14. $6! = 720$ different arrangements are possible for the letters in the 'word' LINCEY.

15. $_{29}C_5 = 118,755$ different subcommittees are possible.

16. $P\left(\begin{array}{c}\text{winning}\\ \text{Pennsylvania's}\\ \text{Cash 5}\end{array}\right) = \frac{1}{_{43}C_5}$

$$= \frac{1}{962,598}$$

$$\approx 0.00000104$$

17. There are 26 ways to select the first character, and 36 ways to select each of the final 7 characters. Therefore, using the Multiplication Rule of Counting, there are $26 \cdot 36^7 = 2.04 \times 10^{12}$ different passwords that are possible for the local area network.

18. It would take the processor

$$\frac{26 \cdot 36^7}{3.0 \times 10^6} \approx 679,156 \text{ seconds (about 7.9 days)}$$

to compute all the passwords.

19. Subjective probability assignment was used. It is not possible to repeat probability experiments to estimate the probability of life on Mars.

20. Since there are two A's, four C's, four G's, and five T's from which to form a DNA sequence, there are $\dfrac{15!}{2!4!4!5!} = 9,459,450$ distinguishable DNA sequences possible using the 15 letters.

21. There are 40 digits, of which 9 are either a 0 or a 1. Therefore,

$$P(\text{guess correctly}) \approx \frac{9}{40} = 0.225.$$

Chapter 6

Discrete Probability Distributions

6.1 Discrete Random Variables

1. A random variable is a numerical measure of the outcome of a probability experiment, so its value is determined by chance.

3. For a discrete probability distribution, each probability must be between 0 and 1 (inclusive) and the sum of the probabilities must equal one.

5. The historical average is based on long-term results. Short-term results can vary dramatically. A batting average of 0.300 means the batter averages 3 hits out of 10 at-bats *in the long run*. It does not mean that he will get 3 hits out of *every* 10 at-bats. A batter in a dry spell is not "due for a hit" nor is a batter with a hitting streak "due for an out". These dry spells and streaks will get washed out in the long run.

7. **(a)**   The number of light bulbs that burn out, X, is a discrete random variable because the value of the random variable results from counting.
    Possible values: $x = 0, 1, 2, 3, ..., 20$

    (b)   The time it takes to fly from New York City to Los Angeles is a continuous random variable because time is measured. If we let the random variable T represent the time it takes to fly from New York City to Los Angeles, the possible values for T are all positive real numbers; that is $t > 0$.

    (c)   The number of hits to a web site in a day is a discrete random variable because the value of the random variable results from counting. If we let the random variable X represent the number of hits, then the possible values of X are $x = 0, 1, 2, ...$

    (d)   The amount of snow in Toronto during the winter is a continuous random variable because the amount of snow is measured. If we let the random variable S represent the amount of snow in Toronto during the winter, the possible values for S are all nonnegative real numbers; that is $s \geq 0$.

9. **(a)**   The amount of rain in Seattle during April is a continuous random variable because the amount of rain is measured. If we let the random variable R represent the amount of rain, the possible values for R are all nonnegative real numbers; that is, $r \geq 0$.

    (b)   The number of fish caught during a fishing tournament is a discrete random variable because the value of the random variable results from counting. If we let the random variable X represent the number of fish caught, the possible values of X are $x = 0, 1, 2, ...$

    (c)   The number of customers arriving at a bank between noon and 1:00 P.M. is a discrete random variable because the value of the random variable results from counting. If we let the random variable X represent the number of customers arriving at the bank between noon and 1:00 P.M., the possible values of X are $x = 0, 1, 2, ...$

    (d)   The time required to download a file from the Internet is a continuous random variable because time is measured. If we let the random variable T represent the time required to download a file, the possible values of T are all positive real numbers; that is, $t > 0$.

11. Yes, because $\sum P(x) = 1$ and $0 \leq P(x) \leq 1$ for all x.

13. No, because $P(50) < 0$.

15. No, because $\sum P(x) = 0.95 \neq 1$.

17. We need the sum of all the probabilities to equal 1. For the given probabilities, we have $0.4 + 0.1 + 0.2 = 0.7$. For the sum of the probabilities to equal 1, the missing probability must be $1 - 0.7 = 0.3$. That is, $P(4) = 0.3$.

19. **(a)** This is a discrete probability distribution because all the probabilities are between 0 and 1 (inclusive) and the sum of the probabilities is 1.

(b)

(c) $\mu_X = \sum [x \cdot P(x)] = 0(0.035) + 1(0.074) + \ldots + 4(0.374) = 2.924 \approx 2.9$

On average, the number of activities that at least one parent of a K-5th grader is involved in is expected to be about 2.9.

(d) $\sigma_X^2 = \sum \left[(x - \mu_X)^2 \cdot P(x) \right]$

$\qquad = (0 - 2.924)^2 (0.035) + (1 - 2.924)^2 (0.074) + \ldots + (4 - 2.924)^2 (0.374)$

$\qquad \approx 1.176$ or about 1.2 activities^2

(e) $\sigma_X = \sqrt{\sigma_X^2} = \sqrt{1.176} \approx 1.085$ or about 1.1 activities.

(f) $P(3) = 0.320$

(g) $P(3 \text{ or } 4) = P(3) + P(4) = 0.320 + 0.374 = 0.694$

21. **(a)** This is a discrete probability distribution because all the probabilities are between 0 and 1 (inclusive) and the sum of the probabilities is 1.

(b)

(c) $\mu_X = \sum [x \cdot P(x)] = 0(0.1677) + 1(0.3354) + \ldots + 5(0.0248) = 1.6273 \approx 1.6$

Over many games, Ichiro is expected to average about 1.6 hits per game.

(d) $\sigma_X^2 = \sum \left[(x - \mu_X)^2 \cdot P(x) \right]$

$\qquad = (0 - 1.6273)^2 (0.1677) + (1 - 1.6273)^2 (0.3354) + \ldots + (5 - 1.6273)^2 (0.0248)$

$\qquad \approx 1.389$

$\qquad \sigma_X = \sqrt{\sigma^2_X} \approx \sqrt{1.389} \approx 1.179$ or about 1.2 hits

(e) $P(2) = 0.2857$

(f) $P(X > 1) = 1 - P(X \leq 1) = 1 - P(0 \text{ or } 1) = 1 - (0.1677 + 0.3354) = 1 - 0.5031 = 0.4969$

23. (a) Total number of World Series
= 17 + 16 + 18 + 33 = 84.

x (games played)	P(x)
4	$\dfrac{17}{84} \approx 0.2024$
5	$\dfrac{16}{84} \approx 0.1905$
6	$\dfrac{18}{84} \approx 0.2143$
7	$\dfrac{33}{84} \approx 0.3929$

(b)

Games Played in the World Series

(c) $\mu_X = \sum \left[x \cdot P(x) \right]$
$\approx 4 \cdot (0.2024) + 5 \cdot (0.1905) + 6 \cdot (0.2143) + 7 \cdot (0.3929)$
≈ 5.7982 or about 5.8 games

The World Series, if played many times, would be expected to last about 5.8 games on average.

(d) $\sigma_X^2 = \sum (x - \mu_x)^2 \cdot P(x)$
$\approx (4 - 5.7982)^2 \cdot 0.2024 + (5 - 5.7982)^2 \cdot 0.1905 + (6 - 5.7982)^2 \cdot 0.2143 + (7 - 5.7982)^2 \cdot 0.3929$
≈ 1.352
$\sigma_X = \sqrt{\sigma_X^2} \approx \sqrt{1.352} \approx 1.163$ or about 1.2 games

25. (a)

x (ideal #)	P(x)	x (ideal #)	P(x)
0	$\dfrac{10}{900} \approx 0.0111$	4	$\dfrac{70}{900} \approx 0.0778$
1	$\dfrac{30}{900} \approx 0.0333$	5	$\dfrac{17}{900} \approx 0.0189$
2	$\dfrac{520}{900} \approx 0.5778$	6	$\dfrac{3}{900} \approx 0.0033$
3	$\dfrac{250}{900} \approx 0.2778$		

(b)

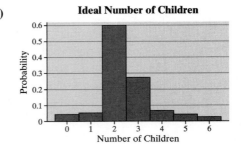

Ideal Number of Children

(c) $\mu_X = \sum \left[x \cdot P(x) \right]$
$\approx 0(0.0111) + 1(0.0333) + 2(0.5778) + \ldots + 6(0.0033) \approx 2.448$

The average ideal number of children is about 2.4 children.

(d) $\sigma_X^2 = \sum (x - \mu_x)^2 \cdot P(x)$
$\approx (0 - 2.448)^2 \cdot 0.0111 + (1 - 2.448)^2 \cdot 0.0333 + \ldots + (6 - 2.448)^2 \cdot 0.0033 \approx 0.689$

$\sigma_X = \sqrt{\sigma_X^2} \approx \sqrt{0.689} \approx 0.830$ or about 0.8 children

27. (a) $P(4) = 0.128$

(b) $P(4 \text{ or } 5) = P(4) + P(5) = 0.128 + 0.060 = 0.188$

(c) $P(X \geq 6) = P(6) + P(7) + P(8) = 0.031 + 0.022 + 0.049 = 0.102$

Section 6.1: *Discrete Random Variables*

(d) $E(X) = \mu_X = \sum x \cdot P(x)$
$= 1(0.292) + 2(0.251) + 3(0.167) + \ldots + 8(0.049)$
$= 2.839$ or about 2.8
We would expect the mother to have had 2.8 live births, on average.

29. $E(X) = \sum x \cdot P(x) = (200)(0.999544) + (200 - 250,000)(0.000456) = \$86.00$
If the company sells many of these policies to 20-year old females, then they will make an average of $86.00 per policy.

31. Let X = the profit from the investment

Profit, x ($)	50,000	10,000	−50,000
Probability	0.2	0.7	0.1

$E(X) = (50,000)(0.2) + (10,000)(0.7) + (-50,000)(0.1) = 12,000$
The expected profit for the investment is $12,000.

33. Let X = player winnings for $5 bet on a single number.

Winnings, x ($)	175	−5
Probability	$\frac{1}{38}$	$\frac{37}{38}$

$E(X) = (175)\left(\frac{1}{38}\right) + (-5)\left(\frac{37}{38}\right) = -\$0.26$
The expected value of the game to the player is a loss of $0.26. If you played the game 1000 times, you would expect to lose $1000 \cdot (\$0.26) = \$260$.

35. (a) $E(X) = \sum x \cdot P(x)$
$= (15,000,000)(0.00000000684) + (200,000)(0.00000028)$
$+ (10,000)(0.000001711) + (100)(0.000153996) + (7)(0.004778961)$
$+ (4)(0.007881463) + (3)(0.01450116) + (0)(0.9726824222)$
≈ 0.30
After many $1 plays, you would expect to win an average of $0.30 per play. That is, you would lose an average of $0.70 per $1 play for a net profit of $-\$0.70$.
(Note: the given probabilities reflect changes made in April 2005 to create larger jackpots that are built up more quickly. It is interesting to note that prior to the change, the expected cash prize was still $0.30)

(b) We need to find the break-even point. That is, the point where we expect to win the same amount that we pay to play. Let x be the grand prize. Set the expected value equation equal to 1 (the cost for one play) and then solve for x.
$E(X) = \sum x \cdot P(x)$
$1 = (x)(0.00000000684) + (200,000)(0.00000028)$
$+ (10,000)(0.000001711) + (100)(0.000153996) + (7)(0.004778961)$
$+ (4)(0.007881463) + (3)(0.01450116) + (0)(0.9726824222)$
$1 = 0.196991659 + 0.00000000684x$
$0.803008341 = 0.00000000684x$
$117,398,880.3 = x$
$118,000,000 \approx x$
The grand prize should be at least $118,000,000 for you to expect a profit after many $1 plays.
(Note: prior to the changes mentioned in part (a), the grand prize only needed to be about $100 million to expect a profit after many $1 plays)

137

(c) No, the size of the grand prize does not affect your chance of winning. Your chance of winning the grand prize is determined by the number of balls that are drawn and the number of balls from which they are drawn. However, the size of the grand prize will impact your expected winnings. A larger grand prize will increase your expected winnings.

37. Answers will vary. The simulations illustrate the Law of Large Numbers.

6.2 The Binomial Probability Distributions

1. A probability experiment is said to be a binomial probability experiment provided:
  a) The experiment consists of a fixed number, n, of trials.
  b) The trials are all independent.
  c) For each trial there are two mutually exclusive (disjoint) outcomes, success or failure.
  d) The probability of success, p, is the same for each trial.

3. If the binomial probability distribution is roughly bell-shaped, then the Empirical Rule can be used to check for unusual observations. As a rule of thumb, the probability distribution for a binomial random variable will be approximately bell-shaped if $np(1-p) \geq 10$.

  In a bell-shaped distribution, about 95% of all observations lie within two standard deviations of the mean. That is, about 95% of the observations lie between $\mu - 2\sigma$ and $\mu + 2\sigma$. An observation would be considered unusual if it is more than two standard deviations above or below the mean because this will occur less than 5% of the time.

5. For a small, fixed value of n, the shape of the binomial probability histogram will be determined by the value of p. The histogram will be skewed right for $p < 0.5$, skewed left for $p > 0.5$, and symmetric (and approximately bell shaped) for $p = 0.5$. The skewness becomes stronger as p approaches either 0 or 1.

7. This is not a binomial experiment because there are more than two possible values for the random variable 'age'.

9. This is a binomial experiment. There is a fixed number of trials ($n = 100$ where each trial corresponds to administering the drug to one of the 100 individuals), the trials are independent (due to random selection), there are two outcomes (favorable or unfavorable response), and the probability of a favorable response is assumed to be constant for each individual.

11. This is not a binomial experiment because the trials (cards) are not independent and the probability of getting an ace changes for each trial (card). Because the cards are not replaced, the probability of getting an ace on the second card depends on what was drawn first.

13. This is not a binomial experiment because the number of trials is not fixed.

15. This is a binomial experiment. There is a fixed number of trials ($n = 100$ where each trial corresponds to selecting one of the 100 parents), the trials are independent (due to random selection), there are two outcomes (spanked or never spanked), and there is a fixed probability (for a given population) that a parent has ever spanked their child.

17. Using $P(x) = {}_nC_x p^x (1-p)^{n-x}$ with $x = 3$, $n = 10$ and $p = 0.4$:

$$P(3) = {}_{10}C_3 \cdot (0.4)^3 \cdot (1-0.4)^{10-3} = \frac{10!}{3!(10-3)!} \cdot (0.4)^3 \cdot (0.6)^7$$

$$= 120 \cdot (0.064) \cdot (0.0279936) \approx 0.2150$$

19. Using $P(x) = {}_nC_x p^x (1-p)^{n-x}$ with $x = 38$, $n = 40$ and $p = 0.99$:

$$P(38) = {}_{40}C_{38} \cdot (0.99)^{38} \cdot (1-0.99)^{40-38} = \frac{40!}{38!(40-38)!} \cdot (0.99)^{38} \cdot (0.01)^2$$

$$= 780 \cdot (0.6825...) \cdot (0.0001) \approx 0.0532$$

21. Using $P(x) = {}_nC_x p^x (1-p)^{n-x}$ with $x = 3$, $n = 8$ and $p = 0.35$:

$$P(3) = {}_8C_3 \cdot (0.35)^3 \cdot (1-0.35)^{8-3} = \frac{8!}{3!(8-3)!} \cdot (0.35)^3 \cdot (0.65)^5$$

$$= 56 \cdot (0.42875)(0.116029...) \approx 0.2786$$

23. Using $n = 9$ and $p = 0.2$:

$$P(X \le 3) = P(0) + P(1) + P(2) + P(3)$$

$$= {}_9C_0 \cdot (0.2)^0 (0.8)^9 + {}_9C_1 \cdot (0.2)^1 (0.8)^8 + {}_9C_2 \cdot (0.2)^2 (0.8)^7 + {}_9C_3 \cdot (0.2)^3 (0.8)^6$$

$$\approx 0.134218 + 0.301990 + 0.301990 + 0.176161$$

$$\approx 0.9144$$

25. Using $n = 7$ and $p = 0.5$:

$$P(X > 3) = P(X \ge 4)$$

$$= P(4) + P(5) + P(6) + P(7)$$

$$= {}_7C_4 \cdot (0.5)^4 (0.5)^3 + {}_7C_5 \cdot (0.5)^5 (0.5)^2 + {}_7C_6 \cdot (0.5)^6 (0.5)^1 + {}_7C_7 \cdot (0.5)^7 (0.5)^0$$

$$= 0.2734375 + 0.1640625 + 0.0546875 + 0.0078125$$

$$= 0.5$$

27. Using $n = 12$ and $p = 0.35$:

$$P(X \le 4) = P(0) + P(1) + P(2) + P(3) + P(4)$$

$$= {}_{12}C_0 \cdot (0.35)^0 (0.65)^{12} + {}_{12}C_1 \cdot (0.35)^1 (0.65)^{11} + {}_{12}C_2 \cdot (0.35)^2 (0.65)^{10}$$

$$+ {}_{12}C_3 \cdot (0.35)^3 (0.65)^9 + {}_{12}C_4 \cdot (0.35)^4 (0.65)^8$$

$$\approx 0.005688 + 0.036753 + 0.108846 + 0.195365 + 0.236692$$

$$\approx 0.5833$$

29. (a)

Distribution

x	$P(x)$
0	0.1176
1	0.3025
2	0.3241
3	0.1852
4	0.0595
5	0.0102
6	0.0007

$x \cdot P(x)$	$(x - \mu_x)^2 \cdot P(x)$
0.0000	0.3812
0.3025	0.1936
0.6483	0.0130
0.5557	0.2667
0.2381	0.2881
0.0510	0.1045
0.0044	0.0129
Σ 1.8000	1.2600

(b) $\mu_X = 1.8$ (from first column in table above and to the right)

$\sigma_X = \sqrt{\sigma_X^2} = \sqrt{1.26} \approx 1.1$ (from second column in table above and to the right)

(c) $\mu_X = n \cdot p = 6 \cdot (0.3) = 1.8$ and $\sigma_X = \sqrt{n \cdot p \cdot (1-p)} = \sqrt{(1.8) \cdot (0.7)} \approx 1.1$

(d)

The distribution is skewed right.

31. (a)

Distribution

x	$P(x)$
0	0.0000
1	0.0001
2	0.0012
3	0.0087
4	0.0389
5	0.1168
6	0.2336
7	0.3003
8	0.2253
9	0.0751

$x \cdot P(x)$	$(x - \mu_x)^2 \cdot P(x)$
0.0000	0.0002
0.0001	0.0034
0.0025	0.0279
0.0260	0.1217
0.1557	0.2944
0.5840	0.3577
1.4016	0.1314
2.1024	0.0188
1.8020	0.3520
0.6758	0.3801
Σ 6.7501	1.6876

(b) $\mu_X = 6.75$ (from first column in table above and to the right)

     $\sigma_X = \sqrt{\sigma_X^2} = \sqrt{1.6876} \approx 1.3$ (from second column in table above and to the right)

(c) $\mu_X = n \cdot p = 9 \cdot (0.75) = 6.75$ and $\sigma_X = \sqrt{n \cdot p \cdot (1 - p)} = \sqrt{(6.75) \cdot (0.25)} \approx 1.3$

(d)

The distribution is skewed left.

33. (a)

Distribution		$x \cdot P(x)$	$(x - \mu_x)^2 \cdot P(x)$
x	$P(x)$	0.0000	0.0250
0	0.0010	0.0098	0.1568
1	0.0098	0.0878	0.3952
2	0.0439	0.3516	0.4690
3	0.1172	0.8204	0.2053
4	0.2051	1.2305	0.0000
5	0.2461	1.2306	0.2049
6	0.2051	0.8204	0.4686
7	0.1172	0.3512	0.3950
8	0.0439	0.0882	0.1568
9	0.0098	0.0100	0.0250
10	0.0010	Σ 5.0005	2.5016

(b) $\mu_X = 5.0$ (from first column in table above and to the right)

$\sigma_X = \sqrt{\sigma_X^2} = \sqrt{2.5016} \approx 1.6$ (from second column in table above and to the right)

(c) $\mu_X = n \cdot p = 10 \cdot (0.5) = 5$ and $\sigma_X = \sqrt{n \cdot p \cdot (1-p)} = \sqrt{5 \cdot (0.5)} = 1.6$

(d)

The distribution is symmetric.

35. (a)   This is a binomial experiment because it satisfies each of the four requirements:
1) There are a fixed number of trials ($n = 15$).
2) The trials are all independent (randomly selected).
3) For each trial, there are only two possible outcomes ('on time' and 'not on time').
4) The probability of "success" (i.e. on time) is the same for all trials ($p = 0.65$).

(b)   We have $n = 15$, $p = 0.65$, and $x = 10$. In the binomial table, we go to the section for $n = 15$ and the column that contains $p = 0.65$. Within the $n = 15$ section, we look for the row $x = 10$.

$P(10) = 0.2123$

There is a 0.2123 probability that in a random sample of 15 such flights, exactly 10 will be on time.

(c)   Here we wish to find $P(X \geq 10)$. Using the complement rule we can write:

$P(X \geq 10) = 1 - P(X < 10) = 1 - P(X \leq 9)$

Now using the cumulative binomial table, we go to the section for $n = 15$ and the column that contains $p = 0.65$. Within the $n = 15$ section, we look for the row $x = 9$.

$P(X \leq 9) = 0.4357$

Therefore, $P(X \geq 10) = 1 - 0.4357 = 0.5643$.

In a random sample of 15 such flights, there is a 0.5643 probability that at least 10 flights will be on time.

(d) $P(X < 10) = P(X \leq 9) = 0.4357$ [from part (c)].

In a random sample of 15 such flights, there is a 0.4357 probability that less than 10 flights will be on time.

(e)   Using the binomial probability table we get:

$P(7 \leq X \leq 10) = P(7) + P(8) + P(9) + P(10) = 0.0710 + 0.1319 + 0.1906 + 0.2123 = 0.6058$

Using the cumulative binomial probability table we get:

$P(7 \leq X \leq 10) = P(X \leq 10) - P(X \leq 6) = 0.6481 - 0.0422 = 0.6059$

In a random sample of 15 such flights, there is a 0.6059 probability that between 7 and 10 flights, inclusive, will be on time.

37. (a)   We have $n = 20$, $p = 0.05$, and $x = 3$. In the binomial table, we go to the section for $n = 20$ and the column that contains $p = 0.05$. Within the $n = 20$ section, we look for the row $x = 3$.

$P(3) = 0.0596$

There is a 0.0596 probability that in a random sample of 20 Clarinex-D users, exactly 3 will have experienced insomnia as a side effect.

(b)   Here we wish to find $P(X \leq 3)$. Using the cumulative binomial table, we go to the section for $n = 20$ and the column that contains $p = 0.05$. Within the $n = 20$ section, we look for the row $x = 3$.

$P(X \leq 3) = 0.9841$

There is a 0.9841 probability that in a random sample of 20 Clarinex-D users, 3 or fewer will have experienced insomnia as a side effect.

(c)   Using the binomial probability table we get:

$P(1 \leq X \leq 4) = P(1) + P(2) + P(3) + P(4)$
$= 0.3774 + 0.1887 + 0.0596 + 0.0133 = 0.6390$

Using the cumulative binomial probability table we get:

$P(1 \leq X \leq 4) = P(X \leq 4) - P(X \leq 0) = 0.9974 - 0.3585 = 0.6389$

There is a 0.6389 probability that in a random sample of 20 Clarinex-D users, between 1 and 4, inclusive, will have experienced insomnia as a side effect.

(d)   Here we are considering $P(X \geq 4)$. Using the cumulative binomial table, we get:

$P(X \geq 4) = 1 - P(X < 4) = 1 - P(X \leq 3) = 1 - 0.9841 = 0.0159$

It would be considered unusual for 4 or more Clarinex-D users in a random sample of 20 to experience insomnia as a side effect since the probability is less than 0.05.

39. (a)   We have $n = 25$, $p = 0.61$, and $x = 20$.

$P(20) = {}_{25}C_{20} \cdot (0.61)^{20} (0.39)^{5} \approx 0.0244$

There is a 0.0244 probability that in a random sample of 25 murders, exactly 20 were cleared.

(b)   The numbers between 16 and 18, inclusive, are 16, 17, and 18. Therefore,

$P(16 \leq X \leq 18) = P(16, 17, \text{ or } 18) = P(16) + P(17) + P(18)$
$= {}_{25}C_{16} \cdot (0.61)^{16} (0.39)^{9} + {}_{25}C_{17} \cdot (0.61)^{17} (0.39)^{8} + {}_{25}C_{18} \cdot (0.61)^{18} (0.39)^{7}$
$\approx 0.156719 + 0.129772 + 0.090212$
≈ 0.3767

There is a 0.3767 probability that in a random sample of 25 murders, between 16 and 18, inclusive, were cleared.

(c) The numbers that are fewer than 10 are 9, 8, 7, 6, 5, 4, 3, 2, 1, and 0. Therefore,

$$P(X < 10) = P(X \le 9)$$
$$= P(0) + P(1) + P(2) + ... + P(9)$$
$$= {}_{25}C_0 \cdot (0.61)^0 (0.39)^{25} + {}_{25}C_1 \cdot (0.61)^1 (0.39)^{24} + ... + {}_{25}C_9 \cdot (0.61)^9 (0.39)^{16}$$
$$\approx 0.000000 + 0.000000 + ... + 0.002316 + 0.006843$$
$$\approx 0.0100$$

It would be considered unusual for fewer than 10 murders in a random sample of 25 murders to be cleared since the probability of this happening is less than 0.05.

41. (a) We have $n = 10$, $p = 0.75$, and $x = 6$. Using the binomial probability table, we get:

$$P(6) = 0.1460$$

There is a 0.1460 probability that in a random sample of 10 adult Americans, exactly 6 would be satisfied with the job the nation's airlines are doing.

(b) Using the cumulative binomial probability table, we get:

$$P(X < 7) = P(X \le 6) = 0.2241$$

There is a 0.2241 probability that in a random sample of 10 adult Americans, fewer than 7 would be satisfied with the job the nation's airlines are doing.

(c) Using the complement rule and the cumulative binomial probability table, we get:

$$P(X \ge 5) = 1 - P(X < 5) = 1 - P(X \le 4) = 1 - 0.0197 = 0.9803$$

There is a 0.9803 probability that in a random sample of 10 adult Americans, 5 or more would be satisfied with the job the nation's airlines are doing.

(d) Using the binomial probability table, we get:

$$P(5 \le X \le 8) = P(5) + P(6) + P(7) + P(8)$$
$$= 0.0584 + 0.1460 + 0.2503 + 0.2816$$
$$= 0.7363$$

Using the cumulative binomial probability table, we get:

$$P(5 \le X \le 8) = P(X \le 8) - P(X \le 4) = 0.7560 - 0.0197 = 0.7363$$

There is a 0.7363 probability that in a random sample of 10 adult Americans, between 5 and 8, inclusive, would be satisfied with the job the nation's airlines are doing.

(e) Using the cumulative binomial probability table, we get:

$$P(X \le 3) = 0.0035$$

If ten adult Americans are randomly selected, it would be unusual for 3 or fewer to be satisfied with the job the nation's major airlines are doing since the probability is less than 0.05. If this happened, we might conclude that airline satisfaction has decreased since 2005.

43. (a) $\hat{p} = \dfrac{17}{20} = 0.85$; based on the sample, we would estimate that 85% of community college males live at home.

(b) Using $n = 20$, $p = 0.53$, we obtain:

$$P(X \ge 17) = P(17) + P(18) + P(19) + P(20)$$
$$= {}_{20}C_{17}(0.53)^{17}(0.47)^3 + {}_{20}C_{18}(0.53)^{18}(0.47)^2 + {}_{20}C_{19}(0.53)^{19}(0.47)^1 + {}_{20}C_{20}(0.53)^{20}(0.47)^0$$
$$\approx 0.002432 + 0.000457 + 0.000054 + 0.000003$$
$$\approx 0.0029$$

There is a 0.0029 probability that in a random sample of 20 male community college students, at least 17 live at home.

(c)   Since this probability is so low, we might conclude that a larger percentage of community college students live at home. That is, community college enrollment appears to be a factor that increases the likelihood of living at home.

45. (a)   We have $n = 100$ and $p = 0.65$.

$\mu_X = n \cdot p = 100(0.65) = 65$ flights; $\sigma_X = \sqrt{np(1-p)} = \sqrt{65(0.35)} = \sqrt{22.75} \approx 4.8$ flights

(b)   We expect that, in a random sample of 100 flights from Orlando to Los Angeles, 65 will be on time.

(c) $P(75) = {}_{100}C_{75} \cdot (0.65)^{75} (0.35)^{25} \approx 0.0090 < 0.05$

Since the probability is less than 0.05, it would be unusual if 75 flights in a random sample of 100 flights were on time.

Alternatively, since $np(1-p) = 22.75 > 10$ the distribution is approximately bell shaped and we can apply the Empirical Rule.

$\mu_X + 2\sigma_X \approx 65 + 2(4.8) = 74.6$

Since 75 is more than 2 standard deviations away from the mean, we would conclude that it would be unusual to observe 75 on-time flights in a sample of 100 flights.

47. (a)   We have $n = 240$ and $p = 0.05$.

$E(X) = \mu_X = n \cdot p = 240(0.05) = 12$ patients with insomnia

We would expect 12 patients from a sample of 240 to experience insomnia as a side effect.

(b)   Since $np(1-p) = 11.4 > 10$, we can use the Empirical Rule to check for unusual observations.

$\sigma_X = \sqrt{np(1-p)} = \sqrt{12(0.95)} \approx 3.4$

20 is above the mean, and we have $\mu_X + 2\sigma_X = 12 + 2(3.4) = 18.8$.

This indicates that 20 is more than two standard deviations above the mean.
It would be unusual to observe 20 patients from a sample of 240 experience insomnia as a side effect because $P(20) < 0.05$. Note: $P(20) = {}_{240}C_{20} \cdot (0.05)^{20} (0.95)^{220} \approx 0.0088$

49. (a)   We have $n = 250$ and $p = 0.61$.

$\mu_X = n \cdot p = 250(0.61) = 152.5$ murder clearances

$\sigma_X = \sqrt{np(1-p)} = \sqrt{250(0.61)(1-0.61)} = \sqrt{59.475} \approx 7.7$ murder clearances

(b)   For a random sample of 250 murders, we would expect 152.5 of them to be cleared by arrest or exceptional means, on average.

(c)   We find the interval of "usual" values by finding the value that is two standard deviations below the mean and the value that is two standard deviations above the mean.

$\mu_X - 2\sigma_X \approx 152.5 - 2(7.7) = 137.1 \approx 137$; $\mu_X + 2\sigma_X \approx 152.5 + 2(7.7) = 167.9 \approx 168$

For a random sample of 250 murders, it would be considered "usual" to clear between 137 and 168.

(d)   Since 170 is above the interval of "usual" values found in part (c), it would be considered unusual to clear more than 170 murders in a random sample of 250 murders.

51.   We have $n = 1000$ and $p = 0.536$.

$\mu_X = n \cdot p = 1000(0.536) = 536$; $\sigma_X = \sqrt{np(1-p)} = \sqrt{1000(0.536)(1-0.536)} = \sqrt{248.704} \approx 15.8$

Since $600 > \mu_X + 2\sigma_X = 536 + 2(15.8) = 567.6$, it would be considered unusual to find 600 searches using Google in a random sample of 1000 U.S. internet searches.

53. We would expect $500,000(0.56) = 280,000$ of the stops to be pedestrians who are nonwhite. Because $500,000(0.56)(1-0.56) = 123,200 > 10$, we can use the Empirical Rule to identify cutoff points for unusual results. The standard deviation number of stops is $\sigma_x = \sqrt{500,000(0.56)(1-0.56)} \approx 351.0$. If the number of stops of nonwhites exceeds $\mu_X + 2\sigma_X = 280,000 + 2(351) = 280,702$, we would say the result is unusual. The actual number of stops is $500,000(0.89) = 445,000$, which is definitely unusual. A potential criticism of this analysis is the use of 0.44 as the proportion of whites, since the actual proportion of whites may be different due to individuals commuting back and forth to the city. Additionally, there could be confounding due to the part of the city where stops are made. If the distribution of nonwhites is not uniform across the city, then location should be taken into account. See Simpson's Paradox in chapter 4.

55. (a), (b), (d), (h) Answers will vary.

    (c) We have $n = 30$, $p = 0.98$, and $x = 29$.
$$P(29) = {}_{30}C_{29} \cdot (0.98)^{29} (0.02)^1 \approx 0.3340$$

    (e) At most 27 means we wish to find $P(X \le 27)$. We can find this using the complement rule and the binomial probability formula.
$$P(X \le 27) = 1 - P(X > 27) = 1 - P(X \ge 28)$$
Now,
$$P(X \ge 28) = P(28) + P(29) + P(30)$$
$$= {}_{30}C_{28} \cdot (0.98)^{28}(0.02)^2 + {}_{30}C_{29} \cdot (0.98)^{29}(0.02)^1 + {}_{30}C_{30} \cdot (0.98)^{30}(0.02)^0$$
$$\approx 0.09883 + 0.33397 + 0.54548$$
$$\approx 0.9783$$
Therefore, $P(X \le 27) = 1 - 0.9783 = 0.0217$.

    (f) $E(X) = \mu_X = n \cdot p = 30(0.98) = 29.4$ males ; simulated means will vary.

    (g) $\sigma_X = \sqrt{np(1-p)} = \sqrt{30(0.98)(0.02)} \approx 0.8$; simulated standard deviations will vary.

57. (a) Since $E(X) = np$ and we want $E(X)$ to be at least 10, we solve
$$np = 10$$
$$n(0.27) = 10$$
$$n = \frac{10}{0.27} \approx 37.04 \rightarrow 38$$
You would need to randomly select at least 38 residents (Since we cannot sample a fractional number, we must sample 38 residents 25 years or older. If we had rounded down to 37, the expected number would have been less than 10, which is not what we required.)

    (b) We would like the probability, $P(X \ge 10)$, to be 0.99 or greater, with $p = 0.27$ and n to be determined. $P(X \ge 10) = 1 - P(X \le 9) = 1 - (P(0) + ... + P(9))$ and using technology we find that:
when $n = 64$, $P(X \ge 10) = 1 - P(X \le 9) = 1 - 0.0107 = 0.9893$;
when $n = 65$, $P(X \ge 10) = 1 - P(X \le 9) = 1 - 0.0089 = 0.9911$.
Thus the minimum number we would need in our sample is 65 residents 25 years and older.

59. Since there are 38 slots on the roulette wheel (18 red, 18 black, and 2 green), the probability of getting a green slot is $p = \dfrac{2}{38} = \dfrac{1}{19}$.

(a) We have $x = 1$ and $r = 1$. $P(1) = {}_0C_0 \cdot \left(\dfrac{1}{19}\right)^1 \left(\dfrac{18}{19}\right)^0 = \dfrac{1}{19} \approx 0.0526$

(b) We have $x = 20$ and $r = 2$. $P(20) = {}_{19}C_1 \cdot \left(\dfrac{1}{19}\right)^2 \left(\dfrac{18}{19}\right)^{18} \approx 0.0199$

(c) We have $x = 30$ and $r = 3$. $P(30) = {}_{29}C_2 \cdot \left(\dfrac{1}{19}\right)^3 \left(\dfrac{18}{19}\right)^{27} \approx 0.0137$

(d) The expected number of trials before observing $r = 3$ green is $E(X) = \dfrac{r}{p} = \dfrac{3}{1/19} = 3 \cdot 19 = 57$.

6.3 The Poisson Probability Distribution

1. A random variable X, measuring the number of successes in a fixed interval, follows a Poisson process provided that the following conditions are met:
   (1) the probability of two or more successes in any sufficiently small subinterval is 0;
   (2) the probability of success is the same for any two intervals of equal length;
   (3) the number of successes in any interval is independent of the number of successes in any other non-overlapping (disjoint) interval.

3. $\lambda = 10$ hits per minute, $t = 5$ minutes (from 7:30 P.M. to 7:35 P.M.).

5. $\lambda = 0.07$ flaws per linear foot, $t = 20$ linear feet of timber.

7. Use $P(x) = \dfrac{\mu^x}{x!} \cdot e^{-\mu}$ to find the following, assuming $\mu = 5$:

(a) $P(6) = \dfrac{5^6}{6!} \cdot e^{-5} = 0.1462$

(b) $P(X < 6) = P(X \le 5) = P(0) + P(1) + P(2) + P(3) + P(4) + P(5)$
$$= \dfrac{5^0}{0!} \cdot e^{-5} + \dfrac{5^1}{1!} \cdot e^{-5} + \dfrac{5^2}{2!} \cdot e^{-5} + \dfrac{5^3}{3!} \cdot e^{-5} + \dfrac{5^4}{4!} \cdot e^{-5} + \dfrac{5^5}{5!} \cdot e^{-5} = 0.6160$$

(c) $P(X \ge 6) = 1 - P(X < 6) = 1 - 0.6160 = 0.3840$

(d) $P(2 \le X \le 4) = P(2) + P(3) + P(4) = \dfrac{5^2}{2!} \cdot e^{-5} + \dfrac{5^3}{3!} \cdot e^{-5} + \dfrac{5^4}{4!} \cdot e^{-5} = 0.4001$

9. Use $P(x) = \dfrac{(\lambda t)^x}{x!} \cdot e^{-\lambda t}$ to find the following, given $\lambda = 0.07$, $t = 10$, and $\lambda t = 0.7$:

(a) $P(4) = \dfrac{0.7^4}{4!} \cdot e^{-0.7} = 0.0050$

(b) $P(X < 4) = P(X \leq 3) = P(X = 0) + P(1) + P(2) + P(3)$

$$= \frac{0.7^0}{0!} \cdot e^{-0.7} + \frac{0.7^1}{1!} \cdot e^{-0.7} + \frac{0.7^2}{2!} \cdot e^{-0.7} + \frac{0.7^3}{3!} \cdot e^{-0.7} = 0.9942$$

(c) $P(X \geq 4) = 1 - P(X < 4) = 1 - 0.9942 = 0.0058$

(d) $P(4 \leq X \leq 6) = P(4) + P(5) + P(6) = \frac{0.7^4}{4!} \cdot e^{-0.7} + \frac{0.7^5}{5!} \cdot e^{-0.7} + \frac{0.7^6}{6!} \cdot e^{-0.7} = 0.0057$

(e) $\mu_X = \lambda t = (0.07)(10) = 0.7$ and $\sigma_X = \sqrt{\lambda t} = \sqrt{0.7} \approx 0.8$

11. We are given that the hits occur at a rate of 1.4 per minute, so $\lambda = 1.4$; and the interval of time we are interested in is 5 minutes (from 7:30 P.M. to 7:35 P.M.), so $t = 5$. Therefore, $\lambda t = (1.4)(5) = 7$.

(a) $P(7) = \frac{7^7}{7!}e^{-7} \approx 0.1490$

On about 15 out of every 100 days, exactly 7 hits will occur between 7:30 P.M. and 7:35 P.M.

(b) $P(X < 7) = P(X \leq 6) = P(0) + ... + P(6)$

$$= \frac{7^0}{0!}e^{-7} + \frac{7^1}{1!}e^{-7} + \frac{7^2}{2!}e^{-7} + \frac{7^3}{3!}e^{-7} + \frac{7^4}{4!}e^{-7} + \frac{7^5}{5!}e^{-7} + \frac{7^6}{6!}e^{-7}$$

$\approx 0.00091 + 0.00638 + 0.02234 + 0.05213 + 0.09123 + 0.12772 + 0.14900$

≈ 0.4497

On about 45 out of every 100 days, fewer than 7 hits will occur between 7:30 P.M. and 7:35 P.M.

(c) Use the Complement Rule: $P(X \geq 7) = 1 - P(X < 7) = 1 - 0.4497 = 0.5503$.

On about 55 out of every 100 days, at least 7 hits will occur between 7:30 P.M. and 7:35 P.M.

13. We are given that FDAL level for insect filth in peanut butter is 0.3 insect fragments per gram, so $\lambda = 0.3$; and the interval we are interested in is 5 grams (the 5-gram sample), so $t = 5$. Therefore, $\lambda t = (0.3)(5) = 1.5$.

(a) $P(2) = \frac{(1.5)^2}{2!}e^{-1.5} \approx 0.2510$

In about 25 out of every hundred 5-gram samples of this peanut butter, exactly 2 insect fragments will be found.

(b) $P(X < 2) = P(X \leq 1) = P(0) + P(1) \approx 0.22313 + 0.33470 \approx 0.5578$

In about 56 out of every hundred 5-gram samples of this peanut butter, fewer than 2 insect fragments will be found.

(c) Use the Complement Rule: $P(X \geq 2) = 1 - P(X < 2) = 1 - 0.5578 = 0.4422$.

In about 44 out of every hundred 5-gram samples of this peanut butter, at least 2 insect fragments will be found.

(d) Use the Complement Rule: $P(X \geq 1) = 1 - P(X < 1) = 1 - P(0) = 1 - 0.2231 = 0.7769$

In about 78 out of every hundred 5-gram samples of this peanut butter, at least 1 insect fragment will be found.

(e)
$$P(X \geq 4) = 1 - P(X < 4) = 1 - P(X \leq 3)$$
$$= 1 - \left[P(0) + P(1) + P(2) + P(3) \right]$$
$$\approx 1 - \left[0.22313 + 0.33470 + 0.25102 + 0.12551 \right]$$
$$= 1 - 0.93436 \approx 0.0656$$

Since $P(X \geq 4) > 0.05$, it would not be unusual for a 5-gram sample of peanut butter to contain four or more insect fragments. About 7 out of every hundred 5-gram samples of this peanut butter will contain at least 4 fragments.

15. (a) We are given that fatalities occur at the rate of 0.03 per 100 million miles. If we think in terms of 100 million mile units, we have $\lambda = 0.03$. The interval we are interested in is 100 million miles, so $t = 1$ (that is, 1 one hundred million mile unit). Therefore, $\lambda t = (0.03)(1) = 0.03$.

$$P(0) = \frac{(0.03)^0}{0!} e^{-0.03} = e^{-0.03} \approx 0.9704$$

In about 97 out of every 100 one hundred million-mile samples, no deaths will occur.

(b) $P(X \geq 1) = 1 - P(X < 1) = 1 - P(0) = 1 - 0.9704 = 0.0296$

In about 3 out of every 100 one hundred million-mile samples, at least one death will occur.

(c) $P(X > 1) = 1 - P(X \leq 1) = 1 - \left[P(0) + P(1) \right]$

$$= 1 - \left[\frac{(0.03)^0}{0!} e^{-0.03} + \frac{(0.03)^1}{1!} e^{-0.03} \right] \approx 1 - 0.9996 = 0.0004$$

In about 4 out of every 10,000 one hundred million-mile samples, more than one death will occur.

17. We are given that major hurricanes strike Florida at a rate of 29 per 108 year interval. That is,
$$\lambda = \frac{29}{108} \approx 0.2685 \text{ major hurricanes per year. The time interval we are interested in is one year, so } t = 1.$$

(a) $P(X = 0) = \dfrac{\left(\frac{29}{108}\right)^0}{0!} e^{-\frac{29}{108}} \approx 0.7645$

(b) $P(X \geq 1) = 1 - P(X = 0) = 1 - 0.7645 = 0.2355$

Since this probability is greater than 0.05 it would not be unusual for Florida to be hit by at least one major hurricane in a single year.

(c) $P(X \geq 3) = 1 - P(X < 3) = 1 - P(X \leq 2)$
$$= 1 - \left[P(0) + P(1) + P(2) \right]$$
$$= 1 - \left[\frac{\left(\frac{29}{108}\right)^0}{0!} e^{-\frac{29}{108}} + \frac{\left(\frac{29}{108}\right)^1}{1!} e^{-\frac{29}{108}} + \frac{\left(\frac{29}{108}\right)^2}{2!} e^{-\frac{29}{108}} \right]$$
$$\approx 1 - 0.9974$$
$$= 0.0026$$

Since this probability is less than 0.05, it would be unusual for Florida to be hit by at least 3 major hurricanes in a single year.

(d) $P(X \geq 2) = 1 - P(X < 2) = 1 - P(X \leq 1)$

$= 1 - \left[P(0) + P(1) \right]$

$= 1 - \left[\dfrac{\left(\frac{29}{108}\right)^0}{0!} e^{-\frac{29}{108}} + \dfrac{\left(\frac{29}{108}\right)^1}{1!} e^{-\frac{29}{108}} \right]$

$\approx 1 - 0.9698$

$= 0.0302$

Since this probability is less than 0.05, it would be unusual for Florida to be hit by at least 2 major hurricanes in a single year. Thus, the 2005 hurricane season in Florida was unusual.

19. (a) and **(b)**

Use technology and the given parameters $\lambda = 0.2$ and $t = 30$ to construct a table:

x	$P(x) = \dfrac{(\lambda t)^x}{x!} \cdot e^{-\lambda t}$	Expected # of restaurants having x arrivals $= 200 \cdot P(x)$	Observed number of restaurants having x arrivals
0	0.0025	0.50	0
1	0.0149	2.97	4
2	0.0446	8.92	5
3	0.0892	17.85	13
4	0.1339	26.77	23
5	0.1606	32.12	25
6	0.1606	32.12	28
7	0.1377	27.54	25
8	0.1033	20.65	27
9	0.0688	13.77	21
10	0.0413	8.26	15
11	0.0225	4.51	5
12	0.0113	2.25	3
13	0.0052	1.04	2
14	0.0022	0.45	2
15	0.0009	0.18	0
16	0.0003	0.07	2

(c) The observed frequencies are lower for fewer cars (x small) and higher for more cars (x large), so the advertising campaign appears to be effective.

21. (a) and (c)

Use technology to construct a table:

x	Observed Frequency	Proportion of years = Frequency ÷ 200	$x \cdot$ proportion of years	$P(x) = \dfrac{(0.61)^x}{x!} \cdot e^{-0.61}$
0	109	0.545	0.000	0.5434
1	65	0.325	0.325	0.3314
2	22	0.110	0.220	0.1011
3	3	0.015	0.045	0.0206
4	1	0.005	0.020	0.0031
		Total =	0.610	

(b) From the data in the table, the mean number of deaths per year is $\mu_X = 0.610$.

(c) Using $\mu = 0.61$ we calculate the Poisson probabilities as shown in the last column.

(d) Answers will vary. These probabilities are very close to the relative frequencies, so the random variable X appears to follow a Poisson distribution.

23. (a) $E(X) = \lambda t = \dfrac{23.6}{100} \cdot (500) = 118$ colds.

(b) – (g) Answers will vary.

6.4 The Hypergeometric Probability Distribution (On CD)

1. In both cases, the random variable X represents the number of successes in n trials of an experiment, where each trial results in one of two mutually exclusive (disjoint) outcomes classified as "success" or "failure". The binomial probability distribution requires that each trial be independent of the others and that the probability of success remain the same for each trial. This is not the case for the hypergeometric probability distribution. For the hypergeometric probability distribution, the population size is finite and there are a fixed number of successes in the population. The sample is obtained without replacement, so the trials are not independent and the probability of success does not remain constant. If the population size is infinite, there is no difference between the hypergeometric distribution and the binomial distribution.

3. n is the sample size and there cannot be more successes than trials. Therefore, we must have $x \le n$. In turn, the population only contains k successes. Therefore, we must also have $x \le k$. If the sample size n is larger than the number of successes in the population (that is, $n > k$), we are still restricted by the fact that there are only k successes in the population. Therefore, the possible values of the random variable must be less than or equal to the smaller of the sample size (n) and the total number of successes in the population (k).

5. This is a hypergeometric probability experiment because:
1) The population is finite ($N = 49$ numbers).
2) There are only two possible outcomes (number matches or does not match).
3) The sample size is $n = 6$ and is drawn without replacement.
There are only 6 matching numbers so $k = 6$. The possible values for X are $x = 0, 1, 2, ..., 6$.

7. This is a hypergeometric probability experiment because:
   1) The population is finite ($N = 250$ numbers).
   2) There are only two possible outcomes (chip is defective or chip is not defective).
   3) The sample size is $n = 20$ and is drawn without replacement.
   There are only 12 chips that are defective so $k = 12$. The possible values for X are $x = 0, 1, 2, ..., 12$.

9. Given $N = 150$, $n = 20$, $k = 30$, and $x = 5$:

$$P(5) = \frac{\left(_{30}C_5\right)\left(_{150-30}C_{20-5}\right)}{_{150}C_{20}} = \frac{\left(_{30}C_5\right)\left(_{120}C_{15}\right)}{_{150}C_{20}} = 0.1856$$

11. Given $N = 230$, $n = 15$, $k = 200$, and $x = 12$:

$$P(12) = \frac{\left(_{200}C_{12}\right)\left(_{230-200}C_{15-12}\right)}{_{230}C_{15}} = \frac{\left(_{200}C_{12}\right)\left(_{30}C_3\right)}{_{230}C_{15}} = 0.1939$$

13. Given $N = 150$, $n = 20$, and $k = 30$:

$$\mu_X = n \cdot \frac{k}{N} = 20 \cdot \frac{30}{150} = 4 \; ;$$

$$\sigma_X = \sqrt{\left(\frac{N-n}{N-1}\right) \cdot n \cdot \frac{k}{N} \cdot \frac{N-k}{N}} = \sqrt{\left(\frac{150-20}{150-1}\right) \cdot 20 \cdot \frac{30}{150} \cdot \frac{150-30}{150}} \approx 1.67$$

15. Given $N = 230$, $n = 15$, and $k = 200$:

$$\mu_X = n \cdot \frac{k}{N} = 15 \cdot \frac{200}{230} \approx 13.0 \; ;$$

$$\sigma_X = \sqrt{\left(\frac{N-n}{N-1}\right) \cdot n \cdot \frac{k}{N} \cdot \frac{N-k}{N}} = \sqrt{\left(\frac{230-15}{230-1}\right) \cdot 15 \cdot \frac{200}{230} \cdot \frac{230-200}{230}} \approx 1.26$$

17. We are given $N = 47$, $n = 6$, and $k = 6$:

   (a)
   $$P(3) = \frac{\left(_6C_3\right)\left(_{47-6}C_{6-3}\right)}{_{47}C_6} = \frac{\left(_6C_3\right)\left(_{41}C_3\right)}{_{47}C_6} = 0.01986$$
   There is a 0.01986 probability of matching 3 numbers.

   (b)
   $$P(4) = \frac{\left(_6C_4\right)\left(_{47-6}C_{6-4}\right)}{_{47}C_6} = \frac{\left(_6C_4\right)\left(_{41}C_2\right)}{_{47}C_6} = 0.001146$$
   There is a 0.001146 probability of matching 4 numbers.

   (c)
   $$P(5) = \frac{\left(_6C_5\right)\left(_{47-6}C_{6-5}\right)}{_{47}C_6} = \frac{\left(_6C_5\right)\left(_{41}C_1\right)}{_{47}C_6} = 0.0000229$$
   There is a 0.0000229 probability of matching 5 numbers.

   (d)
   $$P(6) = \frac{\left(_6C_6\right)\left(_{47-6}C_{6-6}\right)}{_{47}C_6} = \frac{\left(_6C_6\right)\left(_{41}C_0\right)}{_{47}C_6} = 0.000000093$$
   There is a 0.000000093 probability of matching 6 numbers.

(e) $P(\text{win}) = P(3) + P(4) + P(5) + P(6)$
$$= 0.01986 + 0.001146 + 0.0000229 + 0.000000093$$
$$= 0.02103$$

There is a 0.02103 probability of purchasing a winning ticket. This probability is less than 0.05, so it would be considered unusual to purchase a winning ticket.

(f) $\mu_X = n \cdot \dfrac{k}{N} = 6 \cdot \dfrac{6}{47} \approx 0.766$;

$$\sigma_X = \sqrt{\left(\dfrac{N-n}{N-1}\right) \cdot n \cdot \dfrac{k}{N} \cdot \dfrac{N-k}{N}} = \sqrt{\left(\dfrac{47-6}{47-1}\right) \cdot 6 \cdot \dfrac{6}{47} \cdot \dfrac{47-6}{47}} \approx 0.772$$

$E(X) = \mu_x \approx 0.766$. For a randomly selected ticket, we expect to match 0.766 numbers.

19. We are given $N = 250$, $n = 20$, and $k = 12$:

(a) $P(4) = \dfrac{\left(_{12}C_4\right)\left(_{250-12}C_{20-4}\right)}{_{250}C_{20}} = \dfrac{\left(_{12}C_4\right)\left(_{238}C_{16}\right)}{_{250}C_{20}} = 0.0087$

There is a 0.0087 probability of obtaining 4 defective chips.

(b) $P(3) = \dfrac{\left(_{12}C_3\right)\left(_{250-12}C_{20-3}\right)}{_{250}C_{20}} = \dfrac{\left(_{12}C_3\right)\left(_{238}C_{17}\right)}{_{250}C_{20}} = 0.0507$

There is a 0.0507 probability of obtaining 3 defective chips.

(c) $P(0) = \dfrac{\left(_{12}C_0\right)\left(_{250-12}C_{20-0}\right)}{_{250}C_{20}} = \dfrac{\left(_{12}C_0\right)\left(_{238}C_{20}\right)}{_{250}C_{20}} = 0.3590$

There is a 0.3590 probability of obtaining no defective chips.

(d) $P(14) = 0$ because the random variable cannot take on the value 14. The minimum of n and k is 12. Therefore, the possible values for the random variable are $x = 0, 1, 2, \ldots, 12$.

(e) $E(X) = \mu_X = n \cdot \dfrac{k}{N} = 20 \cdot \dfrac{12}{250} = 0.96$

In a random sample of 20 chips, we expect to obtain 0.96 defective chips on average.

21. Let the random variable X represent the number of jurors on the jury who will never convict. From the problem statement, we get $N = 30$, $n = 12$, and $k = 2$.
Here it will take less work if we find the probability of avoiding a hung jury and subtracting from 1. To avoid a hung jury, we need $X = 0$.

$$P(0) = \dfrac{\left(_2C_0\right)\left(_{30-2}C_{12-0}\right)}{_{30}C_{12}} = \dfrac{\left(_2C_0\right)\left(_{28}C_{12}\right)}{_{30}C_{12}} = 0.3517$$

Since, $1 - 0.3517 = 0.6483$, there is a 0.6483 probability of having a hung jury.(Note: Here we are assuming that there will never be a hung jury unless one of the two potential jurors who will never convict is on the jury.)

23. Let the random variable X represent the number of cylinders that pass inspection. From the problem statement we have $N = 200$ and $n = 4$. To accept the shipment, we must obtain 0 defective cylinders. That is, we want $X = 0$.

(a) If 10% are defective, we have $k = (0.1)(200) = 20$.

$$P(0) = \frac{\left({}_{20}C_0\right)\left({}_{200-20}C_{4-0}\right)}{{}_{200}C_4} = \frac{\left({}_{20}C_0\right)\left({}_{180}C_4\right)}{{}_{200}C_4} = 0.6539$$

If 10% are defective, there is a 0.6539 probability that the shipment is accepted.

(b) If 20% are defective, we have $k = (0.2)(200) = 40$.

$$P(0) = \frac{\left({}_{40}C_0\right)\left({}_{200-40}C_{4-0}\right)}{{}_{200}C_4} = \frac{\left({}_{40}C_0\right)\left({}_{160}C_4\right)}{{}_{200}C_4} = 0.4065$$

If 20% are defective, there is a 0.4065 probability that the shipment is accepted.

(c) If 40% are defective, we have $k = (0.4)(200) = 80$.

$$P(0) = \frac{\left({}_{80}C_0\right)\left({}_{200-80}C_{4-0}\right)}{{}_{200}C_4} = \frac{\left({}_{80}C_0\right)\left({}_{120}C_4\right)}{{}_{200}C_4} = 0.1270$$

If 40% are defective, there is a 0.1270 probability that the shipment is accepted.

(d) If 60% are defective, we have $k = (0.6)(200) = 120$.

$$P(0) = \frac{\left({}_{120}C_0\right)\left({}_{200-120}C_{4-0}\right)}{{}_{200}C_4} = \frac{\left({}_{120}C_0\right)\left({}_{80}C_4\right)}{{}_{200}C_4} = 0.0245$$

If 60% are defective, there is a 0.0245 probability that the shipment is accepted.

(e) If 80% are defective, we have $k = (0.8)(200) = 160$.

$$P(0) = \frac{\left({}_{160}C_0\right)\left({}_{200-160}C_{4-0}\right)}{{}_{200}C_4} = \frac{\left({}_{160}C_0\right)\left({}_{40}C_4\right)}{{}_{200}C_4} = 0.0014$$

If 80% are defective, there is a 0.0014 probability that the shipment is accepted.

(f)

Chapter 6 Review Exercises

1. **(a)** The number of students in a randomly selected classroom is a discrete random variable because its value results from counting. If we let the random variable S represent the number of students, the possible values for S are all nonnegative integers (up to the capacity of the room). That is, $s = 0, 1, 2, \ldots$.

   (b) The number of inches of snow that falls in Minneapolis during the winter season is a continuous random variable because its value results from a measurement. If we let the random variable S represent the number of inches of snow, the possible values for S are all nonnegative real numbers. That is, $s \geq 0$.

   (c) The amount of flight time accumulated is a continuous random variable because its value results from a measurement. If we let the random variable H represent the accumulated flight time, the possible values for H are all nonnegative real numbers. That is, $h \geq 0$.

   (d) The number of points scored by the Miami Heat in a game is a discrete random variable because its value results from a count. If we let the random variable X represent the number of points scored by the Miami Heat in a game, the possible values for X are nonnegative integers. That is, $x = 0, 1, 2, \ldots$.

2. **(a)** This is not a valid probability distribution because the sum of the probabilities does not equal 1. ($\sum P(x) = 0.73$)

   (b) This is a valid probability distribution because $\sum P(x) = 1$ and $0 \leq P(x) \leq 1$ for all x.

3. **(a)** Total number of Stanley Cups $= 20 + 17 + 17 + 14 = 68$.

x (games played)	$P(x)$
4	$\frac{20}{68} \approx 0.2941$
5	$\frac{17}{68} = 0.2500$
6	$\frac{17}{68} = 0.2500$
7	$\frac{14}{68} \approx 0.2059$

   (b)

   Stanley Cup Playoff

   (c) $\mu_X = \sum [x \cdot P(x)] \approx 4 \cdot (0.2941) + 5 \cdot (0.25) + 6 \cdot (0.25) + 7 \cdot (0.2059)$
   ≈ 5.368 or about 5.4 games

   The Stanley Cup, if played many times, would be expected to last about 5.4 games.

   (d) $\sigma_X^2 = \sum (x - \mu_x)^2 \cdot P(x)$
   $\approx (4 - 5.368)^2 \cdot 0.2941 + (5 - 5.368)^2 \cdot 0.2500$
   $+ (6 - 5.368)^2 \cdot 0.2500 + (7 - 5.368)^2 \cdot 0.2059$
   ≈ 1.232
   $\sigma_X = \sqrt{\sigma_X^2} \approx \sqrt{1.232} \approx 1.110$ or about 1.1 games

4. Let X represent the amount that the operator makes.

You draw:	Ace	Face Card	7 of Spades	Anything Else
Operator makes, x (\$)	−5	−3	−10	2
Probability, $P(x)$	$\dfrac{1}{13}$	$\dfrac{3}{13}$	$\dfrac{1}{52}$	$\dfrac{35}{52}$

$$E(X)=(-5)\left(\frac{1}{13}\right)+(-3)\left(\frac{3}{13}\right)+(-10)\left(\frac{1}{52}\right)+(2)\left(\frac{35}{52}\right)=\$0.08$$

If many customers play the game, the operator can expect to make \$0.08 per customer.

5. (a) This is a binomial experiment. There are a fixed number of trials ($n=10$) where each trial corresponds to a randomly chosen freshman, the trials are independent, there are only two possible outcomes (graduated or did not graduate within six years), and the probability of a graduation within six years is fixed for all trials ($p=0.54$).

(b) This is not a binomial experiment because the number of trials is not fixed.

6. (a) We have $n=10$, $p=0.08$, and $x=0$.

$$P(0)={}_{10}C_0\cdot(0.08)^0(0.92)^{10}\approx0.4344$$

In about 43% of random samples of 10 females 20–34 years old, there will be 0 females with high serum cholesterol.

(b) $P(2)={}_{10}C_2\cdot(0.08)^2(0.92)^8\approx0.1478$

In about 15% of random samples of 10 females 20–34 years old, there will be 2 females with high serum cholesterol.

(c)
$$\begin{aligned}P(X\geq2)&=1-P(X<1)=1-P(X\leq1)\\&=1-\left[P(0)-P(1)\right]\\&=1-\left[0.4344+0.3777\right]\\&=1-0.8121=0.1879\end{aligned}$$

In about 19% of random samples of 10 females 20–34 years old, there will be at least 2 females with high serum cholesterol.

(d) The probability that 9 will **not** have high cholesterol is the same as the probability that 1 **will** have high cholesterol, which is $P(1)=0.3777$.

In about 38% of random samples of 10 females 20–34 years old, there will be 9 females who *do not* have high serum cholesterol.

(e) $E(X)=\mu_X=n\cdot p=250(0.08)=20$; $\sigma_X=\sqrt{np(1-p)}=\sqrt{20(0.92)}\approx4.3$

(f) No. Since $np\cdot(1-p)=18.4>10$, we can use the Empirical Rule to check for unusual observations. We have that 12 is below the mean and $\mu-2\sigma=20-2\cdot4.3=11.4$. This indicates that 12 is within two standard deviations of the mean, so observing 12 females with high serum cholesterol in a random sample of 250 females 20–34 years old would not be unusual.

7. (a) We have $n=15$ and $p=0.6$.

Using the binomial probability table, we get: $P(10)=0.1859$

(b) Using the cumulative binomial probability table, we get:
$$P(X<5)=P(X\leq4)=0.0093$$

(c) Using the complement rule, we get: $P(X \geq 5) = 1 - P(X < 5) = 1 - 0.0093 = 0.9907$

(d) Using the binomial probability table, we get:

$P(7 \leq X \leq 12) = P(7) + P(8) + P(9) + P(10) + P(11) + P(12)$
$\qquad = 0.1181 + 0.1771 + 0.2066 + 0.1859 + 0.1268 + 0.0634$
$\qquad = 0.8779$

Using the cumulative binomial probability table, we get:

$P(7 \leq X \leq 12) = P(X \leq 12) - P(X \leq 6) = 0.9729 - 0.0950 = 0.8779$

There is a 0.8779 [Tech: 0.8778] probability that in a random sample of 15 U.S. women 18 years old or older, between 7 and 12, inclusive, would feel that the minimum driving age should be 18.

(e) $E(X) = \mu_X = np = 200(0.6) = 120$ women;

$\sigma_X = \sqrt{np(1-p)} = \sqrt{120(0.4)} = \sqrt{48} \approx 6.928$ or about 6.9 women

(f) No; since $np(1-p) = 48 > 10$, we can use the Empirical Rule to check for unusual observations.

$\mu - 2\sigma \approx 120 - 13.8 = 106.2$ and $\mu + 2\sigma \approx 120 + 13.8 = 133.8$

Since 110 is within this range of values, it would not be considered an unusual observation.

8. (a)

Distribution		$x \cdot P(x)$	$(x-\mu_x)^2 \cdot P(x)$
x	$P(x)$		
0	0.00002	0.0000	0.0005
1	0.00037	0.0004	0.0092
2	0.00385	0.0077	0.0615
3	0.02307	0.0692	0.2076
4	0.08652	0.3461	0.3461
5	0.20764	1.0382	0.2076
6	0.31146	1.8688	0.0000
7	0.26697	1.8688	0.2670
8	0.10011	0.8009	0.4005
Σ		6.0000	1.5000

(b) Using the formulas from Section 6.1:

$\mu_X = 6$ (from first column in table above and to the right)

$\sigma_X = \sqrt{\sigma_X^2} = \sqrt{1.5} \approx 1.2$ (from second column in table above and to the right)

Using the formulas from Section 6.2:

$\mu_X = np = 8(0.75) = 6$ and $\sigma_X = \sqrt{np \cdot (1-p)} = \sqrt{8(0.75)(0.25)} \approx 1.2$.

(c)

The distribution is skewed left.

9. If $np \cdot (1-p) \geq 10$, then the Empirical Rule can be used to check for unusual observations.

10. In sampling from large populations without replacement, the trials may be assumed to be independent provided that the sample size is small in relation to the size of the population. As a general rule of thumb, this condition is satisfied if the sample size is less than 5% of the population size.

11. A random variable X, the number of successes in a fixed interval, follows a Poisson process provided that following conditions are met:
   1) the probability of two or more successes in any sufficiently small subinterval is 0;
   2) the probability of success is the same for any two intervals of equal length;
   3) the number of successes in any interval is independent of the number of successes in any other non-overlapping interval.

12. Use $P(x) = \dfrac{(\lambda t)^x}{x!} \cdot e^{-\lambda t}$ with $\lambda = 0.05$ and $t = 30$ so that $\lambda t = 1.5$:

   (a) $P(2) = \dfrac{1.5^2}{2!} \cdot e^{-1.5} = 0.2510$

   (b) $P(X < 2) = P(0) + P(1) = 0.22313 + 0.33470 \approx 0.5578$

   (c) $P(X \geq 2) = 1 - P(X < 2) = 1 - 0.5578 = 0.4422$

   (d) $P(1 \leq X \leq 3) = P(1) + P(2) + P(3) = 0.33470 + 0.25102 + 0.12551 \approx 0.7112$

   (e) $\mu_X = \lambda t = (0.05) \cdot 30 = 1.5$; $\sigma_X = \sqrt{\lambda t} = \sqrt{1.5} \approx 1.2$

13. We need to find $P(X \geq 2)$ where X is the number of defects in a roll. We have $\mu = \lambda \cdot t = \dfrac{1}{500} \cdot (100) = 0.2$

   $$P(X \geq 2) = 1 - P(X \leq 1) = 1 - \left[\dfrac{0.2^0}{0!} \cdot e^{-0.2} + \dfrac{0.2^1}{1!} \cdot e^{-0.2} \right] = 1 - 0.9825 = 0.0175$$

   About 1.75% of the rolls are rejected.

14. We have that maintenance calls occur at the rate of two per month, so $\lambda = 2$. The time interval of interest is one month, so $t = 1$. Therefore, $\lambda t = (2)(1) = 2$.

   $$P(X > 4) = 1 - P(X \leq 4)$$
   $$= 1 - \left[\dfrac{2^0}{0!} \cdot e^{-2} + \dfrac{2^1}{1!} \cdot e^{-2} + \dfrac{2^2}{2!} \cdot e^{-2} + \dfrac{2^3}{3!} \cdot e^{-2} + \dfrac{2^4}{4!} \cdot e^{-2} \right]$$
   $$= 1 - 0.9473 = 0.0527$$

   Since $P(X > 4) > 0.05$, it would not be unusual for the copy machine to require more than 4 maintenance calls in a given month.

15. We have $n = 40$, $p = 0.17$, $x = 12$, and find $\mu_X = n \cdot p = 40(0.17) = 6.8$. Since 12 is above the mean, we compute $P(X \geq 12)$. Using technology we find $P(X \geq 12) \approx 0.0301 < 0.05$, so the result of the survey is unusual. This suggests that emotional abuse may be a factor that increases the likelihood of self-injurious behavior.

Chapter 6 Test

1. **(a)** The number of days of measurable rainfall in Honolulu during a year is a discrete random variable because its value results from counting. If we let the random variable R represent the number of days for which there is measurable rainfall, the possible values for R are integers between 0 and 365, inclusive. That is, $r = 0, 1, 2, \ldots 365$.

   (b) The miles per gallon for a Toyota Prius is a continuous random variable because its value results from a measurement. If we let the random variable M represent the miles per gallon, the possible values for M are all nonnegative real numbers. That is, $m \geq 0$.
   (Note: we need to include the possibility that $m = 0$ since the engine could be idling, in which case it would be getting 0 miles to the gallon.)

   (c) The number of golf balls hit into the ocean on the famous 18th hole at Pebble Beach is a discrete random variable because its value results from a count. If we let the random variable X represent the number of golf balls hit into the ocean, the possible values for X are nonnegative integers. That is, $x = 0, 1, 2, \ldots$

   (d) The weight of a randomly selected robin egg is a continuous random variable because its value results from a measurement. If we let the random variable W represent the weight of a robin's egg, the possible values for W are all nonnegative real numbers. That is, $w > 0$.

2. **(a)** This is a valid probability distribution because $\sum P(x) = 1$ and $0 \leq P(x) \leq 1$ for all x.

   (b) This is not a valid probability distribution because $P(4) < 0$.

3. **(a)** Total number of Wimbledon Men's Single Finals = 18 + 10 + 12 = 40.

x (sets played)	$P(x)$
3	$\frac{18}{40} = 0.45$
4	$\frac{10}{40} = 0.25$
5	$\frac{12}{40} = 0.30$

   (b)

   Wimbledon Men's Singles Finals

   (c) $\mu_X = \sum \left[x \cdot P(x) \right] = 3 \cdot (0.45) + 4 \cdot (0.25) + 5 \cdot (0.30)$
   $= 3.85$ or about 3.9 sets

   The Wimbledon men's single finals, if played many times, would be expected to last about 3.9 sets.

   (d) $\sigma_X^2 = \sum (x - \mu_x)^2 \cdot P(x)$
   $\approx (3 - 3.85)^2 \cdot 0.45 + (4 - 3.85)^2 \cdot 0.25 + (5 - 3.85)^2 \cdot 0.30$
   $= 0.7275$
   $\sigma_X = \sqrt{\sigma_X^2} \approx \sqrt{0.7275} \approx 0.853$ or about 0.9 sets

4. $E(X) = \sum x \cdot P(x) = \$200 \cdot (0.998725) + (-\$99,800) \cdot (0.001275) = \$72.50$.
   If the company issues many \$100,000 1-year policies to 35-year-old males, then they will make an average profit of \$72.50 per male.

5. A probability experiment is said to be a binomial probability experiment provided:
   a) The experiment consists of a fixed number, n, of trials.
   b) The trials are all independent.
   c) For each trial there are two mutually exclusive (disjoint) outcomes, success or failure.
   d) The probability of success, p, is the same for each trial.

6. (a) This is not a binomial experiment because the number of trials is not fixed.

(b) This is a binomial experiment. There are a fixed number of trials ($n = 25$) where each trial corresponding to a randomly chosen property crime, the trials are independent, there are only two possible outcomes (cleared or not cleared), and the probability that the property crime is cleared is fixed for all trials ($p = 0.16$).

7. (a) We have $n = 20$ and $p = 0.20$.
Using the binomial probability table, we get: $P(7) = 0.0545$

(b) Using the cumulative binomial probability table, we get:
$$P(X < 5) = P(X \le 4) = 0.6296$$

(c) Using the complement rule and the cumulative binomial probability table, we get:
$$P(X \ge 3) = 1 - P(X < 3)$$
$$= 1 - P(X \le 2)$$
$$= 1 - 0.2061 = 0.7939$$

(d) Using the binomial probability table, we get:
$$P(5 \le X \le 10) = P(5) + P(6) + P(7) + P(8) + P(9) + P(10)$$
$$= 0.1746 + 0.1091 + 0.0545 + 0.0222 + 0.0074 + 0.0020$$
$$= 0.3698$$

Using the cumulative binomial probability table, we get:
$$P(5 \le X \le 10) = P(X \le 10) - P(X \le 4) = 0.9994 - 0.6296 = 0.3698$$
There is a 0.3698 probability that in a random sample of 20 U.S. households, between 5 and 10, inclusive, would have been tuned into the January 16, 2007, *American Idol* show.

(e) $E(X) = \mu_X = n \cdot p = 500(0.2) = 100$ households;
$$\sigma_X = \sqrt{np(1-p)} = \sqrt{500(0.2)(1-0.2)} = \sqrt{80} \approx 8.9 \text{ households}$$

(f) Yes; since $np(1-p) = 80 > 10$, we can use the Empirical Rule to check for unusual observations.
$$\mu - 2\sigma \approx 100 - 2(8.9) = 82.2 \quad \text{and} \quad \mu + 2\sigma \approx 100 + 2(8.9) = 117.8$$
Since 125 is outside this range of values (i.e. more than 2 standard deviations away from the mean), it would be considered an unusual observation.

8. (a) We have $n = 25$ and $p = 0.35$.
$$P(8) = {}_{25}C_8 \cdot (0.35)^8 (0.65)^{17} = 0.1607$$

(b) $P(X < 4) = P(X \le 3) = P(0) + P(1) + P(2) + P(3)$
$= 0.00002 + 0.00028 + 0.00183 + 0.00755 = 0.00968$
≈ 0.0097

(c) $P(X \ge 5) = 1 - P(X \le 4) = 1 - (P(X \le 3) + P(4)) = 1 - (0.00968 + 0.02236) \approx 0.9680$

(d) The probability that exactly 20 do **not** experience insomnia is the same as the probability that 5 **do**, which is $P(5) = 0.0506$.

(e) $\mu_X = np = 1000 \cdot (0.35) = 350$ users;

$\sigma_X = \sqrt{np \cdot (1-p)} = \sqrt{1000(0.35)(0.65)} \approx 15.083$ or about 15.1 users

(f) No. Since $np(1-p) = 227.5 > 10$, we can use the Empirical Rule to check for unusual observations. $\mu - 2\sigma = 319.8$ and $\mu + 2\sigma = 380.2$. Since 330 is between these values, observing 330 users of Zyban that experience insomnia in a random sample of 1000 users of Zyban would not be unusual.

9. (a)

Distribution		$x \cdot P(x)$	$(x - \mu_x)^2 \cdot P(x)$
x	$P(x)$		
0	0.3277	0	0.3277
1	0.4096	0.4096	0.0000
2	0.2048	0.4096	0.2048
3	0.0512	0.1536	0.2048
4	0.0064	0.0256	0.0576
5	0.0003	0.0016	0.0051
Σ		1.0000	0.8000

(b) Using the formulas from Section 6.1:

$\mu_X = 1$ (from first column in table above and to the right)

$\sigma_X = \sqrt{\sigma_X^2} = \sqrt{0.8} \approx 0.9$ (from second column in table above and to the right)

Using the formulas from Section 6.2:

$\mu_X = np = 5(0.2) = 1$ and $\sigma_X = \sqrt{np \cdot (1-p)} = \sqrt{5(0.2)(0.8)} \approx 0.9$.

(c)

The distribution is skewed right.

10. Use $P(x) = \dfrac{(\lambda t)^x}{x!} \cdot e^{-\lambda t}$ with $\lambda = 0.05$ and $t = 8$ so that $\lambda t = 0.4$:

(a) $P(3) = \dfrac{0.4^3}{3!} \cdot e^{-0.4} = 0.0072$

(b) $P(X < 3) = P(0) + P(1) + P(2) = 0.67032 + 0.26813 + 0.05363 \approx 0.9921$

(c) $P(X \geq 3) = 1 - P(X < 3) = 1 - 0.9921 = 0.0079$

(d) $P(3 \leq X \leq 5) = P(3) + P(4) + P(5) = 0.00715 + 0.00072 + 0.00006 \approx 0.0079$

(e) $\mu_X = \lambda t = (0.05) \cdot 8 = 0.4$; $\sigma_X = \sqrt{\lambda t} = \sqrt{0.4} \approx 0.6$

11. We are given that cars arrive at the rate of 0.41 cars per minute, so $\lambda = 0.41$. The time interval of interest is 10 minutes (4:00 P.M. and 4:10 P.M), so $t = 10$. Therefore, $\lambda t = 0.41(10) = 4.1$.

(a) $P(4) = \dfrac{4.1^4}{4!} \cdot e^{-4.1} \approx 0.1951$

On about 20 of every 100 randomly chosen days, exactly 4 cars will arrive at the bank's drive-through window between 4:00 P.M. and 4:10 P.M.

(b) $P(X < 4) = P(X \le 3) = P(0) + P(1) + P(2) + P(3)$

$\quad = \dfrac{4.1^0}{0!} \cdot e^{-4.1} + \dfrac{4.1^1}{1!} \cdot e^{-4.1} + \dfrac{4.1^2}{2!} \cdot e^{-4.1} + \dfrac{4.1^3}{3!} \cdot e^{-4.1}$

$\quad \approx 0.01657 + 0.06795 + 0.13929 + 0.19037$

$\quad \approx 0.4142$

On about 41 of every 100 randomly chosen days, fewer than 4 cars will arrive at the bank's drive-through window between 4:00 P.M. and 4:10 P.M.

(c) Using the complement rule, we get: $P(X \ge 4) = 1 - P(X < 4) = 1 - 0.4142 = 0.5858$

On about 59 of every 100 randomly chosen days, at least 4 cars will arrive at the bank's drive-through window between 4:00 P.M. and 4:10 P.M.

(d) $\mu_X = \lambda t = (0.41) \cdot 10 = 4.1$ cars

(e) $\sigma_X = \sqrt{\lambda t} = \sqrt{4.1} \approx 2.0$ cars

Chapter 7

The Normal Probability Distribution

Section 7.1

1. For the graph to be that of a probability density function, (1) the area under the graph over all possible values of the random variable must equal 1, and (2) the graph must be greater than or equal to 0 for all possible values of the random variable. That is, the graph of the equation must lie on or above the horizontal axis for all possible values of the random variable.

3. The area under the graph of a probability density function can be interpreted as either:
   (1) The proportion of the population with the characteristic described by the interval; or
   (2) The probability that a randomly selected individual from the population has the characteristic described by the interval.

5. $\mu - \sigma$; $\mu + \sigma$

7. No, the graph cannot represent a normal density function because it is not symmetric.

9. No, the graph cannot represent a normal density function because it crosses below the horizontal axis. That is, it is not always greater than or equal to 0.

11. Yes, the graph can represent a normal density function.

13. The figure presents the graph of the density function with the area we wish to find shaded.

Random Variable (time)

The width of the rectangle is $10 - 5 = 5$ and the height is $\frac{1}{30}$. Thus, the area between 5 and 10 is $5\left(\frac{1}{30}\right) = \frac{1}{6}$. The probability that the friend is between 5 and 10 minutes late is $\frac{1}{6}$.

15. The figure presents the graph of the density function with the area we wish to find shaded.

Random Variable (time)

The width of the rectangle is $30 - 20 = 10$ and the height is $\frac{1}{30}$. Thus, the area between 20 and 30 is $10\left(\frac{1}{30}\right) = \frac{1}{3}$. The probability that the friend is at least 20 minutes late is $\frac{1}{3}$.

17. (a)

   (b) $P(0 \le X \le 0.2) = 1(0.2 - 0) = 0.2$

   (c) $P(0.25 \le X \le 0.6) = 1(0.6 - 0.25) = 0.35$

   (d) $P(X \ge 0.95) = 1(1 - 0.95) = 0.05$

   (e) Answers will vary.

19. The histogram is symmetrical and bell-shaped, so a normal distribution can be used as a model for the variable.

21. The histogram is skewed to the right, so normal distribution cannot be used as a model for the variable.

23. Graph A matches $\mu = 10$ and $\sigma = 2$, and graph B matches $\mu = 10$ and $\sigma = 3$. We can tell because a higher standard deviation makes the graph lower and more spread out.

25. The center is at 2, so $\mu = 2$. The distance to the inflection points is ± 3, so $\sigma = 3$.

27. The center is at 100, so $\mu = 100$. The distance to the inflection points is ± 15, so $\sigma = 15$.

29. (a)

(b)

(c) Interpretation 1: 15.87% of the cell phone plans in the United States are less than $44.00 per month.

Interpretation 2: The probability is 0.1587 that a randomly selected cell phone plan in the United States is less than $44.00 per month.

31. (a)

(b)

(c) Interpretation 1: 2.28% of all full-term babies has a birth weight of at least 4410 grams.

Interpretation 2: The probability is 0.0228 that the birth weight of a randomly chosen full-term baby is at least 4410 grams.

33. (a) Interpretation 1: The proportion of human pregnancies that last more than 280 days is 0.1908.

Interpretation 2: The probability is 0.1908 that a randomly selected human pregnancy lasts more than 280 days.

(b) Interpretation 1: The proportion of human pregnancies that last between 230 and 260 days is 0.3416.

Interpretation 2: The probability is 0.3416 that a randomly selected human pregnancy lasts between 230 and 260 days.

35. (a) $z_1 = \dfrac{x_1 - \mu}{\sigma} = \dfrac{8 - 10}{3} = -\dfrac{2}{3} \approx -0.67$

(b) $z_2 = \dfrac{x_2 - \mu}{\sigma} = \dfrac{12 - 10}{3} = \dfrac{2}{3} \approx 0.67$

(c) The area between z_1 and z_2 is also 0.495.

37. (a), (b)

(c) The normal density function appears to describe the distance Michael hits a pitching wedge fairly accurately. Looking at the graph, the normal curve is a fairly good approximation to the histogram.

Section 7.2

1. The standard normal curve has the following properties:
(1) It is symmetric about its mean $\mu = 0$ and has standard deviation $\sigma = 1$.
(2) Its highest point occurs at $\mu = 0$.
(3) It has inflection points at -1 and 1.
(4) The area under the curve is 1.
(5) The area under the curve to the right of $\mu = 0$ equals the area under the curve to the left of $\mu = 0$. Both equal 0.5.
(6) As Z increases, the graph approaches, but never equals, zero. As Z decreases, the graph approaches, but never equals, zero.
(7) It satisfies the Empirical Rule: Approximately 68% of the area under the standard normal curve is between -1 and 1. Approximately 95% of the area under the standard normal curve is between -2 and 2. Approximately 99.7% of the area under the standard normal curve is between -3 and 3.

3. False. Although the area is very close to 1 (and for all practical purposes we may assume it is approximately 1), it is not exactly 1 because the normal curve continues indefinitely to the right.

5. The standard normal table (Table V) gives the area to the left of the z-score. Thus, we look up each z-score and read the corresponding area. We can also use technology. The areas are:

   (a) The area to the left of $z = -2.45$ is 0.0071.

   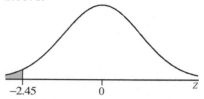

   (b) The area to the left of $z = -0.43$ is 0.3336.

   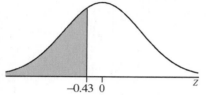

   (c) The area to the left of $z = 1.35$ is 0.9115.

   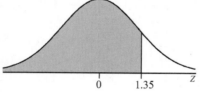

   (d) The area to the left of $z = 3.49$ is 0.9998.

   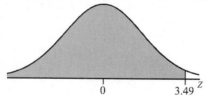

7. The standard normal table (Table V) gives the area to the left of the z-score. Thus, we look up each z-score and read the corresponding area from the table. The area to the right is one minus the area to the left. We can also use technology to find the area. The areas are:

   (a) The area to the right of $z = -3.01$ is $1 - 0.0013 = 0.9987$.

   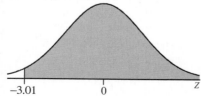

(b) The area to the right of $z = -1.59$ is $1 - 0.0559 = 0.9441$.

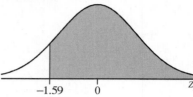

(c) The area to the right of $z = 1.78$ is $1 - 0.9625 = 0.0375$.

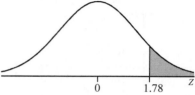

(d) The area to the right of $z = 3.11$ is $1 - 0.9991 = 0.0009$.

9. To find the area between two z-scores using the standard normal table (Table V), we look up the area to the left of each z-score and then we find the difference between these two. We can also use technology to find the area. The areas are:

   (a) The area to the left of $z = -2.04$ is 0.0207, and the area to the left of $z = 2.04$ is 0.9793. So, the area between is $0.9793 - 0.0207 = 0.9586$.

   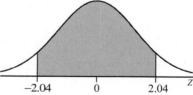

   (b) The area to the left of $z = -0.55$ is 0.2912, and the area to the left of $z = 0$ is 0.5. So, the area between is $0.5 - 0.2912 = 0.2088$.

   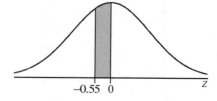

(c) The area to the left of $z = -1.04$ is 0.1492, and the area to the left of $z = 2.76$ is 0.9971. So, the area between is $0.9971 - 0.1492 = 0.8479$

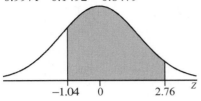

11. (a) The area to the left of $z = -2$ is 0.0228, and the area to the right of $z = 2$ is $1 - 0.9772 = 0.0228$. So, the total area is $0.0228 + 0.0228 = 0.0456$. [Tech: 0.0455]

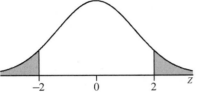

(b) The area to the left of $z = -1.56$ is 0.0594, and the area to the right of $z = 2.56$ is $1 - 0.9948 = 0.0052$. So, the total area is $0.0594 + 0.0052 = 0.0646$.

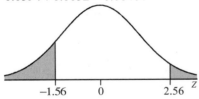

(c) The area to the left of $z = -0.24$ is 0.4052, and the area to the right of $z = 1.20$ is $1 - 0.8849 = 0.1151$. So, the total area is $0.4052 + 0.1151 = 0.5203$.
[Tech: 0.5202]

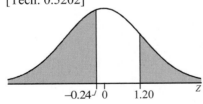

13. (a) The area to the left of $z = -1.34$ is 0.0901, and the area to the left of $z = 2.01$ is 0.9778. So, the area between is $0.9778 - 0.0901 = 0.8877$.

(b) The area to the left of $z = -1.96$ is 0.0250, and the area to the right of $z = 1.96$ is $1 - 0.9750 = 0.0250$. So, the total area is $0.0250 + 0.0250 = 0.0500$.

(c) The area to the left of $z = -2.33$ is 0.0099, and the area to the left of $z = 2.33$ is 0.9901. So, the area between $z = -2.33$ and $z = 2.33$ is $0.9901 - 0.0099 = 0.9802$.

15. The area in the interior of the standard normal table (Table V) that is closest to 0.1000 is 0.1003, corresponding to $z = -1.28$.

17. The area in the interior of the standard normal table (Table V) that is closest to 0.9800 is 0.9798, corresponding to $z = 2.05$.

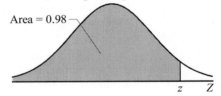

19. The area to the left of the unknown z-score is $1 - 0.25 = 0.75$. The area in the interior of the standard normal table (Table V) that is closest to 0.7500 is 0.7486, corresponding to $z = 0.67$.

21. The area to the left of the unknown z-score is $1 - 0.89 = 0.11$. The area in the interior of the standard normal table (Table V) that is closest to 0.1100 is 0.1093, corresponding to $z = -1.23$.

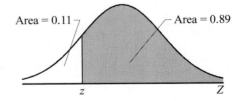

23. The z-scores for the middle 80% are the z-scores for the top and bottom 10%. The area to the left of z_1 is 0.10, and the area to the left of z_2 is 0.90. The area in the interior of the standard normal table (Table V) that is closest to 0.1000 is 0.1003, corresponding to $z_1 = -1.28$. The area in the interior of the standard normal table (Table V) that is closest to 0.9000 is 0.8997, corresponding to $z_2 = 1.28$.

25. The z-scores for the middle 99% are the z-scores for the top and bottom 0.5%. The area to the left of z_1 is 0.005, and the area to the left of z_2 is 0.995. The areas in the interior of the standard normal table (Table V) that are closest to 0.0050 are 0.0049 and 0.0051. So, we use the average of their corresponding z-scores: -2.58 and -2.57, respectively. This gives $z_1 = -2.575$.

The areas in the interior of the standard normal table (Table V) that are closest to 0.9950 are 0.9949 and 0.9951. So, we use the average of their corresponding z-scores: 2.57 and 2.58, respectively. This gives $z_2 = 2.575$.

27. The area to the right of the unknown z-score is 0.05, so the area to the left is $1 - 0.05 = 0.9500$. From the interior of the standard normal table (Table V), we find that the z-scores 1.64 and 1.65 have corresponding areas of 0.9495 and 0.9505, respectively, which are equally close to 0.95. So, we average the two z-scores obtaining $z_{0.05} = 1.645$.

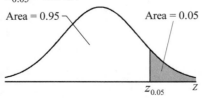

29. The area to the right of the unknown z-score is 0.01, so the area to the left is $1 - 0.01 = 0.99$. The area in the interior of the standard normal table (Table V) that is closest to 0.9900 is 0.9901, so $z_{0.01} = 2.33$.

31. The area to the right of the unknown z-score is 0.20, so the area to the left is $1 - 0.20 = 0.80$. The area in the interior of the standard normal table (Table V) that is closest to 0.8000 is 0.7995, so $z_{0.20} = 0.84$.

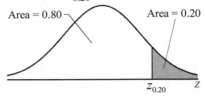

33. From the table, $P(Z < 1.93) = 0.9732$.

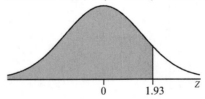

35. From the table, $P(Z < -2.98) = 0.0014$, so $P(Z > -2.98) = 1 - 0.0014 = 0.9986$.

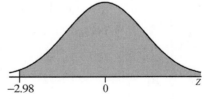

37. From the table, $P(Z < -1.20) = 0.1151$ and $P(Z < 2.34) = 0.9904$. So,
$$P(-1.20 \le Z < 2.34) = 0.9904 - 0.1151$$
$$= 0.8753$$

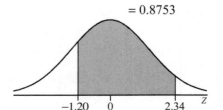

39. From the table, $P(Z < 1.84) = 0.9671$, so
$P(Z \geq 1.84) = 1 - 0.9671 = 0.0329$.

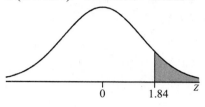

41. From the table, $P(Z \leq 0.72) = 0.7642$

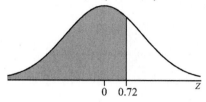

43. From the table, $P(Z < -2.56) = 0.0052$.
Also, $P(Z \leq 1.39) = 0.9177$, which means
$P(Z > 1.39) = 1 - 0.9177 = 0.0823$. So,
$P(Z < -2.56 \text{ or } Z > 1.39) = 0.0052 + 0.0823$
$= 0.0875$

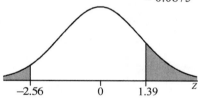

45. From the table, $P(Z < -1.00) = 0.1587$ and
$P(Z < 1.00) = 0.8413$, so
$P(-1 < Z < 1) = 0.8413 - 0.1587$
$= 0.6826$
[Tech: 0.6827]
So, approximately 68% of the data lies within
1 standard deviation of the mean.

Similarly, $P(Z < -2.00) = 0.0228$ and
$P(Z < 2.00) = 0.9772$, so
$P(-2 < Z < 2) = 0.9772 - 0.0228$
$= 0.9544$
[Tech: 0.9545]
So, approximately 95% of the data lies within
2 standard deviations of the mean.
Likewise, $P(Z < -3.00) = 0.0013$ and
$P(Z < 3.00) = 0.9987$, so
$P(-3 < Z < 3) = 0.9987 - 0.0013$
$= 0.9974$
[Tech: 0.9973]
So, approximately 99.7% of the data lies
within 3 standard deviations of the mean.

47. By symmetry, the area to the right of $z = 2.55$
is also 0.0054.

49. By symmetry, the area between $z = 0.53$ and
$z = 1.24$ is also 0.1906.

Section 7.3

1. To find the area under any normal curve:
(1) Draw the curve and shade the desired area.
(2) Convert the values of X to z-scores using
the formula $z = \dfrac{x - \mu}{\sigma}$.
(3) Draw a standard normal curve and shade
the desired area.
(4) Find the area under the standard normal
curve. This area is equal to the area under
the normal curve drawn in Step (1).

3. $z = \dfrac{x - \mu}{\sigma} = \dfrac{35 - 50}{7} \approx -2.14$

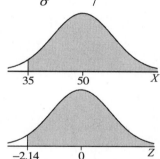

From Table V, the area to the left is 0.0162, so
$P(X > 35) = 1 - 0.0162 = 0.9838$. [Tech: 0.9839]

5. $z = \dfrac{x - \mu}{\sigma} = \dfrac{45 - 50}{7} \approx -0.71$

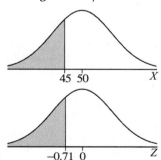

From Table V, the area to the left is 0.2389, so
$P(X \leq 45) = 0.2389$. [Tech: 0.2375]

7. $z_1 = \dfrac{x_1 - \mu}{\sigma} = \dfrac{40 - 50}{7} \approx -1.43$;

$z_2 = \dfrac{x_2 - \mu}{\sigma} = \dfrac{65 - 50}{7} \approx 2.14$

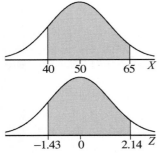

From Table V, the area to the left of $z_1 = -1.43$ is 0.0764 and the area to the left of $z_2 = 2.14$ is 0.9838, so $P(40 < X < 65) = 0.9838 - 0.0764 = 0.9074$. [Tech: 0.9074]

9. $z_1 = \dfrac{x_1 - \mu}{\sigma} = \dfrac{55 - 50}{7} \approx 0.71$;

$z_2 = \dfrac{x_2 - \mu}{\sigma} = \dfrac{70 - 50}{7} \approx 2.86$

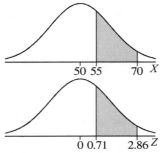

From Table V, the area to the left of $z_1 = 0.71$ is 0.7611 and the area to the left of $z_2 = 2.86$ is 0.9979, so $P(55 \leq X \leq 70) = 0.9979 - 0.7611 = 0.2368$. [Tech: 0.2354]

11. $z_1 = \dfrac{x_1 - \mu}{\sigma} = \dfrac{38 - 50}{7} \approx -1.71$;

$z_2 = \dfrac{x_2 - \mu}{\sigma} = \dfrac{55 - 50}{7} \approx 0.71$

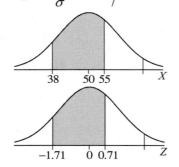

From Table V, the area to the left of $z_1 = -1.71$ is 0.0436 and the area to the left of $z_2 = 0.71$ is 0.7611, so $P(38 < X \leq 55) = 0.7611 - 0.0436 = 0.7175$. [Tech: 0.7192]

13. The figure below shows the normal curve with the unknown value of X separating the bottom 9% of the distribution from the top 91% of the distribution.

From Table V, the area closest to 0.09 is 0.0901. The corresponding z-score is -1.34. So, the 9th percentile for X is $x = \mu + z\sigma = 50 + (-1.34)(7) = 40.62$. [Tech: 40.61]

15. The figure below shows the normal curve with the unknown value of X separating the bottom 81% of the distribution from the top 19% of the distribution.

From Table V, the area closest to 0.81 is 0.8106. The corresponding z-score is 0.88. So, the 81st percentile for X is $x = \mu + z\sigma = 50 + 0.88(7) = 56.16$. [Tech: 56.15]

17. (a) $z = \dfrac{x - \mu}{\sigma} = \dfrac{20 - 21}{1} = -1.00$

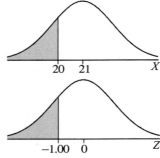

From Table V, the area to the left of $z = -1.00$ is 0.1587, so $P(X < 20) = 0.1587$.

(b) $z = \dfrac{x - \mu}{\sigma} = \dfrac{22 - 21}{1} = 1.00$

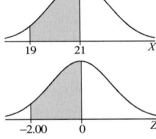

From Table V, the area to the left of
$z = 1.00$ is 0.8413, so $P(X > 22) =$
$1 - 0.8413 = 0.1587$.

(c) $z_1 = \dfrac{x_1 - \mu}{\sigma} = \dfrac{19 - 21}{1} = -2.00$;

$z_2 = \dfrac{x_2 - \mu}{\sigma} = \dfrac{21 - 21}{1} = 0$

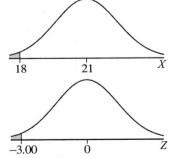

From Table V, the area to the left of
$z_1 = -2.00$ is 0.0228 and the area to the left
of $z_2 = 0$ is 0.5000, so $P(19 \le X \le 21) =$
$0.5000 - 0.0228 = 0.4772$.

(d) $z = \dfrac{x - \mu}{\sigma} = \dfrac{18 - 21}{1} = -3.00$

From Table V, the area to the left of
$z = -3.00$ is 0.0013, so $P(X < 18) =$
0.0013.

Yes, it would be unusual for an egg to
hatch in less than 18 days. Only about 1
egg in 1000 hatches in less than 18 days.

19. (a) $z_1 = \dfrac{x_1 - \mu}{\sigma} = \dfrac{1000 - 1262}{118} \approx -2.22$;

$z_2 = \dfrac{x_2 - \mu}{\sigma} = \dfrac{1400 - 1262}{118} \approx 1.17$

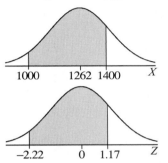

From Table V, the area to the left of
$z_1 = -2.22$ is 0.0132 and the area to the
left of $z_2 = 1.17$ is 0.8790, so
$P(1000 \le X \le 1400) = 0.8790 - 0.0132$
$= 0.8658$. [Tech: 0.8657]

(b) $z = \dfrac{x - \mu}{\sigma} = \dfrac{1000 - 1262}{118} \approx -2.22$

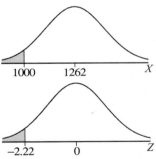

From Table V, the area to the left of
$z = -2.22$ is 0.0132, so $P(X < 1000) =$
0.0132.

(c) $z = \dfrac{x - \mu}{\sigma} = \dfrac{1200 - 1262}{118} \approx -0.53$

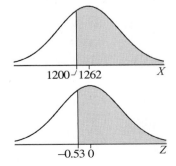

From Table V, the area to the left of
$Z = -0.53$ is 0.2981, so $P(X > 1200) =$
$1 - 0.2981 = 0.7019$. [Tech: 0,7004]
So, the proportion of 18-ounce bags of Chip
Ahoy! cookies that contains more than 1200
chocolate chips is 0.7019, or 70.19%.

(d) $z = \dfrac{x - \mu}{\sigma} = \dfrac{1125 - 1262}{118} \approx -1.16$.

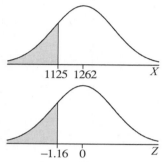

From Table V, the area to the left of
$z = -1.16$ is 0.1230, so $P(X < 1125) =$
0.1230. [Tech: 0.1228] So, the proportion
of 18-ounce bags of Chip Ahoy! cookies
that contains less than 1125 chocolate chips
is 0.1230, or 12.30%.

(e) $z = \dfrac{x - \mu}{\sigma} = \dfrac{1475 - 1262}{118} \approx 1.81$

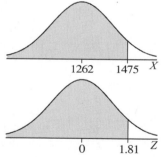

From Table V, the area to the left of
$z = 1.81$ is 0.9649, so $P(X < 1475) =$
0.9649. [Tech: 0.9645] An 18-ounce bag
of Chip Ahoy! Cookies that contains 1475
chocolate chips is at the 96th percentile.

(f) $z = \dfrac{x - \mu}{\sigma} = \dfrac{1050 - 1262}{118} \approx -1.80$.

From Table V, the area to the left of
$z = -1.80$ is 0.0359, so $P(X < 1050) =$
0.0359. [Tech: 0.0362] An 18-ounce bag
of Chip Ahoy! cookies that contains 1050
chocolate chips is at the 4th percentile.

21. (a) $z = \dfrac{x - \mu}{\sigma} = \dfrac{270 - 266}{16} = 0.25$

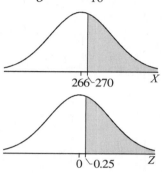

From Table V, the area to the left of
$z = 0.25$ is 0.5987, so $P(X > 270) =$
$1 - 0.5987 = 0.4013$. So, the proportion
of human pregnancies that last more than
270 days is 0.4013.

(b) $z = \dfrac{x - \mu}{\sigma} = \dfrac{250 - 266}{16} = -1$

From Table V, the area to the left of
$z = -1.00$ is 0.1587, so $P(X < 250) =$
0.1587. So, the proportion of human
pregnancies that last less than 250 days is
0.1587, or 15.87%.

(c) $z_1 = \dfrac{x_1 - \mu}{\sigma} = \dfrac{240 - 266}{16} \approx -1.63$;

$z_2 = \dfrac{x_2 - \mu}{\sigma} = \dfrac{280 - 266}{16} \approx 0.88$

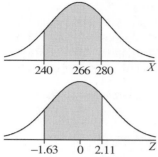

From Table V, the area to the left of $z_1 = -1.63$ is 0.0516 and the area to the left of $z_2 = 0.88$ is 0.8106, so $P(240 \le X \le 280) = 0.8106 - 0.0516 = 0.7590$. [Tech: 0.7571] So, the proportion of human pregnancies lasts between 240 and 280 days is 0.7590, or 75.90%.

(d) $z = \dfrac{x - \mu}{\sigma} = \dfrac{280 - 266}{16} \approx 0.88$

From Table V, the area to the left of $z = 0.88$ is 0.8106, so $P(X > 280) = 1 - 0.8106 = 0.1894$. [Tech: 0.1908]

(e) $z = \dfrac{x - \mu}{\sigma} = \dfrac{245 - 266}{16} \approx -1.31$

From Table V, the area to the left of $z = -1.31$ is 0.0951, so $P(X \le 245) = 0.0951$. [Tech: 0.0947]

(f) $z = \dfrac{x - \mu}{\sigma} = \dfrac{224 - 266}{16} \approx -2.63$

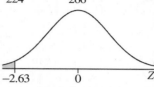

From Table V, the area to the left of $z = -2.63$ is 0.0043, so $P(X < 224) = 0.0043$.

Yes, "very preterm" babies are unusual. Only about 4 births in 1000 are "very preterm."

23. (a) $z = \dfrac{x - \mu}{\sigma} = \dfrac{24.9 - 25}{0.07} \approx -1.43$

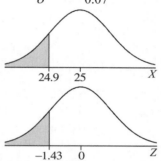

From Table V, the area to the left of $z = -1.43$ is 0.0764, so $P(X < 24.9) = 0.0764$. [Tech: 0.0766] So, the proportion of rods that has a length less than 24.9 cm is 0.0764, or 7.64%.

(b) $z_1 = \dfrac{x_1 - \mu}{\sigma} = \dfrac{24.85 - 25}{0.07} \approx -2.14$;

$z_2 = \dfrac{x_2 - \mu}{\sigma} = \dfrac{25.15 - 25}{0.07} \approx 2.14$

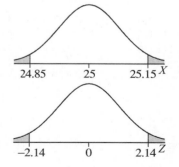

From Table V, the area to the left of $z_1 = -2.14$ is 0.0162, so $P(X < 24.85) = 0.0162$. The area to the left of $z_2 = 2.14$ is 0.9838, so $P(X > 25.15) = 1 - 0.9838 = 0.0162$. So, $P(X < 24.85 \text{ or } X > 25.15) = 2(0.0162) = 0.0324$. [Tech: 0.0321] The proportion of rods that will be discarded is 0.0324, or 3.24%.

(c) The manager should expect to discard $5000(0.0324) = 162$ of the 5000 steel rods.

(d) $z_1 = \dfrac{x_1 - \mu}{\sigma} = \dfrac{24.9 - 25}{0.07} \approx -1.43$;

$z_2 = \dfrac{x_2 - \mu}{\sigma} = \dfrac{25.1 - 25}{0.07} \approx 1.43$

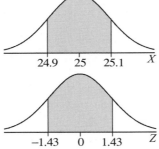

From Table V, the area to the left of $z_1 = -1.43$ is 0.0764 and the area to the left of $z_2 = 1.43$ is 0.9236, so $P(24.9 \le X \le 25.1) = 0.9236 - 0.0764 = 0.8472$. [Tech: 0.8469] So, 0.8472, or 84.72%, of the rods manufactured will be between 24.9 and 25.1 cm. Let n represent the number of rods that must be manufactured. Then, $0.8472n = 10,000$,

so $n = \dfrac{10,000}{0.8472} \approx 11,803.59$. Increase this to the next whole number: 11,804. To meet the order, the manager should manufacture 11,804 rods. [Tech: 11,808 rods]

25. (a) The figure below shows the normal curve with the unknown value of X separating the bottom 17% of the distribution from the top 83%.

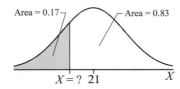

From Table V, the area closest to 0.17 is 0.1711, which corresponds to the z-score -0.95. So, the 17th percentile for incubation times of fertilized chicken eggs is $x = \mu + z\sigma = 21 + (-0.95)(1) \approx 20$ days.

(b) The figure below shows the normal curve with the unknown values of X separating the middle 95% of the distribution from the bottom 2.5% and the top 2.5%.

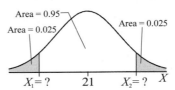

From Table V, the area 0.0250 corresponds to the z-score -1.96. Likewise, the area $0.0250 + .095 = 0.975$ corresponds to the z-score 1.96. Now, $x_1 = \mu + z_1\sigma = 21 + (-1.96)(1) \approx 19$ and $x_2 = \mu + z_2\sigma = 21 + 1.96(1) \approx 23$. Thus, the incubation times that make up the middle 95% of fertilized chicken eggs is between 19 and 23 days.

27. (a) The figure below shows the normal curve with the unknown value of X separating the bottom 30% of the distribution from the top 70%.

From Table V, the area closest to 0.30 is 0.3015, which corresponds to the z-score -0.52. So, the 30th percentile for the number of chocolate chips in an 18-ounce bag of Chips Ahoy! cookies is $x = \mu + z\sigma = 1262 + (-0.52)(118) \approx 1201$ chocolate chips. [Tech: 1200 chocolate chips]

(b) The figure below shows the normal curve with the unknown values of X separating the middle 99% of the distribution from the bottom 0.5% and the top 0.5%.

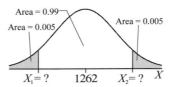

From Table V, the areas 0.0049 and 0.0051 are equally close to 0.005. We average the corresponding z-scores -2.58 and -2.57 to obtain $z_1 = -2.575$. Likewise, the area $0.005 + 0.99 = 0.995$ is equally close 0.9949 and 0.9951. We average the corresponding z-scores 2.57 and 2.58 to obtain $z_2 = 2.575$. Now, $x_1 = \mu + z_1\sigma = 1262 + (-2.575)(118) \approx 958$ and $x_2 = \mu + z_2\sigma = 1262 + 2.575(118) \approx 1566$. So, the number of chocolate chips that make up the middle 99% of 18-ounce bags of Chips Ahoy! cookies is 958 to 1566 chips.

29. (a) $z = \dfrac{x - \mu}{\sigma} = \dfrac{20 - 17}{2.5} = 1.2$

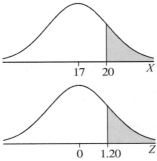

From Table V, the area to the left of $z = 1.20$ is 0.8849, so $P(X > 20) = 1 - 0.8849 = 0.1151$. So, about 11.51% of Speedy Lube's customers receive the service for half price.

(b) The figure below shows the normal curve with the unknown value of X separating the top 3% of the distribution from the bottom 97%.

From Table V, the area closest to 0.97 is 0.9699, which corresponds to the z-score 1.88. So, $x = \mu + z\sigma = 17 + 1.88(2.5) \approx 22$. In order to discount only about 3% of its customers, Speedy Lube should make the guaranteed time limit 22 minutes.

Section 7.4

1. Explanations will vary. One possibility follows: Normal random variables are linearly related to their z-scores (by the formula $X = \mu + Z\sigma$), so the plot of values of X against their expected z-scores should be linear.

3. The plotted points do not lie within the provided bounds, so the sample data do not come from a normally distributed population.

5. The plotted points do not lie within the provided bounds, so the sample data do not come from a normally distributed population.

7. The normal probability plot is roughly linear and all the data lie within the provided bounds, so the sample data could come from a normally distributed population.

9. **(a)** The normal probability plot is roughly linear, and all the data lie within the provided bounds, so the sample data could come from a population that is normally distributed.

(b) $\sum x = 48,895$; $\sum x^2 = 61,211,861$; $n = 40$

$\bar{x} = \dfrac{\sum x}{n} = \dfrac{49,895}{40} \approx 1247.4$ chips ;

$s = \sqrt{\dfrac{\sum x^2 - \dfrac{(\sum x)^2}{n}}{n-1}}$

$= \sqrt{\dfrac{62,635,301 - \dfrac{(49,895)^2}{40}}{40-1}}$

≈ 101.0 chips

(c) $\mu - \sigma \approx \bar{x} - s = 1247.4 - 101.0 = 1146.4$;
$\mu + \sigma \approx \bar{x} + s = 1247.4 + 101.0 = 1348.4$

(d) $z = \dfrac{x - \mu}{\sigma} = \dfrac{1000 - 1247.4}{101.0} \approx -2.45$

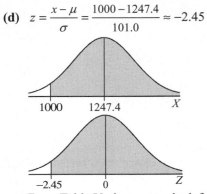

From Table V, the area to the left of
$z = -2.45$ is 0.0071, so $P(X > 1000) =$
$1 - 0.0071 = 0.9929$. [Tech: 0.9928]

(e) $z_1 = \dfrac{x_1 - \mu}{\sigma} = \dfrac{1200 - 1247.4}{101.0} \approx -0.47$;

$z_2 = \dfrac{x_2 - \mu}{\sigma} = \dfrac{1400 - 1247.4}{101.0} \approx 1.51$

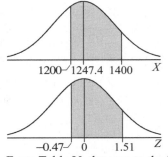

From Table V, the area to the left of
$z_1 = -0.47$ is 0.3192 and the area to the
left of $z_2 = 1.51$ is 0.9345. So,
$P(1200 \le X \le 1400) = 0.9345 - 0.3192 =$
0.6153. [Tech: 0.6152] The proportion
of 18-ounce bags of Chips Ahoy! that
contains between 1200 and 1400 chips is
0.6153, or 61.53%.

11. The normal probability plot is approximately
linear, so the sample data could come from a
normally distributed population.

13. The normal probability plot is not approximately
linear (points lie outside the provided bounds), so
the sample data do not come from a normally
distributed population.

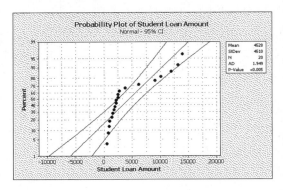

Section 7.5

1. A probability experiment is a binomial
experiment if all the following are true:
   (1) The experiment is performed n
       independent times.
   (2) For each trial, there are two mutually
       exclusive outcomes – success or failure.
   (3) The probability of success, p, is the same
       for each trial of the experiment.

3. We must use a correction for continuity when
using the normal distribution to approximate
binomial probabilities because we are using a
continuous density function to approximate the
probability of a discrete random variable.

5. Approximate $P(X \ge 40)$ by computing the
area under the normal curve to the right of
$x = 39.5$.

7. Approximate $P(X = 8)$ by computing the
area under the normal curve between $x = 7.5$
and $x = 8.5$.

9. Approximate $P(18 \le X \le 24)$ by computing
the area under the normal curve between
$x = 17.5$ and $x = 24.5$.

11. Approximate $P(X > 20) = P(X \ge 21)$ by
computing the area under the normal curve to
the right of $x = 20.5$.

13. Approximate $P(X > 500) = P(X \ge 501)$ by
computing the area under the normal curve to
the right of $x = 500.5$.

15. Using $P(x) = {}_nC_x p^x (1-p)^{n-x}$, with the parameters $n = 60$ and $p = 0.4$, we get

$P(20) = {}_{60}C_{20}(0.4)^{20}(0.6)^{40} \approx 0.0616$. Now $np(1-p) = 60 \cdot 0.4 \cdot (1-0.4) = 14.4 \geq 10$, so the normal approximation can be used, with

$\mu_X = np = 60(0.4) = 24$ and $\sigma_X = \sqrt{np(1-p)}$ $= \sqrt{14.4} \approx 3.795$. With continuity correction we calculate:

$P(20) \approx P(19.5 < X < 20.5)$

$= P\left(\dfrac{19.5 - 24}{\sqrt{14.4}} < Z < \dfrac{20.5 - 24}{\sqrt{14.4}}\right)$

$= P(-1.19 < Z < -0.92)$

$= 0.1788 - 0.1170$

$= 0.0618$ [Tech: 0.0603]

17. Using $P(x) = {}_nC_x p^x (1-p)^{n-x}$, with the parameters $n = 40$ and $p = 0.25$, we get

$P(30) = {}_{40}C_{30}(0.25)^{30}(0.75)^{70} \approx 4.1 \times 10^{-11}$.
Now $np(1-p) = 40 \cdot 0.25 \cdot (1-0.25) = 7.5$, which is below 10, so the normal approximation cannot be used.

19. Using $P(x) = {}_nC_x p^x (1-p)^{n-x}$, with the parameters $n = 75$ and $p = 0.75$, we get

$P(60) = {}_{75}C_{60}(0.75)^{60}(0.25)^{15} \approx 0.0677$. Now $np(1-p) = 75 \cdot 0.75 \cdot (1-0.75) = 14.0625 \geq 10$, so the normal approximation can be used, with

$\mu_X = 75(0.75) = 56.25$ and $\sigma_X = \sqrt{np(1-p)} = \sqrt{14.0625} = 3.75$. With continuity correction we calculate:

$P(60) \approx P(59.5 < X < 60.5)$

$= P\left(\dfrac{59.5 - 56.25}{3.75} < Z < \dfrac{60.5 - 56.25}{3.75}\right)$

$= P(0.87 < Z < 1.13)$

$= 0.8707 - 0.8078$

$= 0.0630$ [Tech: 0.0645]

21. From the parameters $n = 150$ and $p = 0.9$, we get $\mu_X = np = 150 \cdot 0.9 = 135$ and $\sigma_X = \sqrt{np(1-p)} = \sqrt{150 \cdot 0.9 \cdot (1-0.9)} = \sqrt{13.5} \approx 3.674$. Note that $np(1-p) = 13.5 \geq 10$, so the normal approximation to the binomial distribution can be used.

(a) $P(130) \approx P(129.5 < X < 130.5)$

$= P\left(\dfrac{129.5 - 135}{\sqrt{13.5}} < Z < \dfrac{130.5 - 135}{\sqrt{13.5}}\right)$

$= P(-1.50 < Z < -1.22)$

$= 0.1112 - 0.0668$

$= 0.0444$ [Tech: 0.0431]

(b) $P(X \geq 130) \approx P(X \geq 129.5)$

$= P\left(Z \geq \dfrac{129.5 - 135}{\sqrt{13.5}}\right)$

$= P(Z \geq -1.50)$

$= 1 - 0.0668$

$= 0.9332$ [Tech: 0.9328]

(c) $P(X < 125) = P(X \leq 124)$

$\approx P(X \leq 124.5)$

$= P\left(Z \leq \dfrac{124.5 - 135}{\sqrt{13.5}}\right)$

$= P(Z \leq -2.86)$

$= 0.0021$

(d) $P(125 \leq X \leq 135)$

$\approx P(124.5 \leq X \leq 135.5)$

$= P\left(\dfrac{124.5 - 135}{\sqrt{13.5}} < Z < \dfrac{135.5 - 135}{\sqrt{13.5}}\right)$

$= P(-2.86 < Z < 0.14)$

$= 0.5557 - 0.0021$

$= 0.5536$ [Tech: 0.5520]

23. From the parameters $n = 600$ and $p = 0.02$ we get $\mu_X = np = 600 \cdot 0.02 = 12$ and $\sigma_X = \sqrt{np(1-p)} = \sqrt{600 \cdot 0.02 \cdot (1-0.02)} = \sqrt{11.76} \approx 3.429$. Note that $np(1-p) = 11.76 \geq 10$, so the normal approximation to the binomial distribution can be used.

(a) $P(20) \approx P(19.5 \leq X \leq 20.5)$

$= P\left(\dfrac{19.5 - 12}{\sqrt{11.76}} \leq Z \leq \dfrac{20.5 - 12}{\sqrt{11.76}}\right)$

$= P(2.19 \leq Z \leq 2.48)$

$= 0.9934 - 0.9857$

$= 0.0077$ [Tech: 0.0078]

(b) $P(X \le 20) \approx P(X \le 20.5)$

$$= P\left(Z \le \frac{20.5 - 12}{\sqrt{11.76}}\right)$$

$$= P(Z \le 2.48)$$

$$= 0.9934$$

(c) $P(X \ge 22) \approx P(X \ge 21.5)$

$$= P\left(Z \ge \frac{21.5 - 12}{\sqrt{11.76}}\right)$$

$$= P(Z \ge 2.77)$$

$$= 1 - 0.9972$$

$$= 0.0028$$

(d) $P(20 \le X \le 30)$

$$\approx P(19.5 \le X \le 30.5)$$

$$= P\left(\frac{19.5 - 12}{\sqrt{11.76}} \le Z \le \frac{30.5 - 12}{\sqrt{11.76}}\right)$$

$$= P(2.19 \le Z \le 5.39)$$

$$= 1.0000 - 0.9857$$

$$= 0.0143 \quad \text{[Tech: 0.0144]}$$

25. From the parameters $n = 200$ and $p = 0.55$, we get $\mu_X = np = 200 \cdot 0.55 = 110$ and $\sigma_X = \sqrt{np(1-p)} = \sqrt{200 \cdot 0.55 \cdot (1 - 0.55)} = \sqrt{49.5} \approx 7.036$. Note that $np(1-p) = 49.5 \ge 10$, so the normal approximation to the binomial distribution can be used.

(a) $P(X \ge 130) \approx P(X \ge 129.5)$

$$= P\left(Z \ge \frac{129.5 - 110}{\sqrt{49.5}}\right)$$

$$= P(Z \ge 2.77)$$

$$= 1 - 0.9972$$

$$= 0.0028$$

(b) Yes, the result from part (a) contradicts the results of the *Current Population Survey* because the result from part (a) is unusual. Fewer than 3 samples in 1000 will result in 130 or more male students living at home if the true percentage is 55%.

27. From the parameters $n = 150$ and $p = 0.37$, we get $\mu_X = np = 150 \cdot 0.37 = 55.5$ and $\sigma_X = \sqrt{np(1-p)} = \sqrt{150 \cdot 0.37 \cdot (1 - 0.37)} = \sqrt{34.965} \approx 5.913$. Note that $np(1-p) = 34.965 \ge 10$, so the normal approximation to the binomial distribution can be used.

(a) $P(X \ge 75) \approx P(X \ge 74.5)$

$$= P\left(Z \ge \frac{74.5 - 55.5}{\sqrt{34.965}}\right)$$

$$= P(Z \ge 3.21)$$

$$= 1 - 0.9993$$

$$= 0.0007$$

(b) Yes, the result from part (a) contradicts the results of the *Current Population Survey* because the result from part (a) is unusual. Fewer than 1 sample in 1000 will result in 75 or more respondents preferring a male if the true percentage who prefer a male is 37%.

Chapter 7 Review Exercises

1. (a) μ is the center (and peak) of the normal distribution, so $\mu = 60$.

(b) σ is the distance from the center to the points of inflection, so $\sigma = 70 - 60 = 10$.

(c) Interpretation 1: The proportion of values of the random variable to the right of $x = 75$ is 0.0668.
Interpretation 2: The probability that a randomly selected value is greater than $x = 75$ is 0.0668.

(d) Interpretation 1: The proportion of values of the random variable between $x = 50$ and $x = 75$ is 0.7745.
Interpretation 2: The probability that a randomly selected value is between $x = 50$ and $x = 75$ is 0.7745.

2. (a) $z_1 = \dfrac{x_1 - \mu}{\sigma} = \dfrac{18 - 20}{4} = -0.5$

(b) $z_2 = \dfrac{x_2 - \mu}{\sigma} = \dfrac{21 - 20}{4} = 0.25$

(c) The area between z_1 and z_2 is also 0.2912.

3.

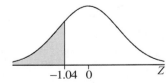

Using the standard normal table, the area to the left of $z = -1.04$ is 0.1492.

4.

Using the standard normal table, the area between $z = -0.34$ and $z = 1.03$ is $0.8485 - 0.3669 = 0.4816$.

5.

$P(Z < 1.19) = 0.8830$

6.

$P(Z \geq 1.61) = 1 - 0.9463 = 0.0537$

7.

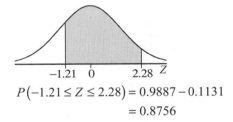

$P(-1.21 \leq Z \leq 2.28) = 0.9887 - 0.1131$
$$= 0.8756$$

8. If the area to the right of z is 0.483 then the area to the left z is $1 - 0.483 = 0.5170$.

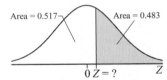

The closest area to this in the interior of the normal tables is 0.5160, corresponding to $z = 0.04$.

9. The z-scores for the middle 92% are the z-scores for the top and bottom 4%. The area to the left of z_1 is 0.04, and the area to the left of z_2 is 0.96. The area in the interior of the standard normal table (Table V) that is closest to 0.0400 is 0.0401, corresponding to $z_1 = -1.75$. The area in the interior of the standard normal table (Table V) that is closest to 0.9600 is 0.9599, corresponding to $z_2 = 1.75$.

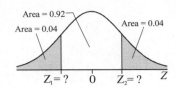

10. The area to the right of the unknown z-score is 0.20, so the area to the left is $1 - 0.20 = 0.80$. The area in the interior of the standard normal table (Table V) that is closest to 0.8000 is 0.7995, corresponding to $z = 0.84$, so $z_{0.20} = 0.84$.

11.

$$z = \frac{55 - 50}{6} \approx 0.83$$

From Table V, the area to the left of $z = 0.83$ is 0.7967, so $P(X > 55) = 1 - 0.7967 = 0.2033$.

[Tech: 0.2023]

12.

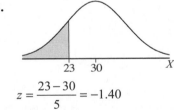

$$z = \frac{23 - 30}{5} = -1.40$$

From Table V, the area to the left of $z = -1.40$ is 0.0808, so $P(X \leq 23) = 0.0808$.

13.

$$z_1 = \frac{65 - 70}{10} = -0.50 \; ; \; z_2 = \frac{85 - 70}{10} = 1.50$$

From Table V, the area to the left of $z = -0.50$ is 0.3085 and the area to the left of $z = 1.50$ is 0.9332, so $P(65 < X < 85) = 0.9332 - 0.3085$
$$= 0.6247.$$

14. (a)

$z = \dfrac{75,000 - 70,000}{4,400} \approx 1.14$

From Table V, the area to the left of $z = 1.14$ is 0.8729, so $P(X \geq 75,000) =$ $1 - 0.8729 = 0.1271$ [Tech: 0.1279]

So, the proportion of tires that will last at least 75,000 miles is 0.1271, or 12.71%.

(b)

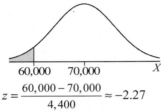

$z = \dfrac{60,000 - 70,000}{4,400} \approx -2.27$

From Table V, the area to the left of $z = -2.27$ is 0.0116, so $P(X \leq 60,000) =$ $= 0.0116$ [Tech: 0.0115] So, the proportion of tires that will last 60,000 miles or less is 0.0116, or 1.16%.

(c)

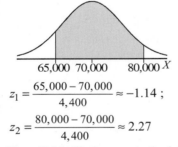

$z_1 = \dfrac{65,000 - 70,000}{4,400} \approx -1.14$;

$z_2 = \dfrac{80,000 - 70,000}{4,400} \approx 2.27$

From Table V, the area to the left of $z_1 = -1.14$ is 0.1271 and the area to the left of $z_2 = 2.27$ is 0.9884, so $P(65,000 \leq X \leq 80,000) = 0.9884 - 0.1271$ $= 0.8613$ [Tech: 0.8606]

(d) The figure below shows the normal curve with the unknown value of X separating the bottom 2% of the distribution from the top 98% of the distribution.

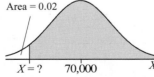

From Table V, 0.0202 is the area closest to 0.02. The corresponding z-score is -2.05.

So, $x = \mu + z\sigma = 70,000 + (-2.05)(4,400)$ $= 60,980$. [Tech: 60,964] In order to warrant only 2% of its tires, Dunlop should advertise its warranty mileage as 60,980 miles.

15. (a)

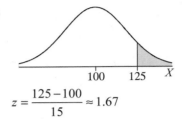

$z = \dfrac{125 - 100}{15} \approx 1.67$

From Table V, the area to the left of $z = 1.67$ is 0.9525, so $P(X > 125) =$ $1 - 0.9525 = 0.0475$ [Tech: 0.0478].

(b)

$z = \dfrac{90 - 100}{15} \approx -0.67$

From Table V, the area to the left of $z = -0.67$ is 0.2514, so $P(X < 90) =$ 0.2514. [Tech: 0.2525]

(c)

$z_1 = \dfrac{110 - 100}{15} \approx 0.67$; $z_2 = \dfrac{140 - 100}{15} \approx 2.67$

From Table V, the area to the left of $z_1 = 0.67$ is 0.7486 and the area to the left of $z_2 = 2.67$ is 0.9962, so $P(110 < X < 140) = 0.9962 - 0.7486$ $= 0.2476$ [Tech: 0.2487] So, the proportion of test takers who score between 110 and 140 is 0.2476, or 24.76%.

(d)

$z = \dfrac{150 - 100}{15} \approx 3.33$

From Table V, the area to the left of $z = 3.33$ is 0.9996, so $P(X > 150) =$ $= 1 - 0.9996 = 0.0004$.

(e) The figure below shows the normal curve with the unknown value of X separating the bottom 98% of the distribution from the top 2% of the distribution.

From Table V, 0.9798 is the area closest to 0.98. The corresponding z-score is 2.05. So, $x = \mu + z\sigma = 100 + 2.05(15) \approx 131$. A score of 131 places a child in the 98th percentile.

(f) The figure below shows the normal curve with the unknown values of X separating the middle 95% of the distribution from the bottom 2.5% and top 2.5% of the distribution.

From Table V, the area 0.0250 corresponds to $z_1 = -1.96$ and the area 0.9750 corresponds to $z_2 = 1.96$. So, $x_1 = \mu + z_1\sigma = 100 + (-1.96)(15) \approx 71$ and $x_2 = \mu + z_2\sigma = 100 + 1.96(15) \approx 129$. Thus, children of normal intelligence scores are between 71 and 129 on the Wechsler Scale.

16. (a)

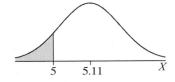

$z = \dfrac{5.25 - 5.11}{0.062} \approx 2.26$

From Table V, the area to the left of $z = 2.26$ is 0.9881, so $P(X > 5.25) = 1 - 0.9881 = 0.0119$. [Tech: 0.0120] So, the proportion of baseballs produced by this factory that are too heavy for use by major league baseball is 0.0119, or 1.19%.

(b)

$z = \dfrac{5 - 5.11}{0.062} \approx -1.77$

From Table V, the area to the left of $z = -1.77$ is 0.0384, so $P(X < 5) = 0.0384$. [Tech: 0.0380] So, the proportion of baseballs produced by this factory that are too light for use by major league baseball is 0.0384, or 3.84%.

(c)

From parts (a) and (b), $z_1 \approx -1.77$ and $z_2 \approx 2.26$. The area to the left of $z_1 = -1.17$ is 0.0384 and the area to the left of $z_2 = 2.26$ is 0.9881, so $P(5 \le X \le 5.25) = 0.9881 - 0.0384 = 0.9497$ [Tech: 0.9500]. So, the proportion of baseballs produced by this factory that can be used by major league baseball is 0.9497, or 94.97%.

(d) From part (c), we know that 94.97% of the baseballs can be used by major league baseball. Let n represent the number of baseballs that must be produced. Then, $0.9497n = 8,000$, so

$$n = \frac{8,000}{0.9497} \approx 8,423.71.$$ Increase this to the next whole number: 8,424. To meet the order, the factory should produce 8,424 baseballs. [Tech: 8421 baseballs]

17. (a) Since $np(1 - p) = 250(0.46)(1 - 0.46) = 62.1 \ge 10$, the normal distribution can be used to approximate the binomial distribution. The parameters are $\mu_X = np = 250(0.46) = 115$ and $\sigma_X = \sqrt{np(1 - p)} = \sqrt{250(0.46)(1 - 0.46)} = \sqrt{62.1} \approx 7.880$

(b) For the normal approximation, we make corrections for continuity 124.5 and 125.5.

$P(125) \approx P(124.5 < X < 125.5)$

$$= P\left(\frac{124.5 - 115}{\sqrt{62.1}} < Z < \frac{125.5 - 115}{\sqrt{62.1}}\right)$$

$$= P(1.21 < Z \leq 1.33)$$

$$= 0.9082 - 0.8869$$

$$= 0.0213 \quad [\text{Tech: } 0.0226]$$

Interpretation: Approximately 2 of every 100 random samples of 250 adult Americans will result in exactly 125 who state that they have read at least 6 books within the past year.

(c) $P(X < 120) = P(X \leq 119)$

For the normal approximation, we make a correction for continuity to 119.5.

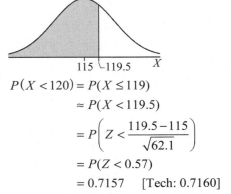

$P(X < 120) = P(X \leq 119)$

$$\approx P(X < 119.5)$$

$$= P\left(Z < \frac{119.5 - 115}{\sqrt{62.1}}\right)$$

$$= P(Z < 0.57)$$

$$= 0.7157 \quad [\text{Tech: } 0.7160]$$

Interpretation: Approximately 72 of every 100 random samples of 250 adult Americans will result in fewer than 120 who state that they have read at least 6 books within the past year.

(d) For the normal approximation, we make a correction for continuity to 139.5.

$P(X \geq 140) \approx P(X > 139.5)$

$$= P\left(Z > \frac{139.5 - 115}{\sqrt{62.1}}\right)$$

$$\approx P(Z > 3.11)$$

$$= 1 - 0.9991$$

$$= 0.0009$$

Interpretation: Approximately 1 of every 1000 random samples of 250 adult Americans will result in 140 or more who state that they have read at least 6 books within the past year.

(e) For the normal approximation, we make corrections for continuity 99.5 and 120.5.

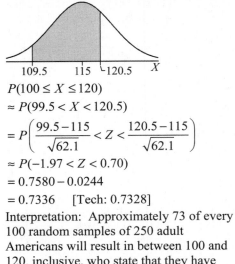

$P(100 \leq X \leq 120)$

$$\approx P(99.5 < X < 120.5)$$

$$= P\left(\frac{99.5 - 115}{\sqrt{62.1}} < Z < \frac{120.5 - 115}{\sqrt{62.1}}\right)$$

$$\approx P(-1.97 < Z < 0.70)$$

$$= 0.7580 - 0.0244$$

$$= 0.7336 \quad [\text{Tech: } 0.7328]$$

Interpretation: Approximately 73 of every 100 random samples of 250 adult Americans will result in between 100 and 120, inclusive, who state that they have read at least 6 books within the past year.

18. The normal probability plot is roughly linear, and all the data lie within the provided bounds, so the sample data could come from a normally distributed population.

19. The plotted points do not appear linear and do not lie within the provided bounds, so the sample data are not from a normally distributed population.

20. The plotted points are not linear and do not lie within the provided bounds, so the data are not from a population that is normally distributed.

21. From the parameters $n = 250$ and $p = 0.20$, we get $\mu_X = np = 250 \cdot 0.20 = 50$ and $\sigma_X = \sqrt{np(1-p)} = \sqrt{250(0.20)(1-0.20)} = \sqrt{40} \approx 6.325$. Note that $np(1-p) = 40 \geq 10$, so the normal approximation to the binomial distribution can be used.

(a) For the normal approximation, we make corrections for continuity to 30.5.

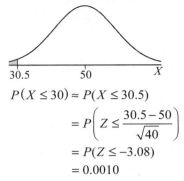

$P(X \le 30) \approx P(X \le 30.5)$

$$= P\left(Z \le \frac{30.5 - 50}{\sqrt{40}} \right)$$

$$= P(Z \le -3.08)$$

$$= 0.0010$$

(b) Yes, the result from part (a) contradicts the *USA Today* "Snapshot" because the result from part (a) is unusual. About 1 sample in 1000 will result in 30 or fewer who do their most creative thinking while driving, if the true percentage is 20%.

22. (a)

(b) $P(0 \le X \le 5) = 0.05 \cdot (5 - 0) = 0.25$

(c) $P(10 \le X \le 18) = 0.05 \cdot (18 - 10) = 0.4$

23. The standard normal curve has the following properties:
  (1) It is symmetric about its mean $\mu = 0$ and has standard deviation $\sigma = 1$.
  (2) Its highest point occurs at $\mu = 0$.
  (3) It has inflection points at -1 and 1.
  (4) The area under the curve is 1.
  (5) The area under the curve to the right of $\mu = 0$ equals the area under the curve to the left of $\mu = 0$, which equals 0.5.
  (6) As Z increases, the graph approaches but never equals zero. As Z decreases, the graph approaches but never equals zero.
  (7) It satisfies the Empirical Rule: Approximately 68% of the area under the standard normal curve is between -1 and 1. Approximately 95% of the area under the standard normal curve is between -2 and 2. Approximately 99.7% of the area under the standard normal curve is between -3 and 3.

24. The graph plots actual observations against expected z-scores, assuming that the data are normally distributed. If the plot is not linear, then we have evidence that that data are not from a normal distribution.

Chapter 7 Test

1. (a) μ is the center (and peak) of the normal distribution, so $\mu = 7$.

  (b) σ is the distance from the center to the points of inflection, so $\sigma = 9 - 7 = 2$.

  (c) Interpretation 1: The proportion of values for the random variable to the left of $x = 10$ is 0.9332.
  Interpretation 2: The probability that a randomly selected value is less than $x = 10$ is 0.9332.

  (d) Interpretation 1: The proportion of values for the random variable between $x = 5$ and $x = 8$ is 0.5328.
  Interpretation 2: The probability that a randomly selected value is between $x = 5$ and $x = 8$ is 0.5328.

2. (a) $z_1 = \dfrac{x_1 - \mu}{\sigma} = \dfrac{48 - 50}{8} = -0.25$

  (b) $z_2 = \dfrac{x_2 - \mu}{\sigma} = \dfrac{60 - 50}{8} = 1.25$

  (c) The area between z_1 and z_2 is also 0.4931.

3.

Using Table V, the area to the left of $z = 2.04$ is 0.9793, so the are to the right of $z = 2.04$ is $1 - 0.9793 = 0.0207$.

4.

Using the Table V, the area to the left of $z = 0.21$ is 0.5832 and the area to the left of $z = 1.69$ is 0.9545. So, $P(0.21 < Z < 1.69) = 0.9545 - 0.5832 = 0.3713$

5. The z-scores for the middle 88% are the z-scores for the top and bottom 6%.

Area = 0.88
Area = 0.06
Area = 0.06
$Z_1 = ?$ 0 $Z_2 = ?$ Z

The area to the left of z_1 is 0.06, and the area to the left of z_2 is 0.94. From the interior of Table V, the areas 0.0606 and 0.0594 are equally close to 0.0600. These areas correspond to $z = -1.55$ and $z = -1.56$, respectively. Splitting the difference we obtain $z_1 = -1.555$. From the interior of Table V, the areas 0.9394 and 0.9406 are equally close to 0.9400. These correspond to $z = 1.55$ and $z = 1.56$, respectively. Splitting the difference, we obtain $z_2 = 1.555$.

6. The area to the right of the unknown z-score is 0.04, so the area to the left is $1 - 0.04 = 0.96$.

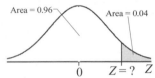

Area = 0.96
Area = 0.04
0 $Z = ?$ Z

The area in the interior of the Table V that is closest to 0.9600 is 0.9599, corresponding to $z = 1.75$, so $z_{0.04} = 1.75$.

7. (a)

17 20 23

(b) $z_1 = \dfrac{22 - 20}{3} \approx 0.67$; $z_2 = \dfrac{27 - 20}{3} \approx 2.33$

20 22 27 X

From Table V, the area to the left of $z_1 = 0.67$ is 0.7486 and the area to the left of $z_2 = 2.33$ is 0.9901, so $P(22 \le X \le 27)$ $= 0.9901 - 0.7486 = 0.2415$. [Tech: 0.2527]

8. (a)

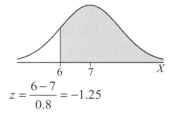

6 7 X

$z = \dfrac{6 - 7}{0.8} = -1.25$

From Table V, the area to the left of $z = -1.25$ is 0.1956, so $P(X \ge 6) =$ $1 - 0.1056 = 0.8944$. So, the proportion of the time that a fully charged iPhone will last at least 6 hours is 0.8944, or 89.44%.

(b)

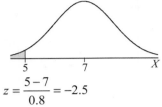

5 7 X

$z = \dfrac{5 - 7}{0.8} = -2.5$

From Table V, the area to the left of $z = -2.50$ is 0.0062, so $P(X < 5) =$ 0.0062. That is, the probability that a fully charged iPhone will last less than 5 hours is 0.0062. This is an unusual result.

(c) The figure below shows the normal curve with the unknown value of X separating the top 5% of the distribution from the bottom 95% of the distribution.

0.95
0.05
7 $X = ?$ X

From the interior of Table V, the areas 0.9495 and 0.9505 are equally close to 0.9500. These areas correspond to $z = 1.64$ and $z = 1.65$, respectively. Splitting the difference, we obtain the z-score 1.645. So, $x = \mu + z\sigma = 7 + 1.645(0.8) \approx 8.3$. The cutoff for the top 5% of all talk times is 8.3 hours.

(d)

7 9 X

$z = \dfrac{9 - 7}{0.8} = 2.5$

From Table V, the area to the left of $z = 2.50$ is 0.9938, so $P(X > 9) =$ $1 - 0.9938 = 0.0062$. So, yes, it would be unusual for the iPhone to last more than 9 hours. Only about 6 out of every 1000 full charges will result in the iPhone lasting more than 9 hours.

9. (a)

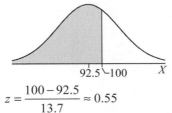

$$z = \frac{100 - 92.5}{13.7} \approx 0.55$$

From Table V, the area to the left of $z = 0.55$ is 0.7088, so $P(X < 6) = 0.7088$. [Tech: 0.7080] So, the proportion of 20- to 29-year-old males whose waist circumferences is less than 100 cm is 0.7088, or 70.88%.

(b)

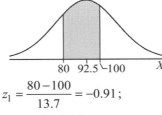

$$z_1 = \frac{80 - 100}{13.7} = -0.91;$$

$$z_2 = \frac{100 - 92.5}{13.7} \approx 0.55$$

From Table V, the area to the left of $z_1 = -0.91$ is 0.1814 and the area to the left of $z_2 = 0.55$ is 0.7088, so

$P(80 \le X \le 100) = 0.7088 - 0.1814 = 0.5274$. [Tech: 0.5272] That is, the probability that a randomly selected 20- to 29-year-old male has a waist circumference between 80 and 100 cm is 0.5274.

(c) The figure that follows shows the normal curve with the unknown values of X separating the middle 90% of the distribution from the bottom 5% and the top 5% of the distribution.

From the interior of Table V, the areas 0.0495 and 0.0505 are equally close to 0.0500. These areas correspond to $z = -1.65$ and $z = -1.64$, respectively. Splitting the difference we obtain $z_1 = -1.645$. So, $x_1 = \mu + z_1\sigma = 92.5 + (-1.645)(13.7) \approx 70$. From the interior of Table V, the areas 0.9495 and

0.9505 are equally close to 0.9500. These areas correspond to $z = 1.64$ and $z = 1.65$, respectively. Splitting the difference we obtain the $z_2 = 1.645$. So, $x_2 = \mu + z_2\sigma = 92.5 + 1.645(13.7) \approx 115$. Therefore, waist circumferences between 70 and 115 cm make up the middle 90% of all waist circumferences.

(d) The figure below shows the normal curve with the unknown value of X separating the bottom 10% of the distribution from the top 90% of the distribution.

From the interior of Table V, the area closest to 0.10 is 0.1003. The corresponding z-score is -1.28. So, $x = \mu + z\sigma = 92.5 + (-1.28)(13.7) \approx 75$. Thus, a waist circumference of 75 cm is at the 10th percentile.

10. $np(1-p) = 500(0.16)(1-0.16) = 67.2 \ge 10$, so the normal distribution can be used to approximate the binomial distribution. The parameters are $\mu_X = np = 500(0.16) = 80$ and $\sigma_X = \sqrt{np(1-p)} = \sqrt{250(0.16)(1-0.16)} = \sqrt{67.2} \approx 8.198$.

(a) For the normal approximation of $P(100)$, we make corrections for continuity 99.5 and 100.5.

$$P(100) \approx P(99.5 < X < 100.5)$$
$$= P\left(\frac{99.5 - 80}{\sqrt{67.2}} < Z < \frac{100.5 - 80}{\sqrt{67.2}}\right)$$
$$= P(2.38 < Z \le 2.50)$$
$$= 0.9938 - 0.9913$$
$$= 0.0025$$

(b) $P(X < 60) = P(X \le 59)$

For the normal approximation, we make a correction for continuity to 59.5.

$P(X < 60) = P(X \le 59)$

$\approx P(X < 69.5)$

$= P\left(Z < \dfrac{59.5 - 80}{\sqrt{67.2}} \right)$

$= P(Z < -2.50)$

$= 0.0062$

11. The plotted points do not appear linear and one point does not lie within the provided bounds, so the sample data do not likely come from a normally distributed population.

12. (a)

(b) $P(20 \le X \le 30) = \dfrac{1}{40}(30 - 20) = 0.25$

(c) $P(X < 15) = P(10 \le X \le 15)$

$= \dfrac{1}{40}(15 - 10)$

$= 0.125$

Chapter 8

Sampling Distributions

Section 8.1

1. The sampling distribution of a statistic (such as the sample mean) is a probability distribution for all possible values of the statistic computed from a sample of size n.

3. standard error; mean

5. The mean of the sampling distribution of $\bar{x}$ is given by $\mu_{\bar{x}} = \mu$ and the standard deviation is given by $\sigma_{\bar{x}} = \dfrac{\sigma}{\sqrt{n}}$.

7. four; to see this, note that $\sigma_{\bar{x}} = \dfrac{\sigma}{\sqrt{4n}} = \dfrac{1}{2} \cdot \dfrac{\sigma}{\sqrt{n}}$.

9. The sampling distribution of $\bar{x}$ would be exactly normal. The mean and standard deviation would be $\mu_{\bar{x}} = \mu = 30$ and $\sigma_{\bar{x}} = \dfrac{\sigma}{\sqrt{n}} = \dfrac{8}{\sqrt{10}} \approx 2.530$.

11. $\mu_{\bar{x}} = \mu = 80$; $\sigma_{\bar{x}} = \dfrac{\sigma}{\sqrt{n}} = \dfrac{14}{\sqrt{49}} = \dfrac{14}{7} = 2$

13. $\mu_{\bar{x}} = \mu = 52$; $\sigma_{\bar{x}} = \dfrac{\sigma}{\sqrt{n}} = \dfrac{10}{\sqrt{21}} \approx 2.182$

15. **a.** The sampling distribution is symmetric about 500, so the mean is $\mu_{\bar{x}} = 500$.

    b. The inflection points are at 480 and 520, so $\sigma_{\bar{x}} = 520 - 500 = 20$ (or $500 - 480 = 20$).

    c. Since $n = 16 \le 30$, the population be must be normal so that the sampling distribution of $\bar{x}$ is normal.

    d. $\sigma_{\bar{x}} = \dfrac{\sigma}{\sqrt{n}}$

    $20 = \dfrac{\sigma}{\sqrt{16}}$

    $\sigma = 20\sqrt{16} = 20(4) = 80$

    The standard deviation of the population from which the sample is drawn is 80.

17. **(a)** Since $\mu = 80$ and $\sigma = 14$, the mean and standard deviation of the sampling distribution of $\bar{x}$ are given by:

    $\mu_{\bar{x}} = \mu = 80$; $\sigma_{\bar{x}} = \dfrac{\sigma}{\sqrt{n}} = \dfrac{14}{\sqrt{49}} = \dfrac{14}{7} = 2$.

    We are not told that the population is normally distributed, but we do have a large sample size ($n = 49 \ge 30$). Therefore, we can use the Central Limit Theorem to say that the sampling distribution of $\bar{x}$ is approximately normal.

    (b) $P(\bar{x} > 83) = P\left(Z > \dfrac{83 - 80}{2}\right)$

    $= P(Z > 1.50)$

    $= 1 - P(Z \le 1.50)$

    $= 1 - 0.9332$

    $= 0.0668$

    If we take 100 simple random samples of size $n = 49$ from a population with $\mu = 80$ and $\sigma = 14$, then about 7 of the samples will result in a mean that is greater than 83.

    (c) $P(\bar{x} \le 75.8) = P\left(Z \le \dfrac{75.8 - 80}{2}\right)$

    $= P(Z \le -2.10)$

    $= 0.0179$

    If we take 100 simple random samples of size $n = 49$ from a population with $\mu = 80$ and $\sigma = 14$, then about 2 of the samples will result in a mean that is less than or equal to 75.8.

    (d) $P(78.3 < \bar{x} < 85.1)$

    $= P\left(\dfrac{78.3 - 80}{2} < Z < \dfrac{85.1 - 80}{2}\right)$

    $= P(-0.85 < Z < 2.55)$

    $= 0.9946 - 0.1977$

    $= 0.7969$ \quad [Tech: 0.7970]

    If we take 100 simple random samples of size $n = 49$ from a population with $\mu = 80$ and $\sigma = 14$, then about 78 of the samples will result in a mean that is between 78.3 and 85.1.

19. (a) The population must be normally distributed. If this is the case, then the sampling distribution of $\bar{x}$ is exactly normal. The mean and standard deviation of the sampling distribution are $\mu_{\bar{x}} = \mu = 64$ and

$$\sigma_{\bar{x}} = \frac{\sigma}{\sqrt{n}} = \frac{17}{\sqrt{12}} \approx 4.907 \ .$$

(b) $P(\bar{x} < 67.3) = P\left(Z < \frac{67.3 - 64}{17/\sqrt{12}}\right)$

$= P(Z < 0.67)$

$= 0.7486$ [Tech: 0.7493]

If we take 100 simple random samples of size $n = 12$ from a population that is normally distributed with $\mu = 64$ and $\sigma = 17$, then about 75 of the samples will result in a mean that is less than 67.3.

(c) $P(\bar{x} \geq 65.2) = P\left(Z \geq \frac{65.2 - 64}{17/\sqrt{12}}\right)$

$= P(Z \geq 0.24)$

$= 1 - P(Z < 0.24)$

$= 1 - 0.5948$

$= 0.4052$ [Tech: 0.4034]

If we take 100 simple random samples of size $n = 12$ from a population that is normally distributed with $\mu = 64$ and $\sigma = 17$ then about 41 of the samples will result in a mean that is greater than or equal to 65.2.

21. (a) $P(X < 260) = P\left(Z < \frac{260 - 266}{16}\right)$

$= P(Z < -0.38)$

$= 0.3520$ [Tech: 0.3538]

If we select a simple random sample of $n = 100$ human pregnancies, then about 35 of the pregnancies would last less than 260 days.

(b) Since the length of human pregnancies is normally distributed, the sampling distribution of $\bar{x}$ is normal with $\mu_{\bar{x}} = 266$

and $\sigma_{\bar{x}} = \frac{16}{\sqrt{20}} \approx 3.578$.

(c) $P(\bar{x} < 260) = P\left(Z < \frac{260 - 266}{16/\sqrt{20}}\right)$

$= P(Z < -1.68)$

$= 0.0465$ [Tech: 0.0468]

If we take 100 simple random samples of size $n = 20$ human pregnancies, then about 5 of the samples will result in a mean gestation period of 260 days or less.

(d) $\mu_{\bar{x}} = \mu = 266$; $\sigma_{\bar{x}} = \frac{\sigma}{\sqrt{n}} = \frac{16}{\sqrt{50}}$

$P(\bar{x} < 260) = P\left(Z < \frac{260 - 266}{16/\sqrt{50}}\right)$

$= P(Z < -2.65)$

$= 0.0040$

If we take 1000 simple random samples of size $n = 50$ human pregnancies, then about 4 of the samples will result in a mean gestation period of 260 days or less.

(e) Answers will vary. Part (d) indicates that this result would be an unusual observation. Therefore, we would conclude that the sample likely came from a population whose mean gestation period is less than 266 days.

(f) $\mu_{\bar{x}} = \mu = 266$; $\sigma_{\bar{x}} = \frac{\sigma}{\sqrt{n}} = \frac{16}{\sqrt{15}}$

$P(256 \leq \bar{x} \leq 276)$

$= P\left(\frac{256 - 266}{16/\sqrt{15}} \leq Z \leq \frac{276 - 266}{16/\sqrt{15}}\right)$

$= P(-2.42 \leq Z \leq 2.42)$

$= 0.9922 - 0.0078$

$= 0.9844$ [Tech: 0.9845]

If we take 100 simple random samples of size $n = 15$ human pregnancies, then about 98 of the samples will result in a mean gestation period between 256 and 276 days, inclusive.

23. (a) $P(X > 95) = P\left(Z > \frac{95 - 90}{10}\right)$

$= P(Z > 0.5)$

$= 1 - P(Z \leq 0.5)$

$= 1 - 0.6915$

$= 0.3085$

If we select a simple random sample of $n = 100$ second grade students, then about 31 of the students would read more than 95 words per minute.

(b) $\mu_{\bar{x}} = \mu = 90$; $\sigma_{\bar{x}} = \dfrac{\sigma}{\sqrt{n}} = \dfrac{10}{\sqrt{12}}$

$$P(\bar{x} > 95) = P\left(Z > \dfrac{95-90}{10/\sqrt{12}}\right)$$
$$= P(Z > 1.73)$$
$$= 1 - P(Z \le 1.73)$$
$$= 1 - 0.9582$$
$$= 0.0418 \quad [\text{Tech: } 0.0416]$$

If we take 100 simple random samples of size $n = 12$ second grade students, then about 4 of the samples will result in a mean reading rate that is more than 95 words per minute.

(c) $\mu_{\bar{x}} = \mu = 90$; $\sigma_{\bar{x}} = \dfrac{\sigma}{\sqrt{n}} = \dfrac{10}{\sqrt{24}}$

$$P(\bar{x} > 95) = P\left(Z > \dfrac{95-90}{10/\sqrt{24}}\right)$$
$$= P(Z > 2.45)$$
$$= 1 - P(Z \le 2.45)$$
$$= 1 - 0.9929$$
$$= 0.0071 \quad [\text{Tech: } 0.0072]$$

If we take 1000 simple random samples of size $n = 24$ second grade students, then about 7 of the samples will result in a mean reading rate that is more than 95 words per minute.

(d) Increasing the sample size decreases the probability that $\bar{x} > 95$. This happens because $\sigma_{\bar{x}}$ decreases as n increases.

(e) No, this result would not be unusual because, if $\mu_{\bar{x}} = \mu = 90$ and $\sigma_{\bar{x}} = \dfrac{\sigma}{\sqrt{n}} = \dfrac{10}{\sqrt{20}}$, then

$$P(\bar{x} > 92.8) = P\left(Z > \dfrac{92.8-90}{10/\sqrt{20}}\right)$$
$$= P(Z > 1.25)$$
$$= 1 - P(Z \le 1.25)$$
$$= 1 - 0.8944$$
$$= 0.1056 \quad [\text{Tech: } 0.1052]$$

If we take 100 simple random samples of size $n = 20$ second grade students, then about 11 of the samples will result in a mean reading rate that is above 92.8 words per minute. This result does not qualify as unusual. This means that the new reading program is not abundantly more effective than the old program.

25. (a) $P(X > 0) = P\left(Z > \dfrac{0 - 0.007233}{0.04135}\right)$
$$= P(Z > -0.17)$$
$$= 1 - P(Z \le -0.17)$$
$$= 1 - 0.4325$$
$$= 0.5675 \quad [\text{Tech: } 0.5694]$$

If we select a simple random sample of $n = 100$ months, then about 57 of the months would have positive rates of return.

(b) $\mu_{\bar{x}} = \mu = 0.007233$; $\sigma_{\bar{x}} = \dfrac{\sigma}{\sqrt{n}} = \dfrac{0.04135}{\sqrt{12}}$

$$P(\bar{x} > 0) = P\left(Z > \dfrac{0 - 0.007233}{0.04135/\sqrt{12}}\right)$$
$$= P(Z > -0.61)$$
$$= 1 - P(Z \le -0.61)$$
$$= 1 - 0.2709$$
$$= 0.7291 \quad [\text{Tech: } 0.7277]$$

If we take 100 simple random samples of size $n = 12$ months, then about 73 of the samples will result in a mean monthly rate that is positive.

(c) $\mu_{\bar{x}} = \mu = 0.007233$; $\sigma_{\bar{x}} = \dfrac{\sigma}{\sqrt{n}} = \dfrac{0.04135}{\sqrt{24}}$

$$P(\bar{x} > 0) = P\left(Z > \dfrac{0 - 0.007233}{0.04135/\sqrt{24}}\right)$$
$$= P(Z > -0.86)$$
$$= 1 - P(Z \le -0.86)$$
$$= 1 - 0.1949$$
$$= 0.8051 \quad [\text{Tech: } 0.8043]$$

If we take 100 simple random samples of size $n = 24$ months, then about 81 of the samples will result in a mean monthly rate that is positive.

(d) $\mu_{\bar{x}} = \mu = 0.007233$; $\sigma_{\bar{x}} = \dfrac{\sigma}{\sqrt{n}} = \dfrac{0.04135}{\sqrt{36}}$

$$P(\bar{x} > 0) = P\left(Z > \dfrac{0 - 0.007233}{0.04135/\sqrt{36}}\right)$$
$$= P(Z > -1.05)$$
$$= 1 - P(Z \le -1.05)$$
$$= 1 - 0.1469$$
$$= 0.8531 \quad [\text{Tech: } 0.8530]$$

If we take 100 simple random samples of size $n = 36$ months, then about 85 of the samples will result in a mean monthly rate that is positive.

(e) Answers will vary. Based on the results of parts (b)–(d), the likelihood of earning a positive rate of return increases as the investment time horizon increases.

27. (a) Without knowing the shape of the distribution, we would need a sample size of at least 30 so we could apply the Central Limit Theorem.

(b) $\mu_{\bar{x}} = \mu = 11.4$; $\sigma_{\bar{x}} = \dfrac{\sigma}{\sqrt{n}} = \dfrac{3.2}{\sqrt{40}}$

$$P(\bar{x} < 10) = P\left(Z < \dfrac{10 - 11.4}{3.2/\sqrt{40}}\right)$$
$$= P(Z < -2.77)$$
$$= 0.0028$$

If we take 1000 simple random samples of size $n = 40$ oil changes, then about 3 of the samples will result in a mean time less than 10 minutes.

29. (a) The sampling distribution of $\bar{x}$ is approximately normal because the sample is large, $n = 50 \geq 30$. From the Central Limit Theorem, as the sample size increases, the sampling distribution of the mean becomes more normal.

(b) Assuming that we are sampling from a population that is exactly at the Food Defect Action Level, $\mu_{\bar{x}} = \mu = 3$ and

$$\sigma_{\bar{x}} = \dfrac{\sigma}{\sqrt{n}} = \dfrac{\sqrt{3}}{\sqrt{50}} = \sqrt{\dfrac{3}{50}} \approx 0.245.$$

(c) $P(\bar{x} \geq 3.6) = P\left(Z \geq \dfrac{3.6 - 3}{\sqrt{3}/\sqrt{50}}\right)$

$$= P(Z \geq 2.45)$$
$$= 1 - 0.9929$$
$$= 0.0071 \quad [\text{Tech: } 0.0072]$$

If we take 1000 simple random samples of size $n = 50$ ten-gram portions of peanut butter, then about 7 of the samples will result in a mean of at least 3.6 insect fragments. This result is unusual. We might conclude that the sample comes from a population with a mean higher than 3 insect fragments per ten-gram portion.

31. (a) No, the variable "weekly time spent watching television" is not likely normally distributed. It is likely skewed right.

(b) Because the sample is large, $n = 40 \geq 30$, the sampling distribution of $\bar{x}$ is approximately normal with $\mu_{\bar{x}} = \mu = 2.35$

and $\sigma_{\bar{x}} = \dfrac{\sigma}{\sqrt{n}} = \dfrac{1.93}{\sqrt{40}} \approx 0.305$.

(c) $P(2 \leq \bar{x} \leq 3)$

$$= P\left(\dfrac{2 - 2.35}{1.93/\sqrt{40}} \leq Z \leq \dfrac{3 - 2.35}{1.93/\sqrt{40}}\right)$$
$$= P(-1.15 \leq Z \leq 2.13)$$
$$= 0.9834 - 0.1251$$
$$= 0.8583 \quad [\text{Tech: } 0.8577]$$

If we take 100 simple random samples of size $n = 40$ adult Americans, then about 86 of the samples will result in a mean time between 2 and 3 hours watching television on a weekday.

(d) $\mu_{\bar{x}} = \mu = 2.35$; $\sigma_{\bar{x}} = \dfrac{\sigma}{\sqrt{n}} = \dfrac{1.93}{\sqrt{35}}$

$$P(\bar{x} \leq 1.89) = P\left(Z \leq \dfrac{1.89 - 2.35}{1.93/\sqrt{35}}\right)$$
$$= P(Z \leq -1.41)$$
$$= 0.0793$$

If we take 100 simple random samples of size $n = 35$ adult Americans, then about 8 of the samples will result in a mean time of 1.89 hours or less watching television on a weekday. This result is not unusual, so this evidence is insufficient to conclude that avid Internet users watch less television.

33. (a) $\mu = \dfrac{\sum x}{N} = \dfrac{232}{6} \approx 38.7$

The population mean age is 38.7 years.

(b) 37, 43; 37, 29; 37, 47; 37, 36; 37, 40; 43, 29; 43, 47; 43, 36; 43, 40; 29, 47; 29, 36; 29, 40; 47, 36; 47, 40; 36, 40

(c) Obtain each sample mean by adding the two ages in a sample and dividing by two.

$$\bar{x} = \dfrac{37 + 43}{2} = 40 \text{ yr}; \quad \bar{x} = \dfrac{37 + 29}{2} = 33 \text{ yr};$$

$$\bar{x} = \dfrac{37 + 47}{2} = 42 \text{ yr}; \quad \bar{x} = \dfrac{37 + 36}{2} = 36.5 \text{ yr},$$

etc.

$\bar{x}$	$P(\bar{x})$	$\bar{x}$	$P(\bar{x})$
32.5	$1/15 \approx 0.0667$	39.5	$1/15 \approx 0.0667$
33	$1/15 \approx 0.0667$	40	$1/15 \approx 0.0667$
34.5	$1/15 \approx 0.0667$	41.5	$2/15 \approx 0.1333$
36	$1/15 \approx 0.0667$	42	$1/15 \approx 0.0667$
36.5	$1/15 \approx 0.0667$	43.5	$1/15 \approx 0.0667$
38	$2/15 \approx 0.1333$	45	$1/15 \approx 0.0667$
38.5	$1/15 \approx 0.0667$		

(d) $\mu_{\bar{x}} = (32.5)\left(\dfrac{1}{15}\right) + (33)\left(\dfrac{1}{15}\right) + \ldots + (45)\left(\dfrac{1}{15}\right)$

≈ 38.7 years

Notice that this is the same value we obtained in part (a) for the population mean.

(e) $P(35.7 \leq \bar{x} \leq 41.7)$

$= \dfrac{1}{15} + \dfrac{1}{15} + \dfrac{2}{15} + \dfrac{1}{15} + \dfrac{1}{15} + \dfrac{1}{15} + \dfrac{2}{15} = \dfrac{9}{15} = 0.6$

(f) for part (b):
37, 43, 29; 37, 43, 47; 37, 43, 36;
37, 43, 40; 37, 29, 47; 37, 29, 36;
37, 29, 40; 37, 47, 36; 37, 47, 40;
37, 36, 40; 43, 29, 47; 43, 29, 36;
43, 29, 40; 43, 47, 36; 43, 47, 40;
43, 36, 40; 29, 47, 36; 29, 47, 40;
29, 36, 40; 47, 36, 40

for part (c):
Obtain each sample mean by adding the three ages in a sample and dividing by three.

$\bar{x} = \dfrac{37+43+29}{3} \approx 36.3$ yr ;

$\bar{x} = \dfrac{37+43+47}{3} \approx 42.3$ yr ;

$\bar{x} = \dfrac{37+43+36}{3} = 38.7$ yr ; etc

$\bar{x}$	$P(\bar{x})$	$\bar{x}$	$P(\bar{x})$
34	$1/20 = 0.05$	39.7	$2/20 = 0.1$
35	$1/20 = 0.05$	40	$2/20 = 0.1$
35.3	$1/20 = 0.05$	41	$1/20 = 0.05$
36	$1/20 = 0.05$	41.3	$1/20 = 0.05$
36.3	$1/20 = 0.05$	42	$1/20 = 0.05$
37.3	$2/20 = 0.1$	42.3	$1/20 = 0.05$
37.7	$2/20 = 0.1$	43.3	$1/20 = 0.05$
38.7	$2/20 = 0.1$		

for part (d):

$\mu_{\bar{x}} = (34)\left(\dfrac{1}{20}\right) + (35)\left(\dfrac{1}{20}\right) + \ldots + (43.3)\left(\dfrac{1}{20}\right)$

≈ 38.7 years

Notice that this is the same value we obtained previously.

for part (e):

$P(35.7 \leq \bar{x} \leq 41.7) = \dfrac{14}{20} = \dfrac{7}{10} = 0.7$

With the larger sample size, the probability of obtaining a sample mean within 3 years of the population mean has increased.

35. (a) – (c) Answers will vary.

(d) $\mu_{\bar{x}} = \mu = 100$; $\sigma_{\bar{x}} = \dfrac{\sigma}{\sqrt{n}} = \dfrac{15}{\sqrt{20}} \approx 3.354$

(e) Answers will vary.

(f) $P(\bar{x} > 108) = P\left(Z > \dfrac{108-100}{15/\sqrt{20}} \right)$

$= P(Z > 2.39)$

$= 1 - P(Z \leq 2.39)$

$= 1 - 0.9916$

$= 0.0084$ [Tech: 0.0085]

(g) Answers will vary.

37. (a) – (b): Answers will vary. Using skewed right with $\mu = 3.92$ and $\sigma = 2.79$, we get:

For (a): $\mu_{\bar{x}} = 3.92$; $\sigma_{\bar{x}} = \dfrac{2.79}{\sqrt{5}} \approx 1.248$

For (b): $\mu_{\bar{x}} = 3.92$; $\sigma_{\bar{x}} = \dfrac{2.79}{\sqrt{10}} \approx 0.882$

(c) Specific results from the applet will vary. We can use the Central Limit Theorem to say that the sampling distribution is approximately normal since we have a large sample (that is, $n \geq 30$).

$\mu_{\bar{x}} = 3.92$; $\sigma_{\bar{x}} = \dfrac{2.79}{\sqrt{50}} \approx 0.395$

(d) Answers will vary. As the sample size increases, we should see the sampling distribution become more normally distributed.

39. (a) Assuming that only one number is selected, the probability distribution will be as follows:

x	$P(x)$
35	$1/38 \approx 0.0263$
-1	$37/38 \approx 0.9737$

(b) $\mu = (35)(0.0263) + (-1)(0.9737) \approx -\$0.05$

$\sigma = \sqrt{(35 - (-.05))^2 (0.0263) + (-1 - (-.05))^2 (0.9737)}$

$\approx \$5.76$

(c) Because the sample size is large, $n = 100 > 30$, the sampling distribution of $\bar{x}$ is approximately normal with $\mu_{\bar{x}} = \mu = -\$0.05$

and $\sigma_{\bar{x}} = \dfrac{\sigma}{\sqrt{n}} = \dfrac{5.76}{\sqrt{100}} \approx \$0.576$.

(d) $P(\bar{x} > 0) = P\left(Z > \dfrac{0 - (-0.05)}{5.76/\sqrt{100}} \right)$

$= P(Z > 0.09)$

$= 1 - P(Z \le 0.09)$

$= 1 - 0.5359$

$= 0.4641$ [Tech: 0.4654]

(e) $\mu_{\bar{x}} = -\$0.05;\ \sigma_{\bar{x}} = \dfrac{5.76}{\sqrt{200}} \approx \$0.407$

$P(\bar{x} > 0) = P\left(Z > \dfrac{0 - (-0.05)}{5.76/\sqrt{200}} \right)$

$= P(Z > 0.12)$

$= 1 - P(Z \le 0.12)$

$= 1 - 0.5478$

$= 0.4522$ [Tech: 0.4511]

(f) $\mu_{\bar{x}} = -\$0.05;\ \sigma_{\bar{x}} = \dfrac{5.76}{\sqrt{1000}} \approx \$0.275$

$P(\bar{x} > 0) = P\left(Z > \dfrac{0 - (-0.05)}{5.76/\sqrt{1000}} \right)$

$= P(Z > 0.27)$

$= 1 - P(Z \le 0.27)$

$= 1 - 0.6064$

$= 0.3936$ [Tech: 0.3918]

(g) The probability of being ahead decreases as the number of games played increases.

Section 8.2

1. $0.44;\ p = \dfrac{220}{500} = 0.44$

3. False; while it is possible for the sample proportion to have the same value as the population proportion, it will not *always* have the same value.

5. The sampling distribution of $\hat{p}$ is approximately normal when $n \le 0.05N$ and $np(1 - p) \ge 10$.

7. $25,000(0.05) = 1250$; the sample size, $n = 500$, is less than 5% of the population size and $np(1 - p) = 500(0.4)(0.6) = 120 > 10$.

The distribution of $\hat{p}$ is approximately normal, with mean $\mu_{\hat{p}} = p = 0.4$ and standard deviation

$\sigma_{\hat{p}} = \sqrt{\dfrac{p(1 - p)}{n}} = \sqrt{\dfrac{0.4(1 - 0.4)}{500}} \approx 0.022$.

9. $25,000(0.05) = 1250$; the sample size, $n = 1000$, is less than 5% of the population size and $np(1 - p) = 1000(0.103)(0.897) = 92.391 > 10$.

The distribution of $\hat{p}$ is approximately normal, with mean $\mu_{\hat{p}} = p = 0.103$ and standard deviation

$\sigma_{\hat{p}} = \sqrt{\dfrac{p(1 - p)}{n}} = \sqrt{\dfrac{0.103(1 - 0.103)}{1000}} \approx 0.010$.

11. (a) $10,000(0.05) = 500$; the sample size, $n = 75$, is less than 5% of the population size and $np(1 - p) = 75(0.8)(0.2) = 12 > 10$. The distribution of $\hat{p}$ is approximately normal, with mean $\mu_{\hat{p}} = p = 0.8$ and standard deviation

$\sigma_{\hat{p}} = \sqrt{\dfrac{p(1 - p)}{n}} = \sqrt{\dfrac{0.8(1 - 0.8)}{75}} \approx 0.046$.

(b) $P(\hat{p} \ge 0.84) = P\left(Z \ge \dfrac{0.84 - 0.8}{\sqrt{0.8(0.2)/75}} \right)$

$= P(Z \ge 0.87)$

$= 1 - P(Z < 0.87)$

$= 1 - 0.8078$

$= 0.1922$ [Tech: 0.1932]

About 19 out of 100 random samples of size $n = 75$ will result in 63 or more individuals (that is, 84% or more) with the characteristic.

(c) $P(\hat{p} \le 0.68) = P\left(Z \le \dfrac{0.68-0.8}{\sqrt{0.8(0.2)/75}} \right)$

$= P(Z \le -2.60)$

$= 0.0047$

About 5 out of 1000 random samples of size $n = 75$ will result in 51 or fewer individuals (that is, 68% or less) with the characteristic.

13. (a) $1,000,000(0.05) = 50,000$; the sample size, $n = 1000$, is less than 5% of the population size and

$np(1-p) = 1000(0.35)(0.65) = 227.5 > 10$.

The distribution of $\hat{p}$ is approximately normal, with mean $\mu_{\hat{p}} = p = 0.35$ and standard deviation

$\sigma_{\hat{p}} = \sqrt{\dfrac{p(1-p)}{n}} = \sqrt{\dfrac{0.35(1-0.35)}{1000}} \approx 0.015.$

(b) $\hat{p} = \dfrac{x}{n} = \dfrac{390}{1000} = 0.39$

$P(X \ge 390) = P(\hat{p} \ge 0.39)$

$= P\left(Z \ge \dfrac{0.39-0.35}{\sqrt{0.53(0.65)/1000}} \right)$

$= P(Z \ge 2.65)$

$= 1 - P(Z < 2.65)$

$= 1 - 0.9960$

$= 0.0040$

About 4 out of 1000 random samples of size $n = 1000$ will result in 390 or more individuals (that is, 39% or more) with the characteristic.

(c) $\hat{p} = \dfrac{x}{n} = \dfrac{320}{1000} = 0.32$

$P(X \le 320) = P(\hat{p} \le 0.32)$

$= P\left(Z \ge \dfrac{0.32-0.35}{\sqrt{0.53(0.65)/1000}} \right)$

$= P(Z \le -1.99)$

$= 0.0233$ [Tech: 0.0234]

About 2 out of 100 random samples of size $n = 1000$ will result in 320 or fewer individuals (that is, 32% or less) with the characteristic.

15. (a) The sample size, $n = 100$, is less than 5% of the population size and

$np(1-p) = 100(0.82)(0.18) = 14.76 > 10$.

The distribution of $\hat{p}$ is approximately normal, with mean $\mu_{\hat{p}} = p = 0.82$ and standard deviation

$\sigma_{\hat{p}} = \sqrt{\dfrac{p(1-p)}{n}} = \sqrt{\dfrac{0.82(1-0.82)}{100}} \approx 0.038$.

(b) $\hat{p} = \dfrac{x}{n} = \dfrac{85}{100} = 0.85$

$P(X \ge 85) = P(\hat{p} \ge 0.85)$

$= P\left(Z \ge \dfrac{0.85-0.82}{\sqrt{0.82(0.18)/100}} \right)$

$= P(Z \ge 0.78)$

$= 1 - P(Z < 0.78)$

$= 1 - 0.7823$

$= 0.2177$ [Tech: 0.2174]

About 22 out of 100 random samples of size $n = 100$ Americans will result in at least 85 (that is, at least 85%) who are satisfied with their lives.

(c) $\hat{p} = \dfrac{x}{n} = \dfrac{75}{100} = 0.75$

$P(X < 76) = P(X \le 75)$

$= P(\hat{p} \le 0.75)$

$= P\left(Z \le \dfrac{0.75-0.82}{\sqrt{0.82(0.18)/100}} \right)$

$= P(Z \le -1.82)$

$= 0.0344$ [Tech: 0.0342]

About 3 out of 100 random samples of size $n = 100$ Americans will result in fewer than 76 individuals (that is, 75% or less) who are satisfied with their lives. This result is unusual.

17. (a) Our sample size, $n = 500$, is less than 5% of the population size and

$np(1-p) = 500(0.26)(0.74) = 96.2 > 10$.

The distribution of $\hat{p}$ is approximately normal, with mean $\mu_{\hat{p}} = p = 0.26$ and standard deviation

$\sigma_{\hat{p}} = \sqrt{\dfrac{p(1-p)}{n}} = \sqrt{\dfrac{0.26(1-0.26)}{500}} \approx 0.020$.

(b) $P(\hat{p} < 0.24) = P\left(Z < \dfrac{0.24 - 0.26}{\sqrt{0.26(0.74)/500}}\right)$

$= P(Z < -1.02)$

$= 0.1539$ [Tech: 0.1540]

About 15 out of 100 random samples of size $n = 500$ adults will result in less than 24% who have no credit cards.

(c) $\hat{p} = \dfrac{x}{n} = \dfrac{150}{500} = 0.3$

$P(X \geq 150) = P(\hat{p} \geq 0.3)$

$= P\left(Z \geq \dfrac{0.3 - 0.26}{\sqrt{0.26(0.74)/500}}\right)$

$= P(Z \geq 2.04)$

$= 1 - P(Z < 2.04)$

$= 1 - 0.9793$

$= 0.0207$

About 2 out of 100 random samples of size $n = 500$ adults will result in more than 150 who have no credit cards. This result is unusual. Thus, it is unusual for 150 of 500 adults to have no credit cards.

19. $\hat{p} = \dfrac{x}{n} = \dfrac{121}{1100} = 0.11$

$P(X \geq 121) = P(\hat{p} \geq 0.11)$

$= P\left(Z \geq \dfrac{0.11 - 0.10}{\sqrt{0.1(0.9)/1100}}\right)$

$= P(Z \geq 1.11)$

$= 1 - P(Z < 1.11)$

$= 1 - 0.8665$

$= 0.1335$ [Tech: 0.1345]

This result is not unusual, so this evidence is insufficient to conclude that the proportion of Americans who are afraid to fly has increased above 0.10.

21. (a) To say the sampling distribution of $\hat{p}$ is approximately normal, we need $np(1-p) \geq 10$ and $n \leq 0.05N$.

With $p = 0.1$, we need

$n(0.1)(1-0.1) \geq 10$

$n(0.1)(0.9) \geq 10$

$n(0.09) \geq 10$

$n \geq 111.11$

Therefore, we need a sample size of 112, or 62 more adult Americans.

(b) With $p = 0.2$, we need

$n(0.2)(1-0.2) \geq 10$

$n(0.2)(0.8) \geq 10$

$n(0.16) \geq 10$

$n \geq 62.5$

Therefore, we need a sample size of 63, or 13 more, adult Americans.

23. A sample of size $n = 20$ households represents more than 5% of the population size $N = 100$ households in the association. Thus, the results from individuals in the sample are not independent of one another.

25. Answers will vary.

Chapter 8 Review Exercises

1. Answers will vary. The sampling distribution of a statistic (such as the sample mean) is a probability distribution for all possible values of the statistic computed from a sample of size n.

2. The sampling distribution of $\bar{x}$ is exactly normal if the population distribution is normal. If the population distribution is not normal, we apply the Central Limit Theorem and say that the distribution of $\bar{x}$ is approximately normal for a sufficiently large n (for finite σ). For our purposes, $n \geq 30$ is considered large enough to apply the theorem.

3. The sampling distribution of $\hat{p}$ is approximately normal if $np(1-p) \geq 10$ and $n \leq 0.05N$.

4. For $\bar{x}$: $\mu_{\bar{x}} = \mu$ and $\sigma_{\bar{x}} = \dfrac{\sigma}{\sqrt{n}}$

For $\hat{p}$: $\mu_{\hat{p}} = p$ and $\sigma_{\hat{p}} = \sqrt{\dfrac{p(1-p)}{n}}$

5. (a) $P(X > 2625) = P\left(Z > \dfrac{2625 - 2600}{50}\right)$

$= P(Z > 0.5)$

$= 1 - P(Z \leq 0.5)$

$= 1 - 0.6915$

$= 0.3085$

If we select a simple random sample of $n = 100$ pregnant women, then about 31 have energy needs of more than 2625 kcal/day. This result is not unusual.

(b) Since the population is normally distributed, the sampling distribution of $\bar{x}$ will be normal, regardless of the sample size. The mean of the distribution is $\mu_{\bar{x}} = \mu = 2600$ kcal, and the standard deviation is

$$\sigma_{\bar{x}} = \frac{\sigma}{\sqrt{n}} = \frac{50}{\sqrt{20}} \approx 11.180 \text{ kcal.}$$

(c) $P(\bar{x} > 2625) = P\left(Z > \dfrac{2625 - 2600}{50/\sqrt{20}} \right)$

$= P(Z > 2.24)$

$= 1 - P(Z \leq 2.24)$

$= 1 - 0.9875$

$= 0.0125$ [Tech: 0.0127]

If we take 100 simple random samples of size $n = 20$ pregnant women, then about 1 of the samples will result in a mean energy need of more than 2625 kcal/day. This result is unusual.

6. (a) Since we have a large sample ($n = 30$), we can use the Central Limit Theorem to say that the sampling distribution of $\bar{x}$ is approximately normal. The mean of the sampling distribution is $\mu_{\bar{x}} = \mu = 0.75$ inch and the standard deviation is

$$\sigma_{\bar{x}} = \frac{\sigma}{\sqrt{n}} = \frac{0.004}{\sqrt{30}} \approx 0.001 \text{ inch.}$$

(b) The quality control inspector will determine the machine needs an adjustment if the sample mean is either less than 0.748 inch or more than 0.752 inch.

$P(\text{needs adjustment})$

$= P(\bar{x} < 0.748) + P(\bar{x} > 0.752)$

$= P\left(Z < \dfrac{0.748 - 0.75}{0.004/\sqrt{30}} \right) + P\left(Z > \dfrac{0.752 - 0.75}{0.004/\sqrt{30}} \right)$

$= P(Z < -2.74) + P(Z > 2.74)$

$= P(Z < -2.74) + \left[1 - P(Z \leq 2.74) \right]$

$= 0.0031 + (1 - 0.9969)$

$= 0.0062$

There is a 0.0062 probability that the quality control inspector will conclude the machine needs an adjustment when, in fact, the machine is correctly calibrated.

7. (a) No, the variable "number of television sets" is not likely normally distributed. It is likely skewed right.

(b) $\bar{x} = \dfrac{102}{40} = 2.55$ televisions per household

(c) Because the sample is large, $n = 40 > 30$, the sampling distribution of $\bar{x}$ is approximately normal with $\mu_{\bar{x}} = \mu = 2.24$ and $\sigma_{\bar{x}} = \dfrac{\sigma}{\sqrt{n}} = \dfrac{1.38}{\sqrt{40}} \approx 0.218$.

$P(\bar{x} \geq 2.55) = P\left(Z \geq \dfrac{2.55 - 2.24}{1.38/\sqrt{40}} \right)$

$= P(Z \geq 1.42)$

$= 1 - P(Z < 1.42)$

$= 1 - 0.9222$

$= 0.0778$ [Tech: 0.0777]

If we take 100 simple random samples of size $n = 40$ households, then about 8 of the samples will result in a mean of 2.55 or more televisions. This result is not unusual, so it does not contradict the results reported by A.C. Nielsen.

8. (a) The sample size, $n = 600$, is less than 5% of the population size and $np(1 - p) = 600(0.72)(0.28) = 120.96 > 10$.

The sampling distribution of $\hat{p}$ is approximately normal, with mean $\mu_{\hat{p}} = p = 0.72$ and standard deviation

$$\sigma_{\hat{p}} = \sqrt{\frac{p(1-p)}{n}} = \sqrt{\frac{0.72(1-0.72)}{600}} \approx 0.018.$$

(b) $P(\hat{p} \leq 0.70) = P\left(Z \leq \dfrac{0.70 - 0.72}{\sqrt{0.72(0.28)/600}} \right)$

$= P(Z \leq -1.09)$

$= 0.1379$ [Tech: 0.1376]

About 14 out of 100 random samples of size $n = 600$ 18- to 29-year-olds will result in no more than 70% who would prefer to start their own business.

(c) $\hat{p} = \dfrac{x}{n} = \dfrac{450}{600} = 0.75$

$P(X \geq 450) = P(\hat{p} \geq 0.75)$

$= P\left(Z \geq \dfrac{0.75 - 0.72}{\sqrt{0.72(0.28)/600}} \right)$

$= P(Z \geq 1.64)$

$= 1 - P(Z < 1.64)$

$= 1 - 0.9495$

$= 0.0505$ [Tech: 0.0509]

About 5 out of 100 random samples of size $n = 600$ 18- to 29-year-olds will result in 450 or more who would prefer to start their own business. Since the probability is approximately 0.05, this result is considered unusual.

9. $\hat{p} = \dfrac{60}{500} = 0.12$; the sample size, $n = 500$, is less than 5% of the population size and $np(1-p) = 500(0.1)(0.9) = 45 > 10$, so the sampling distribution of $\hat{p}$ is approximately normal with mean $\mu_{\hat{p}} = p = 0.1$ and standard deviation

$$\sigma_{\hat{p}} = \sqrt{\frac{p(1-p)}{n}} = \sqrt{\frac{0.1(1-0.1)}{500}} \approx 0.013 .$$

$$P(x \geq 60) = P(\hat{p} \geq 0.12)$$

$$= P\left(Z \geq \frac{0.12 - 0.1}{\sqrt{0.1(0.9)/500}}\right)$$

$$= P(Z \geq 1.49)$$

$$= 1 - P(Z < 1.49)$$

$$= 1 - 0.9319$$

$$= 0.0681 \quad [\text{Tech: } 0.0680]$$

This result is not unusual, so this evidence is insufficient to conclude that the proportion of adults 25 years of age or older with advanced degrees increased.

10. (a) $\hat{p} = \dfrac{192}{638} \approx 0.301$

    (b) We assume that Cabrera' at-bats are independent of one another. Since $np(1-p) = 638(0.273)(1-0.273) \approx 126.6 \geq 10$, the distribution of the sample proportion is normal.

    (c) The sampling distribution of $\hat{p}$ is approximately normal with mean $\mu_{\hat{p}} = p = 0.273$ and standard deviation

    $$\sigma_{\hat{p}} = \sqrt{\frac{p(1-p)}{n}}$$

    $$= \sqrt{\frac{0.273(0.727)}{638}} \approx 0.018$$

(d) $P(\hat{p} \geq 0.301) = P\left(Z \geq \dfrac{0.301 - 0.273}{\sqrt{0.273(0.727)/638}}\right)$

$$= P(Z \geq 1.59)$$

$$= 1 - P(Z < 1.59)$$

$$= 1 - 0.9441$$

$$= 0.0559 \quad [\text{Tech: } 0.0562]$$

If we take 100 simple random samples of size $n = 638$ at-bats, then about 6 of the samples will result in a batting average of 0.301 or greater. This result is not unusual, so Cabrera's year was not all that unusual.

(e) $P(\hat{p} \leq 0.251) = P\left(Z \leq \dfrac{0.251 - 0.273}{\sqrt{0.273(0.727)/638}}\right)$

$$= P(Z \leq -1.25)$$

$$= 0.1056 \quad [\text{Tech: } 0.1061]$$

If we take 100 simple random samples of size $n = 638$ at-bats, then about 11 of the samples will result in a batting average of 0.251 or less. This result is not unusual, so it would not be unusual for Cabrera to have a 0.251 batting average in 2008.

Chapter 8 Test

1. The Central Limit Theorem states that, regardless of the shape of the population, the sampling distribution of $\overline{x}$ becomes approximately normal as the sample size n increases.

2. $\mu_{\overline{x}} = \mu = 50$; $\sigma_{\overline{x}} = \dfrac{\sigma}{\sqrt{n}} = \dfrac{24}{\sqrt{36}} = 4$

3. (a) $P(X > 100) = P\left(Z > \dfrac{100 - 90}{35}\right)$

$$= P(Z > 0.29)$$

$$= 1 - P(Z \leq 0.29)$$

$$= 1 - 0.6141$$

$$= 0.3859 \quad [\text{Tech: } 0.3875]$$

If we select a simple random sample of $n = 100$ batteries of this type, then about 39 batteries would last more than 100 minutes. This result is not unusual.

(b) Since the population is normally distributed, the sampling distribution of $\bar{x}$ will be normal, regardless of the sample size. The mean of the distribution is $\mu_{\bar{x}} = \mu = 90$ minutes, and the standard deviation is

$$\sigma_{\bar{x}} = \frac{\sigma}{\sqrt{n}} = \frac{35}{\sqrt{10}} \approx 11.068 \text{ minutes.}$$

(c) $P(\bar{x} > 100) = P\left(Z > \frac{100-90}{35/\sqrt{10}}\right)$

$= P(Z > 0.90)$

$= 1 - P(Z \le 0.90)$

$= 1 - 0.8159$

$= 0.1841$ [Tech: 0.1831]

If we take 100 simple random samples of size $n = 10$ batteries of this type, then about 18 of the samples will result in a mean charge life of more than 100 minutes. This result is not unusual.

(d) $\mu_{\bar{x}} = \mu = 90$; $\sigma_{\bar{x}} = \frac{\sigma}{\sqrt{n}} = \frac{35}{\sqrt{25}} = 7$

$P(\bar{x} > 100) = P\left(Z > \frac{100-90}{7}\right)$

$= P(Z > 1.43)$

$= 1 - P(Z \le 1.43)$

$= 1 - 0.9263$

$= 0.0764$ [Tech: 0.0766]

If we take 100 simple random samples of size $n = 25$ batteries of this type, then about 8 of the samples will result in a mean charge life of more than 100 minutes.

(e) The probabilities are different because a change in n causes a change in $\sigma_{\bar{x}}$.

4. (a) Since we have a large sample ($n = 45 \ge 30$), we can use the Central Limit Theorem to say that the sampling distribution of $\bar{x}$ is approximately normal. The mean of the sampling distribution is $\mu_{\bar{x}} = \mu = 2.0$ liters and the standard deviation is

$$\sigma_{\bar{x}} = \frac{\sigma}{\sqrt{n}} = \frac{0.05}{\sqrt{45}} \approx 0.007 \text{ liter.}$$

(b) The quality control manager will determine the machine needs an adjustment if the sample mean is either less than 1.98 liters or greater than 2.02 liters.

$P(\text{needs adjustment})$

$= P(\bar{x} < 1.98) + P(\bar{x} > 2.02)$

$= P\left(Z < \frac{1.98-2.0}{0.05/\sqrt{45}}\right) + P\left(Z > \frac{2.02-2.0}{0.05/\sqrt{45}}\right)$

$= P(Z < -2.68) + P(Z > 2.68)$

$= P(Z < -2.68) + \left[1 - P(Z \le 2.68)\right]$

$= 0.0037 + (1 - 0.9963)$

$= 0.0074$ [Tech: 0.0073]

There is a 0.0074 probability that the quality control manager will conclude the machine needs an adjustment even though the machine is correctly calibrated.

5. (a) Our sample size, $n = 300$, is less than 5% of the population size and
$np(1-p) = 300(0.224)(0.776) \approx 52 > 10$.
The distribution of $\hat{p}$ is approximately normal, with mean $\mu_{\hat{p}} = p = 0.224$ and standard deviation

$$\sigma_{\hat{p}} = \sqrt{\frac{p(1-p)}{n}}$$

$$= \sqrt{\frac{0.224(1-0.224)}{300}} \approx 0.024$$

(b) $\hat{p} = \frac{x}{n} = \frac{50}{300} = \frac{1}{6} \approx 0.167$

$P(X \ge 50) = P\left(\hat{p} \ge \frac{1}{6}\right)$

$= P\left(Z \ge \frac{\frac{1}{6} - 0.224}{\sqrt{0.224(0.776)/300}}\right)$

$= P(Z \ge -2.38)$

$= 1 - P(Z < -2.38)$

$= 1 - 0.0087$

$= 0.9913$ [Tech: 0.9914]

About 99 out of 100 random samples of size $n = 300$ adults will result in at least 50 adults (that is, at least 16.7%) who are smokers.

(c) $P(\hat{p} \le 0.18) = P\left(Z \le \frac{0.18 - 0.224}{\sqrt{0.224(0.776)/300}}\right)$

$= P(Z \le -1.83)$

$= 0.0336$ [Tech: 0.0338]

About 3 out of 100 random samples of size $n = 300$ adults will result 18% or less who are smokers. This result would be unusual.

6. (a) For the sample proportion to be normal, the sample size must be large enough to meet the condition $np(1-p) \geq 10$. Since $p = 0.01$, we have:
$$n(0.01)(1-0.01) \geq 10$$
$$0.0099n \geq 10$$
$$n \geq \frac{10}{0.0099} \approx 1010.1$$

Thus, the sample size must be at least 1011 in order to satisfy the condition.

(b) $\hat{p} = \dfrac{x}{n} = \dfrac{9}{1500} = \dfrac{3}{500} = 0.006$

$$P(X < 10) = P(X \leq 9)$$
$$= P(\hat{p} \leq 0.006)$$
$$= P\left(Z \leq \frac{0.006 - 0.01}{\sqrt{0.01(0.99)/1500}}\right)$$
$$= P(Z \leq -1.56)$$
$$= 0.0594 \quad [\text{Tech: } 0.597]$$

About 6 out of 100 random samples of size $n = 1500$ Americans will result in fewer than 10 with peanut or tree nut allergies. This result is not considered unusual.

7. $\hat{p} = \dfrac{82}{1000} = 0.082$; the sample size, $n = 1000$, is less than 5% of the population size and $np(1-p) = 1000(0.07)(0.93) = 65.1 > 10$, so the sampling distribution of $\hat{p}$ is approximately normal with mean $\mu_{\hat{p}} = p = 0.07$ and standard deviation

$$\sigma_{\hat{p}} = \sqrt{\frac{p(1-p)}{n}} = \sqrt{\frac{0.07(1-0.07)}{1000}} \approx 0.008 .$$

$$P(x \geq 82) = P(\hat{p} \geq 0.082)$$
$$= P\left(Z \geq \frac{0.082 - 0.07}{\sqrt{0.07(0.93)/1000}}\right)$$
$$= P(Z \geq 1.49)$$
$$= 1 - P(Z < 1.49)$$
$$= 1 - 0.9319$$
$$= 0.0681 \quad [\text{Tech: } 0.0685]$$

This result is not unusual, so this evidence is insufficient to conclude that the proportion of households with a net worth in excess of $1 million has increased above 7%.

Chapter 9

Estimating the Value of a Parameter Using Confidence Intervals

9.1 The Logic in Constructing Confidence Intervals about a Population Mean Where the Population Standard Deviation is Known

1. The margin of error of a confidence interval of a parameter depends on the level of confidence, the sample size, and the standard deviation of the population.

3. The margin of error decreases as the sample size increases because the Law of Large Numbers states that as the sample size increases the sample mean approaches the value of the population mean.

5. The mean age of the population is a fixed value (i.e., constant), so it is not probabilistic. The 95% level of confidence refers to confidence in the method by which the interval is obtained, not the specific interval. A better interpretation would be: "We are confident that the interval 21.4 years to 28.8 years, obtained by using our method, is one of the 95% of confidence intervals that contains the population mean." The precision of the interval could be increased by increasing the sample size or decreasing the level of confidence.

7. No, a Z-interval should not be constructed because the data are not normal since a point is outside the bounds of the normal probability plot. Also, the data contain an outlier which can be seen in the boxplot.

9. No, a Z-interval should not be constructed because the data are not normal since points are outside the bounds of the normal probability plot. From the boxplot, the data appear to be skewed right.

11. Yes, a Z-interval can be constructed. The plotted points are all within the bounds of the normal probability plot, which also has a generally linear pattern. The boxplot shows that there are no outliers.

13. For a 98% confidence interval, we use $\alpha = 1 - 0.98 = 0.02$, so $z_{\alpha/2} = z_{0.01}$ which is the z-score with area 0.99 below it. The closest area in Table V to 0.9900 is 0.9901 corresponding to $z_{0.01} = 2.33$

15. For a 85% confidence interval we use $\alpha = 1 - 0.85 = 0.15$, so $z_{\alpha/2} = z_{0.075}$ which is the z-score with area 0.9250 below it. The closest area in Table V to 0.9250 is 0.9251 corresponding to $z_{0.075} = 1.44$.

17. The point estimate of the population mean is the midpoint of the confidence interval. We obtain this value by averaging the lower and upper bounds. The margin of error is half the width of the confidence interval. We find this value by subtracting the lower bound from the upper bound and dividing the result by 2.

$$\bar{x} = \frac{18 + 24}{2} = \frac{42}{2} = 21$$

$$E = \frac{24 - 18}{2} = \frac{6}{2} = 3$$

19. The point estimate of the population mean is the midpoint of the confidence interval. We obtain this value by averaging the lower and upper bounds. The margin of error is half the width of the confidence interval. We find this value by subtracting the lower bound from the upper bound and dividing the result by 2.

$$\bar{x} = \frac{5 + 23}{2} = \frac{28}{2} = 14$$

$$E = \frac{23 - 5}{2} = \frac{18}{2} = 9$$

21. (a) For 95% confidence the critical value is $z_{0.025} = 1.96$. Then:

$$\text{Lower bound} = \overline{x} - z_{0.025} \cdot \frac{\sigma}{\sqrt{n}}$$

$$= 34.2 - 1.96 \cdot \frac{5.3}{\sqrt{35}}$$

$$\approx 34.2 - 1.76 = 32.44$$

$$\text{Upper bound} = \overline{x} + z_{0.025} \cdot \frac{\sigma}{\sqrt{n}}$$

$$= 34.2 + 1.96 \cdot \frac{5.3}{\sqrt{35}}$$

$$\approx 34.2 + 1.76 = 35.96$$

(b) $\text{Lower bound} = \overline{x} - z_{0.025} \cdot \frac{\sigma}{\sqrt{n}}$

$$= 34.2 - 1.96 \cdot \frac{5.3}{\sqrt{50}}$$

$$\approx 34.2 - 1.47 = 32.73$$

$$\text{Upper bound} = \overline{x} + z_{0.025} \cdot \frac{\sigma}{\sqrt{n}}$$

$$= 34.2 + 1.96 \cdot \frac{5.3}{\sqrt{50}}$$

$$\approx 34.2 + 1.47 = 35.67$$

Increasing the sample size decreases the margin of error.

(c) For 99% confidence the critical value is $z_{0.005} = 2.575$. Then:

$$\text{Lower bound} = \overline{x} - z_{0.005} \cdot \frac{\sigma}{\sqrt{n}}$$

$$= 34.2 - 2.575 \cdot \frac{5.3}{\sqrt{35}}$$

$$\approx 34.2 - 2.31 = 31.89$$

$$\text{Upper bound} = \overline{x} + z_{0.005} \cdot \frac{\sigma}{\sqrt{n}}$$

$$= 34.2 + 2.575 \cdot \frac{5.3}{\sqrt{35}}$$

$$\approx 34.2 + 2.31 = 36.51$$

Increasing the level of confidence increases the margin of error.

(d) Since a sample size of $n = 15$ is less than 30, we can only compute a confidence interval in this way if the population from which we are sampling is normal.

23. (a) For 96% confidence the critical value is $z_{0.02} = 2.05$. Then:

$$\text{Lower bound} = \overline{x} - z_{0.02} \cdot \frac{\sigma}{\sqrt{n}}$$

$$= 108 - 2.05 \cdot \frac{13}{\sqrt{25}}$$

$$\approx 108 - 5.3 = 102.7$$

$$\text{Upper bound} = \overline{x} + z_{0.02} \cdot \frac{\sigma}{\sqrt{n}}$$

$$= 108 + 2.05 \cdot \frac{13}{\sqrt{25}}$$

$$\approx 108 + 5.3 = 113.3$$

(b) $\text{Lower bound} = \overline{x} - z_{0.02} \cdot \frac{\sigma}{\sqrt{n}}$

$$= 108 - 2.05 \cdot \frac{13}{\sqrt{10}}$$

$$\approx 108 - 8.4 = 99.6$$

$$\text{Upper bound} = \overline{x} + z_{0.02} \cdot \frac{\sigma}{\sqrt{n}}$$

$$= 108 + 2.05 \cdot \frac{13}{\sqrt{10}}$$

$$\approx 108 + 8.4 = 116.4$$

Decreasing the sample size increases the margin of error.

(c) For 88% confidence the critical value is $z_{0.06} = 1.555$. Then:

$$\text{Lower bound} = \overline{x} - z_{0.06} \cdot \frac{\sigma}{\sqrt{n}}$$

$$= 108 - 1.555 \cdot \frac{13}{\sqrt{25}}$$

$$\approx 108 - 4.0 = 104.0$$

$$\text{Upper bound} = \overline{x} + z_{0.06} \cdot \frac{\sigma}{\sqrt{n}}$$

$$= 108 + 1.555 \cdot \frac{13}{\sqrt{25}}$$

$$\approx 108 + 4.0 = 112.0$$

Decreasing the level of confidence decreases the margin of error.

(d) No. Each sample size is too small to insure the $\overline{x}$ sampling distribution is normal.

(e) The outliers would have increased the sample mean, shifting the confidence interval to the right. If there are outliers then we should not use this approach to compute a confidence interval.

25. (a) Flawed; this interpretation implies that the population mean varies rather than the interval. For a given population, the mean number of hours worked is fixed, though typically unknown.

(b) This is the correct interpretation.

(c) Flawed; this interpretation makes an implication statement about individual values rather than the mean.

(d) Flawed; since the sample was of adult Americans, the interpretation should be about the mean number of hours worked by adult Americans, not just about adults in Idaho.

27. Based on the sample data, we are 90% confident that the mean drive-through service time for Taco Bell restaurants is between 161.5 seconds and 164.7 seconds. We are confident that the sample drawn is one of the 90% of the same size that would produce an interval containing the true mean.

29. The precision of the interval could be increased by either increasing the sample size, or lowering the confidence level.

31. (a) For 90% confidence the critical value is $z_{0.05} = 1.645$. Then:

$$\text{Lower bound} = \bar{x} - z_{0.05} \cdot \frac{\sigma}{\sqrt{n}}$$

$$= 0.16 - 1.645 \cdot \frac{0.08}{\sqrt{1200}}$$

$$\approx 0.16 - 0.004 = 0.156 \text{ g/dL}$$

$$\text{Upper bound} = \bar{x} + z_{0.05} \cdot \frac{\sigma}{\sqrt{n}}$$

$$= 0.16 + 1.645 \cdot \frac{0.08}{\sqrt{1200}}$$

$$\approx 0.16 + 0.004 = 0.164 \text{ g/dL}$$

The researcher is 90% confident that the population mean BAC is between 0.156 and 0.164 g/dL for drivers involved in fatal accidents who have a positive BAC value.

(b) Yes, it is possible that the true mean BAC is not captured in the interval from part (a). Using 90% confidence, only 90% of all possible samples of the same size from the population will produce intervals containing the true value of the mean. The interval from part (a) could be one of the 10% that do not capture the true mean.

33. (a) For 95% confidence the critical value is $z_{0.025} = 1.96$. Then:

$$\text{Lower bound} = \bar{x} - z_{0.025} \cdot \frac{\sigma}{\sqrt{n}}$$

$$= 4.113 - 1.96 \cdot \frac{0.110}{\sqrt{900}}$$

$$\approx 4.113 - 0.0072 = 4.1058$$

$$\text{Upper bound} = \bar{x} + z_{0.025} \cdot \frac{\sigma}{\sqrt{n}}$$

$$= 4.113 + 1.96 \cdot \frac{0.110}{\sqrt{900}}$$

$$\approx 4.113 + 0.0072 = 4.1202$$

The EIA is 95% confident that the population mean price per gallon for regular-grade gasoline on July 14, 2008 was between \$4.1058 and \$4.1202.

(b) No; the interval from part (a) is about the mean price for gasoline in the entire nation. A different result in a specific town does not make the interval inaccurate.

35. (a) $\bar{x} = \dfrac{1446 + 743 + 1581 + \ldots + 995}{12}$

$$= \frac{14,451}{12} \approx \$1204.3$$

(b) Yes, the conditions for a Z-interval are satisfied. The plotted points are all within the bounds of the normal probability plot, which also has a generally linear pattern. The boxplot shows that there are no outliers.

(c) For 95% confidence the critical value is $z_{0.025} = 1.96$. Then:

$$\text{Lower bound} = \bar{x} - z_{0.025} \cdot \frac{\sigma}{\sqrt{n}}$$

$$= 1204.3 - 1.96 \cdot \frac{450}{\sqrt{12}}$$

$$\approx 1204.3 - 254.6$$

$$= \$949.7 \quad [\text{Tech: } \$949.6]$$

$$\text{Upper bound} = \bar{x} + z_{0.025} \cdot \frac{\sigma}{\sqrt{n}}$$

$$= 1204.3 + 1.96 \cdot \frac{450}{\sqrt{12}}$$

$$\approx 1204.3 + 254.6$$

$$= \$1458.9$$

The IIHS is 95% confident that the population mean cost of repairs is between \$949.7 and \$1458.9.

(d) For 90% confidence the critical value is $z_{0.05} = 1.645$. Then:

$$\text{Lower bound} = \bar{x} - z_{0.025} \cdot \frac{\sigma}{\sqrt{n}}$$

$$= 1204.3 - 1.645 \cdot \frac{450}{\sqrt{12}}$$

$$\approx 1204.3 - 213.7$$

$$= \$990.6$$

$$\text{Upper bound} = \bar{x} + z_{0.025} \cdot \frac{\sigma}{\sqrt{n}}$$

$$= 1204.3 + 1.645 \cdot \frac{450}{\sqrt{12}}$$

$$\approx 1204.3 + 213.7$$

$$= \$1418.0 \text{ [Tech: \$1417.9]}$$

The IIHS is 90% confident that the population mean cost of repairs is between \$990.6 and \$1418.0.

(e) As the confidence level decreased, the interval width decreased. This is reasonable because, for a fixed standard error, we are less confident that a smaller interval will capture the true mean.

37. (a) $\bar{x} = \dfrac{\sum x}{20} = \dfrac{829}{20} \approx 41.5$ years

(b) Yes. All the data values lie within the bounds on the normal probability plot, indicating that the data could come from a population that is normal. The boxplot does not show any outliers.

(c) For 95% confidence the critical value is $z_{0.025} = 1.96$. Then:

$$\text{Lower bound} = \bar{x} - z_{0.025} \cdot \frac{\sigma}{\sqrt{n}}$$

$$= 41.5 - 1.96 \cdot \frac{8.7}{\sqrt{20}}$$

$$\approx 41.5 - 3.8$$

$$= 37.7 \text{ years [Tech: 37.6 yrs]}$$

$$\text{Upper bound} = \bar{x} + z_{0.025} \cdot \frac{\sigma}{\sqrt{n}}$$

$$= 41.5 + 1.96 \cdot \frac{8.7}{\sqrt{20}}$$

$$\approx 41.5 + 3.8 = 45.3 \text{ years}$$

The agent is 95% confident that the population mean age of buyers in his area who purchase investment property is between 37.7 and 45.3 years.

(d) No, the real estate agent's clients do not appear to differ in age from the general population since the mean age, 39, reported by the National Association of Realtors is contained within the confidence interval.

39. (a) Since the length of dramas (i.e. the population) is not normally distributed, the sample must be large so that the $\bar{x}$ distribution will be approximately normal.

(b) For 99% confidence the critical value is $z_{0.005} = 2.575$. Then:

$$\text{Lower bound} = \bar{x} - z_{0.05} \cdot \frac{\sigma}{\sqrt{n}}$$

$$= 138.3 - 2.575 \cdot \frac{27.3}{\sqrt{30}}$$

$$\approx 138.3 - 12.83$$

$$= 125.47 \text{ minutes}$$

$$\text{Upper bound} = \bar{x} + z_{0.05} \cdot \frac{\sigma}{\sqrt{n}}$$

$$= 138.3 + 2.575 \cdot \frac{27.3}{\sqrt{30}}$$

$$\approx 138.3 + 12.83$$

$$= 151.13 \text{ minutes}$$

The student is 99% confident that the population mean length of a drama is between 125.47 and 151.13 minutes.

41. (a) $\bar{x} = \dfrac{\sum x}{34} = \dfrac{1,589,283}{34} \approx 46,743.6$ miles

(b) The distribution is skewed right.

(c) Because the population is not normally distributed, a large sample is needed to apply the Central Limit Theorem and say that the $\bar{x}$ distribution is approximately normal.

(d) For 99% confidence the critical value is $z_{0.025} = 2.575$. Then:

$$\text{Lower bound} = \bar{x} - z_{0.005} \cdot \frac{\sigma}{\sqrt{n}}$$

$$= 46,743.6 - 2.575 \cdot \frac{19,000}{\sqrt{34}}$$

$$\approx 46,743.6 - 8390.6$$

$$= 38,353.0 \text{ miles}$$

$$\text{[Tech: 38,350.3]}$$

Upper bound $= \bar{x} + z_{0.005} \cdot \dfrac{\sigma}{\sqrt{n}}$

$= 46,743.6 + 2.575 \cdot \dfrac{19,000}{\sqrt{34}}$

$\approx 46,743.6 + 8390.6$

$= 55,134.2$ miles

[Tech: 55,136.9]

The used-car dealer is 99% confident that the mean number of miles on a 4-year-old Hummer H2 is between 38,353.0 and 55,134.2.

(e) For 95% confidence the critical value is $z_{0.025} = 1.96$. Then:

Lower bound $= \bar{x} - z_{0.005} \cdot \dfrac{\sigma}{\sqrt{n}}$

$= 46,743.6 - 1.96 \cdot \dfrac{19,000}{\sqrt{34}}$

$\approx 46,743.6 - 6386.6$

$= 40,357.0$ miles

[Tech: 40,357.1]

Upper bound $= \bar{x} + z_{0.005} \cdot \dfrac{\sigma}{\sqrt{n}}$

$= 46,743.6 + 1.96 \cdot \dfrac{19,000}{\sqrt{34}}$

$\approx 46,743.6 + 6386.6$

$= 53,130.2$ miles

[Tech: 53,130.1]

The used-car dealer is 95% confident that the mean number of miles on a 4-year-old Hummer H2 is between 40,357.0 and 53,130.2.

(f) Decreasing the level of confidence decreases the width of the interval.

(g) No; the used-car dealer only sampled Hummers in the Midwest. Therefore, her results cannot be generalized to the whole United States.

43. For 99% confidence, we use $z_{0.005} = 2.575$. So,

$n = \left(\dfrac{z_{\alpha/2} \cdot \sigma}{E} \right)^2 = \left(\dfrac{2.575 \cdot 13.4}{2} \right)^2 \approx 297.65$,

which we must increase to 298 subjects.

For 95% confidence we use $z_{0.025} = 1.96$. So,

$n = \left(\dfrac{1.96 \cdot 13.4}{2} \right)^2 \approx 172.45$, which we must

increase to 173 subjects.

Decreasing the level of confidence decreases the required sample size.

45. For 95% confidence, we use $z_{0.025} = 1.96$. So,

$n = \left(\dfrac{z_{\alpha/2} \cdot \sigma}{E} \right)^2 = \left(\dfrac{1.96 \cdot 16.6}{1} \right)^2 \approx 1058.59$,

which we must increase to 1059 subjects.

47. (a) For 90% confidence, we use $z_{0.05} = 1.645$. So,

$n = \left(\dfrac{z_{\alpha/2} \cdot \sigma}{E} \right)^2 = \left(\dfrac{1.645 \cdot 19,000}{2000} \right)^2$,

≈ 244.22

which we must increase to 245 vehicles.

(b) For 90% confidence, we use $z_{0.05} = 1.645$. So,

$n = \left(\dfrac{z_{\alpha/2} \cdot \sigma}{E} \right)^2 = \left(\dfrac{1.645 \cdot 19,000}{1000} \right)^2$,

≈ 976.88

which we must increase to 977 vehicles.

(c) Doubling the required accuracy (that is, cutting the margin of error in half) will approximately quadruple the require sample size. This increase is expected because the sample size is inversely proportional to the square of the error. To half the error we must increase the sample size by a factor of $\left(\frac{1}{1/2} \right)^2 = 2^2 = 4$.

49. (a), (b) Answers will vary.

(c) 95% of 20 is $0.95(20) = 19$. We would expect about 19 of the 20 samples to generate confidence intervals that include the population mean. The actual results will vary.

51. (a), (b) Answers will vary.

(c) If these were truly "95% confidence intervals," then we would expect approximately 95% of the 100 samples, or 95 samples, to generate confidence intervals that include the population mean. Actual results will vary.

(d) Since the sample size, $n = 6$, is small and since we are sampling from a non-normal population, the sampling distribution of the sample mean is not normal, so our method for computing a 95% confidence interval is not valid. In other words, it is not true that close to 95% of our intervals contain the population mean.

53. The sample size must be increased by a factor of 4. This is because the sample size, n, is inversely proportional to the square of the error, E. To decrease the error by a factor of $\frac{1}{2}$ we must increase the sample size by a

factor of $\left(\frac{1}{1/2}\right)^2 = 4$.

55. (a) Data Set I: $\bar{x} = \dfrac{\sum x}{n} = \dfrac{793}{8} \approx 99.1$;

Data Set II: $\bar{x} = \dfrac{\sum x}{n} = \dfrac{1982}{20} = 99.1$;

Data Set III: $\bar{x} = \dfrac{\sum x}{n} = \dfrac{2971}{30} \approx 99.0$

(b) For 95% confidence the critical value is $z_{0.025} = 1.96$.

Set I: Lower bound $= \bar{x} - z_{0.025} \cdot \dfrac{\sigma}{\sqrt{n}}$

$= 99.1 - 1.96 \cdot \dfrac{15}{\sqrt{8}}$

$\approx 99.1 - 10.4$

$= 88.7$

Upper bound $= \bar{x} + z_{0.025} \cdot \dfrac{\sigma}{\sqrt{n}}$

$= 99.1 + 1.96 \cdot \dfrac{15}{\sqrt{8}}$

$\approx 99.1 + 10.4$

$= 109.5$

Set II: Lower bound $= \bar{x} - z_{0.025} \cdot \dfrac{\sigma}{\sqrt{n}}$

$= 99.1 - 1.96 \cdot \dfrac{15}{\sqrt{20}}$

$\approx 99.1 - 6.6$

$= 92.5$

Upper bound $= \bar{x} + z_{0.025} \cdot \dfrac{\sigma}{\sqrt{n}}$

$= 99.1 + 1.96 \cdot \dfrac{15}{\sqrt{20}}$

$\approx 99.1 + 6.6$

$= 105.7$

Set III: Lower bound $= \bar{x} - z_{0.025} \cdot \dfrac{\sigma}{\sqrt{n}}$

$= 99.0 - 1.96 \cdot \dfrac{15}{\sqrt{30}}$

$\approx 99.0 - 5.4$

$= 93.6$

[Tech: 93.7]

Upper bound $= \bar{x} + z_{0.025} \cdot \dfrac{\sigma}{\sqrt{n}}$

$= 99.0 + 1.96 \cdot \dfrac{15}{\sqrt{30}}$

$\approx 99.0 + 5.4$

$= 104.4$

(c) As the size of the sample increases, the width of the confidence interval decreases.

(d) Set I: $\bar{x} = \dfrac{\sum x}{n} = \dfrac{703}{8} = 87.9$;

Lower bound $= \bar{x} - z_{0.025} \cdot \dfrac{\sigma}{\sqrt{n}}$

$= 87.9 - 1.96 \cdot \dfrac{15}{\sqrt{8}}$

$\approx 87.9 - 10.4 = 77.5$

Upper bound $= \bar{x} + z_{0.025} \cdot \dfrac{\sigma}{\sqrt{n}}$

$= 87.9 + 1.96 \cdot \dfrac{15}{\sqrt{8}}$

$\approx 87.9 + 10.4 = 98.3$

Set II: $\bar{x} = \dfrac{\sum x}{n} = \dfrac{1892}{20} = 94.6$;

Lower bound $= \bar{x} - z_{0.025} \cdot \dfrac{\sigma}{\sqrt{n}}$

$= 94.6 - 1.96 \cdot \dfrac{15}{\sqrt{20}}$

$\approx 94.6 - 6.6 = 88.0$

Upper bound $= \bar{x} + z_{0.025} \cdot \dfrac{\sigma}{\sqrt{n}}$

$= 94.6 + 1.96 \cdot \dfrac{15}{\sqrt{20}}$

$\approx 94.6 + 6.6 = 101.2$

Set III: $\bar{x} = \dfrac{\sum x}{n} = \dfrac{2881}{30} \approx 96.0$

Lower bound $= \bar{x} - z_{0.025} \cdot \dfrac{\sigma}{\sqrt{n}}$

$= 96.0 - 1.96 \cdot \dfrac{15}{\sqrt{30}}$

$\approx 96.0 - 5.4 = 90.6$

[Tech: 90.7]

Upper bound $= \bar{x} + z_{0.025} \cdot \dfrac{\sigma}{\sqrt{n}}$

$= 96.0 + 1.96 \cdot \dfrac{15}{\sqrt{30}}$

$\approx 96.0 + 5.4 = 101.4$

(e) The confidence intervals for both Data Set II and Data Set III still capture the population mean, 100, with the incorrect entry. The interval from Data Set I does not capture the population mean when the incorrect entry is made. The concept of robustness is illustrated. As the sample size increases the distribution of $\bar{x}$ becomes more normal, making the confidence interval more robust against the effect of the incorrect entry.

57. (a) Answers will vary depending on the results from the applet. You should expect 95% of the intervals to contain the population mean.

(b) Answers will vary depending on the results from the applet.

(c) Answers will vary depending on the results from the applet. You should expect 95% of the intervals to contain the population mean.

(d) Confidence intervals for the samples of size $n = 10$ should be wider than the confidence intervals for the samples of size $n = 50$.

59. (a) Because all subjects were randomly assigned to the treatments, this is a completely randomized design.

(b) The treatment is the smoking cessation program. There are two levels: internet and phone-based intervention, and self-help booklet.

(c) The response variable was abstinence after 12 months (whether or not the subject quit smoking).

(d) The statistics reported are 22.3% of participants in the experimental group reported abstinence and 13.1% of participants in the control group reported abstinence.

(e) $\dfrac{p(1-q)}{q(1-p)} = \dfrac{0.223(1-0.131)}{0.131(1-0.223)} \approx 1.90$; this indicates that abstinence is more likely to occur using the experimental cessation program.

(f) The authors are 95% confident that the population odds ratio is between 1.12 and 3.26.

(g) Answers will vary. One possibility: Smoking cessation is more likely when the Happy Ending Intervention program is used rather than the control method.

9.2 Confidence Intervals for a Population Mean When the Population Standard Deviation is Unknown

1. We can construct a Z-interval if the sample is random, the population from which the sample is drawn is normal or the sample size is large $(n \geq 30)$, and the population standard deviation, σ, is known. A *t*-interval should be constructed if the sample is random, the population from which the sample is drawn is normal, but the population standard deviation, σ, is unknown. Neither interval can be constructed if the sample is not random, the population is not normal and the sample size is small, or when there are outliers.

3. Robust means that the procedure is accurate when there are moderate departures from the requirements, such as normality in the distribution of the population.

5. Similarities: Both the standard normal distribution and the *t*-distribution are probability density functions; both have mean $\mu = 0$, and both are symmetric about their means.
Differences: *t*-distributions vary for different sample sizes while there is only one standard normal distribution. *t*-distributions have longer and thicker tails than the standard normal distribution, which has most of its area between –3 and 3.

7. (a) From the row with df = 25 and the column headed 0.10, we read $t_{0.10} = 1.316$.

(b) From the row with df = 30 and the column headed 0.05, we read $t_{0.05} = 1.697$.

(c) From the row with df = 18 and the column headed 0.01, we read $t = 2.552$. This is the value with area 0.01 to the *right*. By symmetry, the t-value with an area to the *left* of 0.01 is $t_{0.99} = -2.552$.

(d) For a 90% confidence interval, we want the t-value with an area in the right tail of 0.05. With df = 20, we read from Table VI that $t_{0.05} = 1.725$.

9. (a) For 96% confidence, $\alpha/2 = 0.02$. Since $n = 25$, then df = 24. The critical value is $t_{0.02} = 2.172$. Then:

$$\text{Lower bound} = \bar{x} - t_{0.02} \cdot \frac{s}{\sqrt{n}}$$

$$= 108 - 2.172 \cdot \frac{10}{\sqrt{25}}$$

$$\approx 108 - 4.3 = 103.7$$

$$\text{Upper bound} = \bar{x} + t_{0.02} \cdot \frac{s}{\sqrt{n}}$$

$$= 108 + 2.172 \cdot \frac{10}{\sqrt{25}}$$

$$\approx 108 + 4.3 = 112.3$$

(b) Since $n = 10$, then df = 9. The critical value is $t_{0.02} = 2.398$. Then:

$$\text{Lower bound} = \bar{x} - t_{0.02} \cdot \frac{s}{\sqrt{n}}$$

$$= 108 - 2.398 \cdot \frac{10}{\sqrt{10}}$$

$$\approx 108 - 7.6 = 100.4$$

$$\text{Upper bound} = \bar{x} + t_{0.02} \cdot \frac{s}{\sqrt{n}}$$

$$= 108 + 2.398 \cdot \frac{10}{\sqrt{10}}$$

$$\approx 108 + 7.6 = 115.6$$

Decreasing the sample size increases the margin of error.

(c) For 90% confidence, $\alpha/2 = 0.05$. With 24 degrees of freedom, $t_{0.05} = 1.711$. Then:

$$\text{Lower bound} = \bar{x} - t_{0.05} \cdot \frac{s}{\sqrt{n}}$$

$$= 108 - 1.711 \cdot \frac{10}{\sqrt{25}}$$

$$\approx 108 - 3.4 = 104.6$$

$$\text{Upper bound} = \bar{x} + t_{0.05} \cdot \frac{s}{\sqrt{n}}$$

$$= 108 + 1.711 \cdot \frac{10}{\sqrt{25}}$$

$$\approx 108 + 3.4 = 111.4$$

Decreasing the level of confidence decreases the margin of error.

(d) No, because in all cases the sample was small ($n < 30$), so the population must be normally distributed.

11. (a) For 95% confidence, $\alpha/2 = 0.025$. Since $n = 35$, then df = 34. The critical value is $t_{0.025} = 2.032$. Then:

$$\text{Lower bound} = \bar{x} - t_{0.025} \cdot \frac{s}{\sqrt{n}}$$

$$= 18.4 - 2.032 \cdot \frac{4.5}{\sqrt{35}}$$

$$\approx 18.4 - 1.55 = 16.85$$

$$\text{Upper bound} = \bar{x} + t_{0.025} \cdot \frac{s}{\sqrt{n}}$$

$$= 18.4 + 2.032 \cdot \frac{4.5}{\sqrt{35}}$$

$$\approx 18.4 + 1.55 = 19.95$$

(b) Since $n = 50$, then df = 49, but since there is no row in the tables for 49 degrees of freedom we use the closest value, df = 50, instead. The critical value is $t_{0.025} = 2.009$. Then:

$$\text{Lower bound} = \bar{x} - t_{0.025} \cdot \frac{s}{\sqrt{n}}$$

$$= 18.4 - 2.009 \cdot \frac{4.5}{\sqrt{50}}$$

$$\approx 18.4 - 1.28 = 17.12$$

$$\text{Upper bound} = \bar{x} + t_{0.025} \cdot \frac{s}{\sqrt{n}}$$

$$= 18.4 + 2.009 \cdot \frac{4.5}{\sqrt{50}}$$

$$\approx 18.4 + 1.28 = 19.68$$

Increasing the sample size decreases the margin of error.

(c) For 99% confidence, $\alpha/2 = 0.005$. With 34 degrees of freedom, $t_{0.005} = 2.728$. Then:

Lower bound $= \bar{x} - t_{0.005} \cdot \dfrac{s}{\sqrt{n}}$

$= 18.4 - 2.728 \cdot \dfrac{4.5}{\sqrt{35}}$

$\approx 18.4 - 2.08 = 16.32$

Upper bound $= \bar{x} + t_{0.005} \cdot \dfrac{s}{\sqrt{n}}$

$= 18.4 + 2.728 \cdot \dfrac{4.5}{\sqrt{35}}$

$\approx 18.4 + 2.08 = 20.48$

$\left[\text{Tech: } (16.33, 20.48)\right]$

Increasing the level of confidence increases the margin of error.

(d) For a small sample ($n = 15 < 30$), the population must be normally distributed.

13. For 99% confidence, $\alpha/2 = 0.005$. Since $n = 1006$, then df = 1005. There is no row in the table for 1005 degrees of freedom, so we use the closest value, df = 1000, instead. The critical value is $t_{0.005} = 2.581$. Then:

Lower bound $= \bar{x} - t_{0.005} \cdot \dfrac{s}{\sqrt{n}}$

$= 13.4 - 2.581 \cdot \dfrac{16.6}{\sqrt{1006}}$

$\approx 13.4 - 1.35 = 12.05$ books

Upper bound $= \bar{x} - t_{0.005} \cdot \dfrac{s}{\sqrt{n}}$

$= 13.4 + 2.581 \cdot \dfrac{16.6}{\sqrt{1006}}$

$\approx 13.4 + 1.35 = 14.75$ books

We are 99% confident that the population mean number of books read by Americans during 2005 was between 12.05 and 14.75 books.

15. For 95% confidence, $\alpha/2 = 0.025$. Since $n = 81$, then df = 80 and $t_{0.025} = 1.990$. Then:

Lower bound $= \bar{x} - t_{0.025} \cdot \dfrac{s}{\sqrt{n}}$

$= 4.6 - 1.990 \cdot \dfrac{15.9}{\sqrt{81}}$

$\approx 4.6 - 3.52 = 1.08$ days

Upper bound $= \bar{x} + t_{0.025} \cdot \dfrac{s}{\sqrt{n}}$

$= 4.6 + 1.990 \cdot \dfrac{15.9}{\sqrt{81}}$

$\approx 4.6 + 3.52 = 8.12$ days

We are 95% confident that the population mean incubation period of patients with SARS is between 1.08 and 8.12 days.

17. (a) For 90% confidence, $\alpha/2 = 0.05$. Since $n = 547$, then df = 546. There is no row in the table for 546 degrees of freedom, so we use the closest value, df = 1000, instead. The critical value is $t_{0.05} = 1.646$. Then:

Lower bound $= \bar{x} - t_{0.05} \cdot \dfrac{s}{\sqrt{n}}$

$= 45.6 - 1.646 \cdot \dfrac{31.4}{\sqrt{547}}$

$\approx 45.6 - 2.21 = 43.39$ min

Upper bound $= \bar{x} + t_{0.05} \cdot \dfrac{s}{\sqrt{n}}$

$= 45.6 + 1.646 \cdot \dfrac{31.4}{\sqrt{547}}$

$\approx 45.6 + 2.21 = 47.81$ min

Gallup is 90% confident that the population mean commute time of adult Americans employed full-time or part-time is between 43.39 minutes and 47.81 minutes.

(b) Yes; it is possible that the mean commute time is less than 40 minutes since it is possible that the true mean is not captured in the confidence interval. However, it is not very likely since we are 90% confident the true mean commute time is between 43.39 minutes and 47.81 minutes.

19. Using technology, we find $\bar{x} = 3$ days and $s \approx 2.24$ days. For 95% confidence, $\alpha/2 = 0.025$. Since $n = 32$, then df = 31 and $t_{0.025} = 2.040$. Then:

Lower bound $= \bar{x} - t_{0.025} \cdot \dfrac{s}{\sqrt{n}}$

$= 3 - 2.040 \cdot \dfrac{2.24}{\sqrt{32}}$

$\approx 3 - 0.81 = 2.19$ days

Upper bound $= \bar{x} + t_{0.025} \cdot \dfrac{s}{\sqrt{n}}$

$= 3 + 2.040 \cdot \dfrac{2.24}{\sqrt{32}}$

$\approx 3 + 0.81 = 3.81$ days

The researcher is 95% confident that the population mean number of days per week in which adults engage in exercise activities is between 2.19 and 3.81 days.

21. (a) $\bar{x} = \dfrac{\sum x}{n} = \dfrac{93.48}{40} = 2.337$ million shares

(b) Using technology, we find $s \approx 1.217$ million shares. For 90% confidence, $\alpha/2 = 0.05$. Since $n = 40$, then df = 39 and $t_{0.05} = 1.685$. Thus:

Lower bound $= \bar{x} - t_{0.05} \cdot \dfrac{s}{\sqrt{n}}$

$= 2.337 - 1.685 \cdot \dfrac{1.217}{\sqrt{40}}$

$\approx 2.337 - 0.324 = 2.013$

Upper bound $= \bar{x} + t_{0.05} \cdot \dfrac{s}{\sqrt{n}}$

$= 2.337 + 1.685 \cdot \dfrac{1.217}{\sqrt{40}}$

$\approx 2.337 + 0.324 = 2.661$

We are 90% confident that the population mean number of Harley-Davidson shares traded per day in 2007 was between 2.013 and 2.661 million shares.

(c) Using technology, we find $\bar{x} \approx 2.195$ million shares and $s \approx 0.815$ million shares. For 90% confidence, $\alpha/2 = 0.05$. Since $n = 40$, then df = 39 and $t_{0.05} = 1.685$. Thus:

Lower bound $= \bar{x} - t_{0.05} \cdot \dfrac{s}{\sqrt{n}}$

$= 2.195 - 1.685 \cdot \dfrac{0.815}{\sqrt{40}}$

$\approx 2.195 - 0.217 = 1.978$

Upper bound $= \bar{x} + t_{0.05} \cdot \dfrac{s}{\sqrt{n}}$

$= 2.195 + 1.685 \cdot \dfrac{0.815}{\sqrt{40}}$

$\approx 2.195 + 0.217 = 2.412$

We are 90% confident that the population mean number of Harley-Davidson shares traded per day in 2007 was between 1.978 and 2.412 million shares.

(d) The confidence intervals are different because of variation in sampling. The samples have different means and standard deviations that lead to different confidence intervals.

23. (a) Yes. The normal probability plot indicates that the data come from a population that is approximately normal, and the box plot indicates that there are no outliers.

(b) Using technology, $\bar{x} \approx 38.3$ weeks and $s \approx 10.0$ weeks.
For 95% confidence, $\alpha/2 = 0.025$.
Since $n = 12$, then df = 11 and $t_{0.025} = 2.201$.

Lower bound $= \bar{x} - t_{0.025} \cdot \dfrac{s}{\sqrt{n}}$

$= 38.3 - 2.201 \cdot \dfrac{10.0}{\sqrt{12}}$

$\approx 38.3 - 6.4 = 31.9$ weeks

Upper bound $= \bar{x} + t_{0.025} \cdot \dfrac{s}{\sqrt{n}}$

$= 38.3 + 2.201 \cdot \dfrac{10.0}{\sqrt{12}}$

$\approx 38.3 + 6.4 = 44.7$ weeks

We are 95% confident that the population mean age at which a baby first crawls is between 31.9 and 44.7 weeks.

(c) The sample size could be increased in order to increase the accuracy of the interval without changing the confidence level.

25. (a) Yes. The normal probability plot indicates that the data come from a population that is approximately normal, and the box plot indicates that there are no outliers.

(b) Using technology, $\bar{x} = 167.5$ days and $s \approx 21.9$ days.
For 95% confidence, $\alpha/2 = 0.025$.
Since $n = 10$, then df = 9 and $t_{0.025} = 2.262$.

Lower bound $= \bar{x} - t_{0.025} \cdot \dfrac{s}{\sqrt{n}}$

$= 167.5 - 2.262 \cdot \dfrac{21.9}{\sqrt{10}}$

$\approx 167.5 - 15.7 = 151.8$ days

Upper bound $= \bar{x} + t_{0.025} \cdot \dfrac{s}{\sqrt{n}}$

$= 167.5 + 2.262 \cdot \dfrac{21.9}{\sqrt{10}}$

$\approx 167.5 + 15.7 = 183.2$ days

We are 95% confident that the population mean length of the growing season in the Chicago area is between 151.8 and 183.2 days.

(c) The sample size could be increased in order to increase the accuracy of the interval without change the confidence level.

27. (a) From the output:

$\bar{x} = 22.150$; $E = \dfrac{22.209 - 22.091}{2}$

$= 0.059$

(b) We are 95% confident that the population mean age when first married is between 22.091 and 22.209 years.

(c) For 95% confidence, $\alpha / 2 = 0.025$. Since $n = 26{,}540$, then df $= 26{,}539$. There is no row in the table for 26,539 degrees of freedom, so we use df $= 1000$ instead. and $t_{0.025} = 1.962$.

Lower bound $= \bar{x} - t_{0.025} \cdot \dfrac{s}{\sqrt{n}}$

$= 22.150 - 1.962 \cdot \dfrac{4.885}{\sqrt{26{,}540}}$

$\approx 22.150 - 0.059 = 22.091$

Upper bound $= \bar{x} + t_{0.025} \cdot \dfrac{s}{\sqrt{n}}$

$= 22.150 + 1.962 \cdot \dfrac{4.885}{\sqrt{26{,}540}}$

$\approx 22.150 + 0.059 = 22.209$

29. (a)

Home Asking Price,
Lexington, KY

(b) Using technology with the outlier $496,600 included, $\bar{x} \approx \$255{,}383.3$ and $s \approx \$93{,}444.0$.
For 99% confidence, $\alpha / 2 = 0.005$. Since $n = 12$, then df $= 11$ and $t_{0.005} = 3.106$.

Lower bound $= \bar{x} - t_{0.005} \cdot \dfrac{s}{\sqrt{n}}$

$= 255{,}383.3 - 3.106 \cdot \dfrac{93{,}444.0}{\sqrt{12}}$

$\approx 255{,}383.3 - 83{,}784.2$

$= \$171{,}599.1$

Upper bound $= \bar{x} + t_{0.005} \cdot \dfrac{s}{\sqrt{n}}$

$= 255{,}383.3 + 3.106 \cdot \dfrac{93{,}444.0}{\sqrt{12}}$

$\approx 255{,}383.3 + 83{,}784.2$

$= \$339{,}167.5$

[Tech: ($171,604.3, $339,162.3)]

(c) Using technology with the outlier $496,600 removed, $\bar{x} \approx \$233{,}454.5$ and $s \approx \$57{,}074.1$.
Since the outlier was removed, $n = 11$, df $= 10$, and $t_{0.005} = 3.169$.

Lower bound $= \bar{x} - t_{0.005} \cdot \dfrac{s}{\sqrt{n}}$

$= 233{,}454.5 - 3.169 \cdot \dfrac{57{,}074.1}{\sqrt{11}}$

$\approx 233{,}454.5 - 54{,}533.7$

$= \$178{,}920.8$

Upper bound $= \bar{x} + t_{0.005} \cdot \dfrac{s}{\sqrt{n}}$

$= 233{,}454.5 + 3.169 \cdot \dfrac{57{,}074.1}{\sqrt{11}}$

$\approx 233{,}454.5 + 54{,}533.7$

$= \$287{,}988.2$

[Tech: ($178,916.2, $287,992.9)]

(d) The inclusion of the outlier makes the confidence interval wider.

31. (a), (b), (c) Answers will vary.

(d) 95% of 20 is $0.95(20) = 19$. We would expect about 19 of the 20 samples to generate confidence intervals that include the population mean. The actual results will vary.

33. Answers will vary.

9.3 Confidence Intervals for a Population Proportion

1. The best point estimate of the population proportion is the sample proportion, $\hat{p} = \dfrac{x}{n}$.

3. Answers will vary. One possibility follows: By using a prior estimate of p the researcher will get a better estimate of the required sample size, which will be smaller than the "worst-case" sample size given by using no prior estimate of p.

5. $\hat{p} = \dfrac{x}{n} = \dfrac{30}{150} = 0.20$. For 90% confidence, $z_{\alpha/2} = z_{0.05} = 1.645$.

   Lower bound $= \hat{p} - z_{0.05} \cdot \sqrt{\dfrac{\hat{p}(1-\hat{p})}{n}}$

   $= 0.20 - 1.645 \cdot \sqrt{\dfrac{0.2(1-0.2)}{150}}$

   $\approx 0.20 - 0.054 = 0.146$

   Upper bound $= \hat{p} + z_{0.05} \cdot \sqrt{\dfrac{\hat{p}(1-\hat{p})}{n}}$

   $= 0.20 + 1.645 \cdot \sqrt{\dfrac{0.2(1-0.2)}{150}}$

   $\approx 0.20 + 0.054 = 0.254$

7. $\hat{p} = \dfrac{x}{n} = \dfrac{120}{500} = 0.24$. For 99% confidence, $z_{\alpha/2} = z_{0.005} = 2.575$.

   Lower bound $= \hat{p} - z_{0.005} \cdot \sqrt{\dfrac{\hat{p}(1-\hat{p})}{n}}$

   $= 0.24 - 2.575 \cdot \sqrt{\dfrac{0.24(1-0.24)}{500}}$

   $\approx 0.24 - 0.049 = 0.191$

   Upper bound $= \hat{p} + z_{0.005} \cdot \sqrt{\dfrac{\hat{p}(1-\hat{p})}{n}}$

   $= 0.24 + 2.575 \cdot \sqrt{\dfrac{0.24(1-0.24)}{500}}$

   $\approx 0.24 + 0.049 = 0.289$

9. $\hat{p} = \dfrac{x}{n} = \dfrac{860}{1100} \approx 0.782$. For 94% confidence, $z_{\alpha/2} = z_{.03} = 1.88$.

   Lower bound $= \hat{p} - z_{0.03} \cdot \sqrt{\dfrac{\hat{p}(1-\hat{p})}{n}}$

   $= 0.782 - 1.88 \cdot \sqrt{\dfrac{0.782(1-0.782)}{1100}}$

   $\approx 0.782 - 0.023 \approx 0.759$

   Upper bound $= \hat{p} + z_{0.03} \cdot \sqrt{\dfrac{\hat{p}(1-\hat{p})}{n}}$

   $= 0.782 + 1.88 \cdot \sqrt{\dfrac{0.782(1-0.782)}{1100}}$

   $\approx 0.782 + 0.023 \approx 0.805$

11. $\hat{p} = \dfrac{0.249 + 0.201}{2} = \dfrac{0.45}{2} = 0.225$

    $E = \dfrac{0.249 - 0.201}{2} = \dfrac{0.048}{2} = 0.024$

    $x = n \cdot \hat{p} = 1200(0.225) = 270$

13. $\hat{p} = \dfrac{0.509 + 0.462}{2} = \dfrac{0.971}{2} = 0.4855$

    $E = \dfrac{0.509 - 0.462}{2} = \dfrac{0.047}{2} = 0.0235$

    $x = n \cdot \hat{p} = 1680(0.4855) = 815.64 \rightarrow 816$

15. (a) Flawed; no interval has been provided about the population proportion.

    (b) Flawed; this interpretation indicates that the level of confidence is varying.

    (c) This is the correct interpretation.

    (d) Flawed; this interpretation suggests that this interval sets the standard for all the other intervals, which is not true.

17. Rasmussen Reports is 95% confident that the population proportion of adult Americans who dread Valentine's Day is between 0.135 and 0.225.

19. (a) $\hat{p} = \dfrac{x}{n} = \dfrac{47}{863} \approx 0.054$

    (b) The sample size $n = 863$ is less than 5% of the population, and $n\hat{p}(1-\hat{p}) \approx 44.44 \geq 10$.

(c) For 90% confidence, $z_{\alpha/2} = z_{0.05} = 1.645$.
Lower bound

$$= \hat{p} - z_{0.05} \cdot \sqrt{\frac{\hat{p}(1-\hat{p})}{n}}$$

$$= 0.054 - 1.645 \cdot \sqrt{\frac{0.054(1-0.054)}{863}}$$

$$\approx 0.054 - 0.013$$

$$= 0.041 \ [\text{Tech: } 0.042]$$

Upper bound

$$= \hat{p} + z_{0.05} \cdot \sqrt{\frac{\hat{p}(1-\hat{p})}{n}}$$

$$= 0.054 + 1.645 \cdot \sqrt{\frac{0.054(1-0.054)}{863}}$$

$$\approx 0.054 + 0.013 = 0.067$$

(d) We are 90% confident that the population proportion of Lipitor users who will have a headache as a side effect is between 0.041 and 0.067 (i.e., between 4.1% and 6.7%).

21. (a) $\hat{p} = \dfrac{x}{n} = \dfrac{1322}{1979} \approx 0.668$

(b) The sample size $n = 1979$ is less than 5% of the population, and $n\hat{p}(1-\hat{p}) \approx 438.89 \geq 10$.

(c) For 96% confidence, $z_{\alpha/2} = z_{.02} = 2.05$.
Lower bound

$$= \hat{p} - z_{0.02} \cdot \sqrt{\frac{\hat{p}(1-\hat{p})}{n}}$$

$$= 0.668 - 2.05 \cdot \sqrt{\frac{0.668(1-0.668)}{1979}}$$

$$\approx 0.668 - 0.022 = 0.646$$

Upper bound

$$= \hat{p} + z_{0.02} \cdot \sqrt{\frac{\hat{p}(1-\hat{p})}{n}}$$

$$= 0.668 + 2.05 \cdot \sqrt{\frac{0.668(1-0.668)}{1979}}$$

$$\approx 0.668 + 0.022 = 0.690$$

(d) Yes; it is possible that the population proportion is below 60%, because it is possible that the true proportion is not captured in the confidence interval. Since we are 96% confident that the true proportion is between 0.646 and 0.690, it is not likely that the proportion of adult Americans who believe that traditional journalism is out of touch is below 60%.

23. (a) $\hat{p} = \dfrac{x}{n} = \dfrac{322}{2302} \approx 0.140$

(b) The sample size $n = 2302$ is less than 5% of the population, and $n\hat{p}(1-\hat{p}) \approx 277 \geq 10$.

(c) For 90% confidence, $z_{\alpha/2} = z_{0.05} = 1.645$.
Lower bound

$$= \hat{p} - z_{0.05} \cdot \sqrt{\frac{\hat{p}(1-\hat{p})}{n}}$$

$$= 0.140 - 1.645 \cdot \sqrt{\frac{0.140(1-0.140)}{2302}}$$

$$\approx 0.140 - 0.012 = 0.128$$

Upper bound

$$= \hat{p} + z_{0.05} \cdot \sqrt{\frac{\hat{p}(1-\hat{p})}{n}}$$

$$= 0.140 + 1.645 \cdot \sqrt{\frac{0.140(1-0.140)}{2302}}$$

$$\approx 0.140 + 0.012 = 0.152$$

(d) For 98% confidence, $z_{\alpha/2} = z_{.01} = 2.33$.
Lower bound

$$= \hat{p} - z_{0.01} \cdot \sqrt{\frac{\hat{p}(1-\hat{p})}{n}}$$

$$= 0.140 - 2.33 \cdot \sqrt{\frac{0.140(1-0.140)}{2302}}$$

$$\approx 0.140 - 0.017 = 0.123$$

Upper bound

$$= \hat{p} + z_{0.01} \cdot \sqrt{\frac{\hat{p}(1-\hat{p})}{n}}$$

$$= 0.140 + 2.33 \cdot \sqrt{\frac{0.140(1-0.140)}{2302}}$$

$$\approx 0.140 + 0.017 = 0.157$$

(e) Increasing the level of confidence widens the interval.

25. For 99% confidence, $z_{\alpha/2} = z_{.005} = 2.575$.

(a) Using $\hat{p} = 0.69$,

$$n = \hat{p}(1-\hat{p})\left(\frac{z_{\alpha/2}}{E}\right)^2$$

$$= 0.69(1-0.69)\left(\frac{2.575}{0.03}\right)^2$$

$$\approx 1575.9$$

which we must increase to 1576. The researcher needs a sample size of 1576.

(b) Without using a prior estimate,

$$n = 0.25\left(\frac{z_{\alpha/2}}{E}\right)^2 = 0.25\left(\frac{2.575}{0.03}\right)^2$$

$$\approx 1841.8$$

which we must increase to 1842. Without the prior estimate, the researcher would need a sample size of 1842.

27. For 98% confidence, $z_{\alpha/2} = z_{.01} = 2.33$.

(a) Using $\hat{p} = 0.15$,

$$n = \hat{p}(1-\hat{p})\left(\frac{z_{\alpha/2}}{E}\right)^2$$

$$= 0.15(1-0.15)\left(\frac{2.33}{0.02}\right)^2$$

$$\approx 1730.5$$

which we must increase to 1731. The researcher needs a sample size of 1731.

(b) Without using a prior estimate,

$$n = 0.25\left(\frac{z_{\alpha/2}}{E}\right)^2 = 0.25\left(\frac{2.33}{0.02}\right)^2$$

$$\approx 3393.1$$

which we must increase to 3394. Without the prior estimate, the researcher would need a sample size of 3394.

29. For 95% confidence, $z_{\alpha/2} = z_{0.025} = 1.96$.

(a) Using $\hat{p} = 0.48$,

$$n = \hat{p}(1-\hat{p})\left(\frac{z_{\alpha/2}}{E}\right)^2$$

$$= 0.48(1-0.48)\left(\frac{1.96}{0.03}\right)^2$$

$$\approx 1065.4$$

which we must increase to 1066. The commentator needs a sample size of 1066.

(b) $n = 0.25\left(\frac{z_{\alpha/2}}{E}\right)^2 = 0.25\left(\frac{1.96}{0.03}\right)^2$

$$\approx 1067.1$$

which we must increase to 1068. Without the prior estimate, the commentator would need a sample size of 1068.

(c) The results are close because $0.48(1-0.48) = 0.2496$ is very close to 0.25. That is, the prior estimate is very close the conservative estimate of 0.5 that we use when no estimate is available.

31. For 95% confidence, $z_{\alpha/2} = z_{0.025} = 1.96$.

Using $E = 0.03$ and $\hat{p} = 0.69$, we get

$$n = \hat{p}(1-\hat{p})\left(\frac{z_{\alpha/2}}{E}\right)^2$$

$$= 0.69(1-0.69)\left(\frac{1.96}{0.03}\right)^2$$

$$\approx 913.02$$

which we must increase to 914. At least 914 people were included in the survey.

33. Answers may vary. One possibility follows: The confidence interval for the percentage intending to vote for George Bush was $(49-3,\ 49+3) = (46\%,\ 52\%)$. Likewise, the confidence interval for the percentage intending to vote for John Kerry was $(47-3,\ 47+3) = (44\%,\ 50\%)$. Since these intervals overlap, it is possible that the true percentage intending to vote for John Kerry could have been greater than the true percentage intending to vote for George Bush, or vice versa. Hence, the result was too close to call.

35. **(a), (b), (c)** Answers will vary.

(d) As the sample size n increases, the proportion of intervals that capture p gets closer and closer to the level of confidence.

9.4 Confidence Intervals for a Population Standard Deviation

1. The chi-squared distribution is not symmetric but is skewed to the right; the actual shape of the chi-squared distribution depends on the degrees of freedom. As the number of degrees of freedom increases, the chi-squared distribution become more nearly symmetric. The values of χ^2 are nonnegative.

3. After the population is shown to be normal, a confidence interval for the standard deviation is obtained by as follows:
Step 1: Compute the sample variance.
Step 2: Determine the critical values using the desired confidence level, the correct degrees of freedom, and the χ^2 distribution table.

Step 3: Construct the confidence interval for the population variance by using the formulas Lower bound $= \dfrac{(n-1)s^2}{\chi^2_{\alpha/2}}$

and Upper bound $= \dfrac{(n-1)s^2}{\chi^2_{1-\alpha/2}}$

Step 4: Compute the square root of the lower bound and the upper bound to find the confidence interval for the population standard deviation.

5. $df = n-1 = 19$ and, for 90% confidence, $\alpha/2 = 0.05$. From the table, $\chi^2_{0.05} = 30.144$ and $\chi^2_{1-0.05} = \chi^2_{0.95} = 10.117$.

7. $df = n-1 = 22$ and, for 98% confidence, $\alpha/2 = 0.01$. From the table, $\chi^2_{0.01} = 40.289$ and $\chi^2_{1-0.01} = \chi^2_{0.99} = 9.542$.

9. **(a)** $df = n-1 = 19$ and, for 90% confidence, $\alpha/2 = 0.05$.

From the table, $\chi^2_{0.05} = 30.144$ and $\chi^2_{1-0.05} = \chi^2_{0.95} = 10.117$.

Lower bound $= \dfrac{(n-1)s^2}{\chi^2_{\alpha/2}} = \dfrac{19 \cdot 12.6}{30.144} \approx 7.94$

Upper bound $= \dfrac{(n-1)s^2}{\chi^2_{1-\alpha/2}} = \dfrac{19 \cdot 12.6}{10.117} \approx 23.66$

(b) $df = n-1 = 29$ and, for 90% confidence, $\alpha/2 = 0.05$.

From the table, $\chi^2_{0.05} = 42.557$ and $\chi^2_{1-0.05} = \chi^2_{0.95} = 17.708$.

Lower bound $= \dfrac{(n-1)s^2}{\chi^2_{\alpha/2}} = \dfrac{29 \cdot 12.6}{42.557} \approx 8.59$

Upper bound $= \dfrac{(n-1)s^2}{\chi^2_{1-\alpha/2}} = \dfrac{29 \cdot 12.6}{17.708} \approx 20.63$

Increasing the sample size decreases the width of the confidence interval.

(c) $df = n-1 = 19$ and, for 98% confidence, $\alpha/2 = 0.01$.

From the table, $\chi^2_{0.05} = 36.191$ and $\chi^2_{1-0.01} = \chi^2_{0.99} = 7.633$.

Lower bound $= \dfrac{(n-1)s^2}{\chi^2_{\alpha/2}} = \dfrac{19 \cdot 12.6}{36.191} \approx 6.61$

Upper bound $= \dfrac{(n-1)s^2}{\chi^2_{1-\alpha/2}} = \dfrac{19 \cdot 12.6}{7.633} \approx 31.36$

Increasing the level of confidence increases the width of the confidence interval.

11. $df = n-1 = 12-1 = 11$ and, for 95% confidence, $\alpha/2 = 0.025$.

From the table, $\chi^2_{0.025} = 21.920$ and $\chi^2_{1-0.025} = \chi^2_{0.975} = 3.816$.

Lower bound $= \sqrt{\dfrac{(n-1)s^2}{\chi^2_{\alpha/2}}}$

$= \sqrt{\dfrac{11 \cdot (10.00)^2}{21.920}} \approx 7.08$ weeks

Upper bound $= \sqrt{\dfrac{(n-1)s^2}{\chi^2_{1-\alpha/2}}}$

$= \sqrt{\dfrac{11 \cdot (10.00)^2}{3.816}} \approx 16.98$ weeks

Essential Baby can be 95% confident that the population standard deviation of the time at which babies first crawl is between 7.08 and 16.98 weeks.

13. $df = n-1 = 10-1 = 9$ and, for 99% confidence, $\alpha/2 = 0.005$.

From the table, $\chi^2_{0.005} = 23.589$ and $\chi^2_{1-0.005} = \chi^2_{0.995} = 1.735$.

Lower bound $= \sqrt{\dfrac{(n-1)s^2}{\chi^2_{\alpha/2}}}$

$= \sqrt{\dfrac{9 \cdot (21.88)^2}{23.589}} \approx 13.51$ days

Upper bound $= \sqrt{\dfrac{(n-1)s^2}{\chi^2_{1-\alpha/2}}}$

$= \sqrt{\dfrac{9 \cdot (21.88)^2}{1.735}} \approx 49.83$ days

The agricultural researcher can be 99% confident that the population standard deviation of the growing season in Chicago is between 13.51 and 49.83 days.

15. (a) From the probability plot shown below, the data appear to be from a population that is normally distributed.

(b) Using technology, we obtain $s \approx 0.349$ oz.

(c) $df = n-1 = 11$ and, for 90% confidence, $\alpha / 2 = 0.05$. From the table,
$$\chi^2_{1-0.05} = \chi^2_{0.95} = 4.575 \text{ and}$$
$$\chi^2_{0.05} = 19.675 .$$

Lower bound $= \sqrt{\dfrac{(n-1)s^2}{\chi^2_{\alpha/2}}}$

$= \sqrt{\dfrac{11(0.349)^2}{19.675}} \approx 0.261$ oz

Upper bound $= \sqrt{\dfrac{(n-1)s^2}{\chi^2_{1-\alpha/2}}}$

$= \sqrt{\dfrac{11(0.349)^2}{4.575}} \approx 0.541$ oz

The quality-control manager can be 90% confident that the population standard deviation of the number of ounces of peanuts is between 0.261 and 0.541 ounce.

(d) No; we are 90% confident that the population standard deviation is between 0.261 and 0.541 ounces, and so above 0.20 ounces.

17. From the standard normal table, we get $z_{0.975} = -1.96$ and $z_{0.025} = 1.96$. With $v = 100$, we get

$$\chi^2_{0.975} \approx \frac{\left(z_{0.975} + \sqrt{2v-1} \right)^2}{2}$$

$$= \frac{\left(-1.96 + \sqrt{2 \cdot 100 - 1} \right)^2}{2} = 73.772$$

(compared to the table's value of 74.222)

and $\chi^2_{0.025} \approx \dfrac{\left(z_{0.025} + \sqrt{2v-1} \right)^2}{2}$

$$= \frac{\left(1.96 + \sqrt{2 \cdot 100 - 1} \right)^2}{2} = 129.070$$

(compared to the table's value of 129.561).

9.5 Putting It All Together: Which Procedure Do I Use?

1. We construct a *t*-interval when we are estimating the population mean, we do not know the population standard deviation, and the underlying population is normally distributed. If the underlying population is not normally distributed, we can construct a *t*-interval to estimate the population mean provide the sample size is large ($n \geq 30$).

We construct a *Z*-interval when we are estimating the population mean, we know the population standard deviation, and the underlying population is normally distributed. If the underlying population is not normally distributed, we can construct a *Z*-interval to estimate the population mean provided the sample size is large ($n \geq 30$). We also construct a *Z*-interval when we are estimating the population proportion, provided the sample is smaller than 5% of the population and $n\hat{p}(1-\hat{p}) \geq 10$.

3. We construct a *Z*-interval because we are estimating a population mean, we know the population standard deviation, and the underlying population is normally distributed. For 95% confidence the critical value is $z_{0.025} = 1.96$. Then:

Lower bound $= \bar{x} - z_{0.025} \cdot \dfrac{\sigma}{\sqrt{n}}$

$= 60 - 1.96 \cdot \dfrac{20}{\sqrt{14}} \approx 49.5$

Upper bound $= \bar{x} + z_{0.025} \cdot \dfrac{\sigma}{\sqrt{n}}$

$= 60 + 1.96 \cdot \dfrac{20}{\sqrt{14}} \approx 70.5$

5. $\hat{p} = \dfrac{35}{300} \approx 0.117$. The sample size $n = 300$ is less than 5% of the population, and $n\hat{p}(1-\hat{p}) \approx 31.0 \geq 10$, so we can construct a Z-interval. For 99% confidence the critical value is $z_{0.005} = 2.575$. Then:

Lower bound

$$= \hat{p} - z_{0.005} \cdot \sqrt{\dfrac{\hat{p}(1-\hat{p})}{n}}$$

$$= 0.117 - 2.575 \cdot \sqrt{\dfrac{0.117(1-0.117)}{300}}$$

$$\approx 0.117 - 0.048 = 0.069$$

Upper bound

$$= \hat{p} + z_{0.005} \cdot \sqrt{\dfrac{\hat{p}(1-\hat{p})}{n}}$$

$$= 0.117 + 2.575 \cdot \sqrt{\dfrac{0.117(1-0.117)}{300}}$$

$$\approx 0.117 + 0.048 = 0.165 \quad \text{[Tech: 0.164]}$$

7. We construct a t-interval because we are estimating the population mean, we do not know the population standard deviation, and the underlying population is normally distributed. For 90% confidence, $\alpha/2 = 0.05$. Since $n = 12$, then df = 11 and $t_{0.05} = 1.796$. Then:

$$\text{Lower bound} = \overline{x} - t_{0.05} \cdot \dfrac{s}{\sqrt{n}}$$

$$= 45 - 1.796 \cdot \dfrac{14}{\sqrt{12}}$$

$$\approx 45 - 7.26 = 37.74$$

$$\text{Upper bound} = \overline{x} + t_{0.05} \cdot \dfrac{s}{\sqrt{n}}$$

$$= 45 + 1.796 \cdot \dfrac{14}{\sqrt{12}}$$

$$\approx 45 + 7.26 = 52.26$$

9. We construct a t-interval because we are estimating the population mean, we do not know the population standard deviation, and the underlying population is normally distributed. For 99% confidence, $\alpha/2 = 0.005$. Since $n = 40$, then df = 39 and $t_{0.050} = 2.708$. Then:

$$\text{Lower bound} = \overline{x} - t_{0.005} \cdot \dfrac{s}{\sqrt{n}}$$

$$= 120.5 - 2.708 \cdot \dfrac{12.9}{\sqrt{40}}$$

$$\approx 120.5 - 5.52 = 114.98$$

$$\text{Upper bound} = \overline{x} + t_{0.005} \cdot \dfrac{s}{\sqrt{n}}$$

$$= 120.5 + 2.708 \cdot \dfrac{12.9}{\sqrt{40}}$$

$$\approx 120.5 + 5.52 = 126.02$$

11. df $= n - 1 = 12 - 1 = 11$ and, for 90% confidence, $\alpha/2 = 0.05$. From the table, $\chi^2_{0.05} = 19.675$ and $\chi^2_{1-0.05} = \chi^2_{0.95} = 4.575$.

$$\text{Lower bound} = \dfrac{(n-1)s^2}{\chi^2_{\alpha/2}} = \dfrac{11 \cdot 23.7}{19.675} \approx 13.25$$

$$\text{Upper bound} = \dfrac{(n-1)s^2}{\chi^2_{1-\alpha/2}} = \dfrac{11 \cdot 23.7}{4.575} \approx 56.98$$

13. We construct a t-interval because we are estimating the population mean, we do not know the population standard deviation, and the underlying population is normally distributed. For 95% confidence, $\alpha/2 = 0.025$. Since $n = 40$, then df = 39 and $t_{0.025} = 2.023$. Then:

$$\text{Lower bound} = \overline{x} - t_{0.025} \cdot \dfrac{s}{\sqrt{n}}$$

$$= 54 - 2.023 \cdot \dfrac{8}{\sqrt{40}}$$

$$\approx 54 - 2.6 = 51.4 \text{ months}$$

$$\text{Upper bound} = \overline{x} + t_{0.025} \cdot \dfrac{s}{\sqrt{n}}$$

$$= 54 + 2.023 \cdot \dfrac{8}{\sqrt{40}}$$

$$\approx 54 + 2.6 = 56.6 \text{ months}$$

We can be 95% confident that the population of felons convicted of aggravated assault serve a mean sentence between 51.4 and 56.6 months.

15. We construct a Z-interval because we are estimating a population mean, we know the population standard deviation, and the underlying population is normally distributed. For 90% confidence the critical value is $z_{0.05} = 1.645$. Then:

Lower bound $= \bar{x} - z_{0.05} \cdot \dfrac{\sigma}{\sqrt{n}}$

$= 3421 - 1.645 \cdot \dfrac{2583}{\sqrt{100}}$

$\approx 3421 - 424.9 = \$2996.1$

Upper bound $= \bar{x} + z_{0.05} \cdot \dfrac{\sigma}{\sqrt{n}}$

$= 3421 + 1.645 \cdot \dfrac{2583}{\sqrt{100}}$

$\approx 3421 + 424.9 = \$3845.9$

The Internal Revenue Service can be 90% confident that the population mean additional tax owed is between \$2996.1 and \$3845.9.

17. $\hat{p} = \dfrac{567}{1008} \approx 0.563$. The sample size $n = 1008$ is less than 5% of the population, and $n\hat{p}(1-\hat{p}) \approx 248 \geq 10$, so we can construct a Z-interval:
For 90% confidence the critical value is $z_{0.05} = 1.645$. Then:
Lower bound

$= \hat{p} - z_{0.05} \cdot \sqrt{\dfrac{\hat{p}(1-\hat{p})}{n}}$

$= 0.563 - 1.645 \cdot \sqrt{\dfrac{0.563(1-0.563)}{1008}}$

$\approx 0.563 - 0.026 = 0.537$
Upper bound

$= \hat{p} + z_{0.05} \cdot \sqrt{\dfrac{\hat{p}(1-\hat{p})}{n}}$

$= 0.563 + 1.645 \cdot \sqrt{\dfrac{0.563(1-0.563)}{1008}}$

$\approx 0.563 + 0.026 = 0.589$ [Tech: 0.588]
The Gallup Organization can be 90% confident that the population proportion of adult Americans who are worried about having enough money for retirement is between 0.537 and 0.589 (i.e., between 53.7% and 58.9%).

19. The normal probability plot and boxplot show that the data are normal with no outliers. We construct a Z-interval because we are estimating a population mean, and we know the population standard deviation. For 95% confidence the critical value is $z_{0.025} = 1.96$. The data give $n = 15$ and $\bar{x} = 69.85$ inches. Then:

Lower bound $= \bar{x} - z_{0.025} \cdot \dfrac{\sigma}{\sqrt{n}}$

$= 69.85 - 1.96 \cdot \dfrac{2.9}{\sqrt{15}}$

$\approx 69.85 - 1.47 = 68.38$ inches
[Tech: 68.39]

Upper bound $= \bar{x} + z_{0.025} \cdot \dfrac{\sigma}{\sqrt{n}}$

$= 69.85 + 1.96 \cdot \dfrac{2.9}{\sqrt{15}}$

$\approx 69.85 + 1.47 = 71.32$ inches
We are 95% confident that the population mean height of 20- to 29-year-old males is between 68.38 and 71.32 inches.

21. The box plot indicates an outlier in the data, so we cannot compute a confidence interval using either method.

23. The normal probability plot and boxplot show that the data are normal with no outliers. We construct a t-interval because we are estimating the population mean and we do not know the population standard deviation. The data give $n = 15$, $\bar{x} \approx 109.3$ and $s \approx 14.4$. With df = 14, we use $t_{0.025} = 2.145$. Then:

Lower bound $= \bar{x} - t_{0.025} \cdot \dfrac{s}{\sqrt{n}}$

$= 109.3 - 2.145 \cdot \dfrac{14.4}{\sqrt{15}}$

$\approx 109.3 - 8.0$

$= 101.3$ beats per minute
[Tech: 101.4]

Upper bound $= \bar{x} + t_{0.025} \cdot \dfrac{s}{\sqrt{n}}$

$= 109.3 + 2.145 \cdot \dfrac{14.4}{\sqrt{15}}$

$\approx 109.3 + 8.0$

$= 117.3$ beats per minute
We can be 95% confident that the population mean pulse rate for women after 3 minutes of exercise is between 101.3 and 117.3 beats per minute.

Chapter 9 Review Exercises

1. (a) For a 99% confidence interval we want the t-value with an area in the right tail of 0.005. With df $= 17$, we read from the table that $t_{0.005} = 2.898$.

(b) For a 90% confidence interval we want the t-value with an area in the right tail of 0.05. With df $= 26$, we read from the table that $t_{0.05} = 1.706$.

2. (a) df $= n - 1 = 21$ and for 95% confidence, $\alpha/2 = 0.025$. From the table, $\chi^2_{1-0.025} = \chi^2_{0.975} = 10.283$ and $\chi^2_{0.025} = 35.479$.

(b) df $= n - 1 = 11$ and for 99% confidence, $\alpha/2 = 0.005$. From the table, $\chi^2_{1-0.005} = \chi^2_{0.995} = 2.603$ and $\chi^2_{0.005} = 26.757$.

3. In 100 samples, we would expect 95 of the 100 intervals to include the true population mean, 100. Random chance in sampling causes a particular interval to not include the true population mean.

4. In a 90% confidence interval, the 90% represents the proportion of intervals that would contain the parameter of interest (e.g. the population mean, population proportion, or population standard deviation) if a large number of different samples is obtained.

5. If a large number of different samples (of the same size) is obtained, a 90% confidence interval for the population mean will not capture the true value of the population mean 10% of the time.

6. The area to the left of $t = -1.56$ is also 0.0681, because the t-distribution is symmetric about 0.

7. The area under the t-distribution to the right of $t = 2.32$ is greater than the area under the standard normal distribution to the right of $z = 2.32$ because the t-distribution uses s to approximate σ, making it more dispersed than the Z-distribution.

8. The properties of the Student's t-distribution:
  1. It is symmetric around $t = 0$.
  2. It is different for different sample sizes.
  3. The area under the curve is 1; half the area is to the right of 0 and half the area is to the left of 0.
  4. As t gets extremely large, the graph approaches, but never equals, zero. Similarly, as t gets extremely small (negative), the graph approaches, but never equals, zero.
  5. The area in the tails of the t-distribution is greater than the area in the tails of the standard normal distribution.
  6. As the sample size n increases, the distribution (and the density curve) of the t-distribution becomes more like the standard normal distribution.

9. (a) For 90% confidence the critical value is $z_{0.05} = 1.645$. Then:

$$\text{Lower bound} = \bar{x} - z_{0.05} \cdot \frac{\sigma}{\sqrt{n}}$$
$$= 54.8 - 1.645 \cdot \frac{10.5}{\sqrt{20}}$$
$$\approx 54.8 - 3.86 = 50.94$$
$$\text{Upper bound} = \bar{x} + z_{0.05} \cdot \frac{\sigma}{\sqrt{n}}$$
$$= 54.8 + 1.645 \cdot \frac{10.5}{\sqrt{20}}$$
$$\approx 54.8 + 3.86 = 58.66$$

(b)
$$\text{Lower bound} = \bar{x} - z_{0.05} \cdot \frac{\sigma}{\sqrt{n}}$$
$$= 54.8 - 1.645 \cdot \frac{10.5}{\sqrt{30}}$$
$$\approx 54.8 - 3.15 = 51.65$$
$$\text{Upper bound} = \bar{x} + z_{0.05} \cdot \frac{\sigma}{\sqrt{n}}$$
$$= 54.8 + 1.645 \cdot \frac{10.5}{\sqrt{30}}$$
$$\approx 54.8 + 3.15 = 57.95$$

Increasing the sample size decreases the width of the confidence interval.

(c) For 99% confidence the critical value is $z_{0.005} = 2.575$. Then:

Lower bound $= \bar{x} - z_{0.005} \cdot \dfrac{\sigma}{\sqrt{n}}$

$= 54.8 - 2.575 \cdot \dfrac{10.5}{\sqrt{20}}$

$\approx 54.8 - 6.05 = 48.75$

Upper bound $= \bar{x} + z_{0.005} \cdot \dfrac{\sigma}{\sqrt{n}}$

$= 54.8 + 2.575 \cdot \dfrac{10.5}{\sqrt{20}}$

$\approx 54.8 + 6.05 = 60.85$

Increasing the level of confidence increases the width of the confidence interval.

10. (a) For 90% confidence, $\alpha/2 = 0.05$. If $n = 15$, then df = 14. The critical value is $t_{0.05} = 1.761$. Then:

Lower bound $= \bar{x} - t_{0.05} \cdot \dfrac{s}{\sqrt{n}}$

$= 104.3 - 1.761 \cdot \dfrac{15.9}{\sqrt{15}}$

$\approx 104.3 - 7.23 = 97.07$

Upper bound $= \bar{x} + t_{0.05} \cdot \dfrac{s}{\sqrt{n}}$

$= 104.3 + 1.761 \cdot \dfrac{15.9}{\sqrt{15}}$

$\approx 104.3 + 7.23 = 111.53$

(b) If $n = 25$, then df = 24. The critical value is $t_{0.05} = 1.711$. Then:

Lower bound $= \bar{x} - t_{0.05} \cdot \dfrac{s}{\sqrt{n}}$

$= 104.3 - 1.711 \cdot \dfrac{15.9}{\sqrt{25}}$

$\approx 104.3 - 5.44 = 98.86$

Upper bound $= \bar{x} + t_{0.05} \cdot \dfrac{s}{\sqrt{n}}$

$= 104.3 + 1.711 \cdot \dfrac{15.9}{\sqrt{25}}$

$\approx 104.3 + 5.44 = 109.74$

Increasing the sample size decreases the width of the confidence interval.

(c) For 95% confidence, $\alpha/2 = 0.025$. With 14 df, $t_{0.025} = 2.145$. Then:

Lower bound $= \bar{x} - t_{0.025} \cdot \dfrac{s}{\sqrt{n}}$

$= 104.3 - 2.145 \cdot \dfrac{15.9}{\sqrt{15}}$

$\approx 104.3 - 8.81 = 95.49$

Upper bound $= \bar{x} + t_{0.025} \cdot \dfrac{s}{\sqrt{n}}$

$= 104.3 + 2.145 \cdot \dfrac{15.9}{\sqrt{15}}$

$\approx 104.3 + 8.81 = 113.11$

Increasing the level of confidence increases the width of the confidence interval.

(d) df $= n - 1 = 14$ and for 90% confidence, $\alpha/2 = 0.05$. From the table, $\chi^2_{1-0.05} = \chi^2_{0.95} = 6.571$ and $\chi^2_{0.05} = 23.685$.

Lower bound $= \sqrt{\dfrac{(n-1)s^2}{\chi^2_{\alpha/2}}}$

$= \sqrt{\dfrac{14(15.9)^2}{23.685}} = 12.22$

Upper bound $= \sqrt{\dfrac{(n-1)s^2}{\chi^2_{\alpha/2}}}$

$= \sqrt{\dfrac{14(15.9)^2}{6.571}} = 23.21$

11. (a) The size of the sample ($n = 40$) is sufficiently large to apply the Central Limit Theorem and conclude that the sampling distribution of $\bar{x}$ is approximately normal.

(b) We construct a Z-interval because we are estimating a population mean and we know the population standard deviation. For 90% confidence the critical value is $z_{0.05} = 1.645$.

Lower bound $= \bar{x} - z_{0.05} \cdot \dfrac{\sigma}{\sqrt{n}}$

$= 100,294 - 1.645 \cdot \dfrac{4600}{\sqrt{40}}$

$\approx 100,294 - 1196$

$= 99,098$ miles

Upper bound $= \bar{x} + z_{0.05} \cdot \dfrac{\sigma}{\sqrt{n}}$

$= 100,294 + 1.645 \cdot \dfrac{4600}{\sqrt{40}}$

$\approx 100,294 + 1196$

$= 101,490$ miles

Michelin can be 90% confident that the mean mileage for its HydroEdge tire is between 99,098 and 101,490 miles.

(c) For 95% confidence the critical value is $z_{0.05} = 1.96$.

Lower bound $= \bar{x} - z_{0.05} \cdot \dfrac{\sigma}{\sqrt{n}}$

$= 100,294 - 1.96 \cdot \dfrac{4600}{\sqrt{40}}$

$\approx 100,294 - 1426$

$= 98,868$ miles

Upper bound $= \bar{x} + z_{0.05} \cdot \dfrac{\sigma}{\sqrt{n}}$

$= 100,294 + 1.96 \cdot \dfrac{4600}{\sqrt{40}}$

$\approx 100,294 + 1426$

$= 101,720$ miles

Michelin can be 95% confident that the mean mileage for its HydroEdge tire is between 99,868 and 101,720 miles.

(d) For 99% confidence, we use $z_{0.005} = 2.575$. So,

$n = \left(\dfrac{z_{\alpha/2} \cdot \sigma}{E} \right)^2 = \left(\dfrac{2.575 \cdot 4600}{1500} \right)^2 \approx 62.4$

which we must increase to 63. Michelin would require 63 tires to be within 1,500 miles of the mean mileage with 99% confidence.

12. (a) Since the number of emails sent in a day cannot be negative, and one standard deviation to the left of the mean results in a negative number, we expect that the distribution is skewed right.

(b) For 90% confidence, $\alpha/2 = 0.05$. We have $df = n - 1 = 927$. Since there is no row in the table for 927 degrees of freedom, we use $df = 1000$ instead. Thus, $t_{0.05} = 1.646$. Then:

Lower bound $= \bar{x} - t_{0.05} \cdot \dfrac{s}{\sqrt{n}}$

$= 10.4 - 1.646 \cdot \dfrac{28.5}{\sqrt{928}}$

$\approx 10.4 - 1.54$

$= 8.86$ e-mails

Upper bound $= \bar{x} + t_{0.05} \cdot \dfrac{s}{\sqrt{n}}$

$= 10.4 + 1.646 \cdot \dfrac{28.5}{\sqrt{928}}$

$\approx 10.4 + 1.54$

$= 11.94$ e-mails

We are 90% confident that the population mean number of e-mails sent per day is between 8.86 and 11.94.

13. (a) The sample size is probably small due to the difficulty and expense of locating highly trained cyclists and gathering the data.

(b) For 95% confidence, $\alpha/2 = 0.025$. We have $df = n - 1 = 15$, and $t_{0.025} = 2.131$. Then:

Lower bound $= \bar{x} - t_{0.025} \cdot \dfrac{s}{\sqrt{n}}$

$= 218 - 2.131 \cdot \dfrac{31}{\sqrt{16}}$

$\approx 218 - 16.5$

$= 201.5$ kilojoules

Upper bound $= \bar{x} + t_{0.025} \cdot \dfrac{s}{\sqrt{n}}$

$= 218 + 2.131 \cdot \dfrac{31}{\sqrt{16}}$

$\approx 218 + 16.5$

$= 234.5$ kilojoules

The researchers can be 95% confident that the population mean total work performed for the sports-drink treatment group is between 201.5 and 234.5 kilojoules.

(c) Yes; it is possible that the mean total work performed is less than 198 kilojoules since it is possible that the true mean is not captured in the confidence interval. However, it is not very likely since we are 95% confident the true mean total work performed is between 201.5 and 234.5 kilojoules.

(d) For 95% confidence, $\alpha/2 = 0.025$. We have $df = n - 1 = 15$, and $t_{0.025} = 2.131$. Then:

Lower bound $= \bar{x} - t_{0.025} \cdot \dfrac{s}{\sqrt{n}}$

$= 178 - 2.131 \cdot \dfrac{31}{\sqrt{16}}$

$\approx 178 - 16.5$

$= 161.5$ kilojoules

Upper bound $= \bar{x} + t_{0.025} \cdot \dfrac{s}{\sqrt{n}}$

$= 178 + 2.131 \cdot \dfrac{31}{\sqrt{16}}$

$\approx 178 + 16.5$

$= 194.5$ kilojoules

The researchers can be 95% confident that the population mean total work performed for the placebo treatment group is between 161.5 and 194.5 kilojoules.

(e) Yes; it is possible that the mean total work performed is more than 198 kilojoules since it is possible that the true mean is not captured in the confidence interval. However, it is not very likely since we are 95% confident the true mean total work performed is between 161.5 and 194.5 kilojoules.

(f) Yes; our findings support the researchers' conclusion. The confidence intervals do not overlap, so we are confident that the mean total work performed for the sports-drink treatment is greater than the mean total work performed for the placebo treatment.

14. (a) Since the sample size is large ($n \geq 30$), $\bar{x}$ has an approximately normal distribution.

(b) We construct a *t*-interval because we are estimating a population mean and we do not know the population standard deviation. For 95% confidence, $\alpha/2 = 0.025$. Since $n = 60$, then $df = 59$. There is no row in the table for 59 degrees of freedom, so we use $df = 60$ instead. The critical value is $t_{0.025} = 2.000$. Then:

Lower bound $= \bar{x} - t_{0.025} \cdot \dfrac{s}{\sqrt{n}}$

$= 2.27 - 2.000 \cdot \dfrac{1.22}{\sqrt{60}}$

$\approx 2.27 - 0.32$

$= 1.95$ children

Upper bound $= \bar{x} + t_{0.025} \cdot \dfrac{s}{\sqrt{n}}$

$= 2.27 + 2.000 \cdot \dfrac{1.22}{\sqrt{60}}$

$\approx 2.27 + 0.32$

$= 2.59$ children

[Tech: 2.58]

We are 95% confident that couples who have been married for 7 years have a population mean number of children between 1.95 and 2.59.

(c) For 99% confidence, $\alpha/2 = 0.005$. The critical value is $t_{0.005} = 2.660$. Then:

Lower bound $= \bar{x} - t_{0.025} \cdot \dfrac{s}{\sqrt{n}}$

$= 2.27 - 2.660 \cdot \dfrac{1.22}{\sqrt{60}}$

$\approx 2.27 - 0.42$

$= 1.85$ children

Upper bound $= \bar{x} + t_{0.025} \cdot \dfrac{s}{\sqrt{n}}$

$= 2.27 + 2.660 \cdot \dfrac{1.22}{\sqrt{60}}$

$\approx 2.27 + 0.42$

$= 2.69$ children

We are 95% confident that couples who have been married for 7 years have a population mean number of children between 1.85 and 2.69.

15. (a) Using technology, we obtain $\bar{x} \approx 147.3$ cm and $s \approx 28.8$ cm.

(b) Yes. All the data values lie within the bounds on the normal probability plot, indicating that the data should come from a population that is normal. The boxplot does not show any outliers.

(c) We construct a *t*-interval because we are estimating a population mean and we do not know the population standard deviation. For 95% confidence, $\alpha/2 = 0.025$. Since $n = 12$, then df = 11. The critical value is $t_{0.025} = 2.201$. Then:

Lower bound $= \bar{x} - t_{0.025} \cdot \dfrac{s}{\sqrt{n}}$

$$= 147.3 - 2.201 \cdot \dfrac{28.8}{\sqrt{12}}$$

$$\approx 147.3 - 18.3$$

$$= 129.0 \text{ cm}$$

Upper bound $= \bar{x} + t_{0.025} \cdot \dfrac{s}{\sqrt{n}}$

$$= 147.3 + 2.201 \cdot \dfrac{28.8}{\sqrt{12}}$$

$$\approx 147.3 + 18.3$$

$$= 165.6 \text{ cm}$$

We are 95% confident that the population mean diameter of a Douglas fir tree in the western Washington Cascades is between 129.0 and 165.6 centimeters.

(d) From the table, with df = 11 and $\alpha/2 = 0.025$, we find $\chi^2_{1-0.025} = \chi^2_{0.975} = 3.816$ and $\chi^2_{0.025} = 21.920$. Then:

Lower bound $= \sqrt{\dfrac{(n-1)s^2}{\chi^2_{\alpha/2}}}$

$$= \sqrt{\dfrac{11(28.8)^2}{21.920}} \approx 20.4 \text{ cm}$$

Upper bound $= \sqrt{\dfrac{(n-1)s^2}{\chi^2_{\alpha/2}}}$

$$= \sqrt{\dfrac{11(28.8)^2}{3.816}} \approx 48.9 \text{ cm}$$

We are 95% confident that the population standard deviation diameter of a Douglas fir tree in the western Washington Cascades is between 20.4 and 48.9 centimeters.

16. (a) $\hat{p} = \dfrac{x}{n} = \dfrac{58}{678} \approx 0.086$

(b) For 95% confidence, $z_{\alpha/2} = z_{.025} = 1.96$.
Lower bound

$$= \hat{p} - z_{.025} \cdot \sqrt{\dfrac{\hat{p}(1-\hat{p})}{n}}$$

$$= 0.086 - 1.96 \cdot \sqrt{\dfrac{0.086(1-0.086)}{678}}$$

$$\approx 0.086 - 0.021 = 0.065$$

Upper bound

$$= \hat{p} + z_{.025} \cdot \sqrt{\dfrac{\hat{p}(1-\hat{p})}{n}}$$

$$= 0.086 + 1.96 \cdot \sqrt{\dfrac{0.086(1-0.086)}{678}}$$

$$\approx 0.086 + 0.021 = 0.107$$

The Centers for Disease Control can be 95% confident that the population proportion of adult males aged 20–34 years who have hypertension is between 0.065 and 0.107 (i.e. between 6.5% and 10.7%).

(c) $n = \hat{p}(1-\hat{p}) \left(\dfrac{z_{\alpha/2}}{E} \right)^2$

$$= 0.086(1-0.086) \left(\dfrac{1.96}{.03} \right)^2$$

$$\approx 335.5$$

which we must increase to 336. You would need a sample of size 336 for your estimate to be within 3 percentage points of the true proportion, with 95% confidence.

(d) $n = 0.25 \left(\dfrac{z_{\alpha/2}}{E} \right)^2 = 0.25 \left(\dfrac{1.96}{0.03} \right)^2$

$$\approx 1067.1$$

which we must increase to 1068. Without the prior estimate, you would need a sample size of 1068.

Chapter 9 Test

1. The properties of the Student's t-distribution:
   1. It is symmetric around $t = 0$.
   2. It is different for different sample sizes.
   3. The area under the curve is 1; half the area is to the right of 0 and half the area is to the left of 0.
   4. As t gets extremely large, the graph approaches, but never equals, zero. Similarly, as t gets extremely small (negative), the graph approaches, but never equals, zero.
   5. The area in the tails of the t-distribution is greater than the area in the tails of the standard normal distribution.
   6. As the sample size n increases, the distribution (and the density curve) of the t-distribution becomes more like the standard normal distribution.

2. (a) For a 96% confidence interval we want the t-value with an area in the right tail of 0.02. With df = 25, we read from the table that $t_{0.02} = 2.167$.

   (b) For a 98% confidence interval we want the t-value with an area in the right tail of 0.01. With df = 17, we read from the table that $t_{0.01} = 2.567$.

3. (a) df $= n - 1 = 15$ and for 98% confidence, $\alpha/2 = 0.01$. From the table, $\chi^2_{1-0.01} = \chi^2_{0.99} = 5.229$ and $\chi^2_{0.01} = 30.578$.

   (b) df $= n - 1 = 29$ and for 90% confidence, $\alpha/2 = 0.05$. From the table, $\chi^2_{1-0.05} = \chi^2_{0.95} = 17.708$ and $\chi^2_{0.05} = 42.557$.

4. $\bar{x} = \dfrac{125.8 + 152.6}{2} = \dfrac{278.4}{2} = 139.2$

   $E = \dfrac{152.6 - 125.8}{2} = \dfrac{26.8}{2} = 13.4$

5. (a) The size of the sample ($n = 35$) is sufficiently large to apply the Central Limit Theorem and conclude that the sampling distribution of $\bar{x}$ is approximately normal.

   (b) We construct a Z-interval because we are estimating a population mean and we know the population standard deviation.

For 94% confidence the critical value is $z_{0.03} = 1.88$.

Lower bound $= \bar{x} - z_{0.03} \cdot \dfrac{\sigma}{\sqrt{n}}$

$= 325 - 1.88 \cdot \dfrac{31}{\sqrt{35}}$

$\approx 325 - 9.9 = 315.1$ min

Upper bound $= \bar{x} + z_{0.03} \cdot \dfrac{\sigma}{\sqrt{n}}$

$= 325 + 1.88 \cdot \dfrac{31}{\sqrt{35}}$

$\approx 325 + 9.9 = 334.9$ min

Motorola can be 94% confident that the mean talk time for its V505 camera phone before the battery must be recharged is between 315.1 and 334.9 minutes.

(c) For 98% confidence the critical value is $z_{0.01} = 2.33$.

Lower bound $= \bar{x} - z_{0.01} \cdot \dfrac{\sigma}{\sqrt{n}}$

$= 325 - 2.33 \cdot \dfrac{31}{\sqrt{35}}$

$\approx 325 - 12.2 = 312.8$ min

Upper bound $= \bar{x} + z_{0.03} \cdot \dfrac{\sigma}{\sqrt{n}}$

$= 325 + 2.33 \cdot \dfrac{31}{\sqrt{35}}$

$\approx 325 + 12.2 = 337.2$ min

Motorola can be 98% confident that the mean talk time for its V505 camera phone before the battery must be recharged is between 312.8 and 337.2 minutes.

(d) For 95% confidence, we use $z_{0.025} = 1.96$. So,

$n = \left(\dfrac{z_{\alpha/2} \cdot \sigma}{E}\right)^2 = \left(\dfrac{1.96 \cdot 31}{5}\right)^2 \approx 147.67$

which we must increase to 148. Motorola would need to sample 148 V505 camera phones to be within 5 minutes of the population mean talk time before recharging, with 95% confidence.

6. (a) We would expect the distribution to be skewed right. We expect most respondents to have relatively few family members in prison, but there will be some with several (possibly many) family members in prison.

(b) We construct a t-interval because we are estimating a population mean and we do not know the population standard deviation. For 99% confidence, $\alpha/2 = 0.005$. Since $n = 499$, then df $= 498$ and $t_{0.005} = 2.626$. Then:

Lower bound $= \bar{x} - t_{0.005} \cdot \dfrac{s}{\sqrt{n}}$

$= 1.22 - 2.626 \cdot \dfrac{0.59}{\sqrt{499}}$

$\approx 1.22 - 0.069 = 1.151$

[Tech: 1.152]

Upper bound $= \bar{x} + t_{0.005} \cdot \dfrac{s}{\sqrt{n}}$

$= 1.22 + 2.626 \cdot \dfrac{0.59}{\sqrt{499}}$

$\approx 1.22 + 0.069 = 1.289$

[Tech: 1.288]

We are 99% confident that the population mean number of family members in jail is between 1.151 and 1.289.

7. (a) We construct a t-interval because we are estimating a population mean and we do not know the population standard deviation. For 90% confidence, $\alpha/2 = 0.05$. Since $n = 50$, then df $= 49$. There is no row in the table for 49 degrees of freedom, so we use df $= 50$ instead. Thus, $t_{0.05} = 1.676$, and

Lower bound $= \bar{x} - t_{0.05} \cdot \dfrac{s}{\sqrt{n}}$

$= 4.58 - 1.676 \cdot \dfrac{1.10}{\sqrt{50}}$

$\approx 4.58 - 0.261 = 4.319$ yrs

Upper bound $= \bar{x} + t_{0.05} \cdot \dfrac{s}{\sqrt{n}}$

$= 4.58 + 1.676 \cdot \dfrac{1.10}{\sqrt{50}}$

$\approx 4.58 + 0.261 = 4.841$ yrs

We are 90% confident that the population mean time to graduate is between 4.319 years and 4.841 years.

(b) Yes; we are 90% confident that the mean time to graduate is between 4.319 years and 4.841 years. Since the entire interval is above 4 years, our evidence contradicts the belief that it takes 4 years to complete a bachelor's degree.

8. (a) Using technology, we obtain $\bar{x} \approx 57.8$ inches and $s \approx 15.4$ inches.

(b) Yes. The plotted points are generally linear and stay within the bounds of the normal probability plot. The boxplot shows that there are no outliers.

(c) We construct a t-interval because we are estimating a population mean and we do not know the population standard deviation. For 95% confidence, $\alpha/2 = 0.025$. Since $n = 12$, then df $= 11$ and $t_{0.025} = 2.201$. Then:

Lower bound $= \bar{x} - t_{0.025} \cdot \dfrac{s}{\sqrt{n}}$

$= 57.8 - 2.201 \cdot \dfrac{15.4}{\sqrt{12}}$

$\approx 57.8 - 9.8 = 48.0$ in.

[Tech: 47.9]

Upper bound $= \bar{x} + t_{0.025} \cdot \dfrac{s}{\sqrt{n}}$

$= 57.8 + 2.201 \cdot \dfrac{15.4}{\sqrt{12}}$

$\approx 57.8 + 9.8 = 67.6$ in.

The researcher can be 95% confident that the population mean depth of visibility of the Secchi disk is between 48.0 and 67.6 inches.

(d) For 99% confidence, $\alpha/2 = 0.005$. For df $= 11$, $t_{0.005} = 3.106$. Then:

Lower bound $= \bar{x} - t_{0.005} \cdot \dfrac{s}{\sqrt{n}}$

$= 57.8 - 3.106 \cdot \dfrac{15.4}{\sqrt{12}}$

$\approx 57.8 - 13.8 = 44.0$ in.

Upper bound $= \bar{x} + t_{0.005} \cdot \dfrac{s}{\sqrt{n}}$

$= 57.8 + 3.106 \cdot \dfrac{15.4}{\sqrt{12}}$

$\approx 57.8 + 13.8 = 71.6$ in.

The researcher can be 99% confident that the population mean depth of visibility of the Secchi disk is between 44.0 and 71.6 inches.

(e) From the table, with df = 11 and
$\alpha/2 = 0.025$, we find
$\chi^2_{1-0.025} = \chi^2_{0.975} = 3.816$ and
$\chi^2_{0.025} = 21.920$. Then:

Lower bound $= \sqrt{\dfrac{(n-1)s^2}{\chi^2_{\alpha/2}}} = \sqrt{\dfrac{11(15.45)^2}{21.920}}$

≈ 10.9 in.

Upper bound $= \sqrt{\dfrac{(n-1)s^2}{\chi^2_{\alpha/2}}} = \sqrt{\dfrac{11(15.45)^2}{3.816}}$

≈ 26.2 in.

The researcher can be 99% confident that the population standard deviation of the depth of visibility of the Secchi disk is between 10.9 and 26.2 inches.

9. (a) $\hat{p} = \dfrac{x}{n} = \dfrac{1139}{1201} \approx 0.948$

(b) For 99% confidence,
$z_{\alpha/2} = z_{.005} = 2.575$.
Lower bound

$= \hat{p} - z_{.005} \cdot \sqrt{\dfrac{\hat{p}(1-\hat{p})}{n}}$

$= 0.948 - 2.575 \cdot \sqrt{\dfrac{0.948(1-0.948)}{1201}}$

$\approx 0.948 - 0.016 = 0.932$
Upper bound

$= \hat{p} + z_{.005} \cdot \sqrt{\dfrac{\hat{p}(1-\hat{p})}{n}}$

$= 0.948 + 2.575 \cdot \sqrt{\dfrac{0.948(1-0.948)}{1201}}$

$\approx 0.948 + 0.016$

$= 0.964$ [Tech: 0.965]

The EPA can be 99% confident that the population proportion of Americans who live in neighborhoods with acceptable levels of carbon monoxide is between 0.932 and 0.964 (i.e., between 93.2% and 96.4%).

(c) $n = \hat{p}(1-\hat{p})\left(\dfrac{z_{\alpha/2}}{E}\right)^2$

$= 0.948(1-0.948)\left(\dfrac{1.645}{.015}\right)^2$

$= 592.9$

which we must increase to 593. A sample size of 593 would be needed for the estimate to be within 1.5 percentage points with 90% confidence.

(d) $n = 0.25\left(\dfrac{z_{\alpha/2}}{E}\right)^2 = 0.25\left(\dfrac{1.645}{.015}\right)^2$

≈ 3006.7

which we must increase to 3007. Without the prior estimate, a sample size of 3007 would be needed for the estimate to be within 1.5 percentage points with 90% confidence.

Chapter 10

Hypothesis Tests Regarding a Parameter

Section 10.1

1. Type I error: To reject the null hypothesis when it is true.
   Type II error: To not reject the null hypothesis when the alternative hypothesis is true.

3. As we decrease α (the probability of rejecting a true H_0), we are effectively making it less likely that we will reject H_0 since we require stronger evidence against H_0 as α decreases. This means that it is also more likely that we will fail to reject H_0 when H_1 is really true, so β increases. Thus, β increases as α decreases.

5. Answers will vary. One possibility follows:
   In a hypothesis test we make a judgment about the validity of a hypothesis based on the available data. If the data contradicts H_0 then we reject H_0. However, if the available data do not contradict H_0, this does not guarantee that H_0 is true. Consider the court system in the U.S., where suspects are assumed to be innocent until proven guilty. An acquittal does not mean the suspect is innocent, merely that there was not enough evidence to reject the assumption of innocence.

7. False; sample evidence will never prove a null hypothesis is true. We assume the null is true and try to gather enough evidence to say that the null is not true. Failing to reject a null hypothesis does not imply that the hypothesis is actually true, just that there was not enough evidence to reject the assumption that it is true.

9. Right-tailed since $H_1 : \mu > 5$
   Parameter $= \mu$

11. Two-tailed since $H_1 : \sigma \neq 4.2$
    Parameter $= \sigma$

13. Left-tailed since $H_1 : \mu < 120$
    Parameter $= \mu$

15. (a) $H_0 : p = 0.102$, $H_1 : p > 0.102$
    The alternative hypothesis is > because the sociologist believes the percent has increased.

    (b) We make a Type I error if the sample evidence leads us to reject H_0 and conclude that the proportion of registered births to teenage mothers has increased above 0.102 when, in fact, it has not increased above 0.102.

    (c) We make a Type II error if the sample evidence does not lead us to conclude that the proportion of registered births to teenage mothers has increased above 0.102 when, in fact, the proportion of registered births to teenage mothers has increased above 0.102.

17. (a) $H_0 : \mu = \$299,800$; $H_1 : \mu < \$299,800$
    The alternative hypothesis is < because the real estate broker believes the mean price has decreased.

    (b) We make a Type I error if the sample evidence leads us to reject H_0 and conclude that the mean price of a single-family home had decreased below $299,800 when, in fact, it has not decreased below $299,800.

    (c) We make a Type II error if we do not conclude that the mean price of a single-family home has decreased below $299,800, when, in fact, it has decreased below $299,800.

19. (a) $H_0 : \sigma = 0.7$ p.s.i. ; $H_1 : \sigma < 0.7$ p.s.i.
    The alternative hypothesis is < because the quality control manager believes the standard deviation of the required pressure has been reduced.

    (b) We make a Type I error if the sample evidence leads us to reject H_0 and conclude that the standard deviation in the pressure required is less than 0.7 p.s.i when, in fact, the true standard deviation is 0.7 p.s.i.

(c) We make a Type II error if we do not reject the null hypothesis that the standard deviation in the pressure required is 0.7 p.s.i when, in fact, the true standard deviation is less than 0.7 p.s.i.

21. (a) $H_0 : \mu = \$49.94$; $H_1 : \mu \neq \$49.94$

The alternative hypothesis is $\neq$ because no direction of change is given. The researcher feels the mean monthly bill has changed but this could mean an increase or a decrease.

(b) We make a Type I error if the sample evidence leads us to reject H_0 and conclude that the mean monthly cell phone bill is not $49.94 when, in fact, it is $49.94.

(c) We make a Type II error if we do not reject the null hypothesis that the mean monthly cell phone bill is $49.94 when, in fact, it is different than $49.94.

23. There is sufficient evidence to conclude that the proportion of registered births in the U.S. to teenage mothers has increased above 0.102.

25. There is not sufficient evidence to conclude that the mean price of a single-family home has decreased from $299,800.

27. There is not sufficient evidence to conclude that the standard deviation in the pressure required has been reduced from 0.7 p.s.i.

29. There is sufficient evidence to conclude that the mean monthly cell phone bill is different from $49.94.

31. There is not sufficient evidence to conclude that the proportion of registered births to teenage mothers has increased above 0.102.

33. There is sufficient evidence to conclude that the mean price of a single-family home has decreased from $299,800.

35. (a) $H_0 : \mu = 54$ quarts ; $H_1 : \mu > 54$ quarts

(b) Answers may vary. One possibility follows: Congratulations to the marketing department at popcorn.org. After a marketing campaign encouraging people to consume more popcorn, our researchers have determined that the mean annual consumption of popcorn is now greater than 54 quarts, the mean consumption prior to the campaign.

(c) A Type I error has been made, since the true mean consumption has not increased above 54 quarts. The probability of making a Type I error is 0.05.

37. (a) $H_0 : p = 0.152$; $H_1 : p < 0.152$

(b) There is not sufficient evidence to conclude that changes in the DARE program have resulted in a decrease in the proportion of tenth graders who have tried marijuana.

(c) Since we failed to reject a false null hypothesis, a Type II error was committed.

39. Let $\mu =$ the mean increase in gas mileage for cars using the Platinum Gasaver. Then the hypotheses would be $H_0 : \mu = 0$ versus $H_1 : \mu > 0$.

41. Answers will vary. One possibility follows: If you are going to accuse a company of wrongdoing, you should have fairly convincing evidence. In addition, you likely do not want to find out down the road that your accusations were unfounded. Therefore, it is likely more serious to make a Type I error. For this reason, we should probably make the level of significance $\alpha = 0.01$.

Section 10.2

1. The sample must have been obtained using simple random sampling, and either the population from which the sample is selected must be normally distributed or the sample size must be large ($n \geq 30$).

3. When $\alpha = 0.05$ and σ is known, the critical values for a two-tailed test are $-z_{0.025} = -1.96$ and $z_{0.025} = 1.96$.

5. The P-value is the probability of observing a sample statistic as extreme or more extreme than the one observed under the assumption that the null hypothesis is true. A small P-value indicates that the observed data are very unlikely to result from chance variation in samples, so this is evidence that the null hypothesis is not true. More specifically, if the P-value is less than the level of significance α, the null hypothesis is rejected.

7. If the *P*-value is 0.02, then the probability of obtaining a sample mean more than $|z_0|$ standard deviations away from the hypothesized mean of 50, assuming that 50 is the true mean, is 0.02. Decisions regarding whether to reject H_0 will vary. One possibility follows: Since this is an unusual occurrence, we would conclude that our sample data provide strong enough evidence to reject H_0, although this depends on the level of significance that we are using.

9. Answers will vary. Statistical significance typically refers to absolute differences; that is, whether the observed difference is due to chance. Practical significance typically refers to relative differences; that is, whether the observed difference is large enough to cause concern.

11. (a) $z_0 = \dfrac{\bar{x} - \mu_0}{\sigma / \sqrt{n}} = \dfrac{47.1 - 50}{12 / \sqrt{24}} = -1.18$

    (b) This is a left-tailed test, so the critical value is $-z_{0.05} = -1.645$.

    (c)

    (d) Since $-1.18 > -1.645$, the test statistic is not in the critical region. The researcher will not reject the null hypothesis.

13. (a) $z_0 = \dfrac{\bar{x} - \mu_0}{\sigma / \sqrt{n}} = \dfrac{104.8 - 100}{7 / \sqrt{23}} = 3.29$

    (b) This is a two-tailed test, so the critical values are $\pm z_{0.005} = \pm 2.575$.

    (c)

    (d) Since $3.29 > 2.575$, the test statistic is in the critical region. The researcher will reject the null hypothesis.

15. (a) The test statistic is
    $$z_0 = \dfrac{\bar{x} - \mu_0}{\sigma / \sqrt{n}} = \dfrac{18.3 - 20}{3 / \sqrt{18}} = -2.40.$$

This is a left-tailed test, so we get
$$P - \text{value} = P(Z < -2.40) = 0.0082$$
Fewer than 1 sample in 100 will result in a sample mean as extreme or more extreme as the one we obtained *if* the population mean is $\mu = 20$.

(b) Since *P*-value $= 0.0082 < 0.0500 = \alpha$, the researcher will reject the null hypothesis.

17. (a) No, the population does not need to be normally distributed to compute the *P*-value since the sample size is large $(n \geq 30)$.

    (b) The test statistic is
    $$z_0 = \dfrac{\bar{x} - \mu_0}{\sigma / \sqrt{n}} = \dfrac{101.2 - 105}{12 / \sqrt{35}} = -1.87.$$
    This is a two-tailed test so we have
    $$P\text{-value} = 2 \cdot P(Z < -1.87)$$
    $$= 2 \cdot 0.0307 = 0.0614$$
    About 6 samples in 100 will result in a sample mean as extreme or more extreme as the one we obtained *if* the population mean is $\mu = 105$.

    (c) Since *P*-value $= 0.0614 > 0.05 = \alpha$, the researcher will not reject the null hypothesis.

    (d) For a 95% confidence interval, we use
    $$z_{\alpha/2} = z_{(1-0.95)/2} = z_{0.025}.$$

    Lower bound $= \bar{x} - z_{\alpha/2} \cdot \dfrac{\sigma}{\sqrt{n}}$

    $$= 101.2 - 1.96 \cdot \dfrac{12}{\sqrt{35}} \approx 97.22$$

    Upper bound $= \bar{x} + z_{0.025} \cdot \dfrac{\sigma}{\sqrt{n}}$

    $$= 101.2 + 1.96 \cdot \dfrac{12}{\sqrt{35}} \approx 105.18$$

    Since 105 is contained in this interval, we do not reject the null hypothesis.

19. (a) $H_0 : \mu = \$67$ versus $H_1 : \mu > \$67$

    (b) A *P*-value of 0.02 means that there is a 0.02 probability of obtaining a sample mean of \$73 or higher from a population whose mean is \$67. So, if we obtained 100 simple random samples of size $n = 40$ from a population whose mean is \$67, we would expect about 2 of these samples to result in sample means of \$73 or higher.

(c) Since $P\text{-value} = 0.02 < \alpha = 0.05$, we reject the statement in the null hypothesis. There is sufficient evidence to conclude that the mean dollar amount withdrawn from a PayEase ATM is more than the mean amount from a standard ATM. That is, the mean dollar amount withdrawn from a PayEase ATM is more than $67.

21. (a) $H_0 : \mu = 21.2$ versus $H_1 : \mu > 21.2$

(b) $z_0 = \dfrac{\bar{x} - \mu_0}{\sigma / \sqrt{n}} = \dfrac{21.9 - 21.2}{4.9 / \sqrt{200}} = 2.02$

Classical approach:
This is a right-tailed test, so the critical value is $z_\alpha = z_{0.05} = 1.645$ and the critical region lies to the right of our critical value. Since $z_0 = 2.02 > z_{0.05} = 1.645$, the test statistic falls in the critical region. Therefore, we reject the null hypothesis.

P-value approach:
$$P\text{-value} = P(Z > 2.02) = 1 - P(Z \le 2.02)$$
$$= 1 - 0.9783 = 0.0217$$
Since $P\text{-value} = 0.0217 < \alpha = 0.05$, we reject the null hypothesis.

(c) There is sufficient evidence to conclude that students who take the core mathematics curriculum score better on the ACT.

23. (a) The data are all within the confidence bands of the normal probability plot, which also has a generally linear pattern. The boxplot shows that there are no outliers. Therefore, the conditions for a hypothesis test are satisfied.

(b) Hypotheses: $H_0 : \mu = 0.11$ mg/L versus $H_1 : \mu \ne 0.11$ mg/L
We compute the sample mean to be $\bar{x} = 0.1568$ mg/L.

Test Statistic:
$$z_0 = \dfrac{\bar{x} - \mu_0}{\sigma / \sqrt{n}} = \dfrac{0.1568 - 0.11}{0.08 / \sqrt{10}} \approx 1.85$$

Classical approach:
This is a two-tailed test, so the critical values are $\pm z_{\alpha/2} = \pm z_{0.025} = \pm 1.96$. Since $z_0 = 1.85$ is between $-z_{0.05} = -1.96$ and $z_{0.05} = 1.96$, the test statistic does not fall in the critical region. Therefore, we do not reject the null hypothesis.

P-value approach:
$$P\text{-value} = P(Z < -1.85) + P(Z > 1.85)$$
$$= 2 \cdot P(Z < -1.85)$$
$$= 2(0.0322)$$
$$= 0.0644 \text{ [Tech: 0.0643]}$$
Since $P\text{-value} = 0.0644 > \alpha = 0.05$, we do not reject the null hypothesis.

Conclusion: There is not sufficient evidence to indicate that the calcium concentration in rainwater in Chautauqua, New York, has changed since 1990.

25. (a) The data are all within the confidence bands of the normal probability plot, which also has a generally linear pattern. The boxplot shows that there are no outliers. Therefore, the conditions for a hypothesis test are satisfied.

(b) Hypotheses: $H_0 : \mu = 64.05$ ounces versus $H_1 : \mu \ne 64.05$ ounces
We compute the sample mean to be $\bar{x} \approx 64.007$ ounces.
Test Statistic:
$$z_0 = \dfrac{\bar{x} - \mu_0}{\sigma / \sqrt{n}} = \dfrac{64.007 - 64.05}{0.06 / \sqrt{22}} = -3.36$$

Classical approach:
This is a two-tailed test, so the critical values are $\pm z_{\alpha/2} = \pm z_{0.005} = \pm 2.575$.
Since $z_0 = -3.36 < -z_{0.005} = -2.575$, the test statistic falls in the critical region. Thus, we reject the null hypothesis.
P-value approach:
$$P\text{-value} = P(Z < -3.36) + P(Z > 3.36)$$
$$= 2 \cdot P(Z < -3.36)$$
$$= 2(0.0004)$$
$$= 0.0008$$
Since $P\text{-value} = 0.0008 < \alpha = 0.01$, we reject the null hypothesis.

Conclusion: There is sufficient evidence to indicate that the mean amount of juice in each bottle is not 64.05 ounces.

Since the null hypothesis has been rejected, the process should be stopped so the machine can be recalibrated.

(c) Answers will vary. Using $\alpha = 0.1$ means that we will reject a true null 10% of the time. Stopping the machine process to recalibrate, when unnecessary, delays production which can lead to increased costs, lost revenue, and lower profits.

27. (a) Assessed home values are typically skewed right because there are a few homes with very high assessed values. Because the shape of the distribution of assessed home values is not approximately normal, we need a large sample size so that we can use the Central Limit Theorem, which means that the distribution of the sample mean will be approximately normal.

(b) Hypotheses: $H_0 : \mu = \$11.38$ versus $H_1 : \mu < \$11.38$

Test Statistic:

$$z_0 = \frac{\bar{x} - \mu_0}{\sigma / \sqrt{n}} = \frac{11.09 - 11.38}{8.02 / \sqrt{50}} \approx -0.26$$

Classical approach:
We use an $\alpha = 0.05$ level of significance. Since this is a left-tailed test, our critical value is $-z_\alpha = -z_{0.05} = -1.645$. Since $z_0 = -0.26 > -z_{0.05} = -1.645$, the test statistic does not fall in the critical region and we do not reject the null hypothesis.

P-value approach:
P-value $= P(Z < -0.26) = 0.3974$ [Tech: 0.3991]. Since P-value $> \alpha = 0.05$, we do not reject the null hypothesis.

Conclusion: There is not sufficient evidence to indicate that rural households pay less per $1000 assessed valuation than all homes in the United States.

29. (a) The data are not normally distributed and have outliers, so a large sample is needed to conduct a hypothesis about the mean.

(b) Hypotheses: $H_0 : \mu = 24.1$ million shares versus $H_1 : \mu \neq 24.1$ million shares

Test Statistic:

$$z_0 = \frac{\bar{x} - \mu_0}{\sigma / \sqrt{n}} = \frac{21.97 - 24.1}{7.6 / \sqrt{40}} = -1.77$$

Classical approach:
This is a two-tailed test, so the critical values are $\pm z_{\alpha/2} = \pm z_{0.025} = \pm 1.96$. Since $z_0 = -1.77$ is between $-z_{0.025} = -1.96$ and $z_{0.025} = 1.96$, the test statistic does not fall in a critical region. Thus, we do not reject the null hypothesis.

P-value approach:
$$P\text{-value} = P(Z < -1.77) + P(Z > 1.77)$$
$$= 2 \cdot P(Z < -1.77)$$
$$= 2(0.0384)$$
$$= 0.0768 \ [\text{Tech: } 0.0763]$$

Since P-value $= 0.0768 > \alpha = 0.05$, we do not reject the null hypothesis.

Conclusion: There is not sufficient evidence to indicate that the volume of Dell stock is different from what it was in 2002.

31. Hypotheses: $H_0 : \mu = 0.11$ mg/L versus $H_1 : \mu \neq 0.11$ mg/L. From Problem 23, we have $\bar{x} = 0.1568$ mg/L, $\sigma = 0.08$ mg/L, and $n = 10$. Using $\alpha = 0.01$, we have $z_{\alpha/2} = z_{0.005} = 2.575$.

Lower bound $= \bar{x} - z_{\alpha/2} \cdot \dfrac{\sigma}{\sqrt{n}}$

$$= 0.1568 - 2.575 \cdot \frac{0.08}{\sqrt{10}}$$

$$\approx 0.0917 \ [\text{Tech: } 0.0916] \text{ mg/L}$$

Upper bound $= \bar{x} + z_{\alpha/2} \cdot \dfrac{\sigma}{\sqrt{n}}$

$$= 0.1568 + 2.575 \cdot \frac{0.08}{\sqrt{10}}$$

$$\approx 0.2219 \ [\text{Tech: } 0.2220] \text{ mg/L}$$

Since 0.11 lies in this interval, we do not reject the null hypothesis.

Conclusion: There is not sufficient evidence to indicate that the calcium concentration in rainwater in Chautauqua, New York, has changed since 1990.

33. Hypotheses: $H_0 : \mu = 24.1$ million shares versus $H_1 : \mu \neq 24.1$ million shares From Problem 29, $\bar{x} = 21.97$ million shares, $\sigma = 7.6$ million shares, and $n = 40$. Using $\alpha = 0.05$, we have $z_{\alpha/2} = z_{0.025} = 1.96$.

$$\text{Lower bound} = \bar{x} - z_{\alpha/2} \cdot \frac{\sigma}{\sqrt{n}}$$
$$= 21.97 - 1.96 \cdot \frac{7.6}{\sqrt{40}}$$
$$\approx 19.615 \text{ million shares}$$
$$\text{Upper bound} = \bar{x} + z_{\alpha/2} \cdot \frac{\sigma}{\sqrt{n}}$$
$$= 21.97 + 1.96 \cdot \frac{7.6}{\sqrt{40}}$$
$$\approx 24.325 \text{ million shares}$$

Since 24.1 lies in this interval, we do not reject the null hypothesis.

Conclusion: There is not sufficient evidence to indicate that the volume of Dell stock is different from what it was in 2002.

35. (a) $H_0 : \mu = 515$ versus $H_1 : \mu > 515$

(b) The test statistic is
$$z_0 = \frac{\bar{x} - \mu_0}{\sigma/\sqrt{n}} = \frac{519 - 515}{114/\sqrt{1800}} = 1.49$$
Classical approach:
This is a right-tailed test, so our critical value is $z_\alpha = z_{0.10} = 1.28$. Since $z_0 = 1.49 > z_{0.10} = 1.28$, the test statistic falls in the critical region. Thus, we reject the null hypothesis.
P-value approach:
$$P\text{-value} = P(Z > 1.49)$$
$$= 1 - P(Z \leq 1.49)$$
$$= 1 - 0.9319$$
$$= 0.0681 \text{ [Tech: 0.0683]}$$
Since $P\text{-value} = 0.0681 < \alpha = 0.10$, we reject the null hypothesis.
Conclusion: There is sufficient evidence to conclude that the mean score of students taking this review is greater than 515.

(c) Answers will vary. In some states this would be regarded as a highly significant increase, although in most states it is not likely to be thought of as an increase that has any practical significance.

(d) The test statistic is now
$$z_0 = \frac{\bar{x} - \mu_0}{\sigma/\sqrt{n}} = \frac{519 - 515}{114/\sqrt{400}} = 0.70.$$
Classical approach:
This is a right-tailed test, so our critical value is $z_\alpha = z_{0.10} = 1.28$. Since $z_0 = 0.70 < z_{0.10} = 1.28$, the test statistic does not fall in the critical region. Thus, we do not reject the null hypothesis.
P-value approach:
$$P\text{-value} = P(Z > 0.70)$$
$$= 1 - P(Z \leq 0.70)$$
$$= 1 - 0.7580$$
$$= 0.2420 \text{ [Tech: 0.2414]}$$
Since $P\text{-value} = 0.2420 > \alpha = 0.10$, we do not reject the null hypothesis.
Conclusion: There is not sufficient evidence to conclude that the mean score of students taking this review is greater than 515.

The sample size can dramatically alter the conclusion of a hypothesis test. It is often possible to make even slight changes statistically significant by selecting a large enough sample size.

37. (a) Answers will vary.

(b) Since the samples all come from a population that has mean equal to 80, approximately 10% of them, that is 5 out of the 50 samples, should give a sample mean that is in the critical region at a 10% level of significance.

(c) Answers will vary.

(d) Answers will vary. The true mean is 80, so the null hypothesis is correct. Thus, to reject it would be to commit a Type I error.

39. Yes. Because the head of institutional research has access to the entire population, inference is unnecessary. He can say with 100% confidence that the mean age decreased because the mean age in the current semester is less than the mean age is 1995.

Section 10.3

1. A hypothesis regarding a population mean with σ unknown can be tested provided that the sample is obtained using simple random sampling, and the population from which the sample is drawn is either normally distributed with no outliers or the sample size n is larger than 30.

3. For $\alpha = 0.05$ in a two-tailed test with 12 degrees of freedom when σ is unknown, the critical values are $\pm t_{\alpha/2} = \pm t_{0.025} = \pm 2.179$.

5. **(a)** $t_0 = \dfrac{\overline{x} - \mu_0}{s/\sqrt{n}} = \dfrac{47.1 - 50}{10.3/\sqrt{24}} = -1.379$

   (b) This is a left-tailed test with $24 - 1 = 23$ degrees of freedom, so the critical value is $-t_{0.05} = -1.714$.

   (c)

   (d) Since the test statistic is not in the critical region $(-1.379 > -1.714)$, the researcher will not reject the null hypothesis.

7. **(a)** $t_0 = \dfrac{\overline{x} - \mu_0}{s/\sqrt{n}} = \dfrac{104.8 - 100}{9.2/\sqrt{23}} = 2.502$

   (b) This is a two-tailed test with $23 - 1 = 22$ degrees of freedom, so the critical values are $\pm t_{0.005} = \pm 2.819$.

   (c)

   (d) Since the test statistic is not in the critical region $(-2.819 < 2.502 < 2.819)$, the researcher will not reject the null hypothesis.

   (e) Using $\alpha = 0.01$ with 22 degrees of freedom, we have $t_{\alpha/2} = \pm t_{0.005} = \pm 2.819$.

   Lower bound $= \overline{x} - z_{\alpha/2} \cdot \dfrac{\sigma}{\sqrt{n}}$

   $= 104.8 - 2.819 \cdot \dfrac{9.2}{\sqrt{23}}$

   ≈ 99.39

Upper bound $= \overline{x} + z_{\alpha/2} \cdot \dfrac{\sigma}{\sqrt{n}}$

$= 104.8 + 2.819 \cdot \dfrac{9.2}{\sqrt{23}}$

≈ 110.21

Because this confidence interval includes the hypothesized mean of 100, the researcher will not reject the null hypothesis.

9. **(a)** $t_0 = \dfrac{\overline{x} - \mu_0}{s/\sqrt{n}} = \dfrac{18.3 - 20}{4.3/\sqrt{18}} = -1.677$.

   (b)

   (c) This is a left-tailed test with $18 - 1 = 17$ degrees of freedom. The P-value is the area under the t-distribution to the left of the test statistic, $t_0 = -1.677$. Because of symmetry, the area under the distribution to the left of -1.677 equals the area under the distribution to the right of 1.677. From the t-distribution table (Table VI) in the row corresponding to 17 degrees of freedom, since $t = 1.677$ is between 1.333 and 1.740, whose right-tail areas are 0.10 and 0.05, respectively. So, $0.05 < P\text{-value} < 0.10$ [Tech: 0.0559]. Interpretation: If we obtain 100 random samples of size 18, we would expect about 6 of the samples to result in a sample mean of 18.3 or less *if* the population mean is $\mu = 20$.

   (d) Since $P\text{-value} > \alpha = 0.05$, the researcher will not reject the null hypothesis.

11. **(a)** No, because this sample is large ($n \geq 30$).

    (b) $t_0 = \dfrac{\overline{x} - \mu_0}{s/\sqrt{n}} = \dfrac{101.9 - 105}{5.9/\sqrt{35}} = -3.108$.

    (c)

(d) This is a two-tailed test with $35-1=34$ degrees of freedom. The *P*-value of this two tailed test is the area to the left of $t_0 = -3.108$, plus the are to the right of 3.108. From the *t*-distribution table (Table VI) in the row corresponding to 34 degrees of freedom, 3.108 falls between 3.002 and 3.348, whose right-tail areas are 0.0025 and 0.001, respectively. We must double these values in order to get the total area in both tails: 0.005 and 0.002. Thus, $0.002 < P\text{-value} < 0.005$ [Tech: *P*-value = 0.0038].

Interpretation: If we obtain 1000 random samples of size 35, we would expect about 4 of the samples to result in a sample mean as extreme or more extreme than the one observed *if* the population mean is $\mu = 105$.

(e) Since $P\text{-value} < \alpha = 0.010$, we reject the null hypothesis.

13. Hypotheses: $H_0 : \mu = 9.02$ cm^3 versus

$H_1 : \mu < 9.02$ cm^3

Test Statistic:

$t_0 = \dfrac{\bar{x} - \mu_0}{s/\sqrt{n}} = \dfrac{8.10 - 9.02}{0.7/\sqrt{12}} = -4.553$

$\alpha = 0.01$; d.f. $= n - 1 = 12 - 1 = 11$

Classical approach:
Because this is a left-tailed test with 11 degrees of freedom, the critical value is $-t_{0.01} = -2.718$. Since $t_0 = -4.553 < -t_{0.01} = -2.718$, the test statistic falls within the critical region. We reject the null hypothesis.

P-value approach:
This is a left-tailed test with 11 degrees of freedom. The *P*-value is the area under the *t*-distribution to the left of the test statistic, $t_0 = -4.553$. Because of symmetry, the area under the distribution to the left of -4.553 equals the area under the distribution to the right of 4.553. From the *t*-distribution table (Table VI) in the row corresponding to 11 degrees of freedom, 4.553 falls to the right of 4.437, whose right-tail area is 0.0005. So, $P\text{-value} < 0.0005$ [Tech: *P*-value= 0.0004]. Since $P\text{-value} < \alpha = 0.01$, we reject the null hypothesis.

Conclusion: There is sufficient evidence to conclude that the mean hippocampal volume in alcoholic adolescents is less than the normal volume of 9.02 cm^3.

15. Hypotheses: $H_0 : \mu = 703.5$ versus

$H_1 : \mu > 703.5$

Test Statistic:

$t_0 = \dfrac{\bar{x} - \mu_0}{s/\sqrt{n}} = \dfrac{714.2 - 703.5}{83.2/\sqrt{40}} = 0.813$

$\alpha = 0.05$; d.f. $= n - 1 = 40 - 1 = 39$

Classical approach:
Because this is a right-tailed test with 39 degrees of freedom, the critical value is $t_{0.05} = 1.685$. Since $t_0 = 0.813 < t_{0.05} = 1.685$, the test statistic does not fall within the critical region. We do not reject the null hypothesis.

P-value approach:
This is a right-tailed test with 39 degrees of freedom. The *P*-value is the area under the *t*-distribution to the right of the test statistic, $t_0 = 0.813$. From the *t*-distribution table in the row corresponding to 39 degrees of freedom, 0.813 falls between 0.681 and 0.851, whose right-tail areas are 0.25 and 0.20, respectively. So, $0.20 < P\text{-value} < 0.25$ [Tech: *P*-value = 0.2105]. Since $P\text{-value} > \alpha = 0.05$, we do not reject the null hypothesis.

Conclusion: There is not sufficient evidence to conclude that the mean FICO score of high-income individuals is greater than that of the general population. In other words, it is not unlikely to obtain a mean credit score of 714.2 even though the true population mean credit score is 703.4.

17. (a) Hypotheses: $H_0 : \mu = 98.6°$F versus

$H_1 : \mu < 98.6°$F

Test statistic:

$t_0 = \dfrac{\bar{x} - \mu_0}{s/\sqrt{n}} = \dfrac{98.2 - 98.6}{0.7/\sqrt{700}} = -15.119$.

This is a left-tailed test with $n - 1 = 700 - 1 = 699$ degrees of freedom. However, since our *t*-distribution table does not contain a row for 699, we use df = 1000 (the closest one to 699). So, the critical value is $-t_{0.01} = -2.330$. Since $t_0 = -15.119 < \alpha = -2.330$, the test statistic falls in the critical region and we reject the null hypothesis. There is sufficient evidence to conclude that the mean temperature of humans is less than 98.6° F.

(b) This is a left-tailed test with 699 degrees of freedom. The *P*-value is the area under the *t*-distribution to the left of the test statistic $t_0 = -15.119$. Because of symmetry, the area under the distribution to the left of -15.119 equals the area under the distribution to the right of 15.119. From the *t*-distribution table (Table VI) in the row corresponding to 1000 degrees of freedom (the row closest to df = 699), 15.119 falls to the right of 3.300, whose right-tail area is 0.0005. So, *P*-value < 0.0005 [Tech: P-value < 0.0001].

Interpretation: If we obtain 10,000 random samples of size $n = 700$, we would expect only about 1 of the samples to result in a sample mean of 98.2° F or less *if* the population mean is $\mu = 98.6°$ F.

19. Hypotheses: $H_0 : \mu = 40.7$ years versus $H_1 : \mu \neq 40.7$ years

Using $\alpha = 0.05$ with $n - 1 = 32 - 1 = 31$ degrees of freedom, we have $t_{\alpha/2} = t_{0.025} = 2.037$.

$$\text{Lower bound} = \bar{x} - t_{\alpha/2} \cdot \frac{s}{\sqrt{n}}$$

$$= 38.9 - 2.037 \cdot \frac{9.6}{\sqrt{32}}$$

$$= 35.44 \text{ years}$$

$$\text{Upper bound} = \bar{x} + t_{\alpha/2} \cdot \frac{s}{\sqrt{n}}$$

$$= 38.9 + 2.037 \cdot \frac{9.6}{\sqrt{32}}$$

$$= 42.36 \text{ years}$$

Because this interval includes the hypothesized mean 40.7 years, we do not reject the null hypothesis. Thus, there is not sufficient evidence to conclude that the mean age of death-row inmates has change since 2002.

21. (a) The plotted data are all within the bounds of the normal probability plot, which also has a generally linear pattern. The boxplot shows that there are no outliers. Therefore, the conditions for testing the hypothesis are satisfied.

(b) Hypotheses: $H_0 : \mu = 1.68$ inches versus $H_1 : \mu \neq 1.68$ inches.

We compute the sample mean and sample standard deviation to be $\bar{x} = 1.681$ inches and $s \approx 0.0045$ inches.

The test statistic is

$$t_0 = \frac{\bar{x} - \mu_0}{s / \sqrt{n}}$$

$$= \frac{1.681 - 1.68}{0.0045 / \sqrt{12}}$$

$$= 0.770 \text{ [Tech: 0.778]}$$

Classical approach:
This is a two-tailed test with $n - 1 = 12 - 1 = 11$ degrees of freedom, so the critical values are $\pm t_{0.025} = \pm 2.201$. Since $t_0 = 0.770$ is between -2.201 and 2.201, the test statistic does not fall in the critical region. Therefore, we do not reject the null hypothesis.

P-value approach:
This is a two-tailed test with 11 degrees of freedom. The *P*-value of this two tailed test is the area to the right of $t_0 = 0.770$, plus the area to the left of -0.770. From the *t*-distribution table (Table VI) in the row corresponding to 11 degrees of freedom, 0.770 falls between 0.697 and 0.876, whose right-tail areas are 0.25 and 0.20, respectively. We must double these values in order to get the total area in both tails: 0.50 and 0.40. Thus, $0.40 <$ *P*-value < 0.50 [Tech: *P*-value = 0.4529].

Conclusion: There is not sufficient evidence to conclude that the mean diameter of Maxfli XS golf balls is different from 1.68 inches. In other words, there is not sufficient evidence to conclude that the golf balls do not conform. Therefore, we will give Maxfli the benefit of the doubt and assume the balls are conforming.

23. (a) The plotted data are all within the bounds of the normal probability plot, and the boxplot shows that there are no outliers. Therefore, the conditions for a hypothesis test are satisfied.

(b) Hypotheses: $H_0 : \mu = 84.3$ seconds versus $H_1 : \mu < 84.3$ seconds

We compute the sample mean and sample standard deviation to be $\bar{x} = 78$ seconds and $s = 15.21$ seconds.

The test statistic is

$$t_0 = \frac{\bar{x} - \mu_0}{s/\sqrt{n}} = \frac{78 - 84.3}{15.21/\sqrt{10}} = -1.310$$

Classical approach:
This is a left-tailed test with $n-1 = 10-1 = 9$ degrees of freedom, so the critical value is $-t_{0.10} = -1.383$. Since $t_0 = -1.310 > -t_{0.10} = -1.383$, the test statistic is not in the critical region, so we do not reject the null hypothesis.

P-value approach:
From the t-distribution table (Table VI) in the row corresponding to 9 degrees of freedom, 1.310 falls between 1.100 and 1.383, whose right-tail areas are 0.15 and 0.1, respectively. So, $0.1 < P$-value < 0.15 [Tech: 0.1113]. Since P-value $> \alpha = 0.1$, we do not reject the null hypothesis.

Conclusion: There is not sufficient evidence to conclude that the new system results in a mean wait time that is less than 84.3 seconds. In other words, there is not sufficient evidence to conclude that the new system is effective.

25. (a) We assume that there is no difference between actual and predicted earnings, so the null hypothesis is $H_0 : \mu = 0$. We want to gather evidence that shows that the predictions are off target, but we are not indicating whether the analysts, on average, had predictions that were too high or too low, so the alternative hypothesis is $H_1 : \mu \neq 0$.

(b) Since P-value $= 0.046 < \alpha = 0.05$, we reject the statement in the null hypothesis. The evidence suggests that the analyst's earning predictions are not true.

(c) The researcher is supposed to select the level of significance and direction of the alternative hypothesis prior to gathering evidence. This removes the possibility of any researcher bias.

27. The farmer's analysis may be correct, but his data come from only a small sample in a localized area. His farm is probably not representative of the entire country. In other words, the analysis conducted by the farmer only applies to his farm, not the entire United States.

29. (a) Answers will vary. We would expect 50 samples out of 1000 (i.e. 5%) to result in a rejection of the null hypothesis at $\alpha = 0.05$. The probability of a Type I error is $\alpha = 0.05$. Discrepancies might occur if the requirements for hypothesis testing of the mean (e.g. normality) are not met.

(b) Answers will vary. We would expect 50 samples out of 1000 (i.e. 5%) to result in a rejection of the null hypothesis at $\alpha = 0.05$.

Section 10.4

1. A hypothesis about a population proportion can be tested if the sample is obtained using simple random sampling and $np(1-p) \geq 10$ with $n \leq 0.05N$ (the sample size is no more than 5% of the population size).

3. $np_0(1-p_0) = 200 \cdot 0.3(1-0.3) = 42 \geq 10$, so the requirements of the hypothesis test are satisfied.

(a) $\hat{p} = \frac{75}{200} = 0.375$

The test statistic is

$$z_0 = \frac{\hat{p} - p_0}{\sqrt{\frac{p_0(1-p_0)}{n}}} = \frac{0.375 - 0.3}{\sqrt{\frac{0.3(1-0.3)}{200}}} \approx 2.31.$$

This is a right-tailed test, so the critical value is $z_{0.05} = 1.645$. Since $z_0 = 2.31 > z_{0.05} = 1.645$, the test statistic is in the critical region, so we reject the null hypothesis.

(b) P-value $= P(Z > 2.31)$
$= 1 - P(Z \leq 2.31)$
$= 1 - 0.9896$
$= 0.0104$ [Tech: 0.0103]

Since P-value $= 0.0104 < \alpha = 0.05$, we reject the null hypothesis.

5. $np_0(1-p_0)=150\cdot0.55(1-0.55)=37.125\geq10$, so the requirements of the hypothesis test are satisfied.

(a) $\hat{p}=\dfrac{78}{150}=0.52$

The test statistic is

$z_0=\dfrac{\hat{p}-p_0}{\sqrt{\dfrac{p_0(1-p_0)}{n}}}=\dfrac{0.52-0.55}{\sqrt{\dfrac{0.55(1-0.55)}{150}}}\approx-0.74$.

This is a left-tailed test so the critical value is $-z_{0.1}=-1.28$. Since $z_0=-0.74>-z_{0.1}=-1.28$, the test statistic is not in the critical region, so we do not reject the null hypothesis.

(b) $P\text{-value}=P(Z<-0.74)$
$=0.2296$ [Tech: 0.2301]

Since $P\text{-value}=0.2296>\alpha=0.1$, we do not reject the null hypothesis.

7. $np_0(1-p_0)=500\cdot0.9(1-0.9)=45\geq10$, so the requirements of the hypothesis test are satisfied.

(a) $\hat{p}=\dfrac{440}{500}=0.88$

The test statistic is

$z_0=\dfrac{\hat{p}-p_0}{\sqrt{\dfrac{p_0(1-p_0)}{n}}}=\dfrac{0.88-0.9}{\sqrt{\dfrac{0.9(1-0.9)}{500}}}\approx-1.49$

This is a two-tailed test so the critical values are $\pm z_{0.025}=\pm1.96$. Since $z_0=-1.49$ is between $-z_{0.025}=-1.96$ and $z_{0.025}=1.96$, the test statistic is not in the critical region, so we do not reject the null hypothesis.

(b) $P\text{-value}=P(Z<-1.49)+P(Z>1.49)$
$=2\cdot P(-1.49)$
$=2(0.0681)$
$=0.1362$ [Tech: 0.1360]

Since $P\text{-value}=0.1362>\alpha=0.05$, we do not reject the null hypothesis.

9. The hypotheses are $H_0:p=0.019$ versus $H_1:p>0.019$

$np_0(1-p_0)=863\cdot0.019(1-0.019)=16.1\geq10$, so the requirements of the hypothesis test are satisfied. From the survey, $\hat{p}=\dfrac{19}{863}=0.022$.

The test statistic is

$z_0=\dfrac{\hat{p}-p_0}{\sqrt{\dfrac{p_0(1-p_0)}{n}}}=\dfrac{0.022-0.019}{\sqrt{\dfrac{0.019(1-0.019)}{863}}}\approx0.65$.

The level of significance is $\alpha=0.01$.

Classical approach:
This is a right-tailed test, so the critical value is $z_{0.01}=2.33$. Since $z_0=0.65<z_{0.01}=2.33$, the test statistic is not in the critical region, so we do not reject the null hypothesis.

P-value approach:
$P\text{-value}=P(Z>0.65)$
$=1-P(Z\leq0.65)$
$=1-0.7422$
$=0.2578$ [Tech: 0.2582]

Since $P\text{-value}=0.2578>\alpha=0.01$, we do not reject the null hypothesis.

Conclusion: There is not sufficient evidence to conclude that more than 1.9% of Lipitor users experience flulike symptoms as a side effect.

11. Since a majority would constitute a proportion greater than half, the hypotheses are $H_0:p=0.5$ versus $H_1:p>0.5$.

$np_0(1-p_0)=1034(0.5)(1-0.5)=258.5\geq10$, so the requirements of the hypothesis test are satisfied. From the survey, $\hat{p}=\dfrac{548}{1034}\approx0.530$.

The test statistic is

$z_0=\dfrac{\hat{p}-p_0}{\sqrt{\dfrac{p_0(1-p_0)}{n}}}=\dfrac{0.530-0.5}{\sqrt{\dfrac{0.5(1-0.5)}{1034}}}\approx1.93$

The level of significance is $\alpha=0.05$.

Classical approach:
This is a right-tailed test, so the critical value is $z_{0.05}=1.645$. Since $z_0=1.93>z_{0.05}=1.645$, the test statistic falls in the critical region, so we reject the null hypothesis.

P-value approach:
$P\text{-value}=P(Z>1.93)$
$=1-P(Z\leq1.93)$
$=1-0.9732$
$=0.0268$ [Tech: 0.0269]

Since $P\text{-value}=0.0268<\alpha=0.05$, we reject the null hypothesis.

Conclusion: There is sufficient evidence to conclude that a majority of 30- to 64-year-olds in the United States are worried they will outlive their money.

13. The hypotheses are $H_0 : p = 0.38$ versus $H_1 : p < 0.38$.

$np_0(1 - p_0) = 1122(0.38)(1 - 0.38) \approx 264.3 \geq 10$, so the requirements of the hypothesis test are satisfied. From the survey, $\hat{p} = \dfrac{403}{1122} \approx 0.359$.

The test statistic is

$$z_0 = \frac{\hat{p} - p_0}{\sqrt{\dfrac{p_0(1 - p_0)}{n}}} = \frac{0.359 - 0.38}{\sqrt{\dfrac{0.38(1 - 0.38)}{1122}}} \approx -1.45.$$

The level of significance is $\alpha = 0.05$.

Classical approach:
This is a left-tailed test, so the critical value is $-z_{0.05} = -1.645$. Since $z_0 = -1.45 > -z_{0.05} = -1.645$, the test statistic does not fall in the critical region, so we do not reject the null hypothesis.

P-value approach:
P-value $= P(Z < -1.45)$

$= 0.0735$ [Tech: 0.0754]

Since P-value $= 0.0735 < \alpha = 0.05$, we reject the null hypothesis.

Conclusion: There is not sufficient evidence to conclude that the proportion of families with children under the age of 18 who eat dinner together 7 nights a week has decreased since December, 2001.

15. The hypotheses are $H_0 : p = 0.52$ versus $H_1 : p \neq 0.52$.

$np_0(1 - p_0) = 800(0.52)(1 - 0.52) \approx 20.0 \geq 10$, so the requirements of the hypothesis test are satisfied. From the sample, $\hat{p} = \dfrac{256}{800} = 0.32$.

$$z_0 = \frac{\hat{p} - p_0}{\sqrt{\dfrac{p_0(1 - p_0)}{n}}} = \frac{0.32 - 0.52}{\sqrt{\dfrac{0.52(1 - 0.52)}{800}}} \approx -11.32$$

The level of significance is $\alpha = 0.05$.

Classical approach:
This is a two-tailed test, so the critical values are $\pm z_{\alpha/2} = \pm z_{0.025} = \pm 1.96$. Since $z_0 = -11.32 < -z_{0.025} = -1.96$, the test statistic falls in a critical region, so we reject the null hypothesis.

P-value approach:
The test statistic is so far to the left that P-value < 0.0001. Since this P-value is less than the $\alpha = 0.05$ level of significance, we reject the null hypothesis.

Conclusion: There is sufficient evidence to conclude that the proportion of parents with children in high school who feel it is a serious problem that high school students are not being taught enough math and science has changed from what it was in 1994.

17. The hypotheses are $H_0 : p = 0.47$ versus $H_1 : p \neq 0.47$.

From the sample, $\hat{p} = \dfrac{476}{1035} \approx 0.460$.

$n\hat{p}(1 - \hat{p}) = 1035(0.460)(1 - 0.460) \approx 257.1 \geq 10$, so the requirements for constructing a confidence interval are satisfied. For a 95% confidence interval, we have $\alpha = 0.05$, so $z_{\alpha/2} = z_{0.025} = 1.96$.

Lower Bound $= \hat{p} - z_{\alpha/2} \cdot \sqrt{\dfrac{\hat{p}(1 - \hat{p})}{n}}$

$= 0.460 - 1.96 \cdot \sqrt{\dfrac{0.460(1 - 0.460)}{1035}}$

≈ 0.430

Upper Bound $= \hat{p} + z_{\alpha/2} \cdot \sqrt{\dfrac{\hat{p}(1 - \hat{p})}{n}}$

$= 0.460 + 1.96 \cdot \sqrt{\dfrac{0.460(1 - 0.460)}{1035}}$

≈ 0.490

Since 0.47 is contained in this interval, we do not reject the null hypothesis.

Conclusion: There is not sufficient evidence to conclude that parents' attitudes toward the quality of education in the United States has changes since August, 2002.

19. The hypotheses are $H_0 : p = 0.37$ versus $H_1 : p < 0.37$.

$np_0(1 - p_0) = 150(0.37)(1 - 0.37) \approx 35 \geq 10$, so the requirements of the hypothesis test are satisfied. From the survey, $\hat{p} = \dfrac{54}{150} = 0.36$.

$$z_0 = \frac{\hat{p} - p_0}{\sqrt{\dfrac{p_0(1 - p_0)}{n}}} = \frac{0.36 - 0.37}{\sqrt{\dfrac{0.37(1 - 0.37)}{150}}} \approx -0.25$$

The level of significance is $\alpha = 0.05$.

Classical approach:
This is a left-tailed test, so the critical value is $-z_{0.05} = -1.645$. Since $z_0 = -0.25 > -z_{0.05} = -1.645$, the test statistic is not in the critical region, so we do not reject the null hypothesis.

<u>*P*-value approach:</u>

$P\text{-value} = P(Z < -0.25)$

$\qquad = 0.4013 \text{ [Tech: 0.3999]}$

Since $P\text{-value} = 0.4013 > \alpha = 0.05$, we do not reject the null hypothesis.

Conclusion: There is not sufficient evidence to conclude that less than 37% of pet owners speak to their pets on the answering machine or telephone.

21. The hypotheses are $H_0 : p = 0.04$ versus $H_1 : p < 0.04$.

$np_0(1-p_0) = 120 \cdot 0.04(1-0.04) = 4.6 < 10$, so we use small sample techniques. This is a left-tailed test, so we calculate the probability of 3 or fewer successes in 120 binomial trials with $p = 0.04$. Using technology:

$P\text{-value} = P(X \le 3) = 0.2887$

Since this is larger than the $\alpha = 0.05$ level of significance, we do not reject the null hypothesis.

Conclusion: There is not sufficient evidence to support the obstetrician's belief that fewer than 4% of pregnant mothers smoke more than 21 cigarettes per day.

23. The hypotheses are $H_0 : p = 0.096$ versus $H_1 : p > 0.096$.

$np_0(1-p_0) = 80 \cdot 0.096(1-0.096) = 6.9 < 10$, so we use small sample techniques. This is a right-tailed test, so we calculate the probability of 13 or more successes in 80 binomial trials with $p = 0.096$. Using technology:

$P\text{-value} = P(X \ge 13)$

$\qquad = 1 - P(X < 12)$

$\qquad = 1 - 0.9590$

$\qquad = 0.0410$

Since this is less than the $\alpha = 0.1$ level of significance, we reject the null hypothesis.

Conclusion: There is sufficient evidence to support the urban economist's belief that the percentage of Californians who spend more than 60 minutes traveling to work has increased since 2000.

25. (a) Answers will vary. The proportion could be changing due to sampling error – different people are in the sample. The proportion could also be changing because people's attitudes are changing, perhaps due to issues such as the economy, taxes, foreign policy, etc.

(b) The hypotheses are $H_0 : p = 0.45$ versus $H_1 : p < 0.45$.

$np_0(1-p_0) = 1100 \cdot 0.45(1-0.45) = 272.25 \ge 10$, so the requirements of the hypothesis test are satisfied. From the poll, $\hat{p} = \dfrac{484}{1100} = 0.44$. The test statistic is

$z_0 = \dfrac{\hat{p} - p_0}{\sqrt{\dfrac{p_0(1-p_0)}{n}}} = \dfrac{0.44 - 0.45}{\sqrt{\dfrac{0.45(1-0.45)}{1100}}} \approx -0.67$.

$P\text{-value} = P(Z < -0.67)$

$\qquad = 0.2514 \text{ [Tech: 0.2525]}$

Conclusion: The final decision depends on the choice of the level of significance, α. However, this is a very large P-value, so it is very doubtful that we would choose a level of significance larger than it. Therefore, we do not reject the null hypothesis. The proportion of Americans who approved of the job President Bush was doing in February, 2008 was not significantly less than the level in January, 2008.

27. (a) This is an observational study because the researchers simply looked at records. There was no treatment imposed on the individuals in the study.

(b) The study is retrospective because it was conducted after the fact, looking back at historical records.

(c) There were $22 + 80 = 102$ female siblings in the study.

(d) There were 102 female siblings and 92 male siblings included in the study. So, total number of sibling included in the study was $102 + 92 = 194$. The relative frequency for each gender is:

Females: $\dfrac{102}{194} \approx 0.526$

Males: $\dfrac{92}{194} \approx 0.474$

We draw a bar for each gender and represent these relative frequencies by the heights of the bars.

Proportion of Individuals in a Family with a Sibling Who Has Lupus

(e) $\hat{p} = \dfrac{92}{102+92} \approx 0.474$

(f) The hypotheses are $H_0 : p = 0.51$ versus $H_1 : p < 0.51$.

$np_0(1-p_0) = 194(0.51)(1-0.51) \approx 48 \geq 10$,

so the requirements of the hypothesis test are satisfied. The test statistic is

$z_0 = \dfrac{\hat{p} - p_0}{\sqrt{\dfrac{p_0(1-p_0)}{n}}} = \dfrac{0.474 - 0.51}{\sqrt{\dfrac{0.51(1-0.51)}{194}}} \approx -1.00$.

The level of significance is $\alpha = 0.05$.

Classical approach:
This is a left-tailed test, so the critical value is $-z_{0.05} = -1.645$. Since $z_0 = -1.00 > -z_{0.05} = -1.645$, the test statistic is not in the critical region, so we do not reject the null hypothesis.

P-value approach:
$P\text{-value} = P(Z < -1.00)$
$\qquad\qquad = 0.1587$ [Tech: 0.1594]

Since $P\text{-value} = 0.1587 > \alpha = 0.05$, we do not reject the null hypothesis.

Conclusion: There is not sufficient evidence to conclude that the proportion of males in families where a sibling has lupus is less than 0.51, the accepted proportion of males in the general population at birth.

(g) For a 95% confidence interval, we have $\alpha = 0.05$, so $z_{\alpha/2} = z_{0.025} = 1.96$.
Lower Bound

$= \hat{p} - z_{\alpha/2} \cdot \sqrt{\dfrac{\hat{p}(1-\hat{p})}{n}}$

$= 0.474 - 1.96 \cdot \sqrt{\dfrac{0.474(1-0.474)}{194}}$

≈ 0.404

Upper Bound

$= \hat{p} + z_{\alpha/2} \cdot \sqrt{\dfrac{\hat{p}(1-\hat{p})}{n}}$

$= 0.474 + 1.96 \cdot \sqrt{\dfrac{0.474(1-0.474)}{194}}$

≈ 0.544

We are 95% confident that the proportion of males in a family where a sibling has lupus is between 0.404 and 0.544.

Section 10.5

1. To test a hypothesis regarding a population standard deviation, we must have a simple random sample taken from a normally distributed population.

3. **(a)** $\text{df} = 24 - 1 = 23$

$\chi_0^2 = \dfrac{(n-1)s^2}{\sigma_0^2} = \dfrac{23 \cdot (47.2)^2}{(50)^2} = 20.496$

(b) This is a left-tailed test with 23 degrees of freedom and $\alpha = 0.05$, so the critical value is $\chi_{1-\alpha}^2 = \chi_{0.95}^2 = 13.091$.

(c)

(d) No; the researcher will not reject the hull hypothesis because the test statistic is not in the critical region.

5. **(a)** $\text{df} = 18 - 1 = 17$

$\chi_0^2 = \dfrac{(n-1)s^2}{\sigma_0^2} = \dfrac{17 \cdot (2.4)^2}{(1.8)^2} = 30.222$

(b) This is a right-tailed test with 17 degrees of freedom and $\alpha = 0.10$, so the critical value is $\chi_{0.10}^2 = 24.769$.

(c)

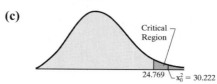

(d) Yes, the researcher will reject H_0 because the test statistic is in the critical region.

7. (a) $df = 12 - 1 = 11$

$$\chi_0^2 = \frac{(n-1)s^2}{\sigma_0^2} = \frac{11 \cdot (4.8)^2}{(4.3)^2} = 13.707$$

(b) This is a two-tailed test with 11 degrees of freedom and $\alpha = 0.05$, so the critical values are $\chi_{0.975}^2 = 3.816$ and $\chi_{0.025}^2 = 21.920$.

(c)

(d) No, the researcher will not reject H_0 because the test statistic is not in the critical region.

9. The hypotheses are $H_0 : \sigma = 4.0$ versus $H_1 : \sigma < 4.0$. The test statistic is

$$\chi_0^2 = \frac{(n-1)s^2}{\sigma_0^2} = \frac{(25-1) \cdot (3.01)^2}{(4.0)^2} = 13.590.$$

Classical Approach: This is a left-tailed test with 24 degrees of freedom and $\alpha = 0.05$, so the critical value is $\chi_{1-\alpha}^2 = \chi_{0.95}^2 = 13.848$. Since the test statistic is in the critical region $(13.590 < 13.848)$, we reject the null hypothesis.

P-value approach: The P-value is the area under the χ^2-distribution with 24 degrees of freedom to the left of $\chi_0^2 = 13.590$. From the χ^2-distribution table (Table VII) in the row corresponding to 24 degrees of freedom, 13.590 is between 12.401 and 13.848. The areas to the right of these critical values are 0.975 and 0.95, respectively. This means the area to the left of these critical values are 0.025 and 0.05, respectively. So, $0.025 < P\text{-value} < 0.05$ [Tech: 0.0446]. Since $P\text{-value} < \alpha = 0.05$, we reject the null hypothesis.

Conclusion: There is sufficient evidence to conclude that the mutual fund has moderate risk.

11. The hypotheses are $H_0 : \sigma = 0.004$ inch versus $H_1 : \sigma < 0.004$ inch. The test statistic is

$$\chi_0^2 = \frac{(n-1)s^2}{\sigma_0^2} = \frac{(25-1) \cdot (0.0025)^2}{(0.004)^2} = 9.375.$$

Classical Approach: This is a left-tailed test with 24 degrees of freedom and $\alpha = 0.01$, so the critical value is $\chi_{1-\alpha}^2 = \chi_{0.99}^2 = 10.856$. Since the test statistic is in the critical region $(9.375 < 10.856)$, we reject the null hypothesis.

P-value approach: The P-value is the area under the χ^2-distribution with 24 degrees of freedom to the left of $\chi_0^2 = 9.375$. From the χ^2-distribution table (Table VII) in the row corresponding to 24 degrees of freedom, 9.375 is to the left of 9.886. The area to the right of this critical value is 0.995, so the area to the left is 0.005. So, $P\text{-value} < 0.005$ [Tech: 0.00339]. Since $P\text{-value} < \alpha = 0.01$, we reject the null hypothesis.

Conclusion: There is sufficient evidence for the manager to conclude that the standard deviation has decreased. The recalibration was effective.

13. The hypotheses are $H_0 : \sigma = 0.08$ versus $H_1 : \sigma \neq 0.08$. The sample size is $n = 10$. We compute the sample standard deviation to be $s \approx 0.0867$. The test statistic is

$$\chi_0^2 = \frac{(n-1)s^2}{\sigma_0^2} = \frac{(10-1) \cdot (0.0867)^2}{(0.08)^2} = 10.571$$

[Tech: 10.567]

Classical Approach: This is a two-tailed test with 9 degrees of freedom and $\alpha = 0.05$, so the critical values are $\chi_{1-\alpha/2}^2 = \chi_{0.975}^2 = 2.700$ and $\chi_{\alpha/2}^2 = \chi_{.025}^2 = 19.023$. Since the test statistic is not in the critical region (10.571 is between 2.700 and 19.023), we do not reject the null hypothesis.

P-value Approach: Using technology, we find $P\text{-value} = 0.6131$. Since this is larger than $\alpha = 0.05$, we do not reject the null hypothesis.

Conclusion: There is not sufficient evidence to conclude that the standard deviation is different from 0.08. The assumption is reasonable.

15. The hypotheses are $H_0 : \sigma = 0.06$ oz versus $H_1 : \sigma \neq 0.06$ oz. The sample size is $n = 22$. We compute the sample standard deviation to be $s \approx 0.045$ ounce. The test statistic is

$$\chi_0^2 = \frac{(n-1)s^2}{\sigma_0^2} = \frac{(22-1) \cdot (0.045)^2}{(0.06)^2} = 11.813$$

[Tech: 11.621]

Classical Approach: This is a two-tailed test with 21 degrees of freedom and $\alpha = 0.05$, so the critical values are $\chi^2_{1-\alpha/2} = \chi^2_{0.975} = 10.283$ and $\chi^2_{\alpha/2} = \chi^2_{0.025} = 35.479$. Since the test statistic is not in the critical region (11.813 is between 10.283 and 35.479), we do not reject the null hypothesis.

P-value Approach: Using technology, we find P-value = 0.1014. Since this is larger than $\alpha = 0.05$, we do not reject the null hypothesis.

Conclusion: There is not sufficient evidence to conclude that the standard deviation is different from 0.06 ounce. The assumption is reasonable.

17. The hypotheses are $H_0 : \sigma = 8.3$ points, $H_1 : \sigma < 8.3$ points. The test statistic is

$$\chi^2_0 = \frac{(n-1)s^2}{\sigma^2_0} = \frac{(25-1)\cdot(6.7)^2}{(8.3)^2} = 15.639.$$

Classical Approach: This is a left-tailed test with 24 degrees of freedom and $\alpha = 0.10$, so the critical value is $\chi^2_{0.90} = 15.659$. Since the test statistic is in the critical region (15.639 < 15.659), we reject the null hypothesis.

P-value approach: The P-value is the area under the χ^2-distribution with 24 degrees of freedom to the left of $\chi^2_0 = 15.639$. From the χ^2-distribution table (Table VII) in the row corresponding to 24 degrees of freedom, 15.639 is between 13.848 and 15.659. The areas to the right of these critical values are 0.95 and 0.90, respectively. This means the area to the left of these critical values are 0.05 and 0.10, respectively. So, $0.05 < P$-value < 0.10 [Tech: 0.0993]. Since P-value $< \alpha = 0.10$, we reject the null hypothesis.

Conclusion: There is sufficient evidence to conclude that Allen Iverson is more consistent than other shooting guards in the NBA.

19. (a) The points lie within the bounds of the normal plot and have a generally linear pattern.

(b) Using technology, we compute the sample standard deviation to be $s = 2.059$ inches.

(c) The hypotheses are $H_0 : \sigma = 2.9$ inches versus $H_1 : \sigma < 2.9$ inches. The test statistic is

$$\chi^2_0 = \frac{(n-1)s^2}{\sigma^2_0} = \frac{(20-1)\cdot(2.059)^2}{(2.9)^2} = 9.578.$$

Classical Approach: This is a left-tailed test with 19 degrees of freedom and $\alpha = 0.01$, so the critical value is $\chi^2_{0.99} = 7.633$. Since the test statistic is not in the critical region (9.578 > 7.633), we do not reject the null hypothesis.

P-value approach: The P-value is the area under the χ^2-distribution with 19 degrees of freedom to the left of $\chi^2_0 = 9.578$. From the χ^2-distribution table (Table VII) in the row corresponding to 19 degrees of freedom, 9.578 is between 8.907 and 10.117. The areas to the right of these critical values are 0.975 and 0.95, respectively. This means that the areas to the left of these critical values are 0.025 and 0.05, respectively. So, $0.025 < P$-value < 0.05 [Tech: 0.0374]. Since P-value $> \alpha = 0.01$, we do not reject the null hypothesis.

Conclusion: There is not sufficient evidence to conclude that the standard deviation of heights of major-league baseball players is less than 2.9 inches.

Section 10.6

1. To test a hypothesis about a population mean, we must use a simple random sample that is either drawn from a normally distributed population, or the sample must have at least 30 subjects. Assuming the prerequisites are met, use a normal model to test a hypothesis if the population standard deviation (or variance) is known, and use a Student's t-distribution if the population standard deviation (or variance) is not known.

3. Hypotheses: $H_0 : \mu = 70$ versus $H_1 : \mu < 70$

Test statistic: $z_0 = \dfrac{\bar{x} - \mu_0}{\sigma / \sqrt{n}} = \dfrac{60 - 70}{20 / \sqrt{14}} = -1.87$

Classical approach:
This is a left-tailed test with $\alpha = 0.1$, so the critical value is $-z_{0.10} = -1.28$. Since $-1.87 < -1.28$, the test statistic falls in the critical region, so we reject the null hypothesis.

P-value approach:
P-value $= P(Z < -1.87) = 0.0307$

Since P-value $= 0.0307 < \alpha = 0.10$, we reject the null hypothesis.

Conclusion:
There is sufficient evidence to conclude that the population mean is less than 70.

5. Hypotheses: $H_0 : p = 0.5$ versus $H_1 : p > 0.5$

$np_0(1 - p_0) = 200 \cdot 0.5(1 - 0.5) = 50 \geq 10$, so the requirements of the hypothesis test are satisfied. From the survey, $\hat{p} = \dfrac{115}{200} = 0.575$.

$z_0 = \dfrac{\hat{p} - p_0}{\sqrt{\dfrac{p_0(1 - p_0)}{n}}} = \dfrac{0.575 - 0.5}{\sqrt{\dfrac{0.5(1 - 0.5)}{200}}} \approx 2.12$

Classical approach:
This is a right-tailed test with $\alpha = 0.05$, so the critical value is $z_{0.05} = 1.645$. The test statistic falls in the critical region ($2.12 > 1.645$), so we reject the null hypothesis.

P-value approach:
P-value $= P(Z > 2.12)$

$= 0.0170$ [Tech: 0.0169]

Since P-value $= 0.0170 < \alpha = 0.05$, we reject the null hypothesis.

Conclusion:
There is sufficient evidence to conclude that more than half of the individuals with a valid driver's license drive an American-made automobile.

7. Hypotheses: $H_0 : \mu = 25$ versus $H_1 : \mu \neq 25$

$t_0 = \dfrac{\bar{x} - \mu_0}{s / \sqrt{n}} = \dfrac{23.8 - 25}{6.3 / \sqrt{15}} = -0.738$

df $= 15 - 1 = 14$

Classical approach:
This is a two-tailed test with $15 - 1 = 14$ degrees of freedom and $\alpha = 0.01$, so the critical values are $\pm t_{0.005} = \pm 2.977$. Since

$t_0 = -0.738$ falls between $-t_{0.005} = -2.977$ and $t_{0.005} = 2.977$, the test statistic is not in the critical region, so we do not reject the null hypothesis.

P-value approach:
This is a two-tailed test with 14 degrees of freedom. The P-value of this two tailed test is the area to the left of $t_0 = -0.738$, plus the area to the right of 0.738. From the t-distribution table (Table VI) in the row corresponding to 14 degrees of freedom, 0.738 falls between 0.692 and 0.868, whose right-tail areas are 0.25 and 0.20, respectively. We must double these values in order to get the total area in both tails: 0.50 and 0.40. Thus, $0.40 < P$-value < 0.50 [Tech: 0.4729]. Since P-value $> \alpha = 0.01$, we do not reject the null hypothesis.

Conclusion:
There is not sufficient evidence to conclude that the population mean is different from 25.

9. Hypotheses: $H_0 : \sigma^2 = 10$ versus $H_1 : \sigma^2 > 10$

Test statistic:

$\chi_0^2 = \dfrac{(n-1)s^2}{\sigma_0^2} = \dfrac{(16-1) \cdot 13.7}{10} = 20.55$

Classical Approach:
This is a right-tailed test with 15 degrees of freedom and $\alpha = 0.05$, so the critical value is $\chi_{0.05}^2 = 24.996$. Since the test statistic does not fall in the critical region ($20.55 < 24.996$), we do not reject the null hypothesis.

P-value approach:
The P-value is the area under the χ^2-distribution with 15 degrees of freedom to the right of $\chi_0^2 = 20.55$. From the χ^2-distribution table (Table VII) in the row corresponding to 15 degrees of freedom, 20.55 is to the left of 22.307. The area to the left of this critical value is 0.10. So, P-value > 0.10 [Tech: 0.1518]. Since P-value $> \alpha = 0.05$, we do not reject the null hypothesis.

Conclusion:
There is not sufficient evidence to conclude that the population variance is greater than 10.

11. Hypotheses: $H_0 : \mu = 100$ versus $H_1 : \mu > 100$

Test Statistic: $t_0 = \dfrac{\bar{x} - \mu_0}{s/\sqrt{n}} = \dfrac{108.5 - 100}{17.9/\sqrt{40}} = 3.003$

Classical approach:
This is a right-tailed test with $40 - 1 = 39$ degrees of freedom and $\alpha = 0.05$, so the critical value is $t_{0.05} = 1.685$. Since $t_0 = 3.003 > t_{0.05} = 1.685$, the test statistic falls in the critical region, so we reject the null hypothesis.

P-value approach:
This is a right-tailed test with 39 degrees of freedom. The *P*-value is the area to the left of $t_0 = 3.003$. From the *t*-distribution table (Table VI) in the row corresponding to 39 degrees of freedom, 3.003 falls between 2.976 and 3.313, whose right-tail areas are 0.0025 and 0.001, respectively. So, $0.0025 < P\text{-value} < 0.001$ [Tech: 0.0023]. Since $P\text{-value} < \alpha = 0.05$, we reject the null hypothesis.

Conclusion:
There is sufficient evidence to conclude that the population mean is more than 100.

13. Hypotheses: $H_0 : \mu = 100$ versus $H_1 : \mu > 100$

Test Statistic: $z_0 = \dfrac{\bar{x} - \mu_0}{\sigma/\sqrt{n}} = \dfrac{104.2 - 100}{15/\sqrt{20}} = 1.25$

Classical approach:
This is a right-tailed test with $\alpha = 0.05$, so the critical values are $z_{0.05} = 1.645$. Since $z_0 = 1.25 < z_{0.05} = 1.645$, the test statistic does not fall in the critical region, so we do not reject the null hypothesis.

P-value approach:
$P\text{-value} = P(Z > 1.25) = 1 - 0.8944 = 0.1056$

[Tech: 0.1052]

Since $P\text{-value} = 0.1056 > \alpha = 0.05$, we do not reject the null hypothesis.

Conclusion:
There is not sufficient evidence to conclude that mothers who listen to Mozart have children with higher IQs.

15. The hypotheses are $H_0 : \sigma = 7000$ psi versus $H_1 : \sigma > 7000$ psi. The test statistic is

$\chi_0^2 = \dfrac{(n-1)s^2}{\sigma_0^2} = \dfrac{(20-1)\cdot(7500)^2}{(7000)^2} = 21.811$.

Classical Approach: This is a right-tailed test with 19 degrees of freedom and $\alpha = 0.01$, so the critical value is $\chi_{0.01}^2 = 36.191$. Since the test statistic is not in the critical region $(21.811 < 36.191)$, we do not reject the null hypothesis.

P-value approach: The *P*-value is the area under the χ^2-distribution with 19 degrees of freedom to the right of $\chi_0^2 = 21.811$. From the χ^2-distribution table (Table VII) in the row corresponding to 19 degrees of freedom, 21.811 falls to the left of 27.204. The area to the right of this critical value is 0.10, so $P\text{-value} > 0.10$ [Tech: 0.2938]. Since $P\text{-value} > \alpha = 0.01$, we do not reject the null hypothesis.

Conclusion: There is not sufficient evidence to conclude that the standard deviation of yield strength exceeds 7000 psi.

17. The hypotheses are $H_0 : \mu = 6.3\%$ versus $H_1 : \mu < 6.3\%$. The test statistic is

$t_0 = \dfrac{\bar{x} - \mu_0}{s/\sqrt{n}} = \dfrac{6.05 - 6.3}{1.75/\sqrt{41}} = -0.915$.

Classical approach: This is a left-tailed test with $41 - 1 = 40$ degrees of freedom and $\alpha = 0.05$, so the critical value is $-t_{0.05} = -1.684$. Since $t_0 = -0.915 > -t_{0.05} = -1.684$, the test statistic does not fall in the critical region, so we do not reject the null hypothesis.

P-value approach: This is a left-tailed test with 40 degrees of freedom, so the *P*-value is the area to the left of $t_0 = -0.915$. From the *t*-distribution table (Table VI) in the row corresponding to 40 degrees of freedom, 0.915 falls between 0.851 and 1.050, whose right-tail areas are 0.20 and 0.15, respectively. So, $0.15 < P\text{-value} < 0.20$ [Tech: 0.1829]. Since $P\text{-value} > \alpha = 0.05$, we do not reject the null hypothesis.

Conclusion: There is not sufficient evidence to conclude that the interest rates are lower than in 2001.

19. Hypotheses: $H_0 : p = 0.23$ versus $H_1 : p \neq 0.23$

$np_0(1 - p_0) = 1026 \cdot 0.23(1 - 0.23) \approx 181.7 \geq 10$,

so the requirements for the hypothesis test are

satisfied. From the survey, $\hat{p} = \dfrac{254}{1026} \approx 0.248$.

The test statistic is

$$z_0 = \dfrac{\hat{p} - p_0}{\sqrt{\dfrac{p_0(1 - p_0)}{n}}} = \dfrac{0.248 - 0.23}{\sqrt{\dfrac{0.23(1 - 0.23)}{1026}}} = 1.37$$

[Tech: 1.34].

Classical approach: This is a two-tailed test with $\alpha = 0.1$, so the critical values are $\pm z_{0.05} = \pm 1.645$. Since $z_0 = 1.37$ is between $-z_{0.05} = -1.645$ and $z_{0.05} = 1.645$, the test statistic does not fall in the critical region, so we do not reject the null hypothesis.

P-value approach:

$$\begin{aligned}
P\text{-value} &= P(Z < -1.37) + P(Z > 1.37) \\
&= 2 \cdot P(Z < -1.37) \\
&= 2 \cdot (0.0853) \\
&= 0.1706 \ [\text{Tech: } 0.1813]
\end{aligned}$$

Since P-value $= 0.1706 > \alpha = 0.1$, we do not reject the null hypothesis.

Conclusion:
There is not sufficient evidence to conclude that the percent of American university undergraduate students with at least one tattoo has changed since 2001.

21. The hypotheses are $H_0 : \mu = 8$ minutes versus $H_1 : \mu < 8$ minutes. The test statistic is

$$t_0 = \dfrac{\overline{x} - \mu_0}{s / \sqrt{n}} = \dfrac{7.34 - 8}{3.2 / \sqrt{49}} = -1.444.$$

Classical approach: This is a left-tailed test with $49 - 1 = 48$ degrees of freedom and $\alpha = 0.01$. However, since our t-distribution table does not contain a row for 48, we use df $= 50$. The critical value is $-t_{0.01} = -2.403$.

Since $t_0 = -1.444 > -t_{0.01} = -2.403$, the test statistic does not fall within the critical region, so we do not reject the null hypothesis.

P-value approach: This is a left-tailed test with 48 degrees of freedom, so the P-value is the area to the left of $t_0 = -1.444$, which is equivalent to the area to the right of 1.444. Since our t-distribution table (Table VI) does not contain a row for 48 degrees of freedom, we use 50. From the row corresponding to 50 degrees of freedom, 1.444 falls between 1.299

and 1.676, whose right-tail areas are 0.10 and 0.05, respectively. So, $0.05 < P$-value < 0.10 [Tech: 0.0777]. Since P-value $> \alpha = 0.01$, we do not reject the null hypothesis.

Conclusion: There is not sufficient evidence to conclude that the mean wait time is less than 8 minutes. In other words, the evidence supports the resident's skepticism.

23. (a) Adding up the frequencies in the table, we find that $n = \sum f = 915$. We divide each frequency by this result to obtain the relative frequencies.

Ideal Number of Children	Relative Frequency
0	$\dfrac{15}{915} \approx 0.0164$
1	$\dfrac{31}{915} \approx 0.0339$
2	0.5738
3	0.2798
4	0.0721
5	0.0109
6	0.0044
7	0.0033
8	0.0011
9	0
10	0.0033
11	0.0011

We draw a bar for each group and represent the relative frequencies by the heights of the bars. The distribution is skewed right.

Ideal Number of Children

(b) The mode is 2 because it is the class with the largest frequency.

(c) To find the mean, we use the formula $\overline{x} = \dfrac{\sum x_i f_i}{\sum f_i}$. To find the standard deviation, we choose to use the computational formula

$$s = \sqrt{s^2} = \sqrt{\dfrac{\sum x_i^2 f_i - \dfrac{\left(\sum x_i f_i\right)^2}{\sum f_i}}{\left(\sum f_i\right) - 1}}.$$

We organize our computations of x_i,

Σf_i, $\Sigma x_i f_i$, and $\Sigma x_i^2 f_i$ in the table that follows:

x_i	f_i	$x_i f_i$	x_i^2	$x_i^2 f_i$
0	15	0	0	0
1	31	31	1	31
2	525	1050	4	2100
3	256	768	9	2304
4	66	264	16	1056
5	10	50	25	250
6	4	24	36	144
7	3	21	49	147
8	1	8	64	64
9	0	0	81	0
10	3	30	100	300
11	1	11	121	121
	$\Sigma f_i = 915$	$\Sigma x_i f_i = 2257$		$\Sigma x_i^2 f_i = 6517$

With the table complete, we compute the sample mean and sample standard deviation:

$$\mu = \frac{\Sigma x_i f_i}{\Sigma f_i} = \frac{2257}{915} \approx 2.47 \text{ children}$$

$$s = \sqrt{s^2} = \sqrt{\frac{\Sigma x_i^2 f_i - \frac{(\Sigma x_i f_i)^2}{\Sigma f_i}}{(\Sigma f_i) - 1}}$$

$$= \sqrt{\frac{6517 - \frac{(2257)^2}{915}}{915 - 1}} \approx 1.02 \text{ children}$$

(d) The distribution is clearly skewed right, not normal. Therefore, a large sample size is necessary in order to apply the Central Limit Theorem.

(e) The hypotheses are $H_0 : \mu = 2.64$ children versus $H_1 : \mu \neq 2.64$ children . The test statistics is

$$t_0 = \frac{\bar{x} - \mu_0}{s/\sqrt{n}} = \frac{2.47 - 2.64}{1.02/\sqrt{915}} = -5.041$$

[Tech: -5.144]

Classical approach: This is a two-tailed test with $915 - 1 = 914$ degrees of freedom and $\alpha = 0.05$. Since our t-distribution table [Table VI] does not contain a row for 914, however, we use 1000 degrees of freedom. The critical values are $\pm t_{0.025} = \pm 1.96$.

Since $t_0 = -5.041 < -t_{0.025} = -1.96$, the test statistic falls within a critical region, so we reject the null hypothesis.

P-value approach: This is a two-tailed test with 914 degrees of freedom, so the P-value is the area to the left of $t_0 = -5.041$, plus the area to the right of 5.041. Again, since our t-distribution table [Table VI] does not contain a row for 914, we use 1000 degrees of freedom. From the row corresponding to 1000 degrees of freedom, 5,041 falls to the right of 3.300, whose right-tail area is 0.0005. We double this to obtain the total area in both tails: $2(0.0005) = 0.001$. So, P-value < 0.001 [Tech: P-value < 0.0001]. Since P-value $< \alpha = 0.05$, we reject the null hypothesis.

Conclusion: There is sufficient evidence to conclude that the mean for the ideal number of children is different from 2.64 children. In other words, the results of the poll indicate that people's beliefs as to the ideal number of children have changed.

Section 10.7

1. A Type II error is made when we fail to reject the null hypothesis when in fact the alternative hypothesis is true.

3. As the true mean gets closer to the hypothesized value of the mean, the probability of making a Type II error increases, so the power of the test decreases. This result is reasonable because sample results consistent with the hypothesized mean are likely to be very similar to sample results consistent with the true mean—it will be difficult to distinguish between the two means based on sample data.

5. **(a)** A Type II error would occur if the sample data led to a conclusion of not rejecting $H_0 : \mu = 50$ when in fact $\mu < 50$.

(b) This is a left-sided test with $\alpha = 0.05$ and $-z_\alpha = -1.645$. We will not reject the null if we get a sample mean greater than $\bar{x} = 50 - 1.645 \cdot \frac{12}{\sqrt{24}} = 45.97$. Therefore, if $\mu = 48$, then

$$P(\text{Type II error}) = P\left(Z > \frac{45.97-48}{12/\sqrt{24}}\right)$$
$$= P(Z > -0.83)$$
$$= 1 - 0.2033$$
$$= 0.7967$$
$$= \beta \ [\text{Tech: } 0.7963]$$

Power $= 1 - \beta = 1 - 0.7967 = 0.2033$

[Tech: 0.2037]

(c) $P(\text{Type II error}) = P\left(Z > \dfrac{45.97-49.2}{12/\sqrt{24}}\right)$
$$= P(Z > -1.32)$$
$$= 1 - 0.0934$$
$$= 0.9066$$
$$= \beta \ [\text{Tech: } 0.9063]$$

Power $= 1 - \beta = 1 - 0.9066 = 0.0934$

[Tech: 0.0937]

7. (a) A Type II error would occur if the sample data led to a conclusion of not rejecting $H_0 : \mu = 100$ when in fact $\mu \neq 100$.

(b) This is a two-sided test with $\alpha = 0.01$ and $\pm z_{\alpha/2} = \pm 2.575$. We will not reject the null if we get a sample mean between
$$\overline{x}_L = 100 - 2.575 \cdot \frac{7}{\sqrt{23}} = 96.24 \text{ and}$$
$$\overline{x}_U = 100 + 2.575 \cdot \frac{7}{\sqrt{23}} = 103.76. \text{ Thus,}$$

$P(\text{Type II error})$
$$= P\left(\frac{96.24-103}{7/\sqrt{23}} < Z < \frac{103.76-103}{7/\sqrt{23}}\right)$$
$$= P(-4.63 < Z < 0.52)$$
$$= 0.6985 - 0$$
$$= 0.6985$$
$$= \beta \ [\text{Tech: } 0.6983]$$
Power $= 1 - 0.6985 = 0.3015$ [Tech: 0.3017]

(c) $P(\text{Type II error})$
$$= P\left(\frac{96.24-101.4}{7/\sqrt{23}} < Z < \frac{103.76-101.4}{7/\sqrt{23}}\right)$$
$$= P(-3.54 < Z < 1.62)$$
$$= 0.9474 - 0$$
$$= 0.9474$$
$$= \beta \ [\text{Tech: } 0.9467]$$
Power $= 1 - 0.9472 = 0.0526$ [Tech: 0.0533]

9. (a) A Type II error would occur if the sample data led to a conclusion of not rejecting $H_0 : \mu = 20$ when in fact $\mu < 20$.

(b) This is a left-sided test with $\alpha = 0.05$ and $-z_\alpha = -1.645$. We will not reject the null if we get a sample mean greater than
$$\overline{x} = 20 - 1.645 \cdot \frac{3}{\sqrt{18}} = 18.84. \text{ Thus,}$$

$$P(\text{Type II error}) = P\left(Z > \frac{18.84-17.4}{3/\sqrt{18}}\right)$$
$$= P(Z > 2.04)$$
$$= 1 - 0.9793$$
$$= 0.0207$$
$$= \beta \ [\text{Tech: } 0.0211]$$

Power $= 1 - \beta = 1 - 0.0207 = 0.9793$

[Tech: 0.9789]

(c) $P(\text{Type II error}) = P\left(Z > \dfrac{18.84-19.2}{3/\sqrt{18}}\right)$
$$= P(Z > -0.51)$$
$$= 1 - 0.3050$$
$$= 0.6950$$
$$= \beta \ [\text{Tech: } 0.6962]$$

Power $= 1 - \beta = 1 - 0.6950 = 0.3050$

[Tech: 0.3038]

11. (a) A Type II error would occur if the sample data led to a conclusion of not rejecting $H_0 : \mu = 105$ when in fact $\mu \neq 105$.

(b) This is a two-sided test with $\alpha = 0.1$ and $\pm z_{\alpha/2} = \pm 1.645$. We will not reject the null if we get a sample mean between
$$\overline{x}_L = 105 - 1.645 \cdot \frac{12}{\sqrt{35}} = 101.66 \text{ and}$$
$$\overline{x}_U = 105 + 1.645 \cdot \frac{12}{\sqrt{35}} = 108.34. \text{ Thus,}$$

$P(\text{Type II error})$
$$= P\left(\frac{101.66-101.8}{12/\sqrt{35}} < Z < \frac{108.34-101.8}{12/\sqrt{35}}\right)$$
$$= P(-0.07 < Z < 3.22)$$
$$= 0.9994 - 0.4721$$
$$= 0.5273$$
$$= \beta \ [\text{Tech: } 0.5262]$$
Power $= 1 - 0.5273 = 0.4727$

[Tech: 0.4738]

(c) P(Type II error)

$$= P\left(\frac{101.66 - 103.1}{12/\sqrt{35}} < Z < \frac{108.34 - 103.1}{12/\sqrt{35}}\right)$$

$$= P(-0.71 < Z < 2.58)$$

$$= 0.9951 - 0.2389$$

$$= 0.7562$$

$$= \beta \quad [\text{Tech: } 0.7557]$$

Power $= 1 - 0.7562 = 0.2438$ [Tech: 0.2443]

13. **(a)** A Type II error would occur if the sample data led to a conclusion of not rejecting $H_0 : \mu = \$67$ when in fact $\mu > \$67$.

(b) This is a right-sided test with $\alpha = 0.05$ and $z_\alpha = 1.645$. We will not reject the null if we get a sample mean less than

$$\bar{x} = 67 + 1.645 \cdot \frac{10}{\sqrt{40}} = \$69.60. \text{ Thus,}$$

$$P(\text{Type II error}) = P\left(Z < \frac{69.60 - 70}{10/\sqrt{40}}\right)$$

$$= P(Z < -0.25)$$

$$= 0.4013$$

$$= \beta \quad [\text{Tech: } 0.4004]$$

Power $= 1 - \beta = 1 - 0.4013 = 0.5987$ [Tech: 0.5996]

(c) $P(\text{Type II error}) = P\left(Z < \frac{69.60 - 72}{10/\sqrt{40}}\right)$

$$= P(Z < -1.52)$$

$$= 0.0643$$

$$= \beta \quad [\text{Tech: } 0.0646]$$

Power $= 1 - \beta = 1 - 0.0643 = 0.9357$ [Tech: 0.9354]

15. **(a)** A Type II error would occur if the sample data led to a conclusion of not rejecting $H_0 : \mu = 515$ when in fact $\mu < 515$.

(b) This is a left-sided test with $\alpha = 0.01$ and $-z_\alpha = -2.33$. We will not reject the null if we get a sample mean greater than

$$\bar{x} = 515 - 2.33 \cdot \frac{112}{\sqrt{20}} = 456.65. \text{ Thus,}$$

$$P(\text{Type II error}) = P\left(Z > \frac{456.65 - 500}{112/\sqrt{20}}\right)$$

$$= P(Z > -1.73)$$

$$= 1 - 0.0418$$

$$= 0.9582$$

$$= \beta \quad [\text{Tech: } 0.9583]$$

Power $= 1 - \beta = 1 - 0.9582 = 0.0418$ [Tech: 0.0417]

(c) $P(\text{Type II error}) = P\left(Z > \frac{456.65 - 480}{112/\sqrt{20}}\right)$

$$= P(Z > -0.93)$$

$$= 1 - 0.1762$$

$$= 0.8238$$

$$= \beta \quad [\text{Tech: } 0.8244]$$

Power $= 1 - \beta = 1 - 0.8238 = 0.1762$ [Tech: 0.1756]

(d) In the table below, we show the probability of making a Type II error, β, and the power of the test, $1 - \beta$, for several different values of the population mean less than 515. We use the values of μ and their corresponding powers to construct the power curve.

μ	β	$1 - \beta$
420	0.0721	0.9279
440	0.2546	0.7454
460	0.5517	0.4483
480	0.8238	0.1762
500	0.9582	0.0418

17. This is a left-sided test with $\alpha = 0.01$ and $-z_\alpha = -2.33$. We will not reject the null if we get a sample mean greater than

$$\bar{x} = 50 - 2.33 \cdot \frac{12}{\sqrt{24}} = 44.29$$

$$P(\text{Type II error}) = P\left(Z > \frac{44.29 - 48}{12/\sqrt{24}}\right)$$

$$= P(Z > -1.51)$$

$$= 1 - 0.0655$$

$$= 0.9345$$

$$= \beta \quad [\text{Tech: } 0.9349]$$

Power $= 1 - \beta = 1 - 0.9345 = 0.0655$ [Tech: 0.0651]. This result is less than the result from problem 5(b). Lowering α has the effect of lowering the power of the test.

19. (a) We will reject H_0 if $Z_0 < -1.645$, i.e., if

$$Z_0 = \frac{\hat{p} - 0.75}{\sqrt{\frac{0.75(1-0.75)}{500}}} < -1.645 \text{, and hence}$$

if $\hat{p} < 0.75 - 1.645\sqrt{\frac{0.75(1-0.75)}{500}} = 0.718$.

$P(\text{Type II error})$

$= P(\hat{p} > 0.718 \text{ given that } p = 0.70)$

$= P\left(Z > \frac{0.718 - 0.70}{\sqrt{0.70(1-0.70)/500}} \right)$

$= P(Z > 0.88)$

$= 1 - 0.8106 = 0.1894 = \beta$ [Tech: 0.1880]

(b) Power $= 1 - \beta = 1 - 0.1894 = 0.8106$
[Tech: 0.8120]

21. Results will vary.

Chapter 10 Review Exercises

1. (a) The hypotheses are $H_0 : \mu = \$4277$
versus $H_1 : \mu > \$4277$.

(b) We make a Type I error if the sample evidence leads us to reject H_0 and believe that the mean outstanding credit card debt per cardholder is more than $4277 when, in fact, it is $4277 (that is, we reject H_0 when in fact H_0 is true).

(c) We make a Type II error if we fail to reject the null hypothesis that the mean outstanding credit card debt per cardholder is $4277 when, in fact, it is more than $4277 (that is, we do not reject H_0 when in fact H_0 is false).

(d) If the null hypothesis is not rejected, this means there is not enough evidence to support the credit counselor's belief that the mean outstanding credit card debt per cardholder is more than $4277.

(e) If the null hypothesis is rejected, this means there is sufficient evidence to support the credit counselor's belief that the mean outstanding credit card debt per cardholder is more than $4277.

2. (a) The hypotheses are $H_0 : p = 0.61$ versus $H_1 : p > 0.61$.

(b) We make a Type I error if the sample evidence leads us to reject H_0 and believe that the percentage of internet users aged 18-24 years old who had downloaded music from the internet is higher than 61% when, in fact, it is 61% (i.e. reject H_0 when H_0 is true).

(c) We make a Type II error if we fail to reject the null hypothesis that the percentage of internet users aged 18-24 years old who had downloaded music from the internet is 61% when, in fact, it is higher than 61% (i.e. do not reject H_0 when in fact H_0 is false).

(d) If the null hypothesis is not rejected, this means there is not enough evidence to support the researcher's suspicion that the percentage of internet users aged 18-24 years old who had downloaded music from the internet is higher than 61%.

(e) If the null hypothesis is rejected, this means there is sufficient evidence to support the researcher's suspicion that the percentage of internet users aged 18-24 years old who had downloaded music from the internet is higher than 61%.

3. $P(\text{Type I Error}) = \alpha = 0.05$

4. $P(\text{Type II Error}) = \beta = 0.113$;
Power $= 1 - \beta = 1 - 0.113 = 0.887$
The power of the test is the probability of rejecting the null hypothesis when the alternative hypothesis is true.

5. (a) $z_0 = \dfrac{\overline{x} - \mu_0}{\sigma/\sqrt{n}} = \dfrac{28.6 - 30}{4.5/\sqrt{12}} = -1.08$.

(b) This is a left-tailed test so the critical value is $-z_{0.05} = -1.645$.

(c)

(d) Since $z_0 = -1.08 > -z_{0.05} = -1.645$, the test statistic is not in the critical region, so the researcher will not reject the null hypothesis.

(e) P-value $= P(Z < -1.08) = 0.1401$ [Tech: 0.1406]. Since P-value $> \alpha = 0.05$, we do not reject the null hypothesis.

6. (a) $z_0 = \dfrac{\bar{x} - \mu_0}{\sigma / \sqrt{n}} = \dfrac{70.6 - 65}{12.3 / \sqrt{23}} = 2.18$

(b) This is a two-tailed test so the critical values are $\pm z_{0.05} = \pm 1.645$.

(c)

Critical Region   Critical Region

-1.645 1.645

(d) Since $z_0 = 2.18 > z_{0.05} = 1.645$, the test statistic falls in a critical region, so we reject the null hypothesis.

(e) $P\text{-value} = P(Z < -2.18) + P(Z > 2.18)$

$\qquad\qquad = 2 \cdot P(Z < -2.18)$

$\qquad\qquad = 2 \cdot (0.0146)$

$\qquad\qquad = 0.0292$ [Tech: 0.0290]

Since $P\text{-value} < \alpha = 0.05$, we reject the null hypothesis.

(f) For 90% confidence the critical value is $z_{0.05} = 1.645$. Then:

Lower bound $= \bar{x} - z_{0.05} \cdot \dfrac{\sigma}{\sqrt{n}}$

$\qquad\qquad = 70.6 - 1.645 \cdot \dfrac{12.3}{\sqrt{23}}$

$\qquad\qquad \approx 66.38$

Upper bound $= \bar{x} + z_{0.05} \cdot \dfrac{\sigma}{\sqrt{n}}$

$\qquad\qquad = 70.6 + 1.645 \cdot \dfrac{12.3}{\sqrt{23}}$

$\qquad\qquad \approx 74.82$

Since $\mu_0 = 65$ does not fall in this interval, we reject the null hypothesis.

7. (a) Since σ is unknown, the test statistic is a *t*-statistic.

$t_0 = \dfrac{\bar{x} - \mu_0}{s / \sqrt{n}} = \dfrac{7.3 - 8}{1.8 / \sqrt{15}} = -1.506$

(b) This is a two-tailed test with 14 degrees of freedom, so the critical values are $\pm t_{0.01} = \pm 2.624$.

(c)

Critical Region   Critical Region

-2.624 2.624

(d) Since $t_0 = -1.506$ falls between $-t_{0.01} = -2.624$ and $t_{0.01} = 2.624$, the test statistic does not fall in a critical region, so we do not reject the null hypothesis.

(e) This is a two-tailed test with 14 degrees of freedom. The *P*-value of this two tailed test is the area to the left of $t_0 = -1.506$, plus the area to the right of 1.506. From the *t*-distribution table (Table VI) in the row corresponding to 14 degrees of freedom, 1.506 falls between 1.345 and 1.761, whose right-tail areas are 0.10 and 0.05, respectively. We must double these values in order to get the total area in both tails: 0.20 and 0.10. Therefore, $0.10 < P\text{-value} < 0.20$ [Tech: 0.1543]. Since $P\text{-value} > \alpha = 0.02$, we do not reject the null hypothesis.

8. (a) Since σ is unknown, the test statistic is a *t*-statistic.

$t_0 = \dfrac{\bar{x} - \mu_0}{s / \sqrt{n}} = \dfrac{3.5 - 3.9}{0.9 / \sqrt{25}} = -2.222$.

(b) This is a left-tailed test with 24 degrees of freedom, so the critical value is $-t_{0.05} = -1.711$.

(c)

Critical Region

-1.711

(d) Since $t_0 = -2.222 < -t_{0.05} = -1.711$, the test statistic falls in a critical region, so we reject the null hypothesis.

(e) This is a left-tailed test with 24 degrees of freedom. The *P*-value is the area to the left of $t_0 = -2.222$, which is equivalent to the area to the right of $t_0 = 2.222$. From the *t*-distribution table (Table VI) in the row corresponding to 24 degrees of freedom, 2.222 falls between 2.172 and 2.492, whose right-tail areas are 0.02 and 0.01, respectively. Therefore, $0.01 < P\text{-value} < 0.02$ [Tech: 0.0180]. Since $P\text{-value} < \alpha = 0.05$, we reject the null hypothesis.

9. $np_0(1-p_0) = 250 \cdot 0.6(1-0.6) = 60 \geq 10$, so the requirements of the hypothesis test are satisfied.

(a) $\hat{p} = \dfrac{165}{250} = 0.66$

The test statistics is

$$z_0 = \frac{\hat{p} - p_0}{\sqrt{\dfrac{p_0(1-p_0)}{n}}} = \frac{0.66 - 0.6}{\sqrt{\dfrac{0.6(1-0.6)}{250}}} = 1.94.$$

This is a right-tailed test with $\alpha = 0.05$, so the critical value is $z_{0.05} = 1.645$. The test statistic fall in the critical region $(1.94 > 1.645)$, so we reject the null hypothesis.

(b) $P\text{-value} = P(Z > 1.94)$

$= 1 - P(Z < 1.94)$

$= 1 - 0.9738$

$= 0.0262$ [Tech: 0.0264]

Since $P\text{-value} = 0.0262 < \alpha = 0.05$, we reject the null hypothesis.

There is sufficient evidence to conclude that $p > 0.6$.

10. $np_0(1-p_0) = 420 \cdot 0.35(1-0.35) \approx 95.6 \geq 10$, so the requirements of the hypothesis test are satisfied.

(a) $\hat{p} = \dfrac{138}{420} \approx 0.329$

The test statistic is

$$z_0 = \frac{\hat{p} - p_0}{\sqrt{\dfrac{p_0(1-p_0)}{n}}} = \frac{0.329 - 0.35}{\sqrt{\dfrac{0.35(1-0.35)}{420}}} = -0.90$$

This is a two-tailed test with $\alpha = 0.01$, so the critical values are $\pm z_{0.005} = \pm 2.575$. Since $z_0 = -0.90$ is between $-z_{0.005} = -2.575$ and $z_{0.005} = 2.575$, the test statistic does not fall in the critical region, so we do not reject the null hypothesis.

(b) $P\text{-value} = P(Z < -0.90) + P(Z > 0.90)$

$= 2 \cdot P(Z < -0.90)$

$= 2(0.1841)$

$= 0.3682$ [Tech: 0.3572]

Since $P\text{-value} > \alpha = 0.01$, we do not reject the null hypothesis.

There is not sufficient evidence to conclude that $p \neq 0.35$.

11. (a) $\chi_0^2 = \dfrac{(n-1)s^2}{\sigma_0^2} = \dfrac{(18-1) \cdot (4.9)^2}{(5.2)^2} = 15.095$

(b) This is a two-tailed test with 17 degrees of freedom and $\alpha = 0.05$, so the critical values are $\chi_{1-\alpha/2}^2 = \chi_{0.975}^2 = 7.564$ and $\chi_{\alpha/2}^2 = \chi_{0.025}^2 = 30.191$.

(c)

(d) No, the researcher will not reject the null hypothesis because the test statistic is not in a critical region (15.095 is between 7.564 and 30.191).

12. (a) $\chi_0^2 = \dfrac{(n-1)s^2}{\sigma_0^2} = \dfrac{24 \cdot (16.5)^2}{(15.7)^2} = 26.508$

(b) This is a right-tailed test with 24 degrees of freedom and $\alpha = 0.10$, so the critical value is $\chi_{0.10}^2 = 33.196$.

(c)

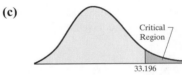

(d) No, the researcher will not reject the null hypothesis because the test statistic is not in the critical region $(15.095 < 33.196)$.

13. The hypotheses are $H_0 : \mu = 0.875$ inch versus $H_1 : \mu > 0.875$ inch. We compute the sample mean and sample standard deviation to be $\bar{x} = 0.876$ inch and $s = 0.005$ inch. The test statistic is $t_0 = \dfrac{\bar{x} - \mu_0}{s/\sqrt{n}} = \dfrac{0.876 - 0.875}{0.005/\sqrt{36}} = 1.2$.

Classical approach: This is a right-tailed test with $n - 1 = 36 - 1 = 35$ degrees of freedom so the critical value is $t_{0.05} = 1.690$. Since $t_0 = 1.2 < t_{0.05} = 1.690$, the test statistic does not fall in the critical region, so we do not reject the null hypothesis.

P-value approach: This is a right-tailed test with 35 degrees of freedom. The P-value is the area to the right of $t_0 = 1.2$. From the t-distribution table (Table VI) in the row corresponding to 35 degrees of freedom, 1.2 falls between 1.052 and 1.306, whose right-tail areas are 0.15 and 0.10, respectively. Thus,

$0.10 < P\text{-value} < 0.15$ [Tech: 0.1191]. Since $P\text{-value} > \alpha = 0.05$, we do not reject the null hypothesis.

Conclusion: There is not sufficient evidence to conclude that the mean distance between the retaining rings is longer than 0.875 inch.

14. The hypotheses are $H_0 : \mu = 7.53$ pounds versus $H_1 : \mu > 7.53$ pounds. The test statistic is $z_0 = \dfrac{\overline{x} - \mu_0}{\sigma / \sqrt{n}} = \dfrac{7.79 - 7.53}{1.15 / \sqrt{50}} = 1.60$.

Classical approach: This is a right-tailed test with $\alpha = 0.01$, so the critical value is $z_{0.01} = 2.33$. Since $z_0 = 1.60 < z_{0.01} = 2.33$, the test statistic is not in the critical region, so we do not reject the null hypothesis. .

P-value approach:
$$
\begin{aligned}
P\text{-value} &= P(Z > 1.60) \\
&= 1 - P(Z < 1.60) \\
&= 1 - 0.9452 \\
&= 0.0548 \ [\text{Tech: } 0.0549]
\end{aligned}
$$

Since $P\text{-value} = 0.0548 > \alpha = 0.05$, we do not reject the null hypothesis.

Conclusion: There is not sufficient evidence that the new diet increases the birth weights of newborns.

15. (a) The hypotheses are $H_0 : \mu = 300$ mg versus $H_1 : \mu > 300$ mg. The test statistic is $t_0 = \dfrac{\overline{x} - \mu_0}{s / \sqrt{n}} = \dfrac{326 - 300}{342 / \sqrt{404}} = 1.528$.

Classical Approach: This is a right-tailed test with $404 - 1 = 403$ degrees of freedom with $\alpha = 0.05$. However, since our t-distribution table (Table VI) does not have a row for df = 403, we use df = 100. The critical value is $t_{0.05} = 1.660$. Since $t_0 = 1.528 < t_{0.05} = 1.660$, the test statistic does not fall in the critical region, so we do not reject the null hypothesis.

P-value approach: This is a right-tailed test with 403 degrees of freedom. The P-value is the area under the t-distribution to the right of the test statistic $t_0 = 1.528$.

From the t-distribution table (Table VI) in the row corresponding to 100 degrees of freedom (since the table does not contain a row for df = 403), 1.528 falls between 1.290 and 1.660, whose right-tail areas are 0.10 and 0.05, respectively. Therefore,

$0.05 < P\text{-value} < 0.10$ [Tech: 0.0636]. Since $P\text{-value} > \alpha = 0.05$, we do not reject the null hypothesis.

Conclusion: There is not sufficient evidence to conclude that mean cholesterol consumption of 20- to 39-year-old males is greater than 300 mg.

(b) Type I error: If the nutritionist rejects the null hypothesis that the mean cholesterol consumption is 300 mg when, in fact, the mean consumption is 300 mg.

Type II error: The nutritionist does not reject the null hypothesis that the mean cholesterol consumption is 300 mg when, in fact, the mean consumption is more than 300 mg.

(c) The probability of making a Type I error is $\alpha = 0.05$.

16. (a) The plotted points are all within the bounds of the normal probability plot, which also has a generally linear pattern. The boxplot shows that there are no outliers. Therefore, the conditions for a hypothesis test are satisfied.

(b) The hypotheses are $H_0 : \mu = \$1400$ versus $H_1 : \mu \neq \$1400$. We compute the sample mean to be $\overline{x} = \$1544.5$. The test statistic is $z_0 = \dfrac{\overline{x} - \mu_0}{\sigma / \sqrt{n}} = \dfrac{1544.5 - 1400}{200 / \sqrt{12}} \approx 2.50$.

Classical approach: This is a two-tailed test with $\alpha = 0.10$, so the critical values are $\pm z_{\alpha/2} = \pm z_{0.05} = \pm 1.645$. Since $z_0 = 2.50 > z_{0.05} = 1.645$, the test statistic falls in a critical region, so we reject the null hypothesis.

P-value approach:
$$
\begin{aligned}
P\text{-value} &= P(Z < -2.50) + P(Z > 2.50) \\
&= 2 \cdot P(Z < -2.50) \\
&= 2 \cdot (0.0062) \\
&= 0.0124 \ [\text{Tech: } 0.0123]
\end{aligned}
$$

Since $P\text{-value} < \alpha = 0.10$, we reject the null hypothesis.

Conclusion: There is sufficient evidence to conclude that the mean price of rent in Cambridge is different from $1400.

17. Since a majority would constitute a proportion greater than half, the hypotheses are $H_0 : p = 0.5$ versus $H_1 : p > 0.5$.

$np_0(1-p_0) = 150(0.5)(1-0.5) = 37.5 \geq 10$, so the requirements of the hypothesis test are satisfied. From the sample, $\hat{p} = \dfrac{81}{150} = 0.54$.

The test statistic is

$$z_0 = \frac{\hat{p} - p_0}{\sqrt{\dfrac{p_0(1-p_0)}{n}}} = \frac{0.54 - 0.5}{\sqrt{\dfrac{0.5(1-0.5)}{150}}} \approx 0.98$$

Classical approach: This is a right-tailed test with $\alpha = 0.05$, so the critical value is $z_{0.05} = 1.645$. Since $z_0 = 0.99 < z_{0.05} = 1.645$, the test statistic does not fall in the critical region, so we do not reject the null hypothesis.

P-value approach:
$$\begin{aligned} P\text{-value} &= P(Z > 0.98) \\ &= 1 - P(Z \leq 0.98) \\ &= 1 - 0.8365 \\ &= 0.1635 \quad [\text{Tech: } 0.1636] \end{aligned}$$

Since P-value $= 0.1635 > \alpha = 0.05$, we do not reject the null hypothesis.

Conclusion: There is not sufficient evidence to conclude that a majority of pregnant women nap at least twice each week.

18. The hypotheses are $H_0 : p = 0.11$ versus $H_1 : p \neq 0.11$.

$np_0(1-p_0) = 1100(0.11)(1-0.11) = 107.69 \geq 10$, so the requirements of the hypothesis test are satisfied. From the survey, $\hat{p} = \dfrac{143}{1100} = 0.13$.

$$z_0 = \frac{\hat{p} - p_0}{\sqrt{\dfrac{p_0(1-p_0)}{n}}} = \frac{0.13 - 0.11}{\sqrt{\dfrac{0.11(1-0.11)}{1100}}} \approx 2.12$$

Classical approach:
This is a two-tailed test with $\alpha = 0.10$, so the critical values are $\pm z_{\alpha/2} = \pm z_{0.05} = \pm 1.645$. Since $z_0 = 2.12 > z_{0.05} = 1.645$, the test statistic falls in a critical region, so we reject the null hypothesis.

P-value approach:
$$\begin{aligned} P\text{-value} &= P(Z < -2.12) + P(Z > 2.12) \\ &= 2 \cdot P(Z < -2.12) \\ &= 2 \cdot (0.0170) \\ &= 0.0340 \end{aligned}$$

Since P-value $= 0.0340 < \alpha = 0.10$, we reject the null hypothesis.

Conclusion: There is sufficient evidence to conclude that the proportion of Americans that have no faith that the federal government is capable of dealing effectively with domestic problems has changed since 1997.

19. The hypotheses are $H_0 : \sigma = 505.6$ grams versus $H_1 : \sigma > 505.6$ grams. The test statistic is $\chi_0{}^2 = \dfrac{(n-1)s^2}{\sigma_0^2} = \dfrac{40 \cdot (840)^2}{(505.6)^2} = 110.409$.

Classical Approach: This is a right-tailed test with $n - 1 = 41 - 1 = 40$ degrees of freedom and $\alpha = 0.01$, so the critical value is $\chi_{0.01}^2 = 63.691$. Since the test statistic falls in the critical region ($\chi_0{}^2 = 110.409 > \chi_{0.01}^2 = 63.691$), we reject the null hypothesis.

P-value approach: The P-value is the area under the χ^2-distribution with 40 degrees of freedom to the left of $\chi_0^2 = 110.409$. From the χ^2-distribution table (Table VII) in the row corresponding to 40 degrees of freedom, 110.409 is to the right of 66.766. The area to the right of this critical value is 0.005. So, P-value < 0.005 [Tech: < 0.0001]. Since P-value $< \alpha = 0.01$, we reject the null hypothesis.

Conclusion: There is sufficient evidence to support the researcher's hypothesis that the variability in the birth weight of preterm babies is more than the variability in the birth weight of full-term babies.

20. (a) This is a right-sided test with $\alpha = 0.01$ and $z_\alpha = 2.33$. We will not reject the null if we get a sample mean less than

$$\bar{x} = 7.53 + 2.33 \cdot \frac{1.15}{\sqrt{50}} = 7.91. \text{ Thus,}$$

$$\begin{aligned} P(\text{Type II error}) &= P\left(Z < \frac{7.91 - 7.59}{1.15/\sqrt{50}}\right) \\ &= P(Z < 1.97) \\ &= 0.9756 \\ &= \beta \quad [\text{Tech: } 0.9751] \end{aligned}$$

(b) Power $= 1 - \beta = 1 - 0.9756 = 0.0244$ [Tech: 0.0249]

21. The hypotheses are $H_0 : p = 0.4$ versus

$H_1 : p > 0.4$.

$np_0(1 - p_0) = 40 \cdot 0.4(1 - 0.4) = 9.6 < 10$, so we use small sample techniques. This is a right-tailed test, so we calculate the probability of 18 or more successes in 40 binomial trials with $p = 0.4$. Using technology:

P-value $= P(X \geq 18)$

$= 1 - P(X \leq 17)$

$= 1 - 0.6885$

$= 0.3115$

Since this is larger than the $\alpha = 0.05$ level of significance, we do not reject the null hypothesis.

Conclusion: There is not sufficient evidence to conclude that the proportion of adolescents who pray daily has increased above 40%.

22. The hypotheses are $H_0 : \mu = 73.2$ versus

$H_1 : \mu < 73.2$. The test statistic is

$t_0 = \dfrac{\bar{x} - \mu_0}{s / \sqrt{n}} = \dfrac{72.8 - 73.2}{12.3 / \sqrt{3851}} = -2.018$.

Classical approach: This is a left-tailed test with $n - 1 = 3851 - 1 = 3850$ degrees of freedom and $\alpha = 0.05$. However, since our t-distribution table (Table VI) does not have a row for df = 3850, we use df = 1000. The critical value is $-t_{0.05} = -1.646$. Since

$t_0 = -2.018 < -t_{0.05} = -1.646$, the test statistic falls in the critical region, so we reject the null hypothesis. *P*-value approach: This is a right-tailed test with 3850 degrees of freedom. The P-value is the area under the t-distribution to the left of the test statistic $t_0 = -2.018$, which is equivalent to the area to the right of 2.018. From the t-distribution table (Table VI) in the row corresponding to 1000 degrees of freedom (since the table does not contain a row for df = 3850), 2.018 falls between 1.962 and 2.056, whose right-tail areas are 0.025 and 0.02, respectively. Thus, $0.02 < P$-value < 0.025 [Tech: 0.0218]. Since P-value $< \alpha = 0.05$, we reject the null hypothesis.

Conclusion: There is sufficient evidence to conclude, from a statistical viewpoint, that the scores on the final exam decreased under the new format. The data are statistically significant, but there is no real practical significance. A difference of 0.4 points (less than ½ of a percentage point) is practically

insignificant when compared to the savings in resources by the university. For all practical purposes, the average scores would be considered the same.

23. Answers will vary.

Chapter 10 Test

1. (a) The hypotheses are $H_0 : \mu = 42.6$ minutes versus $H_1 : \mu > 46.2$ minutes.

(b) There is sufficient evidence to conclude that the mean amount of daily time spent on phone calls and answering or writing emails has increased from what it was in 2006.

(c) Making a Type I error would mean that we rejected the null hypothesis that the mean is 42.6 minutes when, in fact, the mean amount of daily time spent on phone calls and answering or writing emails is 42.6 minutes.

(d) Making a Type II error would mean that we did not reject the null hypothesis that the mean is 42.6 minutes when, in fact, the mean amount of daily time spent on phone calls and answering or writing emails is greater than 42.6 minutes.

2. (a) The hypotheses are $H_0 : \mu = 167.1$ seconds versus $H_1 : \mu < 167.1$ seconds.

(b) By choosing a level of significance of 0.01, we make the probability of making a Type I error small. In other words, the probability of rejecting the null hypothesis that the mean is 167.1 seconds in favor of the alternative hypothesis that the mean is less that 167.1 seconds, when in fact the mean is 167.1 seconds, is small.

(c) The test statistic is

$z_0 = \dfrac{\bar{x} - \mu_0}{\sigma / \sqrt{n}} = \dfrac{163.9 - 167.1}{15.3 / \sqrt{70}} \approx -1.75$.

Classical approach: This is a left-tailed test with $\alpha = 0.01$, so the critical value is $-z_{0.01} = -2.33$. Since $z_0 = -1.75 > -z_{0.01} = -2.33$, the test statistic does not fall in the critical region, so we do not reject the null hypothesis.

P-value approach:

$P\text{-value} = P(Z < -1.75) = 0.0401$

Since $P\text{-value} = 0.0401 > \alpha = 0.01$, we do not reject the null hypothesis.

Conclusion: There is not sufficient evidence to conclude that the drive-through service time has decreased.

3. The hypotheses are $H_0 : \mu = 8$ hours versus $H_1 : \mu < 8$ hours. The test statistic is

$$t_0 = \frac{\bar{x} - \mu_0}{s/\sqrt{n}} = \frac{7.8 - 8}{1.4/\sqrt{151}} = -1.755 .$$

Classical Approach: This is a left-tailed test with $151 - 1 = 150$ degrees of freedom and $\alpha = 0.05$. However, since our *t*-distribution table (Table VI) does not have a row for df = 150, we use df = 100. The critical value is $-t_{0.05} = -1.660$. Since $t_0 = -1.755 < -t_{0.05} = -1.660$, the test statistic falls in the critical region, so we reject the null hypothesis.

P-value approach: This is a left-tailed test with 150 degrees of freedom. The *P*-value is the area under the *t*-distribution to the left of the test statistic $t_0 = -1.755$, which is equivalent to the area to the right of 1.755. From the *t*-distribution table (Table VI) in the row corresponding to 100 degrees of freedom (since the table does not contain a row for df = 150), 1.755 falls between 1.660 and 1.984, whose right-tail areas are 0.05 and 0.025, respectively. Thus, $0.025 < P\text{-value} < 0.05$ [Tech: 0.0406]. Since $P\text{-value} < \alpha = 0.05$, we reject the null hypothesis.

Conclusion: There is sufficient evidence to conclude that postpartum women get less than 8 hours of sleep each night.

4. The hypotheses are $H_0 : \mu = 1.3825$ inches versus $H_1 : \mu \neq 1.3825$ inches.

We compute the sample mean and sample standard deviation and obtain $\bar{x} = 1.3826$ inches and $s \approx 0.00033$ inch.

Using $\alpha = 0.05$ with $n - 1 = 10 - 1 = 9$ degrees of freedom, we have $t_{\alpha/2} = t_{0.025} = 2.262$.

$$\text{Lower bound} = \bar{x} - t_{\alpha/2} \cdot \frac{s}{\sqrt{n}}$$

$$= 1.3826 - 2.262 \cdot \frac{0.00033}{\sqrt{10}}$$

$$\approx 1.3824 \text{ inches}$$

$$\text{Upper bound} = \bar{x} + t_{\alpha/2} \cdot \frac{s}{\sqrt{n}}$$

$$= 1.3826 + 2.262 \cdot \frac{0.00033}{\sqrt{10}}$$

$$\approx 1.3828 \text{ inches}$$

Because this interval includes the hypothesized mean 1.3825 inches, we do not reject the null hypothesis. There is not sufficient evidence to conclude that the mean is different from 1.3825 inches. Therefore, we presume that the part has been manufactured to specifications.

5. The hypotheses are $H_0 : p = 0.6$ versus $H_1 : p > 0.6$.

$np_0(1 - p_0) = 1561(0.6)(1 - 0.6) = 374.64 \geq 10$, so the requirements of the hypothesis test are satisfied. From the sample, $\hat{p} = \frac{954}{1561} \approx 0.611$.

The test statistic is

$$z_0 = \frac{\hat{p} - p_0}{\sqrt{\frac{p_0(1 - p_0)}{n}}} = \frac{0.611 - 0.6}{\sqrt{\frac{0.6(1 - 0.6)}{1561}}} \approx 0.89$$

[Tech: 0.90]

Classical approach: We use $\alpha = 0.05$. This is a right-tailed test, so the critical value is $z_{0.05} = 1.645$. Since $z_0 = 0.89 < z_{0.05} = 1.645$, the test statistic does not fall in the critical region, so we do not reject the null hypothesis.

P-value approach:

$$P\text{-value} = P(Z > 0.89)$$

$$= 1 - P(Z \leq 0.89)$$

$$= 1 - 0.8133$$

$$= 0.1867 \ [\text{Tech: 0.1843}]$$

Since $P\text{-value} = 0.1867 > \alpha = 0.05$, we do not reject the null hypothesis.

Conclusion: There is not sufficient evidence to conclude that a supermajority of Americans did not feel that the United States would need to fight Japan in their lifetimes.

6. The hypotheses are $H_0 : \mu = 0$ pounds versus $H_1 : \mu > 0$ pounds. The test statistic is

$$t_0 = \frac{\bar{x} - \mu_0}{s/\sqrt{n}} = \frac{1.6 - 0}{5.4/\sqrt{79}} \approx 2.634 .$$

Classical Approach: This is a right-tailed test with $79 - 1 = 78$ degrees of freedom and $\alpha = 0.05$. However, since our *t*-distribution table (Table VI) does not have a row for df = 78, we use df = 80. The critical value is $t_{0.05} = 1.664$. Since $t_0 = 2.634 > t_{0.05} = 1.664$, the test statistic falls in the critical region, so we reject the null hypothesis.

P-value approach: This is a right-tailed test with $79 - 1 = 78$ degrees of freedom. The *P*-value is the area under the *t*-distribution to the right of the test statistic $t_0 = 2.634$. From the *t*-distribution table (Table VI) in the row corresponding to 80 degrees of freedom (since the table does not contain a row for df = 78), 2.634 falls between 2.374 and 2.639, whose right-tail areas are 0.01 and 0.005, respectively. Thus, $0.005 < P\text{-value} < 0.01$ [Tech: 0.0051]. Since $P\text{-value} < \alpha = 0.05$, we reject the null hypothesis.

Conclusion: There is sufficient evidence to suggest that the diet is effective. However, loosing 1.6 kg of weight over the course of a year does not seem to have much practical significance.

7. The hypotheses are $H_0 : p = 0.37$ versus $H_1 : p > 0.37$.

$np_0(1 - p_0) = 30 \cdot 0.37(1 - 0.37) = 6.993 < 10$, so we use small sample techniques. This is a right-tailed test, so we calculate the probability of 16 or more successes in 30 binomial trials with $p = 0.37$. Using technology:

$$P\text{-value} = P(X \geq 16)$$
$$= 1 - P(X \leq 15)$$
$$= 1 - 0.9499$$
$$= 0.0501$$

Since this is smaller than the $\alpha = 0.10$ level of significance, we reject the null hypothesis.

Conclusion: There is sufficient evidence to conclude that the proportion of 20- to 24-year-olds who live on their own and do not have a land line is greater than 0.37.

8. The hypotheses are $H_0 : \sigma = 18\%$ versus $H_1 : \sigma < 18\%$. The test statistic is

$$\chi_0^2 = \frac{(n-1)s^2}{\sigma_0^2} = \frac{(10-1) \cdot (16)^2}{(18)^2} \approx 7.111.$$

Classical Approach: This is a left-tailed test with 9 degrees of freedom and $\alpha = 0.05$, so the critical value is $\chi_{1-\alpha}^2 = \chi_{0.95}^2 = 3.325$. Since the test statistic is not in the critical region ($7.111 > 3.325$), we do not reject the null hypothesis.

P-value approach: The *P*-value is the area under the χ^2-distribution with 9 degrees of freedom to the left of $\chi_0^2 = 7.111$. From the χ^2-distribution table (Table VII) in the row corresponding to 9 degrees of freedom, 7.111 is between 4.168 and 14.684. The areas to the right of these critical values are 0.90 and 0.10, respectively. This means the area to the left of these critical values are 0.10 and 0.90, respectively. So, $0.10 < P\text{-value} < 0.90$ [Tech: 0.3744]. Since $P\text{-value} > \alpha = 0.05$, we do not reject the null hypothesis.

Conclusion: There is not sufficient evidence to conclude that the investment manager's portfolio is less risky than the market.

9. (a) This is a left-sided test with $\alpha = 0.01$ and $z_\alpha = 2.33$. We will not reject the null if we get a sample mean greater than $\bar{x} = 167.1 - 2.33 \cdot \frac{15.3}{\sqrt{70}} = 162.84$. Thus,

$$P(\text{Type II error}) = P\left(Z > \frac{162.84 - 166.5}{15.3/\sqrt{70}}\right)$$
$$= P(Z > -2.00)$$
$$= 1 - P(Z < -2.00)$$
$$= 1 - 0.0228$$
$$= 0.9772$$
$$= \beta \quad [\text{Tech: } 0.9774]$$

(b) $\text{Power} = 1 - \beta = 1 - 0.9772 = 0.0228$ [Tech: 0.0226]

Chapter 11

Inferences on Two Samples

Section 11.1

1. independent

3. For the null hypothesis, define $H_0 : \mu_d = 0$. Since the researcher wants to show that the mean of population 1, μ_1, is less than the mean of population 2, μ_2, in matched pair data, the difference $\mu_1 - \mu_2$ should be negative. Thus, define $H_1 : \mu_d < 0$ with $d_i = X_i - Y_i$.

   Alternatively, we could define $H_1 : \mu_d > 0$ with $d_i = Y_i - X_i$.

5. Since the members of the two samples are married to each other, the sampling is dependent.

7. Because the 80 students are randomly allocated to one of two groups, the sampling is independent.

9. Because the 29 patients with schizophrenia who were examined did not determine the 29 healthy patients who were examined, the sampling is independent.

11. **(a)** We measure differences as $d_i = X_i - Y_i$.

Obs	1	2	3	4	5	6	7
X_i	7.6	7.6	7.4	5.7	8.3	6.6	5.6
Y_i	8.1	6.6	10.7	9.4	7.8	9.0	8.5
d_i	−0.5	1.0	−3.3	−3.7	0.5	−2.4	−2.9

   (b) $\sum d_i = (-0.5) + 1.0 + (-3.3) + (-3.7) + 0.5$
   $+ (-2.4) + (-2.9) = -11.3$

   so $\bar{d} = \dfrac{\sum d_i}{n} = \dfrac{-11.3}{7} \approx -1.614$

   $\sum d_i^2 = (-0.5)^2 + (1.0)^2 + (-3.3)^2 + (-3.7)^2$
   $+ (0.5)^2 + (-2.4)^2 + (-2.9)^2 = 40.25$

   so

   $s_d = \sqrt{\dfrac{\sum d_i^2 - \dfrac{\left(\sum d_i\right)^2}{n}}{n-1}} = \sqrt{\dfrac{40.25 - \dfrac{(-11.3)^2}{7}}{7-1}}$

   ≈ 1.915

(c) The hypotheses are $H_0 : \mu_d = 0$ versus $H_1 : \mu_d < 0$. The level of significance is $\alpha = 0.05$. The test statistic is

$$t_0 = \dfrac{\bar{d}}{\dfrac{s_d}{\sqrt{n}}} = \dfrac{-1.614}{\dfrac{1.915}{\sqrt{7}}} \approx -2.230.$$

Classical approach: Since this is a left-tailed test with 6 degrees of freedom, the critical value is $-t_{0.05} = -1.943$. Since the test statistic $t_0 \approx -2.230$ is less than the critical value $-t_{0.05} = -1.943$ (i.e., since the test statistic fall within the critical region), we reject H_0.

P-value approach: The *P*-value for this left-tailed test is the area under the *t*-distribution with 6 degrees of freedom to the left of the test statistic $t_0 = -2.230$, which by symmetry is equal to the area to the right of $t_0 = 2.230$. From the *t*-distribution table (Table VI) in the row corresponding to 6 degrees of freedom, 2.230 falls between 1.943 and 2.447 whose right-tail areas are 0.05 and 0.025, respectively. So, $0.025 < P\text{-value} < 0.05$. [Tech: *P*-value = 0.0336.] Because the *P*-value is less than the level of significance $\alpha = 0.05$, we reject H_0.

Conclusion: There is sufficient evidence at the $\alpha = 0.05$ level of significance to reject the null hypothesis that $\mu_d = 0$ and conclude that $\mu_d < 0$.

(d) For $\alpha = 0.05$ and df = 6, $t_{\alpha/2} = t_{0.025} = 2.447$. Then:

Lower bound:

$\bar{d} - t_{0.025} \cdot \dfrac{s_d}{\sqrt{n}} = -1.614 - 2.447 \cdot \dfrac{1.915}{\sqrt{7}}$

≈ -3.39

Upper bound:

$\bar{d} + t_{0.025} \cdot \dfrac{s_d}{\sqrt{n}} = -1.614 + 2.447 \cdot \dfrac{1.915}{\sqrt{7}}$

≈ 0.16

We can be 95% confident that the mean difference is between −3.39 and 0.16.

13. (a) These are matched-pair data because two measurements (A and B) are taken on the same round.

(b) We measure differences as $d_i = A_i - B_i$.

Obs	1	2	3	4	5	6
A	793.8	793.1	792.4	794.0	791.4	792.4
B	793.2	793.3	792.6	793.8	791.6	791.6
d_i	0.6	−0.2	−0.2	0.2	−0.2	0.8

Obs	7	8	9	10	11	12
A	791.4	792.3	789.6	794.4	790.9	793.5
B	791.6	792.4	788.5	794.7	791.3	793.5
d_i	0.1	−0.1	1.1	−0.3	−0.4	0

We compute the mean and standard deviation of the differences and obtain $\bar{d} \approx 0.1167$ feet per second, rounded to four decimal places, and $s_d \approx 0.4745$ feet per second, rounded to four decimal places. The hypotheses are $H_0 : \mu_d = 0$ versus $H_1 : \mu_d \neq 0$. The level of significance is $\alpha = 0.01$. The test statistic is

$$t_0 = \frac{\bar{d}}{\frac{s_d}{\sqrt{n}}} = \frac{0.1167}{\frac{0.4745}{\sqrt{12}}} \approx 0.852.$$

Classical approach: Since this is a two-tailed test with 11 degrees of freedom, the critical values are $\pm t_{0.005} = \pm 3.106$. Since the test statistic $t_0 \approx 0.852$ falls between the critical values $\pm t_{0.005} = \pm 3.106$ (i.e., since the test statistic does not fall within the critical regions), we do not reject H_0.

P-value approach: The P-value for this two-tailed test is the area under the t-distribution with 11 degrees of freedom to the right of the test statistic $t_0 = 0.853$ plus the area to the left of $t_0 = -0.853$. From the t-distribution table, in the row corresponding to 11 degrees of freedom, 0.852 falls between 0.697 and 0.876 whose right-tail areas are 0.25 and 0.20, respectively. We must double these values in order to get the total area in both tails: 0.50 and 0.40. So, $0.50 > P\text{-value} > 0.40$. [Tech: P-value = 0.4125.] Because the P-value is greater than the level of significance $\alpha = 0.01$, we do not reject H_0.

Conclusion: There is not sufficient evidence at the $\alpha = 0.01$ level of significance to conclude that there is a difference in the measurements of velocity between device A and device B.

(c) For $\alpha = 0.01$ and df = 11, $t_{\alpha/2} = t_{0.005} = 3.106$. Then:

Lower bound:

$$\bar{d} - t_{0.005} \cdot \frac{s_d}{\sqrt{n}} = 0.1167 - 3.106 \cdot \frac{0.4745}{\sqrt{12}}$$
$$\approx -0.31$$

Upper bound:

$$\bar{d} + t_{0.005} \cdot \frac{s_d}{\sqrt{n}} = 0.1167 + 3.106 \cdot \frac{0.4745}{\sqrt{12}}$$
$$\approx 0.54$$

We are 99% confident that the population mean difference in measurement is between −0.31 and 0.54 feet per second.

(d)

A–B

Yes. Since a difference of 0 is located in the middle 50%, the boxplot supports that there is no difference in measurements.

15. (a) It is important to take the measurements on the same date in order to control for any role that season might have on lake clarity.

(b) We measure differences as $d_i = Y_i - X_i$.

Obs	1	2	3	4	5	6	7	8
X_i	38	58	65	74	56	36	56	52
Y_i	52	60	72	72	54	48	58	60
d_i	14	2	7	−2	−2	12	2	8

We compute the mean and standard deviation of the differences and obtain $\bar{d} = 5.125$ inches and $s_d \approx 6.0813$ inches, rounded to four decimal places. The hypotheses are $H_0 : \mu_d = 0$ versus $H_1 : \mu_d > 0$. The level of significance is $\alpha = 0.05$. The test statistic is

$$t_0 = \frac{\bar{d}}{\frac{s_d}{\sqrt{n}}} = \frac{5.125}{\frac{6.0813}{\sqrt{8}}} \approx 2.384.$$

Classical approach: Since this is a right-tailed test with 7 degrees of freedom, the critical value is $t_{0.05} = 1.895$. Since the test statistic $t_0 \approx 2.384$ falls to the right of the critical value $t_{0.05} = 1.895$ (i.e., since the test statistic falls within the critical region), we reject H_0.

<u>*P*-value approach</u>: The *P*-value for this right-tailed test is the area under the *t*-distribution with 7 degrees of freedom to the right of the test statistic $t_0 = 2.384$.

From the *t*-distribution table, in the row corresponding to 7 degrees of freedom, 2.384 falls between 2.365 and 2.517 whose right-tail areas are 0.025 and 0.02, respectively. So, $0.02 < P\text{-value} < 0.025$. [Tech: *P*-value = 0.0243]. Because the *P*-value is less than the level of significance $\alpha = 0.05$, we reject H_0.

<u>Conclusion</u>: There is sufficient evidence at the $\alpha = 0.05$ level of significance to conclude that there has been improvement in the clarity of the water in the 5-year period.

(c)

5 Years Minus Initial Measure

Yes. Since the majority of differences fall above 0, the boxplot supports that the lake is becoming clearer.

17. We measure differences as $d_i = Y_i - X_i$.

Obs	1	2	3	4	5	6	7
X_i	70.3	67.1	70.9	66.8	72.8	70.4	71.8
Y_i	74.1	69.2	66.9	69.2	68.9	70.2	70.4
d_i	3.8	2.1	−4.0	2.4	−3.9	−0.2	−1.4

Obs	8	9	10	11	12	13
X_i	70.1	69.9	70.8	70.2	70.4	72.4
Y_i	69.3	75.8	72.3	69.2	68.6	73.9
d_i	−0.8	5.9	1.5	−1.0	−1.8	1.5

We compute the mean and standard deviation of the differences and obtain $\bar{d} \approx 0.3154$ inches, rounded to four decimal places, and $s_d \approx 2.8971$ inches, rounded to four decimal places. The hypotheses are $H_0 : \mu_d = 0$ versus $H_1 : \mu_d > 0$. The level of significance is $\alpha = 0.1$. The test statistic is

$$t_0 = \frac{\bar{d}}{\frac{s_d}{\sqrt{n}}} = \frac{0.3154}{\frac{2.8971}{\sqrt{13}}} \approx 0.393.$$

<u>Classical approach</u>: Since this is a right-tailed test with 12 degrees of freedom, the critical value is $t_{0.10} = 1.356$. Since the test statistic $t_0 \approx 0.393$ does not fall to the right of the

critical value $t_{0.10} = 1.356$ (i.e., since the test statistic falls outside the critical region), we do not reject H_0.

<u>*P*-value approach</u>: The *P*-value for this right-tailed test is the area under the *t*-distribution with 12 degrees of freedom to the right of the test statistic $t_0 = 0.393$. From the *t*-distribution table, in the row corresponding to 12 degrees of freedom, 0.393 falls to the left of 0.695, whose right-tail area is 0.25. So, $P\text{-value} > 0.25$ [Tech: *P*-value = 0.3508]. Because the *P*-value is greater than the level of significance $\alpha = 0.10$, we do not reject H_0.

<u>Conclusion</u>: No, there is not sufficient evidence at the $\alpha = 0.10$ level of significance to conclude that sons are taller than their fathers.

19. We measure differences as $d_i = \text{diamond} - \text{steel}$.

Specimen	1	2	3	4	5	6	7	8	9
Steel ball	50	57	61	71	68	54	65	51	53
Diamond	52	56	61	74	69	55	68	51	56
d_i	2	−1	0	3	1	1	3	0	3

We compute the mean and standard deviation of the differences and obtain $\bar{d} \approx 1.3333$, rounded to four decimal places, and $s_d = 1.5$. The hypotheses are $H_0 : \mu_d = 0$ versus $H_1 : \mu_d \neq 0$. For $\alpha = 0.05$ and df $= 8$, $t_{\alpha/2} = t_{0.025} = 2.306$. Then:

Lower bound:

$$\bar{d} - t_{0.025} \cdot \frac{s_d}{\sqrt{n}} = 1.3333 - 2.306 \cdot \frac{1.5}{\sqrt{9}} \approx 0.2$$

Upper bound:

$$\bar{d} + t_{0.025} \cdot \frac{s_d}{\sqrt{n}} = 1.3333 + 2.306 \cdot \frac{1.5}{\sqrt{9}} \approx 2.5$$

We can be 95% confident that the population mean difference in hardness reading is between 0.2 and 2.5. This interval does not include 0, so we reject H_0.

<u>Conclusion</u>: There is sufficient evidence to conclude that the two indenters produce different hardness readings.

21. (a) It is important to randomly select whether the student would first be tested with normal or impaired vision in order to control for any "learning" that may occur in using the simulator.

(b) We measure differences as $d_i = Y_i - X_i$.

Subject	1	2	3	4	5
Normal, X_i	4.47	4.24	4.58	4.65	4.31
Impaired, Y_i	5.77	5.67	5.51	5.32	5.83
d_i	1.30	1.43	0.93	0.67	1.52

Subject	6	7	8	9
Normal, X_i	4.80	4.55	5.00	4.79
Impaired, Y_i	5.49	5.23	5.61	5.63
d_i	0.69	0.68	0.61	0.84

We compute the mean and standard deviation of the differences and obtain $\bar{d} \approx 0.9633$ seconds, rounded to four decimal places, and $s_d \approx 0.3576$ seconds, rounded to four decimal places. For $\alpha = 0.05$ and df $= 8$, $t_{\alpha/2} = t_{0.025} = 2.306$.

Lower bound:

$$\bar{d} - t_{0.025} \cdot \frac{s_d}{\sqrt{n}} = 0.9633 - 2.306 \cdot \frac{0.3576}{\sqrt{9}}$$

$$\approx 0.688$$

Upper bound:

$$\bar{d} + t_{0.025} \cdot \frac{s_d}{\sqrt{n}} = 0.9633 + 2.306 \cdot \frac{0.3576}{\sqrt{9}}$$

$$\approx 1.238$$

We can be 95% confident that the population mean difference in reaction time when teenagers are driving impaired from when driving normally is between 0.688 and 1.238 seconds. This interval does not include 0, so we reject H_0.

Conclusion: There is sufficient evidence to conclude that there is a difference in braking time with impaired vision and normal vision.

23. (a) Matching by driver and car for the two different octane levels is important as a means of controlling the experiment. Drivers and cars behave differently, so this matching reduces variability in miles per gallon that is attributable to the driver's driving style.

(b) Conducting the experiment on a closed track allows for control so that all cars and drivers can be put through the exact same driving conditions.

(c) No, neither variable is normally distributed. Both have at least one point outside the bounds of the normal probability plot.

(d) Yes, the differences in mileages appear to the approximately normally distributed since all of the points fall within the boundaries of the probability plot.

(e) From the MINITAB printout, $\bar{d} \approx 5.091$ miles and $s_d \approx 14.876$ miles, where the differences are 92 octane minus 87 octane. The hypotheses are $H_0 : \mu_d = 0$ versus $H_1 : \mu_d > 0$, where $d_i = 92$ Oct $- 87$ Oct. The level of significance is $\alpha = 0.05$. The test statistic is $t_0 \approx 1.14$.

Classical approach: Since this is a right-tailed test with 10 degrees of freedom, the critical value is $t_{0.05} = 1.812$. Since the test statistic $t_0 \approx 1.14$ does not fall to the right of the critical value $t_{0.05} = 1.812$ (i.e., since the test statistic does not fall within the critical region), we fail to reject H_0.

P-value approach: From the MINITAB printout, we find that P-value ≈ 0.141. Because the P-value is greater than the level of significance $\alpha = 0.05$, we do not reject H_0.

Conclusion: We would expect to get the results we obtained in about 14 out of 100 samples if the statement in the null hypothesis were true. Our result is not unusual. Thus, there is not sufficient evidence at the $\alpha = 0.05$ level of significance to conclude that cars get better mileage when using 92-octain gasoline than when using 87-octane gasoline.

Section 11.2

1. To test a hypothesis regarding the difference of two means with unknown population standard deviations, (1) the samples must be obtained using simple random sampling, (2) the samples must be independent, and (3) the populations from which the samples are drawn must be normally distributed or the sample sizes must be large (i.e., $n_1 \geq 30$ and $n_2 \geq 30$).

3. (a) $H_0 : \mu_1 = \mu_2$ versus $H_1 : \mu_1 \neq \mu_2$. The level of significance is $\alpha = 0.05$. Since the sample size of both groups is 15, we use $n_1 - 1 = 14$ degrees of freedom. The test statistic is

$$t_0 = \frac{(\bar{x}_1 - \bar{x}_2) - (\mu_1 - \mu_2)}{\sqrt{\frac{s_1^2}{n_1} + \frac{s_2^2}{n_2}}}$$

$$= \frac{(15.3 - 14.2) - 0}{\sqrt{\frac{3.2^2}{15} + \frac{3.5^2}{15}}} \approx 0.898$$

Classical approach: Since this is a two-tailed test with 14 degrees of freedom, the critical values are $\pm t_{0.025} = \pm 2.145$. Since the test statistic $t_0 \approx 0.898$ is between than the critical values $-t_{0.025} = -2.145$ and $t_{0.025} = 2.145$ (i.e., since the test statistic does not fall within the critical region), we do not reject H_0.

P-value approach: The P-value for this two-tailed test is the area under the t-distribution with 14 degrees of freedom to the right of $t_0 = 0.898$ plus the area to the left of -0.898. From the t-distribution table in the row corresponding to 14 degrees of freedom, 0.898 falls between 0.868 and 1.076 whose right-tail areas are 0.20 and 0.15, respectively. We must double these values in order to get the total area in both tails: 0.40 and 0.30. So, $0.30 < P\text{-value} < 0.40$ [Tech: P-value = 0.3767]. Because the P-value is greater than the level of significance $\alpha = 0.05$, we do not reject H_0.

Conclusion: There is not sufficient evidence at the $\alpha = 0.05$ level of significance to conclude that the population means are different.

(b) For a 95% confidence interval with df = 14, we use $t_{\alpha/2} = t_{0.025} = 2.145$. Then:

Lower bound:

$$(\bar{x}_1 - \bar{x}_2) - t_{\alpha/2} \cdot \sqrt{\frac{s_1^2}{n_1} + \frac{s_2^2}{n_2}}$$

$$= (15.3 - 14.2) - 2.145 \cdot \sqrt{\frac{3.2^2}{15} + \frac{3.5^2}{15}}$$

$$= -1.53 \text{ [Tech: } -1.41]$$

Upper bound:

$$(\bar{x}_1 - \bar{x}_2) + t_{\alpha/2} \cdot \sqrt{\frac{s_1^2}{n_1} + \frac{s_2^2}{n_2}}$$

$$= (15.3 - 14.2) + 2.145 \cdot \sqrt{\frac{3.2^2}{15} + \frac{3.5^2}{15}}$$

$$\approx 3.73 \text{ [Tech: 3.61]}$$

We can be 95% confident that the mean difference is between -1.53 and 3.73 [Tech: between -1.41 and 3.61].

5. (a) $H_0 : \mu_1 = \mu_2$ versus $H_1 : \mu_1 > \mu_2$. The level of significance is $\alpha = 0.10$. Since the smaller sample size is $n_2 = 18$, we use $n_2 - 1 = 18 - 1 = 17$ degrees of freedom. The test statistic is

$$t_0 = \frac{(\bar{x}_1 - \bar{x}_2) - (\mu_1 - \mu_2)}{\sqrt{\frac{s_1^2}{n_1} + \frac{s_2^2}{n_2}}}$$

$$= \frac{(50.2 - 42.0) - 0}{\sqrt{\frac{6.4^2}{25} + \frac{9.9^2}{18}}} \approx 3.081$$

Classical approach: Since this is a right-tailed test with 17 degrees of freedom, the critical value is $t_{0.10} = 1.333$. Since the test statistic $t_0 \approx 3.081$ is to the right of the critical value (i.e., since the test statistic falls within the critical region), we reject H_0.

P-value approach: The P-value for this right-tailed test is the area under the t-distribution with 17 degrees of freedom to the right of $t_0 = 3.081$. From the t-distribution table, in the row corresponding to 17 degrees of freedom, 3.081 falls between 2.898 and 3.222 whose right-tail areas are 0.005 and 0.0025, respectively. Thus, $0.0025 < P\text{-value} < 0.005$ [Tech: P-value = 0.0024]. Because the P-value is less than the level of significance $\alpha = 0.10$, we reject H_0.

Conclusion: There is sufficient evidence at the $\alpha = 0.01$ level of significance to conclude that $\mu_1 > \mu_2$.

(b) For a 90% confidence interval with df $= 17$, we use $t_{\alpha/2} = t_{0.05} = 1.740$. Then:

Lower bound:

$$(\bar{x}_1 - \bar{x}_2) - t_{\alpha/2} \cdot \sqrt{\frac{s_1^2}{n_1} + \frac{s_2^2}{n_2}}$$

$$= (50.2 - 42.0) - 1.740 \cdot \sqrt{\frac{6.4^2}{25} + \frac{9.9^2}{18}}$$

$$\approx 3.57 \text{ [Tech: 3.67]}$$

Upper bound:

$$(\bar{x}_1 - \bar{x}_2) + t_{\alpha/2} \cdot \sqrt{\frac{s_1^2}{n_1} + \frac{s_2^2}{n_2}}$$

$$= (50.2 - 42.0) + 1.740 \cdot \sqrt{\frac{6.4^2}{25} + \frac{9.9^2}{18}}$$

$$\approx 12.83 \text{ [Tech: 12.73]}$$

We can be 90% confident that the mean difference is between 3.57 and 12.83 [Tech: between 3.67 and 12.73].

7. $H_0 : \mu_1 = \mu_2$ versus $H_1 : \mu_1 < \mu_2$. The level of significance is $\alpha = 0.02$. Since the smaller sample size is $n_2 = 25$, we use $n_2 - 1 = 25 - 1 = 24$ degrees of freedom. The test statistic is

$$t_0 = \frac{(\bar{x}_1 - \bar{x}_2) - (\mu_1 - \mu_2)}{\sqrt{\frac{s_1^2}{n_1} + \frac{s_2^2}{n_2}}} = \frac{(103.4 - 114.2) - 0}{\sqrt{\frac{12.3^2}{32} + \frac{13.2^2}{25}}}$$

$$\approx -3.158$$

Classical approach: Since this is a left-tailed test with 24 degrees of freedom, the critical value is $-t_{0.02} = -2.172$. Since the test statistic $t_0 \approx -3.158$ is to the left of the critical value (i.e., since the test statistic falls within the critical region), we reject H_0.

P-value approach: The P-value for this left-tailed test is the area under the t-distribution with 24 degrees of freedom to the left of $t_0 \approx -3.158$, which is equivalent to the area to the right of $t_0 \approx 3.158$. From the t-distribution table in the row corresponding to 24 degrees of freedom, 3.158 falls between 3.091 and 3.467 whose right-tail areas are 0.0025 and 0.001, respectively. Thus, $0.001 < P\text{-value} < 0.0025$ [Tech: P-value = 0.0013]. Because the P-value is less than the level of significance $\alpha = 0.02$, we reject H_0.

Conclusion: There is sufficient evidence at the $\alpha = 0.02$ level of significance to conclude that $\mu_1 < \mu_2$.

9. (a) $H_0 : \mu_{CC} = \mu_{NT}$ versus $H_1 : \mu_{CC} > \mu_{NT}$. The level of significance is $\alpha = 0.01$. Since the smaller sample size is $n_2 = 268$, we use $n_2 - 1 = 268 - 1 = 267$ degrees of freedom. The test statistic is

$$t_0 = \frac{(\bar{x}_1 - \bar{x}_2) - (\mu_1 - \mu_2)}{\sqrt{\frac{s_1^2}{n_1} + \frac{s_2^2}{n_2}}} = \frac{(5.43 - 4.43) - 0}{\sqrt{\frac{1.162^2}{268} + \frac{1.015^2}{1145}}}$$

$$\approx 12.977$$

Classical approach: Since this is a right-tailed test with 267 degrees of freedom. However, since our t-distribution table does not contain a row for 267, we use df $= 100$. Thus, the critical value is $t_{0.01} = 2.364$. Since the test statistic $t_0 \approx 12.977$ is to the right of the critical value (i.e., since the test statistic falls within the critical region), we reject H_0.

P-value approach: The P-value for this right-tailed test is the area under the t-distribution with 267 degrees of freedom to the right of $t_0 = 12.977$. From the t-distribution table, in the row corresponding to 100 degrees of freedom (since the table does not contain a row for df $= 267$), 12.977 falls to the right of 3.390 whose right-tail area is 0.0005. Thus, $P\text{-value} < 0.0005$ [Tech: P-value < 0.0001]. Because the P-value is less than the level of significance $\alpha = 0.01$, we reject H_0.

Conclusion: There is sufficient evidence at the $\alpha = 0.01$ level of significance to conclude that $\mu_{CC} > \mu_{NT}$. That is, the evidence suggests that the mean time to graduate for students who first start in community college is longer than the mean time to graduate for those who do not transfer.

(b) For a 95% confidence interval with 100 degrees of freedom (since the table does not contain a row for df $= 267$), we use $t_{\alpha/2} = t_{0.025} = 1.984$. Then:

Lower bound:

$$(\bar{x}_1 - \bar{x}_2) - t_{\alpha/2} \cdot \sqrt{\frac{s_1^2}{n_1} + \frac{s_2^2}{n_2}}$$

$$= (5.43 - 4.43) - 1.984 \cdot \sqrt{\frac{1.162^2}{268} + \frac{1.015^2}{1145}}$$

$$\approx 0.847 \text{ [Tech: 0.848]}$$

Upper bound:

$$(\overline{x}_1 - \overline{x}_2) + t_{\alpha/2} \cdot \sqrt{\frac{s_1^2}{n_1} + \frac{s_2^2}{n_2}}$$

$$= (5.43 - 4.43) + 1.984 \cdot \sqrt{\frac{1.162^2}{268} + \frac{1.015^2}{1145}}$$

$$\approx 1.153 \ \text{[Tech: 1.152]}$$

We can be 95% confident that the mean additional time to graduate for students who start in community college is between 0.847 and 1.153 years [Tech: between 0.848 and 1.152 years].

(c) No, the results from parts (a) and (b) do not imply that community college *causes* one to take extra time to earn a bachelor's degree. This is observational data. Community college students may be working more hours, which does not allow them to take additional classes.

11. (a) This is an observational study, since no treatment is imposed. The researcher did not influence the data.

(b) Though we do not know if the population is normally distributed, the samples are independent with sizes that are sufficiently large ($n_{\text{arrival}} = n_{\text{departure}} = 35$).

(c) $H_0: \mu_{\text{arrival}} = \mu_{\text{departure}}$ versus $H_1: \mu_{\text{arrival}} \neq \mu_{\text{departure}}$. The level of significance is $\alpha = 0.05$. Since the sample size of both groups is 35, we use $n_{\text{arrival}} - 1 = 34$ degrees of freedom. The test statistic is

$$t_0 = \frac{(\overline{x}_{\text{arrival}} - \overline{x}_{\text{departure}}) - (\mu_{\text{arrival}} - \mu_{\text{departure}})}{\sqrt{\frac{s_{\text{arrival}}^2}{n_{\text{arrival}}} + \frac{s_{\text{departure}}^2}{n_{\text{departure}}}}}$$

$$= \frac{(269 - 260) - 0}{\sqrt{\frac{53^2}{35} + \frac{34^2}{35}}} \approx 0.846$$

Classical approach: Since this is a two-tailed test with 34 degrees of freedom, the critical values are $\pm t_{0.025} = \pm 2.032$. Since the test statistic $t_0 \approx 0.846$ is between the critical values (i.e., since the test statistic does not fall within the critical regions), we do not reject H_0.

P-value approach: The *P*-value for this two-tailed test is the area under the *t*-distribution with 34 degrees of freedom to the right of $t_0 \approx 0.846$ plus the area to the left of $t_0 \approx -0.846$. From the *t*-distribution table in the row corresponding to 34 degrees of freedom, 0.846 falls between 0.682 and 0.852 whose right-tail areas are 0.25 and 0.20, respectively. We must double these values in order to get the total area in both tails: 0.50 and 0.40. So, $0.40 < P\text{-value} < 0.50$ [Tech: *P*-value = 0.4013]. Since the *P*-value is greater than the level of significance $\alpha = 0.05$, we do not reject H_0.

Conclusion: There is not sufficient evidence at the $\alpha = 0.05$ level of significance to conclude that travelers walk at different speeds depending on whether they are arriving or departing an airport.

(d) For a 95% confidence interval with df = 34, we use $t_{\alpha/2} = t_{0.025} = 2.032$. Then:

Lower bound:

$$(\overline{x}_{\text{arrival}} - \overline{x}_{\text{departure}}) - t_{\alpha/2} \cdot \sqrt{\frac{s_{\text{arrival}}^2}{n_{\text{arrival}}} + \frac{s_{\text{departure}}^2}{n_{\text{departure}}}}$$

$$= (269 - 260) - 2.032 \cdot \sqrt{\frac{53^2}{35} + \frac{34^2}{35}}$$

$$\approx -12.6 \ \text{[Tech: } -12.3]$$

Upper bound:

$$(\overline{x}_{\text{arrival}} - \overline{x}_{\text{departure}}) + t_{\alpha/2} \cdot \sqrt{\frac{s_{\text{arrival}}^2}{n_{\text{arrival}}} + \frac{s_{\text{departure}}^2}{n_{\text{departure}}}}$$

$$= (269 - 260) + 2.032 \cdot \sqrt{\frac{53^2}{35} + \frac{34^2}{35}}$$

$$\approx 30.6 \ \text{[Tech: 30.3]}$$

We can be 95% confident that the mean difference in walking speed between passengers arriving and departing an airport is between -12.6 and 30.6 feet per minute [Tech: between -12.3 and 30.3].

13. (a) Yes, Welch's *t*-test can be used. We can treat each sample as a simple random sample of all mixtures of each type. The samples were obtained independently. We are told that a normal probability plot indicates that the data could come from a population that is normal, with no outliers.

(b) $H_0 : \mu_{301} = \mu_{400}$ versus $H_1 : \mu_{301} < \mu_{400}$.
The level of significance is $\alpha = 0.05$.
The sample statistics for the data are
$\bar{x}_{301} \approx 3668.9$, $s_{301} \approx 458.5$, and $n_{301} = 9$,
$\bar{x}_{400} = 4483$, $s_{400} \approx 473.7$, and $n_{400} = 10$.
Since the smaller sample size is $n_{301} = 9$,
we use $n_{301} - 1 = 8$ degrees of freedom.
The critical t-value is $t_{0.05} = 1.860$. The
test statistic is

$$t_0 = \frac{(\bar{x}_{301} - \bar{x}_{400}) - (\mu_{301} - \mu_{400})}{\sqrt{\dfrac{s_{301}^2}{n_{301}} + \dfrac{s_{400}^2}{n_{400}}}}$$

$$= \frac{(3668.9 - 4483) - 0}{\sqrt{\dfrac{458.5^2}{9} + \dfrac{473.7^2}{10}}} \approx -3.804$$

Classical approach: Since this is a left-
tailed test with 8 degrees of freedom, the
critical value is $-t_{0.05} = -1.860$. Since
the test statistic $t_0 \approx -3.804$ is to the left
of the critical value (i.e., since the test
statistic falls within the critical region),
we reject H_0 .

P-value approach: The *P*-value for this left-
tailed test is the area under the t-distribution
with 8 degrees of freedom to the left of
$-t_0 = -3.804$, which is equivalent to the
area to the right of $t_0 = 3.804$. From the t-
distribution table in the row corresponding
to 8 degrees of freedom, 3.804 falls
between 3.355 and 3.833 whose right-tail
areas are 0.005 and 0.0025, respectively.
Thus, $0.0025 < P\text{-value} < 0.005$ [Tech:
P-value = 0.0007]. Because the *P*-value is
less than the level of significance $\alpha = 0.05$,
we reject H_0 .

Conclusion: There is sufficient evidence
at the $\alpha = 0.05$ level of significance to
conclude that mixture 67-0-400 is stronger
than mixture 67-0-301.

(c) For a 90% confidence interval with df = 8,
we use $t_{\alpha/2} = t_{0.05} = 1.860$. Then:

Lower bound:

$$(\bar{x}_{400} - \bar{x}_{301}) - t_{\alpha/2} \cdot \sqrt{\frac{s_{400}^2}{n_{400}} + \frac{s_{301}^2}{n_{301}}}$$

$$= (4483 - 3668.9) - 1.860 \cdot \sqrt{\frac{473.7^2}{10} + \frac{458.5^2}{9}}$$

$$\approx 416.1 \quad [\text{Tech: 442}]$$

Upper bound:

$$(\bar{x}_{400} - \bar{x}_{301}) + t_{\alpha/2} \cdot \sqrt{\frac{s_{400}^2}{n_{400}} + \frac{s_{301}^2}{n_{301}}}$$

$$= (4483 - 3668.9) + 1.860 \cdot \sqrt{\frac{473.7^2}{10} + \frac{458.5^2}{9}}$$

$$\approx 1212 \quad [\text{Tech: 1187}]$$

We can be 90% confident that the mean
strength of mixture 67-0-400 is between
416 and 1212 pounds per square inch
stronger than the mean strength of mixture
67-0-301 [Tech: between 442 and 1187].

(d) The five number summaries follow:
Mixture 67-0-301:
2940, 3150, 3830, 4060, 4090
Mixture 67-0-400:
3730, 4120, 4475, 4890, 5220
[Minitab: 3730, 4108, 4475, 4923, 5220]

Yes, the boxplots supports that mixture
67-0-400 is stronger than mixture 67-0-301.

15. $H_0 : \mu_{\text{carpeted}} = \mu_{\text{uncarpeted}}$ versus
$H_1 : \mu_{\text{carpeted}} > \mu_{\text{uncarpeted}}$. The level of
significance is $\alpha = 0.05$. The sample statistics
for the data are $\bar{x}_1 = 11.2$, $s_1 \approx 2.6774$, and
$n_1 = 8$, and $\bar{x}_2 = 9.7875$, $s_2 \approx 3.2100$, and
$n_2 = 8$. Since the sample size of both groups is
8, we use $n_{\text{carpeted}} - 1 = 7$ degrees of freedom.
The test statistic is

$$t_0 = \frac{(\bar{x}_{\text{carpeted}} - \bar{x}_{\text{uncarpeted}}) - (\mu_{\text{carpeted}} - \mu_{\text{uncarpeted}})}{\sqrt{\dfrac{s_{\text{carpeted}}^2}{n_{\text{carpeted}}} + \dfrac{s_{\text{uncarpeted}}^2}{n_{\text{uncarpeted}}}}}$$

$$= \frac{(11.2 - 9.7875) - 0}{\sqrt{\dfrac{2.6774^2}{8} + \dfrac{3.2100^2}{8}}} \approx 0.956$$

Classical approach: Since this is a right-tailed
test with 7 degrees of freedom, the critical
value is $t_{0.05} = 1.895$. Since the test
statistic $t_0 \approx 0.954$ is not to the right of the
critical value (i.e., since the test statistic does
not fall within the critical region), we do not
reject H_0 .

P-value approach: The *P*-value for this right-tailed test is the area under the *t*-distribution with 7 degrees of freedom to the right of $t_0 \approx 0.956$. From the *t*-distribution table in the row corresponding to 7 degrees of freedom, 0.956 falls between 0.896 and 1.119 whose right-tail areas are 0.20 and 0.15, respectively. So, $0.15 < P\text{-value} < 0.20$ [Tech: *P*-value = 0.1780]. Since the *P*-value is greater than the level of significance $\alpha = 0.05$, we do not reject H_0.

Conclusion: There is not sufficient evidence at the $\alpha = 0.05$ level of significance to conclude that carpeted rooms have more bacteria than uncarpeted rooms.

17. Since both sample sizes are sufficiently large ($n_{AL} = n_{NL} = 30$), we can use Welch's *t*-test. The hypotheses are $H_0 : \mu_{AL} = \mu_{NL}$ versus $H_1 : \mu_{AL} > \mu_{NL}$. The level of significance is $\alpha = 0.05$. We are given the summary statistics $\bar{x}_{AL} = 6.0$ and $s_{AL} = 3.5$, $\bar{x}_{NL} = 4.3$, and $s_{NL} = 2.6$. Since the sample size of both groups is 30, we use $n_{AL} - 1 = 29$ degrees of freedom. The test statistic is

$$t_0 = \frac{(\bar{x}_{AL} - \bar{x}_{NL}) - (\mu_{AL} - \mu_{NL})}{\sqrt{\dfrac{s_{AL}^2}{n_{AL}} + \dfrac{s_{NL}^2}{n_{NL}}}}$$

$$= \frac{(6.0 - 4.3) - 0}{\sqrt{\dfrac{3.5^2}{30} + \dfrac{2.6^2}{30}}} \approx 2.136$$

Classical approach: Since this is a right-tailed test with 29 degrees of freedom, the critical value is $t_{0.05} = 1.699$. Since the test statistic $t_0 \approx 2.136$ falls to the right of the critical value 1.699 (i.e., since the test statistic falls within the critical region), we reject H_0.

P-value approach: The *P*-value for this right-tailed test is the area under the *t*-distribution with 29 degrees of freedom to the right of $t_0 \approx 2.136$. From the *t*-distribution table, in the row corresponding to 29 degrees of freedom, 2.136 falls between 2.045 and 2.150 whose right-tail areas are 0.025 and 0.02, respectively. So, $0.02 < P\text{-value} < 0.025$ [Tech: *P*-value = 0.0187]. Since the *P*-value is less than the level of significance $\alpha = 0.05$, we reject H_0.

Conclusion: There is sufficient evidence at the $\alpha = 0.05$ level of significance to conclude that games played with a designated hitter result in more runs.

19. For this 90% confidence level with $df = 40 - 1 = 39$, we use $t_{\alpha/2} = t_{0.05} = 1.685$. Then:

Lower bound:

$$(\bar{x}_{\text{no children}} - \bar{x}_{\text{children}}) - t_{\alpha/2} \cdot \sqrt{\frac{s_{\text{no children}}^2}{n_{\text{no children}}} + \frac{s_{\text{children}}^2}{n_{\text{children}}}}$$

$$= (5.62 - 4.10) - 1.685 \cdot \sqrt{\frac{2.43^2}{40} + \frac{1.82^2}{40}}$$

$$\approx 0.71 \quad [\text{Tech: } 0.72]$$

Upper bound:

$$(\bar{x}_{\text{no children}} - \bar{x}_{\text{children}}) + t_{\alpha/2} \cdot \sqrt{\frac{s_{\text{no children}}^2}{n_{\text{no children}}} + \frac{s_{\text{children}}^2}{n_{\text{children}}}}$$

$$= (5.62 - 4.10) + 1.685 \cdot \sqrt{\frac{2.43^2}{40} + \frac{1.82^2}{40}}$$

$$\approx 2.33 \quad [\text{Tech: } 2.32]$$

We can be 90% confident that the mean difference in daily leisure time between adults without children and those with children is between 0.71 and 2.33 hours [Tech: between 0.72 and 2.33 hours]. Since the confidence interval does not include zero, we can conclude that there is a significance difference in the leisure time of adults without children and those with children.

21. (a) The hypotheses are $H_0 : \mu_{\text{men}} = \mu_{\text{women}}$ versus $H_1 : \mu_{\text{men}} < \mu_{\text{women}}$.

(b) From the MINITAB printout, *P*-value = 0.0051. Because the *P*-value is less than the level of significance $\alpha = 0.01$, the researcher will reject H_0. Thus, there is sufficient evidence at the $\alpha = 0.01$ level of significance to conclude that the mean step pulse of men is lower than the mean step pulse of women.

(c) From the MINITAB printout, the 95% confidence interval is:
Lower bound: -10.7; Upper bound: -1.5
We are 95% confident that the mean step pulse of men is between 1.5 and 10.7 beats per minute lower than the mean step pulse of women

23. (a) This is an experiment using a completely randomized design.

(b) The response variable is scores on the final exam. The treatments are online homework system versus old-fashioned paper-and-pencil homework.

(c) Factors that are controlled in the experiment are the teacher, the text, the syllabus, the tests, the meeting time, the meeting location.

(d) In this case, the assumption is that the students "randomly" enrolled in the course.

(e) The hypotheses are $H_0 : \mu_{\text{fall}} = \mu_{\text{spring}}$ versus $H_1 : \mu_{\text{fall}} > \mu_{\text{spring}}$. The level of significance is $\alpha = 0.05$. We are given the summary statistics $\overline{x}_{\text{fall}} = 73.6$, $s_{\text{fall}} = 10.3$, and $n_{\text{fall}} = 27$, and $\overline{x}_{\text{spring}} = 67.9$, $s_{\text{spring}} = 12.4$, and $n_{\text{spring}} = 25$. Since the smaller sample size is $n_{\text{spring}} = 25$, we use $n_{\text{spring}} - 1 = 24$ degrees of freedom. The test statistic is

$$t_0 = \frac{(\overline{x}_{\text{fall}} - \overline{x}_{\text{spring}}) - (\mu_{\text{fall}} - \mu_{\text{spring}})}{\sqrt{\dfrac{s_{\text{fall}}^2}{n_{\text{fall}}} + \dfrac{s_{\text{spring}}^2}{n_{\text{spring}}}}}$$

$$= \frac{(73.6 - 67.9) - 0}{\sqrt{\dfrac{10.3^2}{27} + \dfrac{12.4^2}{25}}} \approx 1.795$$

Classical approach: Since this is a right-tailed test with 24 degrees of freedom, the critical values are $t_{0.05} = 1.711$. Since the test statistic $t_0 \approx 1.795$ falls to the right of the critical value 1.711 (i.e., since the test statistic falls within the critical region), we reject H_0.

P-value approach: The P-value for this right-tailed test is the area under the t-distribution with 24 degrees of freedom to the right of $t_0 \approx 1.795$. From the t-distribution table, in the row corresponding to 24 degrees of freedom, 1.795 falls between 1.711 and 2.064 whose right-tail areas are 0.05 and 0.025, respectively. So, $0.025 < P$-value < 0.05 [Tech: P-value = 0.0395]. Since the P-value is less than the level of significance $\alpha = 0.05$, we reject H_0.

Conclusion: There is sufficient evidence at the $\alpha = 0.05$ level of significance to conclude that final exam scores in the fall semester were higher than final exam scores in the spring semester. It would appear to be the case that the online homework system helped in raising final exam scores.

(f) Answers will vary. One possibility follows: One factor that may confound the results is that the weather is pretty lousy at the end of the fall semester, but pretty nice at the end of the spring semester. If "spring fever" kicked in for the spring semester students, then they probably studied less for the final exam.

Section 11.3

1. When testing hypotheses regarding the difference of two proportions, we have the null hypothesis $p_1 = p_2$. A pooled estimate of p is the best point estimate of this common population proportion. However, when constructing confidence intervals for the difference of two population proportions no assumption about their equality is made, so the sample proportions are not pooled.

3. To test a hypothesis regarding two population proportions from independent samples, the samples must be two independent random samples. In addition, $n_1 \hat{p}_1 (1 - \hat{p}_1) \geq 10$ and $n_2 \hat{p}_2 (1 - \hat{p}_2) \geq 10$, and neither sample can be more than 5% of the population from which it was drawn (i.e., $n_1 \leq 0.05 N_1$ and $n_2 \leq 0.05 N_2$).

5. (a) The hypotheses are $H_0 : p_1 = p_2$ versus $H_1 : p_1 > p_2$.

(b) The two sample estimates are
$$\hat{p}_1 = \frac{x_1}{n_1} = \frac{368}{541} \approx 0.6802 \text{ and}$$
$$\hat{p}_2 = \frac{x_2}{n_2} = \frac{351}{593} \approx 0.5919.$$

The pooled estimate is
$$\hat{p} = \frac{x_1 + x_2}{n_1 + n_2} = \frac{368 + 351}{541 + 593} \approx 0.6340.$$

The test statistic is

$$z_0 = \frac{\hat{p}_1 - \hat{p}_2}{\sqrt{\hat{p}(1-\hat{p})}\sqrt{\dfrac{1}{n_1} + \dfrac{1}{n_2}}}$$

$$= \frac{0.6802 - 0.5919}{\sqrt{0.6340(1-0.6340)}\sqrt{\dfrac{1}{541} + \dfrac{1}{593}}}$$

$$\approx 3.08$$

(c) This is a right-tailed test, so the critical value is $z_\alpha = z_{0.05} = 1.645$.

(d) P-value $= P(z_0 \geq 3.08)$

$$= 1 - 0.9990 = 0.0010$$

Since $z_0 = 3.08 > z_{0.05} = 1.645$ and P-value $= 0.0010 < \alpha = 0.05$, we reject H_0. There is sufficient evidence to conclude that $p_1 > p_2$.

7. (a) The hypotheses are $H_0 : p_1 = p_2$ versus $H_1 : p_1 \neq p_2$.

(b) The two sample estimates are

$$\hat{p}_1 = \frac{x_1}{n_1} = \frac{28}{254} \approx 0.1102 \text{ and}$$

$$\hat{p}_2 = \frac{x_2}{n_2} = \frac{36}{301} \approx 0.1196.$$

The pooled estimate is

$$\hat{p} = \frac{x_1 + x_2}{n_1 + n_2} = \frac{28 + 36}{254 + 301} \approx 0.1153.$$

The test statistic is

$$z_0 = \frac{\hat{p}_1 - \hat{p}_2}{\sqrt{\hat{p}(1-\hat{p})}\sqrt{\dfrac{1}{n_1} + \dfrac{1}{n_2}}}$$

$$= \frac{0.1102 - 0.1196}{\sqrt{0.1153(1-0.1153)}\sqrt{\dfrac{1}{254} + \dfrac{1}{301}}}$$

$$\approx -0.35 \quad [\text{Tech:} -0.34]$$

(c) This is a two-tailed test, so the critical values are $\pm z_{\alpha/2} = \pm z_{0.025} = \pm 1.96$.

(d) P-value $= 2 \cdot P(z_0 \leq -0.35)$

$$= 2 \cdot 0.3632$$

$$= 0.7264 \quad [\text{Tech: } 0.7307]$$

Since $z_0 = -0.35$ falls between $-z_{0.025} = -1.96$ and $z_{0.025} = 1.96$ and P-value $= 0.7264 > \alpha = 0.05$, we do not reject H_0. There is not sufficient evidence to conclude that $p_1 \neq p_2$.

---

9. We have $\hat{p}_1 = \dfrac{x_1}{n_1} = \dfrac{368}{541} \approx 0.6802$ and $\hat{p}_2 = \dfrac{x_2}{n_2} = \dfrac{421}{593} \approx 0.7099$. For a 90% confidence level, we use $\pm z_{0.05} = \pm 1.645$. Then:

Lower Bound: $(\hat{p}_1 - \hat{p}_2) - z_{\alpha/2} \cdot \sqrt{\dfrac{\hat{p}_1(1-\hat{p}_1)}{n_1} + \dfrac{\hat{p}_2(1-\hat{p}_2)}{n_2}}$

$$= (0.6802 - 0.7099) - 1.645 \cdot \sqrt{\frac{0.6802(1-0.6802)}{541} + \frac{0.7099(1-0.7099)}{593}}$$

$$\approx -0.075$$

Upper Bound: $(\hat{p}_1 - \hat{p}_2) + z_{\alpha/2} \cdot \sqrt{\dfrac{\hat{p}_1(1-\hat{p}_1)}{n_1} + \dfrac{\hat{p}_2(1-\hat{p}_2)}{n_2}}$

$$= (0.6802 - 0.7099) + 1.645 \cdot \sqrt{\frac{0.6802(1-0.6802)}{541} + \frac{0.7099(1-0.7099)}{593}}$$

$$\approx 0.015$$

11. We have $\hat{p}_1 = \dfrac{x_1}{n_1} = \dfrac{28}{254} \approx 0.1102$ and $\hat{p}_2 = \dfrac{x_2}{n_2} = \dfrac{36}{301} \approx 0.1196$. For a 95% confidence interval we use

$\pm z_{0.025} = \pm 1.96$. Then:

Lower Bound: $(\hat{p}_1 - \hat{p}_2) - z_{\alpha/2} \cdot \sqrt{\dfrac{\hat{p}_1(1-\hat{p}_1)}{n_1} + \dfrac{\hat{p}_2(1-\hat{p}_2)}{n_2}}$

$\qquad = (0.1102 - 0.1196) - 1.96 \cdot \sqrt{\dfrac{0.1102(1-0.1102)}{254} + \dfrac{0.1196(1-0.1196)}{301}}$

$\qquad \approx -0.063$

Upper Bound: $(\hat{p}_1 - \hat{p}_2) + z_{\alpha/2} \cdot \sqrt{\dfrac{\hat{p}_1(1-\hat{p}_1)}{n_1} + \dfrac{\hat{p}_2(1-\hat{p}_2)}{n_2}}$

$\qquad = (0.1102 - 0.1196) + 1.96 \cdot \sqrt{\dfrac{0.1102(1-0.1102)}{254} + \dfrac{0.1196(1-0.1196)}{301}}$

$\qquad \approx 0.044$

---

13. (a) The hypotheses are $H_0 : p_A = p_B$ versus $H_1 : p_A \neq p_B$.

(b) $\hat{p}_A = \dfrac{45+14}{45+19+14+23} = \dfrac{59}{101} \approx 0.5842$

$\hat{p}_B = \dfrac{45+19}{45+19+14+23} = \dfrac{64}{101} \approx 0.6337$

$z_0 = \dfrac{|f_{12} - f_{21}| - 1}{\sqrt{f_{12} + f_{21}}} = \dfrac{|19-14| - 1}{\sqrt{19+14}} \approx 0.70$

(c) This is a two-tailed test, so the critical values are $\pm z_{\alpha/2} = \pm z_{0.025} = \pm 1.96$.

(d) P-value $= 2 \cdot P(z_0 \geq 0.70)$

$\qquad = 2 \cdot [1 - P(z_0 \leq 0.70)]$

$\qquad = 2 \cdot [1 - 0.7580]$

$\qquad = 2 \cdot [0.2420]$

$\qquad = 0.4840$

Since $z_0 = 0.70$ falls between $-z_{0.025} = -1.96$ and $z_{0.025} = 1.96$ and since P-value $= 0.4840 > \alpha = 0.05$, we do not reject H_0. There is not sufficient evidence to conclude that $p_A \neq p_B$.

15. We first verify the requirements to perform the hypothesis test: (1) Each sample can be thought of as a simple random sample; (2) We have $x_1 = 107$, $n_1 = 710$, $x_2 = 67$, and $n_2 = 611$, so $\hat{p}_1 = \dfrac{x_1}{n_1} = \dfrac{107}{710} \approx 0.1507$ and $\hat{p}_2 = \dfrac{x_2}{n_2} = \dfrac{67}{611} \approx 0.1097$. Thus, $n_1\hat{p}_1(1-\hat{p}_1)$ $= 710(0.1507)(1-0.1507) \approx 91 \geq 10$ and $n_2\hat{p}_2(1-\hat{p}_2) = 611(0.1097)(1-0.1097) \approx 60$ ≥ 10; and (3) Each sample is less than 5% of the population. Thus, the requirements are met, and we can conduct the test.

The hypotheses are $H_0 : p_1 = p_2$ versus $H_1 : p_1 > p_2$. From before, the two sample estimates are $\hat{p}_1 \approx 0.1507$ and $\hat{p}_2 \approx 0.1097$. The pooled estimate is

$\hat{p} = \dfrac{x_1 + x_2}{n_1 + n_2} = \dfrac{107 + 67}{710 + 611} \approx 0.1317$.

The test statistic is

$z_0 = \dfrac{\hat{p}_1 - \hat{p}_2}{\sqrt{\hat{p}(1-\hat{p})}\sqrt{1/n_1 + 1/n_2}}$

$\quad = \dfrac{0.1507 - 0.1097}{\sqrt{0.1317(1-0.1317)}\sqrt{1/710 + 1/611}}$

$\quad \approx 2.20 \quad \text{[Tech: 2.19]}$

Classical approach: This is a right-tailed test, so the critical value is $z_\alpha = z_{0.05} = 1.645$. Since $z_0 > z_{0.05}$ (the test statistic lies within the critical region), we reject H_0.

P-value approach: P-value $= P(z_0 \ge 2.20) =$ $1 - 0.9861 = 0.0139$ [Tech: 0.0143]. Since P-value $< \alpha = 0.05$, we reject H_0.

Conclusion: There is sufficient evidence at the $\alpha = 0.05$ level of significance to conclude that a higher proportion of subjects in the treatment group (taking Prevnar) experienced fever as a side effect than in the control (placebo) group.

17. We first verify the requirements to perform the hypothesis test: (1) Each sample is a simple random sample; (2) we have $x_{1945} = 363$, $n_{1945} = 1100$, $x_{2007} = 396$, and $n_{2007} = 1100$, so

$$\hat{p}_{1945} = \frac{x_{1945}}{n_{1945}} = \frac{363}{1100} = 0.33 \text{ and}$$

$$\hat{p}_{2007} = \frac{x_{2007}}{n_{2007}} = \frac{396}{1100} = 0.36. \text{ Thus,}$$

$n_{1945}\hat{p}_{1945}(1 - \hat{p}_{1945}) = 1100(0.33)(1 - 0.33) \approx$ $243 \ge 10$ and $n_{2007}\hat{p}_{2007}(1 - \hat{p}_{2007}) =$ $1100(0.36)(1 - 0.36) \approx 253 \ge 10$; and (3) each sample is less than 5% of the population. Thus, the requirements are met, so we can conduct the test.

The hypotheses are $H_0 : p_{1945} = p_{2007}$ versus $H_1 : p_{1945} \ne p_{2007}$. From before, the two sample estimates are $\hat{p}_{1945} = 0.33$ and $\hat{p}_{2007} = 0.36$. The pooled estimate is

$$\hat{p} = \frac{x_{1945} + x_{2007}}{n_{1945} + n_{2007}} = \frac{363 + 396}{1100 + 1100} = 0.345.$$

The test statistic is

$$z_0 = \frac{\hat{p}_{1945} - \hat{p}_{2007}}{\sqrt{\hat{p}(1 - \hat{p})}\sqrt{1/n_{1945} + 1/n_{2007}}}$$

$$= \frac{0.33 - 0.36}{\sqrt{0.345(1 - 0.345)}\sqrt{1/1100 + 1/1100}}$$

$$\approx -1.48$$

Classical approach: This is a two-tailed test, so the critical values are $\pm z_{\alpha/2} = \pm z_{0.025} = \pm 1.96$. Since the test statistic $z_0 = -1.48$ lies between $-z_{0.025} = -1.96$ and $z_{0.025} = 1.96$ (the test statistic does not fall in the critical region), we do not reject H_0.

P-value approach: P-value $= 2 \cdot P(z_0 \le -1.48)$ $= 2 \cdot 0.0694 = 0.1388$ [Tech: 0.1389] Since P-value $> \alpha = 0.05$, we do not reject H_0.

Conclusion: There is not sufficient evidence at the $\alpha = 0.05$ level of significance to conclude that the proportion of adult Americans who were abstainers in 1945 is different from the proportion of abstainers in 2007.

19. $H_0 : p_m = p_f$ versus $H_1 : p_m \ne p_f$. We have $x_m = 181$, $n_m = 1205$, $x_f = 143$, and $n_f = 1097$, so

$$\hat{p}_m = \frac{x_m}{n_m} = \frac{181}{1205} \approx 0.1502 \text{ and } \hat{p}_f = \frac{x_f}{n_f} = \frac{143}{1097} \approx 0.1304. \text{ For a 95\% confidence interval we use}$$

$\pm z_{0.025} = \pm 1.96$. Then:

Lower Bound: $(\hat{p}_m - \hat{p}_f) - z_{\alpha/2} \cdot \sqrt{\dfrac{\hat{p}_m(1 - \hat{p}_m)}{n_m} + \dfrac{\hat{p}_f(1 - \hat{p}_f)}{n_f}}$

$$= (0.1502 - 0.1304) - 1.96 \cdot \sqrt{\frac{0.1502(1 - 0.1502)}{1205} + \frac{0.1304(1 - 0.1304)}{1097}}$$

$$\approx -0.009$$

Upper Bound: $(\hat{p}_m - \hat{p}_f) + z_{\alpha/2} \cdot \sqrt{\dfrac{\hat{p}_m(1 - \hat{p}_m)}{n_m} + \dfrac{\hat{p}_f(1 - \hat{p}_f)}{n_f}}$

$$= (0.1502 - 0.1304) + 1.96 \cdot \sqrt{\frac{0.1502(1 - 0.1502)}{1205} + \frac{0.1304(1 - 0.1304)}{1097}}$$

$$\approx 0.048$$

We are 95% confident that the difference in the proportion of males and females that have at least one tattoo is between −0.009 and 0.048. Because the interval includes zero, we do not reject the null hypothesis. There is no significant difference in the proportion of males and females that have tattoos.

21. (a) We first verify the requirements to perform the hypothesis test: (1) Each sample can be thought of as a simple random sample; (2) we have $x_C = 50$, $n_C = 1655$, $x_p = 31$, and $n_p = 1652$, so

$$\hat{p}_C = \frac{50}{1655} \approx 0.0302 \text{ and } \hat{p}_p = \frac{31}{1652} \approx 0.0188. \text{ Thus, } n_C\hat{p}_C(1-\hat{p}_C) = 1655(0.0302)(1-0.0302) \approx 48 \geq 10$$

and $n_p\hat{p}_p(1-\hat{p}_p) = 1652(0.0188)(1-0.0188) \approx 30 \geq 10$; and (3) each sample is less than 5% of the population. So, the requirements are met, so we can conduct the test.

The hypotheses are $H_0 : p_C = p_p$ versus $H_1 : p_C > p_p$. From before, the two sample estimates are

$$\hat{p}_C \approx 0.0302 \text{ and } \hat{p}_p \approx 0.0188. \text{ The pooled estimate is } \hat{p} = \frac{x_C + x_p}{n_C + n_p} = \frac{50+31}{1655+1652} \approx 0.0245.$$

The test statistic is

$$z_0 = \frac{\hat{p}_C - \hat{p}_p}{\sqrt{\hat{p}(1-\hat{p})}\sqrt{1/n_C + 1/n_p}} = \frac{0.0302 - 0.0188}{\sqrt{0.0245(1-0.0245)}\sqrt{\frac{1}{1655}+\frac{1}{1652}}} \approx 2.12 \quad [\text{Tech: 2.13}]$$

Classical approach: This is a right-tailed test, so the critical value is $z_{0.05} = 1.645$. Since $z_0 > z_{0.05} = 1.645$ (the test statistic falls in the critical region), we reject H_0.

P-value approach: P-value $= P(z_0 \geq 2.12)$
$= 1 - 0.9830 = 0.0170$ [Tech: 0.0166]. Since this P-value is less than the $\alpha = 0.05$ level of significance, we reject H_0.

Conclusion: There is sufficient evidence at the $\alpha = 0.05$ level of significance to conclude that the proportion of individuals taking Clarinex and experiencing dry mouth is greater than that of those taking a placebo.

(b) No, the difference between the experimental group and the control group is not practically significant. Both Clarinex and the placebo have fairly low proportions of individuals that experience dry mouth.

23. (a) The choices were randomly rotated in order to remove any potential nonsampling error due to the respondent hearing the word "right" or "wrong" first.

(b) We have $x_{2003} = 1086$, $n_{2003} = 1508$, $x_{2008} = 573$, and $n_{2008} = 1508$, so $\hat{p}_{2003} = \frac{x_{2003}}{n_{2003}} = \frac{1086}{1508} \approx 0.7202$

and $\hat{p}_{2008} = \frac{x_{2008}}{n_{2008}} = \frac{573}{1508} = 0.3800$. For a 90% confidence interval we use $\pm z_{0.05} = \pm 1.645$. Then:

Lower Bound: $(\hat{p}_{2003} - \hat{p}_{2008}) - z_{\alpha/2} \cdot \sqrt{\frac{\hat{p}_{2003}(1-\hat{p}_{2003})}{n_{2003}} + \frac{\hat{p}_{2008}(1-\hat{p}_{2008})}{n_{2008}}}$

$= (0.7202 - 0.3800) - 1.645 \cdot \sqrt{\frac{0.7202(1-0.7202)}{1508} + \frac{0.3800(1-0.3800)}{1508}}$

≈ 0.312

Upper Bound: $(\hat{p}_{2003} - \hat{p}_{2008}) + z_{\alpha/2} \cdot \sqrt{\frac{\hat{p}_{2003}(1-\hat{p}_{2003})}{n_{2003}} + \frac{\hat{p}_{2008}(1-\hat{p}_{2008})}{n_{2008}}}$

$= (0.7202 - 0.3800) + 1.645 \cdot \sqrt{\frac{0.7202(1-0.7202)}{1508} + \frac{0.3800(1-0.3800)}{1508}}$

≈ 0.368

We are 90% confident that the difference in the proportion of adult Americans who believe the United States made the right decision to use military force in Iraq from 2003 to 2008 is between 0.312 and 0.318. The attitude regarding the decision to go to war changed substantially.

25. (a) This is a dependent sample because the same person answered both questions.

(b) We first verify the requirements to perform McNemar's test. The samples are dependent and were obtained randomly. The total number of individuals who meet one condition but do not meet the other condition, is $448 + 327 = 775$, which is greater than 10. . Thus, the requirements are met, so we can conduct the test.

The hypotheses are $H_0 : p_{\text{no seat belt}} = p_{\text{smoke}}$ verses $H_1 : p_{\text{no seat belt}} \neq p_{\text{smoke}}$.

The test statistic is

$$z_0 = \frac{|f_{12} - f_{21}| - 1}{\sqrt{f_{12} + f_{21}}} = \frac{|448 - 327| - 1}{\sqrt{448 + 327}} \approx 4.31$$

Classical approach: The critical value for an $\alpha = 0.05$ level of significance is $z_{\alpha/2} = z_{0.025} = 1.96$. Since $z_0 = 4.31 > z_{0.025} = 1.96$ (the test statistic falls in the critical region), we reject H_0.

P-value approach: The P-value is two times the area under the standard normal distribution to the right of the test statistic, $z_0 = 4.31$. The test statistic is so large that $P\text{-value} < 0.0001$. Since this P-value is less than the $\alpha = 0.05$ level of significance, we reject H_0.

Conclusion: There is sufficient evidence at the $\alpha = 0.05$ level of significance to conclude that there is a difference in the proportion who do not use a seatbelt and the proportion who smoke. The sample proportion of individuals who do not wear a seat belt is $\hat{p}_{\text{no seat belt}} = \frac{67 + 327}{3029} \approx 0.13$, while the proportion of individuals who smoke is $\hat{p}_{\text{smoke}} = \frac{67 + 448}{3029} \approx 0.17$. So, smoking appears to be the more popular hazardous activity.

27. We first verify the requirements to perform McNemar's test. The samples are dependent and were obtained randomly. The total number of words meeting one condition but not meeting the other condition, is $385 + 456 = 841$, which is greater than 10. Thus, the requirements are met, so we can conduct the test.

The hypotheses are $H_0 : p_{\text{NN}} = p_{\text{RN}}$ verses $H_1 : p_{\text{NN}} \neq p_{\text{RN}}$. The test statistic is

$$z_0 = \frac{|f_{12} - f_{21}| - 1}{\sqrt{f_{12} + f_{21}}} = \frac{|385 - 456| - 1}{\sqrt{385 + 456}} \approx 2.41$$

Classical approach: The critical value for $\alpha = 0.05$ is $z_{\alpha/2} = z_{0.025} = 1.96$. Since $z_0 = 2.41 > z_{0.025} = 1.96$ (the test statistic falls in the critical region), we reject H_0.

P-value approach: The P-value is two times the area under the standard normal distribution to the right of the test statistic, $z_0 = 2.41$.

$$\begin{aligned} P\text{-value} &= 2 \cdot P(x > 2.41) \\ &= 2 \cdot [1 - P(x < 2.41)] \\ &= 2 \cdot (1 - 0.9920) \\ &= 0.0160 \end{aligned}$$

Since this P-value is less than the $\alpha = 0.05$ level of significance, we reject H_0.

Conclusion: There is sufficient evidence at the $\alpha = 0.05$ level of significance to conclude that there is a difference in the proportion of words not recognized by the two systems. The sample proportion of words recognized by the neural network is

$$\hat{p}_{NN} = \frac{9326 + 385}{9326 + 385 + 456 + 29} \approx 0.95,$$

while the sample proportion of words recognize by the remapped network is

$$\hat{p}_{RN} = \frac{9326 + 456}{9326 + 385 + 456 + 29} \approx 0.96.$$ So, it appears that the remapped network recognizes more words than the neural network. That is, the neural network appears to make more errors.

29. To answer this question, we test to see if the difference in the proportion of gun owners in May 2007 and the proportion of gun owners in October 2004 is statistically significant. We will use a $\alpha = 0.05$ level of significance.

We first verify the requirements to perform the hypothesis test: (1) Each sample can be thought of as a simple random sample; (2) we have $x_{2007} = 340$, $n_{2007} = 1134$, $x_{2004} = 363$, and $n_{2004} = 1134$, so $\hat{p}_{2007} = \frac{340}{1134} \approx 0.2998$ and $\hat{p}_{2004} = \frac{363}{1134} \approx 0.3201$. Therefore,

$n_{2007} \hat{p}_{2007} \left(1 - \hat{p}_{2007}\right) = 1134(0.2998)(1 - 0.2998)$

$\approx 238 \geq 10$ and $n_{2004} \hat{p}_{2004} \left(1 - \hat{p}_{2004}\right) =$

$1134(0.3201)(1 - 0.3201) \approx 247 \geq 10$; and (3) each sample is less than 5% of the population. Thus, the requirements are met, so we can conduct the test. The hypotheses are

$H_0 : p_{2007} = p_{2004}$ versus $H_1 : p_{2007} > p_{2004}$.
From before, the two sample estimates are
$\hat{p}_{2007} \approx 0.2998$ and $\hat{p}_{2004} \approx 0.3201$. The pooled estimate is

$\hat{p} = \dfrac{x_{2007} + x_{2004}}{n_{2007} + n_{2004}} = \dfrac{340 + 363}{1134 + 1134} \approx 0.3100$.

The test statistic is

$z_0 = \dfrac{\hat{p}_{2007} - \hat{p}_{2004}}{\sqrt{\hat{p}(1-\hat{p})}\sqrt{1/n_{2007} + 1/n_{2004}}}$

$= \dfrac{0.2998 - 0.3201}{\sqrt{0.3100(1-0.3100)}\sqrt{1/1134 + 1/1134}}$

≈ -1.05 [Tech: -1.04]

Classical approach: This is a left-tailed test, so the critical value for a $\alpha = 0.05$ level of significance is $-z_{0.05} = -1.645$. Since
$z_0 = -1.04 > -z_{0.05} = -1.645$ (the test statistic does not fall in the critical region), we do not reject H_0.

P-value approach: P-value = $P(z_0 \leq -1.05) = 0.1469$ [Tech: 0.1482]. Since this P-value is greater than the $\alpha = 0.05$ level of significance, we do not reject H_0.

Conclusion: There is not sufficient evidence at the $\alpha = 0.05$ level of significance to conclude that the proportion of gun owners in 2007 is less than the proportion of gun owners in 2004. Therefore, the headline is not accurate.

31. (a) $n = n_1 = n_2$

$= \left[\hat{p}_1(1 - \hat{p}_1) + \hat{p}_2(1 - \hat{p}_2)\right]\left(\dfrac{z_{\alpha/2}}{E}\right)^2$

$= \left[0.219(1 - 0.219) + 0.197(1 - 0.197)\right]\left(\dfrac{1.96}{0.03}\right)^2$

≈ 1405.3

We increase this result to 1406.

(b) $n = n_1 = n_2$

$= 0.5\left(\dfrac{z_{\alpha/2}}{E}\right)^2 = 0.5\left(\dfrac{1.96}{0.03}\right)^2 = 2134.2$

We increase this result to 2135.

33. (a) This is an experiment using a completely randomized design.

(b) The response variable is whether the subject contracted polio or not.

(c) The treatments are the vaccine and the placebo.

(d) A placebo is an innocuous medication that looks, tastes, and smells like the experimental treatment.

(e) Because the incidence rate of polio is low, a large number of subjects is needed so that we are guaranteed a sufficient number of successes.

(f) To answer this question, we test to see if there is a significant difference in the proportion of subjects from the experiment group who contracted polio and the proportion of subjects from the control group who contracted polio. We will use a $\alpha = 0.01$ level of significance.

We first verify the requirements to perform the hypothesis test: (1) Each sample can be thought of as a simple random sample; (2) we have $x_1 = 33$, $n_1 = 200,000$, $x_2 = 115$, and $n_2 = 200,000$, so

$\hat{p}_1 = \dfrac{x_1}{n_1} = \dfrac{33}{200,000} = 0.000165$ and

$\hat{p}_2 = \dfrac{x_2}{n_2} = \dfrac{115}{200,000} = 0.000575$.

Therefore, $n_1 \hat{p}_1 \left(1 - \hat{p}_1\right) =$
$200,000(0.000165)(1 - 0.000165) \approx 33 \geq 10$
and $n_2 \hat{p}_2 \left(1 - \hat{p}_2\right) =$
$200,000(0.000575)(1 - 0.000575) \approx 115 \geq 10$;
and (3) each sample is less than 5% of the population. Thus, the requirements are met, so we can conduct the test.

The hypotheses are $H_0 : p_1 = p_2$ versus $H_1 : p_1 < p_2$. From before, the two sample estimates are $\hat{p}_1 = 0.000165$ and $\hat{p}_2 = 0.000575$. The pooled estimate is

$\hat{p} = \dfrac{x_1 + x_2}{n_1 + n_2} = \dfrac{33 + 115}{200,000 + 200,000} = 0.00037$.

The test statistic is

$$z_0 = \frac{\hat{p}_1 - \hat{p}_2}{\sqrt{\hat{p}(1-\hat{p})}\sqrt{1/n_1 + 1/n_2}}$$

$$= \frac{0.000165 - 0.000575}{\sqrt{0.00037(1-0.00037)}\sqrt{\dfrac{1}{200,000} + \dfrac{1}{200,000}}}$$

$$\approx -6.74$$

Classical approach: This is a left-tailed test, so the critical value is $-z_{0.01} = -2.33$. Since $z_0 = -6.74 < -z_{0.05} = -2.33$ (the test statistic falls in the critical region), we reject H_0.

P-value approach:
P-value $= P(z_0 \le -6.74) < 0.0001$. Since this P-value is less than the $\alpha = 0.01$ level of significance, we reject H_0.

Conclusion: There is sufficient evidence at the $\alpha = 0.01$ level of significance to conclude that the proportion of children in the experimental group who contracted polio is less than the proportion of children in the control group who contracted polio.

Section 11.4

1. The test for comparing two population standard deviations is not robust, so even small departures from normality adversely affect the test, making it unreliable. So, in the test for comparing two population standard deviations, the samples must come from populations that are normally distributed.

3. $F_{0.05,9,10} = 3.02$

5. $\alpha = 0.05$, so $\dfrac{\alpha}{2} = 0.025$. Thus, $F_{0.025,6,8} = 4.65$

and $F_{0.975,6,8} = \dfrac{1}{F_{0.025,8,6}} = \dfrac{1}{5.60} \approx 0.18$.

7. $F_{0.90,25,20} = \dfrac{1}{F_{0.10,20,25}} = \dfrac{1}{1.72} \approx 0.58$

9. $F_{0.05,45,15} = \dfrac{F_{0.05,40,15} + F_{0.05,50,15}}{2}$

$= \dfrac{2.20 + 2.18}{2}$

$= 2.19$

11. The hypotheses are $H_0 : \sigma_1 = \sigma_2$ versus $H_1 : \sigma_1 \ne \sigma_2$. The test statistic is

$$F_0 = \frac{s_1^2}{s_2^2} = \frac{3.2^2}{3.5^2} \approx 0.84.$$

Classical approach: This is a two-tailed test with $n_1 = n_2 = 16$ and $\alpha = 0.05$. So, $\dfrac{\alpha}{2} = 0.025$, and the critical values are $F_{0.025,15,15} = 2.86$ and

$$F_{0.975,15,15} = \frac{1}{F_{0.025,15,15}} = \frac{1}{2.86} \approx 0.35.$$ Since $F_0 = 0.84$ is between $F_{0.975,15,15} = 0.35$ and $F_{0.025,15,15} = 2.86$ (the test statistic does not fall in the critical regions), we do not reject H_0.

P-value approach: Using technology, we find P-value ≈ 0.7330. Since this P-value is greater than the $\alpha = 0.05$ level of significance, we do no reject H_0.

Conclusion: There is not sufficient evidence at the $\alpha = 0.05$ level of significance to conclude that $\sigma_1 \ne \sigma_2$.

13. The hypotheses are $H_0 : \sigma_1 = \sigma_2$ versus $H_1 : \sigma_1 > \sigma_2$. The test statistic is

$$F_0 = \frac{s_1^2}{s_2^2} = \frac{9.9^2}{6.4^2} \approx 2.39.$$

Classical approach: This is a right-tailed test with $n_1 = 26$, $n_2 = 19$, and $\alpha = 0.01$. So, the critical value is $F_{0.01,25,18}$, which we approximate by $F_{0.01,25,20} = 2.84$. Since $F_0 = 2.39 < F_{0.01,25,20} = 2.84$ the test statistic does not fall in the critical region, we do not reject H_0.

P-value approach: Using technology, we find P-value ≈ 0.0303. Since this P-value is greater than the $\alpha = 0.01$ level of significance, we do no reject H_0.

Conclusion: There is not sufficient evidence at the $\alpha = 0.01$ level of significance to conclude that $\sigma_1 > \sigma_2$.

15. The hypotheses are $H_0 : \sigma_1 = \sigma_2$ versus $H_1 : \sigma_1 < \sigma_2$. The test statistic is

$$F_0 = \frac{s_1^2}{s_2^2} = \frac{8.3^2}{13.2^2} \approx 0.40.$$

Classical approach: This is a left-tailed test with $n_1 = 51$, $n_2 = 26$, and $\alpha = 0.10$. So, the critical value is $F_{0.90, 50, 25} = \dfrac{1}{F_{0.10, 25, 50}} = \dfrac{1}{1.53} \approx 0.65$.

Since $F_0 = 0.40 < F_{0.90, 50, 25} = 0.65$ (the test statistic falls in a critical region), we reject H_0.

P-value approach: Using technology, we find P-value ≈ 0.0026. Since this P-value is less than the $\alpha = 0.10$ level of significance, we reject H_0.

Conclusion: There is sufficient evidence at the $\alpha = 0.1$ level of significance to conclude that $\sigma_1 < \sigma_2$.

17. The hypotheses are $H_0 : \sigma_1 = \sigma_2$ versus $H_1 : \sigma_1 \neq \sigma_2$. The test statistic is

$$F_0 = \frac{s_1^2}{s_2^2} = \frac{12.5^2}{12.7^2} \approx 0.97.$$

Classical approach: This is a two-tailed test with $n_1 = 55$, $n_2 = 60$, and $\alpha = 0.05$. The critical values are $F_{0.025, 54, 59}$ and $F_{0.975, 54, 59}$, which we approximate by $F_{0.025, 50, 50} = 1.75$

and $F_{0.975, 54, 59} = \dfrac{1}{F_{0.025, 59, 54}} \approx \dfrac{1}{F_{0.025, 60, 50}} =$

$\dfrac{1}{1.72} \approx 0.58$. Since $F_0 = 0.97$ falls in between $F_{0.975, 54, 59} \approx 1.75$ and $F_{0.975, 54, 59} \approx 0.58$ (the test statistic does not fall in the critical region), we do not reject H_0.

P-value approach: Using technology, we find P-value ≈ 0.9087. Since this P-value is greater than the $\alpha = 0.05$ level of significance, we do not reject H_0.

Conclusion: There is not sufficient evidence at the $\alpha = 0.05$ level of significance to conclude that the standard deviation in the treatment group is different from the standard deviation in the control group.

19. The hypotheses are $H_0 : \sigma_1 = \sigma_2$ versus $H_1 : \sigma_1 > \sigma_2$. The test statistic is

$$F_0 = \frac{s_1^2}{s_2^2} = \frac{17.07^2}{6.48^2} \approx 6.94.$$

Classical approach: This is a right-tailed test with $n_1 = n_2 = 65$ and $\alpha = 0.01$. So, the critical value is $F_{0.01, 64, 64}$, which we approximate by $F_{0.01, 60, 50} = 1.91$. Since $F_0 \approx 6.94 > F_{0.01, 60, 50} = 1.91$ (the test statistic falls in the critical region), we reject H_0.

P-value approach: Using technology, we find P-value < 0.0001. Since this P-value is less than the $\alpha = 0.01$ level of significance, we reject H_0.

Conclusion: There is sufficient evidence at the $\alpha = 0.01$ level of significance to conclude that the treatment group had a higher standard deviation for serum retinal concentration than did the control group.

21. (a) The hypotheses are $H_0 : \sigma_1 = \sigma_2$ versus $H_1 : \sigma_1 < \sigma_2$, where $\sigma_1 =$ the standard deviation of waiting time for a single line. We compute the sample standard deviations to be $s_1 \approx 0.574$ minutes and $s_2 \approx 1.012$ minutes. The test statistic is

$$F_0 = \frac{s_1^2}{s_2^2} = \frac{0.574^2}{1.012^2} \approx 0.32.$$

Classical approach: This is a left-tailed test with $n_1 = n_2 = 20$ and $\alpha = 0.05$. The critical value is $F_{0.95, 19, 19}$. We approximate this by $F_{0.95, 20, 20} = \dfrac{1}{F_{0.05, 20, 20}} = \dfrac{1}{2.12} \approx 0.47$.

Since $F_0 = 0.32 < F_{0.95, 20, 20} \approx 0.47$ (the test statistic falls in the critical region), we reject H_0.

P-value approach: Using technology, we find P-value ≈ 0.0087. Since this P-value is less than the $\alpha = 0.05$ level of significance, we reject H_0.

Conclusion: There is sufficient evidence at the $\alpha = 0.05$ level of significance to conclude that the standard deviation of waiting time in a single line is less than the standard deviation for wait time in the multiple lines.

(b) We compute the five-number summary for each variable:
Single: 1.1, 1.9, 2.2, 2.8, 3.1
Multiple: 0.9, 1.8, 2.45, 2.9, 4.6
[Minitab: 0.9, 1.7, 2.45, 2.9, 4.6]

The boxplot indicates that there is much greater variability in the waiting times for multiple lines than for a single line.

23. (a) The normal probability plots are roughly linear, so we conclude that the data could come from normal distributions. Thus, this requirement of the F-test is satisfied.

(b) Since P-value = 0.0844 is larger than the $\alpha = 0.05$ level of significance, the nursing student should conclude that there is not sufficient evidence to conclude that men have more variability in their systolic blood pressure than women have.

(c) We find the sample standard deviations to be $s_f \approx 10.346$ and $s_m \approx 14.604$. The test statistic is $F_0 = \dfrac{s_m^2}{s_f^2} = \dfrac{14.604^2}{10.346^2} \approx 1.99$. We test the hypotheses $H_0 : \sigma_m = \sigma_f$ versus $H_1 : \sigma_m > \sigma_f$. This is a right-tailed test with df = 19 for the numerator and df = 16 for the denominator. Using Table VIII, we only approximate these by 20 df and 15 df, respectively. Since $F_{0.10,20,15} = 1.92$ and $F_{0.05,20,15} = 2.33$, the P-value should be close to 0.10 which is consistent with the given P-value (0.0844).

Section 11.5

1. Answers will vary. When deciding whether a sample is dependent or independent, one word to look for is *matched*, which indicates that the sampling is dependent. Obviously, if the word *independent* is in the problem, the sampling

method is likely independent. Also, look for language that implies that the two variables are measured on related items, which would mean the sampling is dependent.

3. The completely randomized design is analyzed using the inferential techniques for comparing two means from independent samples. The response variable is qualitative. There are two levels of treatment.

5. We first verify the requirements to perform McNemar's test. The samples are dependent because the different creams were applied to two warts on the same individual. We can think of the sample as a random sample. The total number of individuals for which one treatment was successful while the other was not is $9 + 13 = 22$, which is greater than 10. Thus, the requirements are met, so we can conduct the test.

The hypotheses are $H_0 : p_A = p_B$ verses $H_1 : p_A \neq p_B$. The test statistic is

$$z_0 = \frac{|f_{12} - f_{21}| - 1}{\sqrt{f_{12} + f_{21}}} = \frac{|9 - 13| - 1}{\sqrt{9 + 13}} \approx 0.64$$

Classical approach: The critical value for $\alpha = 0.05$ is $z_{\alpha/2} = z_{0.025} = 1.96$. Since $z_0 = 0.64 < z_{0.025} = 1.96$ (the test statistic does not fall in the critical region), we do not reject H_0.

P-value approach: The P-value is two times the area under the standard normal distribution to the right of the test statistic, $z_0 = 0.64$.

P-value $= 2 \cdot P(x > 0.64)$
$\qquad = 2 \cdot [1 - P(x < 0.64)]$
$\qquad = 2 \cdot (1 - 0.7389)$
$\qquad = 0.5222$

Since this P-value is greater than $\alpha = 0.05$, we do not reject H_0.

Conclusion: There is not sufficient evidence at the $\alpha = 0.05$ level of significance to conclude that there is a difference in the proportions.

7. The hypotheses are $H_0 : \sigma_1 = \sigma_2$ versus $H_1 : \sigma_1 > \sigma_2$. We assume that all requirements to conduct the test are satisfied. The test statistic is $F_0 = \dfrac{s_1^2}{s_2^2} = \dfrac{12^2}{10^2} = 1.44$.

Classical approach: This is a right-tailed test with $n_1 = 31$, $n_2 = 51$, and $\alpha = 0.05$. So, the critical value is $F_{0.05,30,50} = 1.69$. Since

$F_0 = 1.44 < F_{0.05,30,50} = 1.69$ (the test statistic does not fall in the critical region), we do not reject H_0.

P-value approach: Using technology, we find P-value ≈ 0.1249. Since this P-value is greater than the $\alpha = 0.05$ level of significance, we do no reject H_0.

Conclusion: There is not sufficient evidence at the $\alpha = 0.05$ level of significance to conclude that $\sigma_1 > \sigma_2$.

9. We verify the requirements to perform the hypothesis test: (1) We are told that the samples are random samples; (2) We have $x_1 = 40$, $n_1 = 135$, $x_2 = 60$, and $n_2 = 150$, so

$\hat{p}_1 = \dfrac{40}{135} \approx 0.2963$ and $\hat{p}_2 = \dfrac{60}{150} = 0.4$. Thus,

$n_1 \hat{p}_1 (1 - \hat{p}_1) = 135(0.2963)(1 - 0.2963) \approx 28 \geq$ 10 and $n_2 \hat{p}_2 (1 - \hat{p}_2) = 150(0.4)(1 - 0.4) = 36 \geq$ 10; and (3) We assume each sample is less than 5% of the population. So, the requirements are met, and we can conduct the test.

The hypotheses are $H_0 : p_1 = p_2$ versus $H_1 : p_1 < p_2$. From before, the two sample estimates are $\hat{p}_1 \approx 0.2963$ and $\hat{p}_2 = 0.4$.

The pooled estimate is $\hat{p} = \dfrac{40 + 60}{135 + 150} \approx 0.3509$.

The test statistic is

$z_0 = \dfrac{\hat{p}_1 - \hat{p}_2}{\sqrt{\hat{p}(1-\hat{p})}\sqrt{\dfrac{1}{n_1} + \dfrac{1}{n_2}}}$

$= \dfrac{0.2963 - 0.4}{\sqrt{0.3509(1-0.3509)}\sqrt{\dfrac{1}{135} + \dfrac{1}{150}}} \approx -1.83$

Classical approach:
This is a left-tailed test, so the critical value for $\alpha = 0.05$ is $-z_{0.05} = -1.645$. Since

$z_0 = -1.83 < -z_{0.05} = -1.645$ (the test statistic falls in the critical region), we reject H_0.

P-value approach: The P-value is the area under the standard normal distribution to the left of the test statistic, $z_0 = -1.83$.

P-value $= P(z_0 \leq -1.83)$

$= 0.0336$ [Tech: 0.0335]

Since the P-value $= 0.0336 < \alpha = 0.05$, we reject H_0.

Conclusion: There is sufficient evidence at the $\alpha = 0.05$ level of significance to conclude that $p_1 < p_2$.

11. $H_0 : \mu_1 = \mu_2$ versus $H_1 : \mu_1 \neq \mu_2$. We assume that all requirements to conduct the test are satisfied. The level of significance is $\alpha = 0.01$. The smaller sample size is $n_1 = 41$, so we use $n_1 - 1 = 41 - 1 = 40$ degrees of freedom. The test statistic is

$t_0 = \dfrac{(\bar{x}_1 - \bar{x}_2) - (\mu_1 - \mu_2)}{\sqrt{\dfrac{s_1^2}{n_1} + \dfrac{s_2^2}{n_2}}} = \dfrac{(125.3 - 130.8) - 0}{\sqrt{\dfrac{8.5^2}{41} + \dfrac{7.3^2}{50}}}$

≈ -3.271

Classical approach: Since this is a two-tailed test with 40 degrees of freedom, the critical values are $\pm t_{0.005} = \pm 2.704$. Since

$t_0 \approx -3.271 < -t_{0.005} = -2.704$ (the test statistic falls in the critical region), we reject H_0.

P-value approach: The P-value for this two-tailed test is the area under the t-distribution with 40 degrees of freedom to the left of $t_0 = -3.271$ plus the area to the right of 3.271. From the t-distribution table in the row corresponding to 40 degrees of freedom, 3.271 falls between 2.971 and 3.307 whose right-tail areas are 0.0025 and 0.001, respectively. We must double these values in order to get the total area in both tails: 0.005 and 0.002. So, $0.002 < $ P-value < 0.005 [Tech: P-value = 0.0016]. Because the P-value is less than the level of significance $\alpha = 0.05$, we reject H_0.

Conclusion: There is sufficient evidence at the $\alpha = 0.05$ level of significance to conclude that there is a difference in the population means.

13. These are matched-pair data because two measurements, X_i and Y_i, are taken on the same individual. We measure differences as $d_i = Y_i - X_i$.

Individual	1	2	3	4	5
X_i	40	32	53	48	38
Y_i	38	33	49	48	33
d_i	−2	1	−4	0	−5

We compute the mean and standard deviation of the differences and obtain $\bar{d} = -2$ and

$s_d \approx 2.5495$, rounded to four decimal places. The hypotheses are $H_0 : \mu_d = 0$ versus $H_1 : \mu_d < 0$. The level of significance is $\alpha = 0.10$. The test statistic is

$$t_0 = \frac{\overline{d}}{\frac{s_d}{\sqrt{n}}} = \frac{-2}{\frac{2.5495}{\sqrt{5}}} \approx -1.754.$$

Classical approach: Since this is a left-tailed test with 4 degrees of freedom, the critical value is $-t_{0.10} = -1.533$. Since $t_0 \approx -1.754 < -t_{0.10} = -1.533$ (the test statistic falls within the critical region), we reject H_0.

P-value approach: The P-value for this left-tailed test is the area under the t-distribution with 4 degrees of freedom to the left of the test statistic $t_0 = -1.754$, which is equivalent to the area to the right of 1.754. From the t-distribution table, in the row corresponding to 4 degrees of freedom, 1.754 falls between 1.533 and 2.132 whose right-tail areas are 0.10 and 0.05, respectively. So, $0.10 > P\text{-value} > 0.05$ [Tech: P-value = 0.0771]. Because the P-value is less than the level of significance $\alpha = 0.10$, we reject H_0.

Conclusion: There is sufficient evidence at the $\alpha = 0.10$ level of significance to conclude that there is a difference in the measure of the variable before and after a treatment. That is, there is sufficient evidence to conclude that the treatment is effective.

15. (a) Collision claims tend to be skewed right because there are a few very large collision claims relative to the majority of claims.

    (b) The hypotheses were $H_0 : \mu_{30-59} = \mu_{20-24}$ versus $H_1 : \mu_{30-59} < \mu_{20-24}$. All requirements are satisfied to conduct the test for two independent population means. The level of significance is $\alpha = 0.05$. The sample sizes are both $n = 40$, so we use $n - 1 = 40 - 1 = 39$ degrees of freedom. The test statistic is

$$t_0 = \frac{(\overline{x}_1 - \overline{x}_2) - (\mu_1 - \mu_2)}{\sqrt{\frac{s_1^2}{n_1} + \frac{s_2^2}{n_2}}}$$

$$= \frac{(3669 - 4586) - 0}{\sqrt{\frac{2029^2}{40} + \frac{2302^2}{40}}} \approx -1.890$$

Classical approach: Since this is a left-tailed test with 39 degrees of freedom, the critical value is $-t_{0.05} = -1.685$. Since $t_0 = -1.890 < -t_{0.05} = -1.685$ (the test statistic falls in the critical region), we reject H_0.

P-value approach: The P-value for this left-tailed test is the area under the t-distribution with 39 degrees of freedom to the left of $t_0 = -1.890$. From the t-distribution table in the row corresponding to 39 degrees of freedom, 1.890 falls between 1.685 and 2.023 whose right-tail areas are 0.05 and 0.025, respectively. So, $0.025 < P\text{-value} < 0.05$ [Tech: P-value = 0.0313]. Because the P-value is less than the level of significance $\alpha = 0.05$, we reject H_0.

Conclusion: There is sufficient evidence at the $\alpha = 0.05$ level of significance to conclude that the mean collision claim of a 30- to 59-year-old is less than the mean claim of a 20- to 24-year-old. Given that 20- to 24-year-olds tend to claim more for each accident, it makes sense to charge them more for coverage.

17. We first verify the requirements to perform McNemar's test (for dependent population proportions). The samples are dependent because the two questions were asked of the same individual. We can think of the sample as a random sample. The total number of individuals answering favorably for one question while answering unfavorably for the other question is $123 + 70 = 193$, which is greater than 10. Thus, the requirements are met, so we can conduct the test.

The hypotheses are $H_0 : p_{HE} = p_{HA}$ verses $H_1 : p_{HE} \neq p_{HA}$. The test statistic is

$$z_0 = \frac{|f_{12} - f_{21}| - 1}{\sqrt{f_{12} + f_{21}}} = \frac{|123 - 70| - 1}{\sqrt{123 + 70}} \approx 3.74$$

Classical approach: The critical value for $\alpha = 0.05$ is $z_{\alpha/2} = z_{0.025} = 1.96$. Since $z_0 = 3.74 > z_{0.025} = 1.96$ (the test statistic falls in the critical region), we reject H_0.

P-value approach: The P-value is two times the area under the standard normal distribution to the right of the test statistic, $z_0 = 3.74$.

P-value $= 2 \cdot P(x > 3.74)$

$\qquad = 2 \cdot [1 - P(x < 3.74)]$

$\qquad = 2 \cdot (1 - 0.9999)$

$\qquad = 0.0002$

Since P-value $= 0.0002 < \alpha = 0.05$, we reject H_0.

Conclusion: There is sufficient evidence at the $\alpha = 0.05$ level of significance to conclude that there is a difference in the proportions. It seems that the proportion of individuals who are healthy differs from the proportion of individuals who are happy. The data imply that more people are happy even though they are unhealthy.

19. These are matched-pair data because two prices, W_i and T_i, are taken on the same product. We measure differences as $d_i = W_i - T_i$. We are told in the problem to assume the conditions for conducting the test are satisfied.

Product	1	2	3	4	5
W_i	2.94	3.16	4.74	2.84	1.72
T_i	2.94	3.79	4.74	2.49	1.97
d_i	0	−0.63	0	0.35	−0.25

Product	6	7	8	9	10
W_i	1.72	3.77	7.44	3.77	5.98
T_i	1.72	3.99	7.99	3.64	5.99
d_i	0	−0.22	−0.55	0.13	−0.01

We compute the mean and standard deviation of the differences and obtain $\bar{d} = -0.118$ and $s_d \approx 0.3001$, rounded to four decimal places.

The hypotheses are $H_0 : \mu_d = 0$ versus $H_1 : \mu_d \neq 0$. The level of significance is $\alpha = 0.05$. The test statistic is

$$t_0 = \frac{\bar{d}}{\frac{s_d}{\sqrt{n}}} = \frac{-0.118}{\frac{0.3001}{\sqrt{10}}} \approx -1.243.$$

Classical approach: Since this is a two-tailed test with 9 degrees of freedom, the critical value is $\pm t_{\alpha/2} = \pm t_{0.025} = \pm 2.262$. Since $t_0 = -1.243$ falls between $-t_{0.025} = -2.262$ and $t_{0.025} = 2.262$ (the test statistic does not fall within the critical region), we do not reject H_0.

P-value approach: The P-value for this two-tailed test is the area under the t-distribution with 9 degrees of freedom to the left of the test statistic $t_0 = -1.243$ plus the area to the right of 1.243. From the t-distribution table in the row corresponding to 9 degrees of freedom, 1.243 falls between 1.100 and 1.383 whose right-tail areas are 0.15 and 0.10, respectively. We must double these values in order to get the total area in both tails: 0.30 and 0.20. So, $0.30 > P$-value > 0.30. [Tech: P-value $= 0.2451$]. Because the P-value is greater than the level of significance $\alpha = 0.05$, we do not reject H_0.

Conclusion: There is not sufficient evidence at the $\alpha = 0.05$ level of significance to conclude that there is a difference in the pricing of health and beauty products at Wal-Mart and Target.

---

21. The hypotheses were $H_0 : p_{>100K} = p_{<100K}$ versus $H_1 : p_{>100K} \neq p_{<100K}$. We have $x_{>100K} = 710$, $n_{>100K} = 1205$, $x_{<100K} = 695$, and $n_{<100K} = 1310$, so $\hat{p}_{>100K} = \frac{x_{>100K}}{n_{>100K}} = \frac{710}{1205} \approx 0.5892$ and

$\hat{p}_{<100K} = \frac{x_{<100K}}{n_{<100K}} = \frac{695}{1310} = 0.5305$. For a 95% confidence interval we use $\pm z_{0.025} = \pm 1.96$. Then:

Lower Bound: $(\hat{p}_{>100K} - \hat{p}_{<100K}) - z_{\alpha/2} \cdot \sqrt{\frac{\hat{p}_{>100K}(1-\hat{p}_{>100K})}{n_{>100K}} + \frac{\hat{p}_{<100K}(1-\hat{p}_{<100K})}{n_{<100K}}}$

$= (0.5892 - 0.5305) - 1.96 \cdot \sqrt{\frac{0.5892(1-0.5892)}{1205} + \frac{0.5305(1-0.5305)}{1310}} \approx 0.020$

Upper Bound: $(\hat{p}_{>100K} - \hat{p}_{<100K}) + z_{\alpha/2} \cdot \sqrt{\frac{\hat{p}_{>100K}(1-\hat{p}_{>100K})}{n_{>100K}} + \frac{\hat{p}_{<100K}(1-\hat{p}_{<100K})}{n_{<100K}}}$

$= (0.5892 - 0.5305) + 1.96 \cdot \sqrt{\frac{0.5892(1-0.5892)}{1205} + \frac{0.5305(1-0.5305)}{1310}} \approx 0.097$

We are 95% confident that the difference in the proportion of individuals who believe it is morally wrong for unwed women to have children (for individuals who earned more that $100,000 versus individuals who earned less that $100,000) is between 0.020 and 0.097. Because the confidence interval does not include 0, there is sufficient evidence at the $\alpha = 0.05$ level of significance to conclude that there is a difference in the proportions. It appears that a higher proportion of individuals who earn over $100,000 per year feel it is morally wrong for unwed women to have children.

23. (a) Yes, the both plots indicate that the data could come from populations that are normally distributed since all of the points fall within the boundaries of the probability plots.

(b) The five number summaries follow:
Value: 6.02, 6.09, 8.735, 9.54, 12.53
[Minitab: 6.02, 6.073, 8.735, 10.083, 12.53]

$$\text{Lower fence} = Q_1 - 1.5(Q_3 - Q_1)$$
$$= 6.09 - 1.5(9.54 - 6.09)$$
$$= 0.915 \quad [\text{Tech}: -1.063]$$

$$\text{Upper fence} = Q_3 + 1.5(Q_3 - Q_1)$$
$$= 9.54 + 1.5(9.54 - 6.09)$$
$$= 14.715 \quad [\text{Tech}: 17.966]$$

There are no outliers in the value funds data.

Growth: 6.93, 7.57, 8.365, 8.63, 9.33
[Minitab: 6.93, 7.56, 8.365, 8.728, 9.33]

$$\text{Lower fence} = Q_1 - 1.5(Q_3 - Q_1)$$
$$= 7.57 - 1.5(8.63 - 7.57)$$
$$= 5.98 \quad [\text{Tech}: 5.808]$$

$$\text{Upper fence} = Q_3 - 1.5(Q_3 - Q_1)$$
$$= 8.63 + 1.5(8.63 - 7.57)$$
$$= 10.22 \quad [\text{Tech}: 10.480]$$

There are no outliers in the growth funds data.

Three-year Rate of Return

So, there are no outliers in either data set.

(c) We verify the requirements to perform Welch's *t*-test. Each sample is a simple random sample. The samples were obtained independently. The normal probability plot indicates that the data could come from a population that is normal, with no outliers. Thus, the requirements are met, so we can conduct the test. The hypotheses are $H_0 : \mu_v = \mu_g$ versus $H_1 : \mu_v \neq \mu_g$. The level of significance is $\alpha = 0.05$. The sample statistics for the data are $\bar{x}_v = 8.653$, $s_v \approx 2.2952$, and $n_v = 10$, $\bar{x}_g = 8.206$, $s_g \approx 0.7256$, and $n_g = 10$. Since the two sample sizes are equal $n_v = n_g = 10$, we use $n_v - 1 = 10 - 1 = 9$ degrees of freedom. The critical *t*-values are $\pm t_{0.025} = \pm 2.262$.

The test statistic is
$$t_0 = \frac{(\bar{x}_v - \bar{x}_g) - (\mu_v - \mu_g)}{\sqrt{\dfrac{s_v^2}{n_v} + \dfrac{s_g^2}{n_g}}}$$
$$= \frac{(8.653 - 8.206) - 0}{\sqrt{\dfrac{2.2952^2}{10} + \dfrac{0.7256^2}{10}}} \approx 0.587$$

Classical approach: Since this is a two-tailed test with 9 degrees of freedom, the critical values are $\pm t_{0.025} = \pm 2.262$. Since $t_0 = 0.587$ is between $-t_{0.025} = -2.262$ and $t_{0.025} = 2.262$ (the test statistic does not fall within the critical region), we do not reject H_0.

P-value approach: The *P*-value for this two-tailed test is the area under the *t*-distribution with 9 degrees of freedom to the right of $t_0 = 0.587$, plus the area to the left of -0.587. From the *t*-distribution table in the row corresponding to 9 degrees of freedom, 0.587 falls to the left of 0.703 whose right-tail area is 0.25. We must double this value in order to get the total area in both tails: 0.50. Thus, *P*-value > 0.50 [Tech: *P*-value $= 0.5872$]. Because the *P*-value is larger than the level of significance $\alpha = 0.05$, we do not reject H_0.

Conclusion: There is not sufficient evidence at the $\alpha = 0.05$ level of significance to conclude that the mean rate of return of value funds is different from the mean rate of return of growth funds.

(d) The hypotheses are $H_0 : \sigma_v = \sigma_g$ versus $H_1 : \sigma_v \neq \sigma_g$. The test statistic is

$$F_0 = \frac{s_v^2}{s_g^2} = \frac{2.2952^2}{0.7256^2} \approx 10.01.$$

Classical approach: This is a two-tailed test with $n_v = 10$, $n_g = 10$, and $\alpha = 0.05$. The critical values are $F_{0.025,9,9} = 4.03$ and

$$F_{0.975,9,9} = \frac{1}{F_{0.025,9,9}} = \frac{1}{4.03} \approx 0.25. \text{ Since}$$

$F_0 = 10.01 > F_{0.025,9,9} = 4.03$ (the test statistic falls in the critical region), we reject H_0.

P-value approach: Using technology, we find P-value ≈ 0.0021. Since this P-value is less than the $\alpha = 0.05$ level of significance, we reject H_0.

Conclusion: There is sufficient evidence at the $\alpha = 0.05$ level of significance to conclude that the standard deviation of the value funds is different from the standard deviation of the growth funds. In particular, the value funds appear to be more volatile than the growth funds.

(e) Answers may vary. Since there does not appear to be a difference in mean rates of return, risk should be used as a determining factor. Using risk as the criteria for investing, the growth fund appears to be the better option.

Chapter 11 Review Exercises

1. This is dependent since the members of the two samples are matched by diagnosis.

2. This is independent because the subjects are randomly selected from two distinct populations.

3. (a) $F_{0.05,8,9} = 3.23$

(b) The level of significance is $\alpha = 0.05$, so $\frac{\alpha}{2} = 0.025$. Thus, $F_{0.025,10,5} = 6.62$ and

$$F_{0.975,10,5} = \frac{1}{F_{0.025,5,10}} = \frac{1}{4.24} \approx 0.24.$$

4. (a) We compute each difference, $d_i = X_i - Y_i$.

Obs	1	2	3	4	5	6
X_i	34.2	32.1	39.5	41.8	45.1	38.4
Y_i	34.9	31.5	39.5	41.9	45.5	38.8
d_i	−0.7	0.6	0	−0.1	−0.4	−0.4

(b) Using technology, $\bar{d} \approx -0.1667$ and $s_d \approx 0.4502$, rounded to four decimal places.

(c) The hypotheses are $H_0 : \mu_d = 0$ versus $H_1 : \mu_d < 0$. The level of significance is $\alpha = 0.05$. The test statistic is

$$t_0 = \frac{\bar{d}}{\frac{s_d}{\sqrt{n}}} = \frac{-0.1667}{\frac{0.4502}{\sqrt{6}}} \approx -0.907.$$

Classical approach: Since this is a left-tailed test with 5 degrees of freedom, the critical value is $-t_{0.05} = -2.015$. Since the test statistic $t_0 = -0.907 > -t_{0.05} = -2.015$ (the test statistic does not fall within the critical region), we do not reject H_0.

P-value approach: The P-value for this left-tailed test is the area under the t-distribution with 5 degrees of freedom to the left of the test statistic $t_0 \approx -0.907$, which is equivalent to the area to the right of 0.907. From the t-distribution table in the row corresponding to 5 degrees of freedom, 0.907 falls between 0.727 and 0.920 whose right-tail areas are 0.25 and 0.20, respectively. So, $0.20 < P$-value < 0.25 [Tech: P-value = 0.2030]. Because the P-value is greater than the $\alpha = 0.05$ level of significance, we do not reject H_0.

Conclusion: There is not sufficient evidence at the $\alpha = 0.05$ level of significance to conclude that the mean difference is less than zero.

(d) For 98% confidence, we use $\alpha = 0.02$. With 5 degrees of freedom, we have $t_{\alpha/2} = t_{0.01} = 3.365$. Then:

$$\text{Lower bound} = \bar{d} - t_{0.01} \cdot \frac{s_d}{\sqrt{n}}$$

$$= -0.1667 - 3.365 \cdot \frac{0.4502}{\sqrt{6}}$$

$$\approx -0.79$$

$$\text{Upper bound} = \bar{d} + t_{0.01} \cdot \frac{s_d}{\sqrt{n}}$$

$$= -0.167 + 3.365 \cdot \frac{0.450}{\sqrt{6}}$$

$$\approx 0.45$$

We can be 98% confident that the mean difference is between -0.79 and 0.45.

5. (a) $H_0 : \mu_1 = \mu_2$ versus $H_1 : \mu_1 \neq \mu_2$. The level of significance is $\alpha = 0.1$. Since the smaller sample size is $n_2 = 8$, we use $n_2 - 1 = 7$ degrees of freedom. The test statistic is

$$t_0 = \frac{(\bar{x}_1 - \bar{x}_2) - (\mu_1 - \mu_2)}{\sqrt{\frac{s_1^2}{n_1} + \frac{s_2^2}{n_2}}} = \frac{(32.4 - 28.2) - 0}{\sqrt{\frac{4.5^2}{13} + \frac{3.8^2}{8}}}$$

$$\approx 2.290$$

<u>Classical approach</u>: Since this is a two-tailed test with 7 degrees of freedom, the critical values are $\pm t_{0.05} = \pm 1.895$. Since the test statistic $t_0 \approx 2.290$ is to the right of the critical value 1.895 (the test statistic falls within a critical region), we reject H_0.

<u>P-value approach</u>: The P-value for this two-tailed test is the area under the t-distribution with 7 degrees of freedom to the right of $t_0 = 2.290$ plus the area to the left of $-t_0 = -2.290$. From the t-distribution table in the row corresponding to 7 degrees of freedom, 2.290 falls between 1.895 and 2.365 whose right-tail areas are 0.05 and 0.025, respectively. We must double these values in order to get the total area in both tails: 0.10 and 0.05. Thus, $0.05 < P\text{-value} < 0.10$ [Tech: P-value = 0.0351]. Because the P-value is less than the $\alpha = 0.10$ level of significance, we reject H_0.

<u>Conclusion</u>: There is sufficient evidence at the $\alpha = 0.1$ level of significance to conclude that $\mu_1 \neq \mu_2$.

(b) For a 90% confidence interval with df $= 7$, we use $t_{\alpha/2} = t_{0.05} = 1.895$. Then:

Lower bound

$$= (\bar{x}_1 - \bar{x}_2) - t_{\alpha/2} \cdot \sqrt{\frac{s_1^2}{n_1} + \frac{s_2^2}{n_2}}$$

$$= (32.4 - 28.2) - 1.895 \cdot \sqrt{\frac{4.5^2}{13} + \frac{3.8^2}{8}}$$

$$\approx 0.73 \quad [\text{Tech: } 1.01]$$

Upper bound

$$= (\bar{x}_1 - \bar{x}_2) + t_{\alpha/2} \cdot \sqrt{\frac{s_1^2}{n_1} + \frac{s_2^2}{n_2}}$$

$$= (32.4 - 28.2) + 1.895 \cdot \sqrt{\frac{4.5^2}{13} + \frac{3.8^2}{8}}$$

$$\approx 7.67 \quad [\text{Tech: } 7.39]$$

We are 90% confident that the population mean difference is between 0.73 and 7.67. [Tech: between 1.01 and 7.39].

(c) The hypotheses are $H_0 : \sigma_1 = \sigma_2$ versus $H_1 : \sigma_1 \neq \sigma_2$. The test statistic is

$$F_0 = \frac{s_1^2}{s_2^2} = \frac{4.5^2}{3.8^2} \approx 1.40.$$

<u>Classical approach</u>: This is a two-tailed test with $n_1 = 13$, $n_2 = 8$, and $\alpha = 0.05$. The critical values are $F_{0.025,12,7}$, which we approximate by $F_{0.025,10,7} = 4.76$ and

$$F_{0.975,12,7} = \frac{1}{F_{0.025,7,12}} = \frac{1}{3.61} \approx 0.28. \text{ Since}$$

$F_0 = 1.40$ falls between $F_{0.975,12,7} \approx 0.28$ and $F_{0.025,12,7} \approx 4.76$ (the test statistic does not fall in the critical region), we do not reject H_0.

<u>P-value approach</u>: Using technology, we find P-value ≈ 0.6734. Since this P-value is greater than the $\alpha = 0.05$ level of significance, we do not reject H_0.

<u>Conclusion</u>: There is not sufficient evidence at the $\alpha = 0.05$ level of significance to conclude that the standard deviation in population 1 is different from the standard deviation in population 2.

6. (a) $H_0 : \mu_1 = \mu_2$ versus $H_1 : \mu_1 > \mu_2$. The level of significance is $\alpha = 0.01$. Since the smaller sample size is $n_2 = 41$, we use $n_2 - 1 = 40$ degrees of freedom. The test statistic is

$$t_0 = \frac{(\bar{x}_1 - \bar{x}_2) - (\mu_1 - \mu_2)}{\sqrt{\frac{s_1^2}{n_1} + \frac{s_2^2}{n_2}}}$$

$$= \frac{(48.2 - 45.2) - 0}{\sqrt{\frac{8.4^2}{45} + \frac{10.3^2}{41}}} \approx 1.472$$

Classical approach: Since this is a right-tailed test with 40 degrees of freedom, the critical value is $t_{0.01} = 2.423$. Since the test statistic $t_0 \approx 1.472 < t_{0.01} = 2.423$ (the test statistic does not fall within the critical region), we do not reject H_0.

P-value approach: The P-value for this right-tailed test is the area under the t-distribution with 40 degrees of freedom to the right of $t_0 \approx 1.472$. From the t-distribution table in the row corresponding to 40 degrees of freedom, 1.472 falls between 1.303 and 1.684 whose right-tail areas are 0.10 and 0.05, respectively. Thus, $0.10 > $ P-value > 0.05 [Tech: P-value = 0.0726]. Because the P-value is greater than the $\alpha = 0.01$ level of significance, we do not reject H_0.

Conclusion: There is not sufficient evidence at the $\alpha = 0.01$ level of significance to conclude that the mean of population 1 is larger than the mean of population 2.

(b) The hypotheses are $H_0 : \sigma_1 = \sigma_2$ versus $H_1 : \sigma_1 < \sigma_2$. The test statistic is

$$F_0 = \frac{s_1^2}{s_2^2} = \frac{8.4^2}{10.3^2} \approx 0.67.$$

Classical approach: This is a left-tailed test with $n_1 = 45$, $n_2 = 41$, and $\alpha = 0.01$.

The critical value is $F_{0.99, 44, 40} = \frac{1}{F_{0.01, 40, 44}}$

$\approx \frac{1}{F_{0.01, 40, 50}} \approx \frac{1}{2.01} \approx 0.50$. Since

$F_0 = 0.67 > F_{0.99, 44, 40} \approx 0.50$ (the test

statistic does not fall in the critical region), we do not reject H_0.

P-value approach: Using technology, we find P-value ≈ 0.0940. Since this P-value is greater than the $\alpha = 0.01$ level of significance, we do not reject H_0.

Conclusion: There is not sufficient evidence at the $\alpha = 0.01$ level of significance to conclude that the standard deviation in population 1 is less than the standard deviation in population 2.

7. The hypotheses are $H_0 : p_1 = p_2$ versus $H_1 : p_1 \neq p_2$. The level of significance is $\alpha = 0.05$. The two sample estimates are

$\hat{p}_1 = \frac{451}{555} \approx 0.8126$ and $\hat{p}_2 = \frac{510}{600} = 0.85$. The

pooled estimate is $\hat{p} = \frac{451 + 510}{555 + 600} \approx 0.8320$.

The test statistic is

$$z_0 = \frac{\hat{p}_1 - \hat{p}_2}{\sqrt{\hat{p}(1 - \hat{p})}\sqrt{\frac{1}{n_1} + \frac{1}{n_2}}}$$

$$= \frac{0.8126 - 0.85}{\sqrt{0.8320(1 - 0.8320)}\sqrt{\frac{1}{555} + \frac{1}{600}}} \approx -1.70$$

Classical approach: This is a two-tailed test, so the critical values for $\alpha = 0.05$ are $\pm z_{\alpha/2} = \pm z_{0.025} = \pm 1.96$. Since $z_0 = -1.70$ is between $-z_{0.025} = -1.96$ and $z_{0.025} = 1.96$ (the test statistic does not fall in the critical region), we do not reject H_0.

P-value approach: The P-value is two times the area under the standard normal distribution to the left of the test statistic, $z_0 = -1.70$.

P-value $= 2 \cdot P(z \leq -1.70) = 2(0.0446)$

$= 0.0892$ [Tech: 0.0895]

Since the P-value is greater than the $\alpha = 0.05$ level of significance, we do not reject H_0.

Conclusion: There is not sufficient evidence at the $\alpha = 0.05$ level of significance to conclude that the proportion in population 1 is different from the proportion in population 2.

8. (a) Since the same individual is used for both measurements, the sampling method is dependent.

(b) We measure differences as $d_i = A_i - H_i$.

Student	1	2	3	4	5
Height, H_i	59.5	69	77	59.5	74.5
Arm span, A_i	62	65.5	76	63	74
$d_i = A_i - H_i$	2.5	−3.5	−1	3.5	−0.5

Student	6	7	8	9	10
Height, H_i	63	61.5	67.5	73	69
Arm span, A_i	66	61	69	70	71
$d_i = A_i - H_i$	3	−0.5	1.5	−3	2

We compute the mean and standard deviation of the differences and obtain $\bar{d} = 0.4$ inches and $s_d \approx 2.4698$ inches, rounded to four decimal places. The hypotheses are $H_0 : \mu_d = 0$ versus $H_1 : \mu_d \neq 0$. The level of significance is $\alpha = 0.05$. The test statistic is

$$t_0 = \frac{\bar{d}}{\frac{s_d}{\sqrt{n}}} = \frac{0.4}{\frac{2.4698}{\sqrt{10}}} \approx 0.512 .$$

<u>Classical approach</u>: Since this is a two-tailed test with 9 degrees of freedom, the critical values are $\pm t_{0.025} = \pm 2.262$. Since $t_0 = 0.512$ falls between $-t_{0.025} = -2.262$ and $t_{0.025} = 2.262$ (the test statistic does not fall within the critical regions), we do not reject H_0.

<u>P-value approach</u>: The P-value for this two-tailed test is the area under the t-distribution with 9 degrees of freedom to the right of the test statistic $t_0 = 0.512$ plus the area to the left of $-t_0 = -0.512$. From the t-distribution table in the row corresponding to 9 degrees of freedom, 0.512 falls to the left of 0.703 whose right-tail area is 0.25. We must double this value in order to get the total area in both tails: 0.50. So, P-value > 0.50 [Tech: P-value = 0.6209]. Because the P-value is greater than the $\alpha = 0.05$ level of significance, we do not reject H_0.

<u>Conclusion</u>: There is not sufficient evidence at the $\alpha = 0.05$ level of significance to conclude that an individual's arm span is different from the individual's height. That is, the sample evidence does not contradict the belief that arm span and height are the same.

9. (a) The sampling method is independent since the cars selected for the McDonald's sample has no bearing on the cars chosen for the Wendy's sample.

(b) The hypotheses are $H_0 : \mu_{McD} = \mu_W$ versus $H_1 : \mu_{McD} \neq \mu_W$. We compute the means and standard deviations, rounded to four decimal places when necessary, of both samples and obtain $\bar{x}_{McD} = 133.601$, $s_{McD} \approx 39.6110$, $n_{McD} = 30$, and $\bar{x}_W \approx 219.1444$, $s_W \approx 102.8459$, and $n_W = 27$. Since the smaller sample size is $n_W = 27$, we use $n_W - 1 = 26$ degrees of freedom.

The test statistic is

$$t_0 = \frac{(\bar{x}_{McD} - \bar{x}_W) - (\mu_{McD} - \mu_W)}{\sqrt{\frac{s_{McD}^2}{n_{McD}} + \frac{s_W^2}{n_W}}}$$

$$= \frac{(133.601 - 219.1444) - 0}{\sqrt{\frac{39.6110^2}{30} + \frac{102.8459^2}{27}}} \approx -4.059$$

<u>Classical approach</u>: Since this is a two-tailed test with 26 degrees of freedom, the critical values are $\pm t_{0.05} = \pm 1.706$. Since the test statistic $t_0 \approx -4.059$ is to the left of -1.706 (the test statistic falls within a critical region), we reject H_0.

<u>P-value approach</u>: The P-value for this two-tailed test is the area under the t-distribution with 26 degrees of freedom to the left of $t_0 \approx -4.059$ plus the area to the right of 4.059. From the t-distribution table in the row corresponding to 26 degrees of freedom, 4.059 falls to the right of 3.707 whose right-tail area is 0.0005. We must double this value in order to get the total area in both tails: 0.001. So, P-value < 0.001 [Tech: P-value = 0.0003]. Since the P-value is less than $\alpha = 0.1$, we reject H_0.

<u>Conclusion</u>: There is sufficient evidence at the $\alpha = 0.1$ level of significance to conclude that the population mean wait times are different for McDonald's and Wendy's drive-through windows.

(c) We compute the five-number summary for each restaurant:

McDonald's: 71.37, 111.84, 127.13, 147.28, 246.59 [Minitab: 71.37, 109.14, 127.13, 148.23, 246.59]

Wendy's: 71.02, 133.56, 190.91, 281.90, 471.62 [Minitab: 71.0, 133.6, 190.9, 281.9, 471.6]

The boxplots are shown to the right. The boxplots support the results from part (b). Based on the boxplot, it would appear to be the case that the wait time at McDonald's is less than the wait time at Wendy's drive-through windows.

---

10. (a) We first verify the requirements to perform the hypothesis test: (1) Each sample can be thought of as a simple random sample; (2) We have $x_{\exp} = 27$, $n_{\exp} = 696$, $x_2 = 49$, and $n_2 = 678$, so

$$\hat{p}_{\exp} = \frac{27}{696} \approx 0.0388 \text{ and } \hat{p}_2 = \frac{49}{678} \approx 0.0723.$$ So, $n_{\exp}\hat{p}_{\exp}\left(1 - \hat{p}_{\exp}\right) = 696(0.0388)(1 - 0.0388) \approx 26 \geq 10$

and $n_{\text{control}}\hat{p}_{\text{control}}\left(1 - \hat{p}_{\text{control}}\right) = 678(0.0723)(1 - 0.0723) \approx 45 \geq 10$; and (3) Each sample is less than 5% of the population. Thus, the requirements are met, and we can conduct the test. The hypotheses are $H_0 : p_{\exp} = p_{\text{control}}$ versus $H_1 : p_{\exp} < p_{\text{control}}$. From before, the two sample estimates are $\hat{p}_{\exp} \approx 0.0388$

and $\hat{p}_{\text{control}} \approx 0.0723$. The pooled estimate is $\hat{p} = \dfrac{x_{\exp} + x_{\text{control}}}{n_{\exp} + n_{\text{control}}} = \dfrac{27 + 49}{696 + 678} \approx 0.0553$. The test statistic

is $z_0 = \dfrac{\hat{p}_{\exp} - \hat{p}_{\text{control}}}{\sqrt{\hat{p}(1 - \hat{p})}\sqrt{\dfrac{1}{n_{\exp}} + \dfrac{1}{n_{\text{control}}}}} = \dfrac{0.0388 - 0.0723}{\sqrt{0.0553(1 - 0.0553)}\sqrt{\dfrac{1}{696} + \dfrac{1}{678}}} \approx -2.71$.

<u>Classical approach:</u> This is a left-tailed test, so the critical value is $-z_{0.01} = -2.33$. Since $z_0 = -2.71 < -z_{0.01} = -2.33$ (the test statistic falls in the critical region), we reject H_0.

<u>P-value approach:</u> P-value $= P(z_0 \leq -2.71) = 0.0034$ [Tech: 0.0033]. Since this P-value is less than the $\alpha = 0.01$ level of significance, we reject H_0.

<u>Conclusion:</u> There is sufficient evidence at the $\alpha = 0.01$ level of significance to conclude that a lower proportion of women in the experimental group experienced a bone fracture than in the control group.

(b) For a 95% confidence interval we use $\pm z_{0.025} = \pm 1.96$. Then:

$$\text{Lower bound} = (\hat{p}_{\exp} - \hat{p}_{\text{control}}) - z_{\alpha/2} \cdot \sqrt{\dfrac{\hat{p}_{\exp}(1 - \hat{p}_{\exp})}{n_{\exp}} + \dfrac{\hat{p}_{\text{control}}(1 - \hat{p}_{\text{control}})}{n_{\text{control}}}}$$

$$= (0.0388 - 0.0723) - 1.96 \cdot \sqrt{\dfrac{0.0388(1 - 0.0388)}{696} + \dfrac{0.0723(1 - 0.0723)}{678}} \approx -0.06$$

$$\text{Upper bound} = (\hat{p}_{\exp} - \hat{p}_{\text{control}}) + z_{\alpha/2} \cdot \sqrt{\dfrac{\hat{p}_{\exp}(1 - \hat{p}_{\exp})}{n_{\exp}} + \dfrac{\hat{p}_{\text{control}}(1 - \hat{p}_{\text{control}})}{n_{\text{control}}}}$$

$$= (0.0388 - 0.0723) + 1.96 \cdot \sqrt{\dfrac{0.0388(1 - 0.0388)}{696} + \dfrac{0.0723(1 - 0.0723)}{678}} \approx -0.01$$

We can be 95% confident that the difference in the proportion of women who experienced a bone fracture between the experimental and control group is between -0.06 and -0.01.

(c) This is a completely randomized design with two treatments: 5 mg of Actonel versus the placebo.

(d) A double-blind experiment is one in which neither the subject nor the individual administering the treatment knows to which group (experimental or control) the subject belongs.

11. We first verify the requirements to perform McNemar's test. The samples are dependent because the different question were asked of the same individual. The sample is a random sample. The total number of individuals for which one response was favorable while the other was not favorable is $65 + 45 = 110$, which is greater than 10. Thus, the requirements are met, so we can conduct the test.

The hypotheses are $H_0 : p_J = p_V$ verses $H_1 : p_J \neq p_V$. The test statistic is

$$z_0 = \frac{|f_{12} - f_{21}| - 1}{\sqrt{f_{12} + f_{21}}} = \frac{|65 - 45| - 1}{\sqrt{65 + 45}} \approx 1.81$$

Classical approach: The critical value for $\alpha = 0.05$ is $z_{\alpha/2} = z_{0.025} = 1.96$. Since $z_0 = 1.81 < z_{0.025} = 1.96$ (the test statistic does not fall within the critical region), we do not reject H_0.

P-value approach: The P-value is two times the area under the standard normal distribution to the right of the test statistic, $z_0 = 1.81$.

$$\begin{aligned} P\text{-value} &= 2 \cdot P(x > 1.81) \\ &= 2 \cdot [1 - P(x < 1.81)] \\ &= 2 \cdot (1 - 0.9649) \\ &= 0.0702 \end{aligned}$$

Since this P-value is greater than $\alpha = 0.05$, we do not reject H_0.

Conclusion: There is not sufficient evidence at the $\alpha = 0.05$ level of significance to conclude that there is a difference in the proportions. It seems that the proportion of individuals who feel serving on a jury is a civic duty is the same as the proportion who feel voting is a civic duty.

12. (a) $n = n_1 = n_2$

$$= \left[\hat{p}_1(1 - \hat{p}_1) + \hat{p}_2(1 - \hat{p}_2) \right]\left(\frac{z_{\alpha/2}}{E} \right)^2$$

$$= \left[0.188(1 - 0.188) + 0.205(1 - 0.205) \right]\left(\frac{1.645}{0.02} \right)^2$$

$$\approx 2135.3$$
which we must increase to 2136.

(b) $n = n_1 = n_2 = 0.5\left(\frac{z_{\alpha/2}}{E} \right)^2 = 0.5\left(\frac{1.645}{0.02} \right)^2$

≈ 3382.5
which we must increase to 3383.

13. The hypotheses are $H_0 : \sigma_{McD} = \sigma_W$ versus $H_1 : \sigma_{McD} < \sigma_W$. From problem 9, we know $s_{McD} \approx 39.6110$ and $s_W \approx 102.8459$. The test statistic is $F_0 = \frac{s_{McD}^2}{s_W^2} = \frac{39.6110^2}{102.8459^2} \approx 0.15$.

Classical approach: This is a left-tailed test with $n_{McD} = 30$ and $n_W = 27$. The critical values is $F_{0.95,29,26}$, which we approximate by

$$F_{0.95,25,25} = \frac{1}{F_{0.05,25,25}} = \frac{1}{1.96} \approx 0.51. \text{ Since}$$

$F_0 = 0.15 < F_{0.95,29,26} \approx 0.51$ (the test statistic falls in a critical region), we reject H_0.

P-value approach: Using technology, we find P-value < 0.0001. Since this P-value is less than the $\alpha = 0.05$ level of significance, we reject H_0.

Conclusion: There is sufficient evidence at the $\alpha = 0.05$ level of significance to conclude that the standard deviation in wait time at Wendy's is more than the standard deviation in wait time at McDonald's.

14. From problem 8, we have $\bar{d} = 0.4$ and $s_d \approx 2.4698$. For 95% confidence, we use $\alpha = 0.05$, so with 9 degrees of freedom, we have $t_{\alpha/2} = t_{0.025} = 2.262$. Then:

$$\text{Lower bound} = \bar{d} - t_{0.025} \cdot \frac{s_d}{\sqrt{n}}$$
$$= 0.4 - 2.262 \cdot \frac{2.4698}{\sqrt{10}} \approx -1.37$$

$$\text{Upper bound} = \bar{d} + t_{0.025} \cdot \frac{s_d}{\sqrt{n}}$$
$$= 0.4 + 2.262 \cdot \frac{2.4698}{\sqrt{10}} \approx 2.17$$

We can be 95% confident that the mean difference between height and arm span is between -1.37 and 2.17. The interval contains zero, so we conclude that there is not sufficient evidence at the $\alpha = 0.05$ level of significance to reject the claim that arm span and height are equal.

15. From problem 9, we have $\bar{x}_{McD} = 133.601$, $s_{McD} \approx 39.6110$, $n_{McD} = 30$, $\bar{x}_W \approx 219.1444$, $s_W \approx 102.8459$, and $n_W = 27$. For a 95% confidence interval with 26 degrees of freedom, we use $t_{\alpha/2} = t_{0.025} = 2.056$. Then:

$$\text{Lower bound} = (\bar{x}_{McD} - \bar{x}_W) - t_{\alpha/2} \cdot \sqrt{\frac{s_{McD}^2}{n_{McD}} + \frac{s_W^2}{n_W}} = (133.601 - 219.1444) - 2.056 \cdot \sqrt{\frac{39.6110^2}{30} + \frac{102.8459^2}{27}}$$

$$\approx -128.869 \quad [\text{Tech:} -128.4]$$

$$\text{Upper bound} = (\bar{x}_{McD} - \bar{x}_W) + t_{\alpha/2} \cdot \sqrt{\frac{s_{McD}^2}{n_{McD}} + \frac{s_W^2}{n_W}} = (133.601 - 219.1444) + 2.056 \cdot \sqrt{\frac{39.6110^2}{30} + \frac{102.8459^2}{27}}$$

$$\approx -42.218 \quad [\text{Tech:} -42.67]$$

We can be 95% confident that the mean difference in wait times between McDonald's and Wendy's drive-through windows is between −128.869 and −42.218 seconds [Tech: between −128.4 and −42.67].

Answer will vary. One possibility is that a marketing campaign could be initiated by McDonald's touting the fact that wait times are up to 2 minutes less at McDonald's.

Chapter 11 Test

1. This is independent because the subjects are randomly selected from two distinct populations.

2. This is dependent since the members of the two samples are matched by type of crime committed.

3. (a) We compute $d_i = X_i - Y_i$.

Obs	1	2	3	4	5	6	7
X_i	18.5	21.8	19.4	22.9	18.3	20.2	23.1
Y_i	18.3	22.3	19.2	22.3	18.9	20.7	23.9
d_i	0.2	−0.5	0.2	0.6	−0.6	−0.5	−0.8

(b) Using technology, $\bar{d} = -0.2$ and $s_d \approx 0.5260$, rounded to four decimal places.

(c) The hypotheses are $H_0 : \mu_d = 0$ versus $H_1 : \mu_d \neq 0$. The level of significance is $\alpha = 0.01$. The test statistic is

$$t_0 = \frac{\bar{d}}{\frac{s_d}{\sqrt{n}}} = \frac{-0.2}{\frac{0.5260}{\sqrt{7}}} \approx -1.006 .$$

Classical approach: Since this is a two-tailed test with 6 degrees of freedom, the critical values are $\pm t_{0.005} = \pm 3.707$. Since $t_0 \approx -1.006$ falls between $-t_{0.005} = -3.707$ and $t_{0.005} = 3.707$ (the test statistic does not fall within the critical regions), we do not reject H_0.

P-value approach: The *P*-value for this two-tailed test is the area under the *t*-distribution with 6 degrees of freedom to the left of the test statistic $t_0 = -1.006$ plus the area to the right of $t_0 = 1.006$. From the *t*-distribution table in the row corresponding to 6 degrees of freedom, 1.006 falls between 0.906 and 1.134 whose right-tail areas are 0.20 and 0.15, respectively. We must double these values in order to get the total area in both tails: 0.40 and 0.30. So, $0.30 < P\text{-value} < 0.40$ [Tech: *P*-value = 0.3532]. Because the *P*-value is greater than the $\alpha = 0.01$ level of significance, we do not reject H_0.

Conclusion: There is not sufficient evidence at the $\alpha = 0.01$ level of significance to conclude that the population mean difference is different from zero.

(d) For $\alpha = 0.05$ and 7 degrees of freedom, $t_{\alpha/2} = t_{0.025} = 2.447$. Then:

$$\text{Lower bound} = \bar{d} - t_{0.025} \cdot \frac{s_d}{\sqrt{n}}$$

$$= -0.2 - 2.447 \cdot \frac{0.5260}{\sqrt{7}} \approx -0.69$$

$$\text{Upper bound} = \bar{d} - t_{0.025} \cdot \frac{s_d}{\sqrt{n}}$$

$$= -0.2 + 2.447 \cdot \frac{0.5260}{\sqrt{7}} \approx 0.29$$

We are 95% confident that the population mean difference is between −0.69 and 0.29.

4. (a) $H_0 : \mu_1 = \mu_2$ versus $H_1 : \mu_1 \neq \mu_2$. The level of significance is $\alpha = 0.1$. Since the smaller sample size is $n_1 = 24$, we use $n_1 - 1 = 23$ degrees of freedom. The test statistic is

$$t_0 = \frac{(\bar{x}_1 - \bar{x}_2) - (\mu_1 - \mu_2)}{\sqrt{\dfrac{s_1^2}{n_1} + \dfrac{s_2^2}{n_2}}}$$

$$= \frac{(104.2 - 110.4) - 0}{\sqrt{\dfrac{12.3^2}{24} + \dfrac{8.7^2}{27}}} \approx -2.054$$

Classical approach: Since this is a two-tailed test with 23 degrees of freedom, the critical values are $\pm t_{0.05} = \pm 1.714$. Since the test statistic $t_0 \approx -2.054$ is to the left of the critical value -1.714 (the test statistic falls within a critical region), we reject H_0.

P-value approach: The P-value for this two-tailed test is the area under the t-distribution with 23 degrees of freedom to the left of $t_0 = -2.054$ plus the area to the right of $t_0 = 2.054$. From the t-distribution table in the row corresponding to 23 degrees of freedom, 2.054 falls between 1.714 and 2.069 whose right-tail areas are 0.05 and 0.025, respectively. We must double these values in order to get the total area in both tails: 0.10 and 0.05. Thus, $0.05 < P\text{-value} < 0.10$ [Tech: P-value = 0.0464]. Because the P-value is less than the $\alpha = 0.1$ level of significance, we reject H_0.

Conclusion: There is sufficient evidence at the $\alpha = 0.1$ level of significance to conclude that the means are different.

(b) For a 95% confidence interval with 23 degrees of freedom, we use $t_{\alpha/2} = t_{0.025} = 2.069$. Then:

Lower bound

$$= (\bar{x}_1 - \bar{x}_2) - t_{\alpha/2} \cdot \sqrt{\frac{s_1^2}{n_1} + \frac{s_2^2}{n_2}}$$

$$= (104.2 - 110.4) - 2.069 \cdot \sqrt{\frac{12.3^2}{24} + \frac{8.7^2}{27}}$$

$$\approx -12.44 \quad [\text{Tech}: -12.3]$$

Upper bound:

$$= (\bar{x}_1 - \bar{x}_2) + t_{\alpha/2} \cdot \sqrt{\frac{s_1^2}{n_1} + \frac{s_2^2}{n_2}}$$

$$= (104.2 - 110.4) + 2.069 \cdot \sqrt{\frac{12.3^2}{24} + \frac{8.7^2}{27}}$$

$$\approx 0.04 \quad [\text{Tech}: -0.10]$$

We are 95% confident that the population mean difference is between -12.44 and 0.04 [Tech: between -12.3 and -0.10].

(c) The hypotheses are $H_0 : \sigma_1 = \sigma_2$ versus $H_1 : \sigma_1 > \sigma_2$. The test statistic is

$$F_0 = \frac{s_1^2}{s_2^2} = \frac{12.3^2}{8.7^2} \approx 2.00.$$

Classical approach: This is a right-tailed test with $n_1 = 24$, $n_2 = 27$, and $\alpha = 0.10$. The critical value is $F_{0.10, 23, 26}$, which we approximate by $F_{0.10, 25, 25} = 1.68$. Since $F_0 = 2.00 > F_{0.10, 23, 26} \approx 1.68$ (the test statistic falls in the critical region), we reject H_0.

P-value approach: Using technology, we find P-value ≈ 0.0448. Since this P-value is less than the $\alpha = 0.05$ level of significance, we reject H_0.

Conclusion: There is sufficient evidence at the $\alpha = 0.1$ level of significance to conclude that the standard deviation in population 1 is greater than the standard deviation in population 2.

5. (a) $H_0 : \mu_1 = \mu_2$ versus $H_1 : \mu_1 < \mu_2$. The level of significance is $\alpha = 0.05$. Since the smaller sample size is $n_2 = 8$, we use $n_2 - 1 = 7$ degrees of freedom. The test statistic is

$$t_0 = \frac{(\bar{x}_1 - \bar{x}_2) - (\mu_1 - \mu_2)}{\sqrt{\dfrac{s_1^2}{n_1} + \dfrac{s_2^2}{n_2}}}$$

$$= \frac{(96.6 - 98.3) - 0}{\sqrt{\dfrac{3.2^2}{13} + \dfrac{2.5^2}{8}}} \approx -1.357$$

Classical approach: Since this is a left-tailed test with 7 degrees of freedom, the critical value is $-t_{0.05} = -1.895$. Since $t_0 = -1.375 > -t_{0.05} = -1.895$ (the test statistic does not fall within the critical region), we do not reject H_0.

P-value approach: The P-value for this left-tailed test is the area under the t-distribution with 7 degrees of freedom to the left of $t_0 = -1.375$, which is equivalent to the area to the right of $t_0 = 1.375$. From the t-distribution table in the row corresponding to 7 degrees of freedom, 1.375 falls between 1.119 and 1.415 whose right-tail areas are 0.15 and 0.10, respectively. Thus, $0.10 < P\text{-value} < 0.15$ [Tech: P-value = 0.0959]. Because the P-value is greater than the $\alpha = 0.05$ level of significance, we do not reject H_0.

Conclusion: There is not sufficient evidence at the $\alpha = 0.05$ level of significance to conclude that the mean of population 1 is less than the mean of population 2.

(b) The hypotheses are $H_0 : \sigma_1 = \sigma_2$ versus $H_1 : \sigma_1 \neq \sigma_2$. The test statistic is

$$F_0 = \frac{s_1^2}{s_2^2} = \frac{3.2^2}{2.5^2} \approx 1.64.$$

Classical approach: This is a two-tailed test with $n_1 = 13$, $n_2 = 8$, and $\alpha = 0.05$. The critical values are $F_{0.025,12,7}$, which we approximate by $F_{0.025,10,7} = 4.76$ and

$$F_{0.975,12,7} = \frac{1}{F_{0.025,7,12}} = \frac{1}{3.61} \approx 0.28.$$ Since

$F_0 = 1.64$ is between $F_{0.025,12,7} \approx 4.76$ and $F_{0.975,12,7} \approx 0.28$ (the test statistic does not fall in the critical region), we do not reject H_0.

P-value approach: Using technology, we find $P\text{-value} \approx 0.5243$. Since this P-value is greater than the $\alpha = 0.05$ level of significance, we do not reject H_0.

Conclusion: There is not sufficient evidence at the $\alpha = 0.05$ level of significance to conclude that the standard deviation in population 1 is different from the standard deviation in population 2.

6. The hypotheses are $H_0 : p_1 = p_2$ versus $H_1 : p_1 < p_2$. The level of significance is $\alpha = 0.05$. The two sample estimates are $\hat{p}_1 = \frac{156}{650} = 0.24$ and $\hat{p}_2 = \frac{143}{550} = 0.26$. The pooled estimate is $\hat{p} = \frac{156+143}{650+550} \approx 0.2492$.

The test statistic is

$$z_0 = \frac{\hat{p}_1 - \hat{p}_2}{\sqrt{\hat{p}(1-\hat{p})}\sqrt{\dfrac{1}{n_1} + \dfrac{1}{n_2}}}$$

$$= \frac{0.24 - 0.26}{\sqrt{0.2492(1-0.2492)}\sqrt{\dfrac{1}{650} + \dfrac{1}{550}}} \approx -0.80$$

Classical approach: This is a left-tailed test, so the critical value is $-z_{0.05} = -1.645$. Since $z_0 = -0.80 > -z_{0.05} = -1.645$ (the test statistic does not fall in the critical region), we do not reject H_0.

P-value approach: The P-value is the area under the standard normal distribution to the left of the test statistic, $z_0 = -0.80$. P-value = $P(z \leq -0.80) = 0.2119$ [Tech: 0.2124]. Since the P-value is greater than the $\alpha = 0.05$ level of significance, we do not reject H_0.

Conclusion: There is not sufficient evidence at the $\alpha = 0.05$ level of significance to conclude that the proportion in population 1 is less than the proportion in population 2.

7. (a) Since the dates selected for the Texas sample have no bearing on the dates chosen for the Illinois sample, the testing method is independent.

(b) Because the sample sizes are small, both samples must come from populations that are normally distributed.

(c) We compute the five-number summary for each restaurant:

Texas: 4.11, 4.555, 4.735, 5.12, 5.22 [Minitab: 4.11, 4.508, 4.735, 5.13, 5.22]

Illinois: 4.22, 4.40, 4.505, 4.64, 4.75 [Minitab: 4.22, 4.39, 4.505, 4.6525, 4.75]

The boxplots that follow indicate that rain in Chicago, Illinois has a lower pH than rain in Houston, Texas.

Illinois

Texas

4.0   4.5   5.0

Acidity of Rain (pH)

(d) The hypotheses are $H_0 : \mu_{\text{Texas}} = \mu_{\text{Illinois}}$ versus $H_1 : \mu_{\text{Texas}} \neq \mu_{\text{Illinois}}$. We compute the summary statistics, rounded to four decimal places when necessary: $\bar{x}_{\text{Texas}} = 4.77$, $s_{\text{Texas}} \approx 0.3696$, $n_{\text{Texas}} = 12$, $\bar{x}_{\text{Illinois}} \approx 4.5093$, $s_{\text{Illinois}} \approx 0.1557$, and $n_{\text{Illinois}} = 14$. Since the smaller sample size is $n_{\text{Texas}} = 12$, we use $n_{\text{Texas}} - 1 = 11$ degrees of freedom. The test statistic is

$$t_0 = \frac{(\bar{x}_{\text{Texas}} - \bar{x}_{\text{Illinois}}) - (\mu_{\text{Texas}} - \mu_{\text{Illinois}})}{\sqrt{\frac{s_{\text{Texas}}^2}{n_{\text{Texas}}} + \frac{s_{\text{Illinois}}^2}{n_{\text{Illinois}}}}}$$

$$= \frac{(4.77 - 4.5093) - 0}{\sqrt{\frac{0.3696^2}{12} + \frac{0.1557^2}{14}}} \approx 2.276$$

Classical approach: Since this is a two-tailed test with 11 degrees of freedom, the critical values are $\pm t_{0.025} = \pm 2.201$. Since the test statistic $t_0 \approx 2.276$ is to the right of 2.201 (the test statistic falls within a critical region), we reject H_0.

P-value approach: The P-value for this two-tailed test is the area under the t-distribution with 11 degrees of freedom to the right of $t_0 \approx 2.276$ plus the area to the left of $-t_0 \approx -2.276$. From the t-distribution table in the row corresponding to 11 degrees of freedom, 2.276 falls between 2.201 and 2.328 whose right-tail areas are 0.025 and 0.02, respectively. We must double these values in order to get the total area in both tails: 0.05 and 0.04. So, $0.04 < P\text{-value} < 0.05$ [Tech: P-value = 0.0387]. Since the P-value is less than $\alpha = 0.05$, we reject H_0.

Conclusion: There is sufficient evidence at the $\alpha = 0.05$ level of significance to conclude that the acidity of the rain near Houston, Texas is different from the acidity of rain near Chicago, Illinois.

8. (a) This is a matched-pairs design because the measurements in each sample were taken on the same individuals.

(b) The hypotheses are $H_0 : \mu_d = 0$ versus $H_1 : \mu_d > 0$. The level of significance is $\alpha = 0.05$. The test statistic is

$$t_0 = \frac{\bar{d}}{s_d / \sqrt{n}} = \frac{12.61}{4.90 / \sqrt{110}} \approx 26.991.$$

Classical approach: This is a right-tailed test with 109 degrees of freedom. Since our t-distribution table does not have a row for 109 degrees of freedom, we use 100 degrees of freedom instead, and obtain the critical value $t_{0.05} = 1.660$. Since $t_0 = 26.991 > t_{0.05} = 1.660$ (the test statistic fall within the critical region), we reject H_0.

P-value approach: The P-value for this right-tailed test is the area under the t-distribution with 109 degrees of freedom to the right of the test statistic $t_0 = 26.991$. From the t-distribution table in the row corresponding to 100 degrees of freedom (since our table does not have a row for df = 109), 26.991 falls to the right of 3.390 whose right-tail area is 0.0005. So, $< P\text{-value} < 0.0005$ [Tech: P-value < 0.0001]. Because the P-value is less than the $\alpha = 0.05$ level of significance, we reject H_0.

Conclusion: There is sufficient evidence at the $\alpha = 0.05$ level of significance to conclude that there is a positive BMI reduction 10 months after gastric bypass surgery.

(c) For 90% confidence and df = 100 (since df = 109 not in our table), we use $t_{\alpha/2} = t_{0.05} = 1.660$. Then:

$$\text{Lower bound} = \bar{d} - t_{0.05} \cdot \frac{s_d}{\sqrt{n}}$$

$$= 12.61 - 1.660 \cdot \frac{4.90}{\sqrt{110}}$$

$$\approx 11.834 \quad [\text{Tech: } 11.835]$$

$$\text{Upper bound} = \bar{d} + t_{0.05} \cdot \frac{s_d}{\sqrt{n}}$$

$$= 12.61 + 1.660 \cdot \frac{4.90}{\sqrt{110}}$$

$$\approx 13.386 \quad [\text{Tech: } 13.385]$$

We can be 90% confident that the mean BMI reduction is between 11.83 and 13.39.

9. (a) This is a completely randomized design. The subjects were randomly divided into two groups.

(b) The response variable is whether the subjects get dry mouth or not.

(c) We verify the requirements to perform the hypothesis test: (1) Each sample can be thought of as a simple random sample; (2) We have $x_{exp} = 77$, $n_{exp} = 553$, $x_{control} = 34$, and $n_{control} = 373$, so

$$\hat{p}_{exp} = \frac{x_{exp}}{n_{exp}} = \frac{77}{553} \approx 0.1392 \text{ and}$$

$$\hat{p}_{control} = \frac{x_{control}}{n_{control}} = \frac{34}{373} \approx 0.0912. \text{ So,}$$

$$n_{exp}\hat{p}_{exp}\left(1 - \hat{p}_{exp}\right) = 553(0.1392)(1 - 0.1392)$$

$\approx 66 \geq 10$ and $n_{control}\hat{p}_{control}\left(1 - \hat{p}_{control}\right) =$

$373(0.0912)(1 - 0.0912) \approx 31 \geq 10$; and (3) Each sample is less than 5% of the population. Thus, the requirements are met, and we can conduct the test. The hypotheses are $H_0 : p_{exp} = p_{control}$ and $H_1 : p_{exp} > p_{control}$. From before, the two sample estimates are $\hat{p}_{exp} \approx 0.1392$ and $\hat{p}_{control} \approx 0.0912$. The pooled estimate is

$$\hat{p} = \frac{x_{exp} + x_{control}}{n_{exp} + n_{control}} = \frac{77 + 34}{553 + 373} \approx 0.1199.$$

The test statistic is

$$z_0 = \frac{\hat{p}_{exp} - \hat{p}_{control}}{\sqrt{\hat{p}(1 - \hat{p})}\sqrt{\dfrac{1}{n_{exp}} + \dfrac{1}{n_{control}}}}$$

$$= \frac{0.1392 - 0.0912}{\sqrt{0.1199(1 - 0.1199)}\sqrt{\dfrac{1}{553} + \dfrac{1}{373}}}$$

≈ 2.21

Classical approach: This is a right-tailed test, so the critical value is $z_{0.05} = 1.645$. Since $z_0 = 2.21 > z_{0.05} = 1.645$ (the test statistic falls in the critical region), we reject H_0.

P-value approach: The P-value is the area under the standard normal curve to the right of $z_0 = 2.21$:

$$P\text{-value} = P(z_0 \geq 2.21)$$
$$= 1 - P(z_0 < 2.21)$$
$$= 1 - 0.9864$$
$$= 0.0136$$

Since this P-value is less than the $\alpha = 0.05$ level of significance, we reject H_0.

Conclusion: There is sufficient evidence at the $\alpha = 0.05$ level of significance to conclude that a higher proportion of subjects in the experimental group experienced dry mouth than in the control group.

---

10. We have $x_F = 644$, $n_F = 2800$, $x_M = 840$, and $n_M = 2800$, so $\hat{p}_F = \dfrac{x_F}{n_F} = \dfrac{644}{2800} = 0.23$ and

$\hat{p}_M = \dfrac{x_M}{n_M} = \dfrac{840}{2800} = 0.3$. The hypotheses are $H_0 : p_M = p_F$ and $H_1 : p_M \neq p_F$. For a 90% confidence interval we use $\pm z_{0.05} = \pm 1.645$. Then:

Lower bound:

$$(\hat{p}_M - \hat{p}_F) - z_{\alpha/2} \cdot \sqrt{\frac{\hat{p}_M(1 - \hat{p}_M)}{n_M} + \frac{\hat{p}_F(1 - \hat{p}_F)}{n_F}} = (0.3 - 0.23) - 1.645 \cdot \sqrt{\frac{0.3(1 - 0.3)}{2800} + \frac{0.23(1 - 0.23)}{2800}} \approx 0.051$$

Upper bound:

$$(\hat{p}_M - \hat{p}_F) + z_{\alpha/2} \cdot \sqrt{\frac{\hat{p}_M(1 - \hat{p}_M)}{n_M} + \frac{\hat{p}_F(1 - \hat{p}_F)}{n_F}} = (0.3 - 0.23) + 1.645 \cdot \sqrt{\frac{0.3(1 - 0.3)}{2800} + \frac{0.23(1 - 0.23)}{2800}} \approx 0.089$$

We are 90% confident that the difference in the proportions for males and females for which hypnosis led to quitting smoking is between 0.051 and 0.089. Because the confidence interval does not include 0, we reject the null hypothesis. There is sufficient evidence at the $\alpha = 0.1$ level of significance to conclude that the proportion of males and females for which hypnosis led to quitting smoking is different.

11. We first verify the requirements to perform McNemar's test. The samples are dependent because the different questions were posed to the same individual. The sample is a random sample. The total number of individuals for which one response was favorable while the other was not favorable is $2{,}549 + 11{,}685 = 14{,}234$, which is greater than 10. Thus, the requirements are met, so we can conduct the test.

The hypotheses are $H_0 : p_A = p_{DP}$ verses $H_1 : p_A \neq p_{DP}$. The test statistic is

$$z_0 = \frac{|f_{12} - f_{21}| - 1}{\sqrt{f_{12} + f_{21}}} = \frac{|2{,}549 - 11{,}685| - 1}{\sqrt{2{,}549 + 11{,}685}} \approx 76.6$$

Classical approach: The critical value for $\alpha = 0.01$ is $z_{\alpha/2} = z_{0.005} = 2.575$. Since $z_0 = 76.6 > z_{0.005} = 2.575$ (the test statistic falls within the critical region), we reject H_0.

P-value approach: The P-value is two times the area under the standard normal distribution to the right of the test statistic, $z_0 = 76.6$. The test statistic is so large that P-value < 0.0001.
Since this P-value is less than $\alpha = 0.01$, we reject H_0.

Conclusion: There is sufficient evidence at the $\alpha = 0.01$ level of significance to conclude that the proportion of individuals who favor the death penalty is different from the proportion who favor abortion.

12. (a) $n = n_1 = n_2$

$$= \left[\hat{p}_1 (1 - \hat{p}_1) + \hat{p}_2 (1 - \hat{p}_2) \right] \left(\frac{z_{\alpha/2}}{E} \right)^2$$

$$= \left[0.322(1 - 0.322) + 0.111(1 - 0.111) \right] \left(\frac{1.96}{0.04} \right)^2$$

$$\approx 761.1$$

which we must increase to 762.

(b) $n = n_1 = n_2$

$$= 0.5 \left(\frac{z_{\alpha/2}}{E} \right)^2 = 0.5 \left(\frac{1.96}{0.04} \right)^2 = 1200.5$$

which we must increase to 1201.

13. We have $n_p = 72$, $\bar{x}_p = 1.9$, $s_p = 0.22$, $n_n = 75$, $\bar{x}_n = 0.8$, and $s_n = 0.21$. The smaller sample size is $n_p = 72$, so we have $n_p - 1 = 71$ degrees of freedom. Since our t-distributions table does not have a row for df $= 71$, we will use df $= 70$, For a 95% confidence interval with 70 degrees of freedom, we use $t_{\alpha/2} = t_{0.025} = 1.994$. Then:

Lower bound

$$= (\bar{x}_p - \bar{x}_n) - t_{\alpha/2} \cdot \sqrt{\frac{s_p^2}{n_p} + \frac{s_n^2}{n_n}}$$

$$= (1.9 - 0.8) - 1.994 \cdot \sqrt{\frac{0.22^2}{72} + \frac{0.21^2}{75}}$$

$$\approx 1.03$$

Upper bound

$$= (\bar{x}_p - \bar{x}_n) + t_{\alpha/2} \cdot \sqrt{\frac{s_p^2}{n_p} + \frac{s_n^2}{n_n}}$$

$$= (1.9 - 0.8) + 1.994 \cdot \sqrt{\frac{0.22^2}{72} + \frac{0.21^2}{75}}$$

$$\approx 1.17$$

We can be 95% confident that the mean difference in weight of the placebo group versus the Naltrexone group is between 1.03 and 1.17 pounds. Because the confidence interval does not contain 0, we reject H_0.

There is sufficient evidence to conclude that Naltrexone is effective in preventing weight gain among individuals who quit smoking. Answers will vary regarding practical significance, but one must ask, "Do I want to take a drug so that I can keep about 1 pound off ?" Probably not.

14. The hypotheses are $H_0 : \sigma_{\text{Texas}} = \sigma_{\text{Illinois}}$ versus $H_1 : \sigma_{\text{Texas}} \neq \sigma_{\text{Illinois}}$. From problem 7, we know $s_{\text{Texas}} \approx 0.3696$, $n_{\text{Texas}} = 12$, $s_{\text{Illinois}} \approx 0.1557$, and $n_{\text{Illinois}} = 14$. The test statistic is $F_0 = \frac{s_{\text{Texas}}^2}{s_{\text{Illinois}}^2} = \frac{0.3696^2}{0.1557^2} \approx 5.63$.

Classical approach: This is a two-tailed test with $n_{\text{Texas}} = 12$ and $n_{\text{Illinois}} = 14$. The critical values are $F_{0.025, 12, 14} \approx F_{0.025, 12, 15} = 2.96$ and

$$F_{0.975, 12, 14} = \frac{1}{F_{0.025, 14, 12}} \approx \frac{1}{F_{0.025, 15, 12}} = \frac{1}{3.18} \approx 0.31.$$

Since $F_0 = 5.63 > F_{0.025,12,14} \approx 2.96$ (the test statistic falls in a critical region), we reject H_0.

<u>*P*-value approach</u>: Using technology, we find *P*-value ≈ 0.0044. Since this *P*-value is less than the $\alpha = 0.05$ level of significance, we reject H_0.

<u>Conclusion:</u> There is sufficient evidence at the $\alpha = 0.05$ level of significance to conclude that the standard deviation in the acidity of rain near Houston, Texas is different from the standard deviation in the acidity of rain near Chicago, Illinois.

Chapter 12
Inference on Categorical Data

12.1 Goodness of Fit Test

1. *Goodness of Fit* is a good choice because the procedures are for testing whether sample data are a good fit with a hypothesized distribution.

3. To conduct a goodness of fit test, the sample data (obtained by random sampling) must satisfy two requirements: (1) all expected frequencies must be greater than or equal to 1, and (2) no more than 20% of the expected frequencies should be less than 5.

5. Each expected count is $n \cdot p_i$ where $n = 500$. This gives expected counts of $500(0.2) = 100$, $500(0.1) = 50$, $500(0.45) = 225$, and $500(0.25) = 125$.

p_i	0.2	0.1	0.45	0.25
E_i	100	50	225	125

7. (a)

O_i	E	$(O_i - E_i)^2$	$\dfrac{(O_i - E_i)^2}{E_i}$
30	25	25	1
20	25	25	1
28	25	9	0.36
22	25	9	0.36
		$\chi_0^2 =$	2.72

(b) df = 4 − 1 = 3

(c) With 3 degrees of freedom, the critical value is $\chi_{0.05}^2 = 7.815$.

(d) The test statistic is not in the (right-tailed) critical region (that is, $\chi_0^2 < \chi_{0.05}^2$), so we do not reject H_0. There is not sufficient evidence at the 5% level of significance to conclude that any one of the proportions is different from the others.

9. (a)

O_i	E_i	$(O_i - E_i)^2$	$\dfrac{(O_i - E_i)^2}{E_i}$
1	1.6	0.36	0.225
38	25.6	153.76	6.006
132	153.6	466.56	3.038
440	409.6	924.16	2.256
389	409.6	424.36	1.036
		$\chi_0^2 =$	12.561

(b) df = 5 − 1 = 4

(c) With 4 degress of freedom, the critical value is $\chi_{0.05}^2 = 9.488$.

(d) The test statistic is in the (right-tailed) critical region (that is, $\chi_0^2 > \chi_{0.05}^2$), so we reject H_0. There is sufficient evidence at the 5% level of significance to conclude that the random variable X is not binomial with $n = 4$ and $p = 0.8$.

11. Using $\alpha = 0.05$, we want to test
H_0: The color of plain M&Ms follows the manufacturer's stated distribution
vs.
H_1: The color of plain M&Ms follows a different distribution.
There are 400 candies in the bag. To determine the expected counts, multiply 400 by the given percentage for each color. For example, the expected count of brown M&Ms would be $400(0.13) = 52$. We summarize the observed and expected counts in the following table:

	O_i	E_i	$(O_i - E_i)^2$	$\dfrac{(O_i - E_i)^2}{E_i}$
Brown	61	52	81	1.558
Yellow	64	56	64	1.143
Red	54	52	4	0.077
Blue	61	96	1225	12.760
Orange	96	80	256	3.2
Green	64	64	0	0
			$\chi_0^2 =$	18.738

Since all the expected cell counts are greater than or equal to 5, the requirements for the goodness-of-fit test are satisfied.

Classical approach: The critical value, with df $= 6 - 1 = 5$, is $\chi^2_{0.05} = 11.071$. Since the test statistic is in the critical region ($\chi^2_0 > \chi^2_{0.05}$), we reject the null hypothesis.

P-value approach: Using the chi-square table, we find the row that corresponds to 5 degrees of freedom. The value of 18.738 is greater than 16.750, which has an area under the chi-square distribution of 0.005 to the right. Therefore, we have P-value < 0.005. Since P-value $< \alpha$, we reject the null hypothesis. [Note: using technology, we find P-value $= 0.0022 < \alpha$ so the null hypothesis is rejected.]

Conclusion: There is enough evidence to conclude that the distribution of colors of plain M&Ms differs from that claimed by M&M/Mars.

13. **(a)** Answers will vary. Because of the seriousness of the situation, we would want to have strong evidence that there is fraudulent activity. Therefore, we would want to select a small value for α, such as $\alpha = 0.01$.

(b) There are 200 digits being checked. To determine the expected counts, multiply 200 by the percentage for each digit given by Benford's Law. For example, the expected count of the digit 1 would be $200(0.301) = 60.2$. We summarize the observed and expected counts in the following table:

	O_i	E_i	$(O_i - E_i)^2$	$\dfrac{(O_i - E_i)^2}{E_i}$
1	36	60.2	585.64	9.728
2	32	35.2	10.24	0.291
3	28	25	9	0.360
4	26	19.4	43.56	2.245
5	23	15.8	51.84	3.281
6	17	13.4	12.96	0.967
7	15	11.6	11.56	0.997
8	16	10.2	33.64	3.298
9	7	9.2	4.84	0.526
			$\chi^2_0 =$	21.693

Since all the expected cell counts are greater than or equal to 5, the requirements for the goodness-of-fit test are satisfied. Our hypotheses are:
H_0 : the digits obey Benford's law
H_1 : the digits do not obey Benford's law

Classical approach: The critical value, with df $= 9 - 1 = 8$, is $\chi^2_{0.01} = 20.090$. Since the test statistic is in the critical region ($\chi^2_0 > \chi^2_{0.01}$), we reject the null hypothesis.

P-value approach: Using the chi-square table, we find the row that corresponds to 8 degrees of freedom. The value of 21.693 is greater than 20.090, which has an area under the chi-square distribution of 0.01 to the right. Therefore, we have P-value < 0.01. Since P-value $< \alpha$, we reject the null hypothesis. [Note: using technology, we find P-value $= 0.006 < \alpha$ and reject H_0]

Conclusion: There is enough evidence to reject the null hypothesis. The digits do not appear to obey Benford's law.

(c) Answers will vary. Based on the results from part (b), it appears that the employee is guilty of embezzlement at the $\alpha = 0.01$ level of significance.

15. **(a)** To determine the expected counts, multiply 2068 by the given percentages. For example, the expected count for Multiple Locations would be $2068(0.57) = 1178.76$. We summarize the observed and expected counts in the following table:

	O_i	E_i	$(O_i - E_i)^2$	$\dfrac{(O_i - E_i)^2}{E_i}$
Mult. Locations	1036	1178.76	20,380.4176	17.2897
Head	864	641.08	49,693.3264	77.5150
Neck	38	62.04	577.9216	9.3153
Thorax	83	124.08	1687.5664	13.6006
Ab/Lum/ Spine	47	62.04	226.2016	3.6461
			$\chi^2_0 \approx$	121.367

Since all the expected cell counts are greater than or equal to 5, the requirements for the goodness-of-fit test are satisfied. Our hypotheses are

H_0 : the distribution of fatal injuries for those not wearing helmets is the same as for all riders.

H_1 : the distributions are not the same

Classical approach: The critical value, with df $= 5 - 1 = 4$, is $\chi_{0.05}^2 = 9.488$. Since the test statistic is in the critical region ($\chi_0^2 > \chi_{0.05}^2$), we reject the null hypothesis.

P-value approach: Using the chi-square table, we find the row that corresponds to 4 degrees of freedom. The value of 121.367 is greater than 14.860, which has an area under the chi-square distribution of 0.005 to the right. Therefore, we have P-value < 0.005 [Tech: P-value < 0.001]. Since P-value $< \alpha$, we reject the null hypothesis.

Conclusion: There is sufficient evidence to conclude that the distribution of fatal injuries for those not wearing helmets is different than for all riders.

(b) The observed count for head injuries is much higher than expected, while the observed count for all other fatal injuries are lower than expected. We might conclude that motorcycle fatalities from head injuries occur more frequently for riders not wearing helmets.

17. (a) We obtain the number of students from each group attending by multiplying each percent by 100:

Group 1: $100(0.84) = 84$

Group 2: $100(0.84) = 84$

Group 3: $100(0.84) = 84$

Group 4: $100(0.81) = 81$

If seat location has no effect, we expect 83% of each group to attend. Since there were 100 students in each group, we expect 83 of the 100 students in each group to attend. We summarize the observed and expected counts in the following table:

	O_i	E_i	$(O_i - E_i)^2$	$\dfrac{(O_i - E_i)^2}{E_i}$
1	84	83	1	0.0120
2	84	83	1	0.0120
3	84	83	1	0.0120
4	81	83	4	0.0482
			$\chi_0^2 \approx$	0.084

Since all the expected cell counts are greater than or equal to 5, the requirements for the goodness-of-fit test are satisfied. Our hypotheses are

H_0 : there is no difference in attendance patterns among the 4 groups

H_1 : there are differences in attendance patterns among the 4 groups.

Classical approach: The critical value, with df $= 4 - 1 = 3$, is $\chi_{0.05}^2 = 7.815$. Since the test statistic is not in the critical region ($\chi_0^2 < \chi_{0.05}^2$), we do not reject the null hypothesis.

P-value approach: Using the chi-square table, we find the row that corresponds to 3 degrees of freedom. The value of 0.084 [Tech: 0.081] is less than 0.115, which has an area under the chi-square distribution of 0.99 to the right. Therefore, we have P-value > 0.99 [Tech: P-value $= 0.994$]. Since P-value $> \alpha$, we do not reject the null hypothesis.

Conclusion: There is not sufficient evidence to conclude that there are differences among the groups in terms of attendance patterns.

(b) We obtain the number of students from each group attending by multiplying each percent by 100:

Group 1: $100(0.84) = 84$

Group 2: $100(0.81) = 81$

Group 3: $100(0.78) = 78$

Group 4: $100(0.76) = 76$

If seat location has no effect, we expect 80% of each group to attend. Since there were 100 students in each group, we expect 80 of the 100 students in each group to attend. We summarize the observed and expected counts in the following table:

	O_i	E_i	$(O_i - E_i)^2$	$\dfrac{(O_i - E_i)^2}{E_i}$
1	84	80	16	0.2
2	81	80	1	0.0125
3	78	80	4	0.05
4	76	80	16	0.2
			$\chi_0^2 \approx$	0.463

Since all the expected cell counts are greater than or equal to 5, the requirements for the goodness-of-fit test are satisfied. Our hypotheses are

H_0 : there is no difference in attendance patterns among the 4 groups

H_1 : there is a difference in attendance patterns among the 4 groups.

Classical approach: The critical value, with df = 4 − 1 = 3, is $\chi_{0.05}^2 = 7.815$. Since the test statistic is not in the critical region ($\chi_0^2 < \chi_{0.05}^2$), we do not reject the null hypothesis.

P-value approach: Using the chi-square table, we find the row that corresponds to 3 degrees of freedom. The value of 0.463 [Tech: 0.461] is less than 0.584, which has an area under the chi-square distribution of 0.90 to the right. Therefore, we have *P*-value > 0.90 [Tech: *P*-value = 0.927]. Since *P*-value > α , we do not reject the null hypothesis.

Conclusion: There is not sufficient evidence to conclude that there are differences among the groups in terms of attendance patterns.

It is curious that Group 1's attendance rate stayed the same, and the farther a group's original position is located from the front of the room, the more the attendance rate for the group decreases.

(c) We obtain the number of students from each group in the top 20% by multiplying each percent by 100:

Group 1: $100(0.25) = 25$

Group 2: $100(0.20) = 20$

Group 3: $100(0.15) = 15$

Group 4: $100(0.19) = 19$

If seat location has no effect, we expect 20% of each group to be in the top 20%.

Since there were 100 students in each group, we expect 20 of the 100 students in each group to be in the top 20%. We summarize the observed and expected counts in the following table:

	O_i	E_i	$(O_i - E_i)^2$	$\dfrac{(O_i - E_i)^2}{E_i}$
1	25	20	25	1.25
2	20	20	0	0
3	15	20	25	1.25
4	19	20	1	0.05
			$\chi_0^2 =$	2.55

Since all the expected cell counts are greater than or equal to 5, the requirements for the goodness-of-fit test are satisfied. Our hypotheses are

H_0 : there is no difference in the number of students in the top 20% of the class by group.

H_1 : there is a difference in the number of students in the top 20% of the class by group.

Classical approach: The critical value, with df = 4 − 1 = 3, is $\chi_{0.05}^2 = 7.815$. Since the test statistic is not in the critical region ($\chi_0^2 < \chi_{0.05}^2$), we do not reject the null hypothesis.

P-value approach: Using the chi-square table, we find the row that corresponds to 3 degrees of freedom. The value of 2.55 [Tech: 2.57] is less than 6.251, which has an area under the chi-square distribution of 0.10 to the right. Therefore, we have *P*-value > 0.10 [Tech: *P*-value = 0.463]. Since *P*-value > α , we do not reject the null hypothesis.

Conclusion: There is not sufficient evidence to conclude that there is a difference in the number of students in the top 20% of the class by group.

(d) Though not statistically significant, the group located in the front had both better attendance and a larger number of students in the top 20%. Given a choice, one should choose to sit in the front (though other factors, such as study habits, also affect student success).

19. We summarize the observed and expected counts in the following table:

	O_i	E_i	$(O_i - E_i)^2$	$\dfrac{(O_i - E_i)^2}{E_i}$
Jan	40	$\frac{500}{12}$	2.7778	0.0667
Feb	38	$\frac{500}{12}$	13.4444	0.3227
Mar	41	$\frac{500}{12}$	0.4444	0.0107
Apr	40	$\frac{500}{12}$	2.7778	0.0667
May	42	$\frac{500}{12}$	0.1111	0.0027
Jun	41	$\frac{500}{12}$	0.4444	0.0107
Jul	45	$\frac{500}{12}$	11.1111	0.2667
Aug	44	$\frac{500}{12}$	5.4444	0.1307
Sep	44	$\frac{500}{12}$	5.4444	0.1307
Oct	43	$\frac{500}{12}$	1.7778	0.0427
Nov	39	$\frac{500}{12}$	7.1111	0.1707
Dec	43	$\frac{500}{12}$	1.7778	0.0427
			$\chi_0^2 =$	1.264

Since all the expected cell counts are greater than or equal to 5, the requirements for the goodness-of-fit test are satisfied. Our hypotheses are:

H_0 : distribution of birth months is uniform
H_1 : distribution of birth months is not uniform

Classical approach: The critical value, with df = 12 − 1 = 11, is $\chi_{0.05}^2 = 19.675$. Since the test statistic is not in the critical region ($\chi_0^2 < \chi_{0.05}^2$), we do not reject the null hypothesis.

P-value approach: Using the chi-square table, we find the row that corresponds to 11 degrees of freedom. The value of 1.264 is less than 2.603, which has an area under the chi-square distribution of 0.995 to the right. Therefore, we have P-value > 0.995 . Since P-value $> \alpha$, we do not reject the null hypothesis.
[Note: using technology, we find P-value $= 0.9998 > \alpha = 0.05$ so the null hypothesis is not rejected.]

Conclusion: There is not enough evidence to reject the null hypothesis. The birth months appear to occur with equal frequency.

21. We summarize the observed and expected counts in the following table:

	O_i	E_i	$(O_i - E_i)^2$	$\dfrac{(O_i - E_i)^2}{E_i}$
Sun	39	$\frac{300}{7} = 42.857$	14.878	0.347
Mon	40	42.857	8.163	0.190
Tue	30	42.857	165.306	3.857
Wed	40	42.857	8.163	0.190
Thu	41	42.857	3.449	0.080
Fri	49	42.857	37.735	0.880
Sat	61	42.857	329.163	7.680
			$\chi_0^2 =$	13.227

Since all the expected cell counts are greater than or equal to 5, the requirements for the goodness-of-fit test are satisfied. Our hypotheses are:

H_0 : the distribution of pedestrian fatalities is uniformly distributed during the week
H_1 : the distribution of pedestrian fatalities is not uniformly distributed during the week

Classical approach: The critical value, with df = 7 − 1 = 6, is $\chi_{0.05}^2 = 12.592$. Since the test statistic is in the critical region ($\chi_0^2 > \chi_{0.05}^2$), we reject the null hypothesis.

P-value approach: Using the chi-square table, we find the row that corresponds to 6 degrees of freedom. The value of 13.227 is greater than 12.592, which has an area under the chi-square distribution of 0.05 to the right. Therefore, we have P-value < 0.05 [Tech: P-value $= 0.040$]. Since P-value $< \alpha$, we reject the null hypothesis.

Conclusion: There is enough evidence to indicate that pedestrian deaths are not uniformly distributed over the days of the week. We might conclude that fewer pedestrian deaths occur on Tuesdays, and more on Saturdays.

23. (a)

	Rel. Freq.	Obs. Freq.	Exp. Freq.
K-3	0.329	15	$25(0.329) = 8.225$
4-8	0.393	7	$25(0.393) = 9.825$
9-12	0.278	3	$25(0.278) = 6.95$

(b) Since all the expected cell counts are greater than or equal to 5, the requirements for the goodness-of-fit test are satisfied. Our hypotheses are:

H_0 : grade distribution is the same as the national distribution

H_1 : grade distribution is not the same as the national distribution.

$$\chi_0^2 = \frac{(15-8.225)^2}{8.225} + \frac{(7-9.825)^2}{9.825}$$
$$+ \frac{(3-6.95)^2}{6.95}$$
$$\approx 8.638$$

Classical approach: The critical value, with df = 3 − 1 = 2, is $\chi_{0.05}^2 = 5.991$. Since the test statistic is in the critical region ($\chi_0^2 > \chi_{0.05}^2$), we reject the null hypothesis.

P-value approach: Using the chi-square table, we find the row that corresponds to 2 degrees of freedom. The value of 8.638 is greater than 7.378, which has an area under the chi-square distribution of 0.025 to the right. Therefore, we have P-value < 0.025 [Tech: P-value = 0.013]. Since P-value < α , we reject the null hypothesis.

Conclusion: There is enough evidence to indicate that the grade distribution of home-schooled students in the social worker's district is different from the national distribution. We might conclude that a greater proportion of students in grades K-3 are home-schooled in this district than nationally.

25. (a) Answers will vary.

(b) Each of the five numbers should be equally likely so the proportions should all be 20%.

(c) Answers will vary.

27. (a) For 240 randomly selected births, we expect $240(0.071) = 17.04$ to result in low birth weight, and $240(1-0.071) = 222.96$ to not result in low birth weight.

(b) We summarize the observed and expected counts

	O_i	E_i	$(O_i - E_i)^2$	$\dfrac{(O_i - E_i)^2}{E_i}$
Low	22	17.04	24.6016	1.4438
Not Low	218	222.96	24.6016	0.1103
	160		$\chi_0^2 \approx$	1.554

Our hypotheses are:

$H_0 : p = 0.071$

$H_1 : p \neq 0.071$

Classical approach: The critical value, with df = 2 − 1 = 1, is $\chi_{0.05}^2 = 3.841$. Since the test statistic is not in the critical region ($\chi_0^2 < \chi_{0.05}^2$), we do not reject H_0 .

P-value approach: Using the chi-square table, we find the row that corresponds to 1 degree of freedom. The value of 1.554 is less than 2.706, which has an area under the chi-square distribution of 0.10 to the right. Therefore, we have P-value > 0.10 [Tech: P-value = 0.213]. Since P-value > α , we do not reject the null hypothesis.

Conclusion: There is not enough evidence to conclude that the percentage of low-birth-weight babies is higher for mothers 35–39 years old.

(c) Note that

$$n \cdot p_0 (1 - p_0) = 240(0.071)(1 - 0.071)$$
$$\approx 15.83 \geq 10$$

so the requirements of the hypothesis test are satisfied.

The hypotheses are $H_0 : p = 0.071$,

$H_1 : p > 0.071$. From the sample data,

$\hat{p} = \dfrac{22}{240} \approx 0.0917$. The test statistic is

$$z_0 = \frac{\hat{p} - p_0}{\sqrt{p_0(1-p_0)/n}}$$
$$= \frac{0.092 - 0.071}{\sqrt{0.071(1-0.071)/240}}$$
$$= 1.27$$

Classical approach: This is a right-tailed test so the critical value is $z_{0.05} = 1.645$. The test statistic is not in the critical region ($z_0 < z_{0.05}$) so we do not reject the null hypothesis.

P-value approach: Since this is a right-tailed test,

P-value $= P(Z > 1.27) = 0.1020$

[Tech: P-value $= 0.1063$]. Since P-value $> \alpha$, we do not reject the null hypothesis.

Conclusion: There is not enough evidence to conclude that the percentage of low-birth-weight babies is higher for mothers 35–39 years old.

29. (a) There are $229 + 211 + \ldots + 0 + 1 = 576$ rockets that landed in London. We obtain the probabilities by dividing each frequency by 576 (obtaining the relative frequencies).

# of rockets, x	$P(x)$
0	$\dfrac{229}{576} \approx 0.3976$
1	$\dfrac{211}{576} \approx 0.3663$
2	$\dfrac{93}{576} \approx 0.1615$
3	$\dfrac{35}{576} \approx 0.0608$
4	$\dfrac{7}{576} \approx 0.0122$
5	$\dfrac{0}{576} = 0$
6	$\dfrac{0}{576} = 0$
7	$\dfrac{1}{576} \approx 0.0017$

$\mu = 0\left(\frac{229}{576}\right) + 1\left(\frac{211}{576}\right) + \ldots + 6(0) + 7\left(\frac{1}{576}\right)$
≈ 0.9323

The mean number of hits in a region was approximately 0.9323.

(b) The requirements for conducting a goodness-of-fit test are not satisfied because all the expected frequencies will not be greater than or equal to 1, and more than 20% of the expected frequencies are less than 5 (37.5%).

(c) Using the model $P(x) = \dfrac{0.9323^x}{x!} e^{-0.9323}$, the probability distribution is given by

x	$P(x)$
0	0.3936
1	0.3670
2	0.1711
3	0.0532
4 or more	0.0151

(d) We determine the expected number of regions that will be hit by x rockets by multiplying the number of rockets (576) by the probability of a region being hit by x rockets, $P(x) = \dfrac{0.9323^x}{x!} e^{-0.9323}$.

x	Expected number of rocket hits
0	$576 \cdot \dfrac{0.9323^0}{0!} e^{-0.9323} = 226.741$
1	$576 \cdot \dfrac{0.9323^1}{1!} e^{-0.9323} = 211.390$
2	$576 \cdot \dfrac{0.9323^2}{2!} e^{-0.9323} = 98.540$
3	$576 \cdot \dfrac{0.9323^3}{3!} e^{-0.9323} = 30.623$
4 or more	$576 \cdot \left(\dfrac{0.9323^4}{4!} + \ldots + \dfrac{0.9323^7}{7!}\right) e^{-0.9323} = 8.706$

(e) We summarize the observed and expected counts:

	O_i	E_i	$(O_i - E_i)^2$	$\dfrac{(O_i - E_i)^2}{E_i}$
0	229	226.741	5.1031	0.0225
1	211	211.390	0.1521	0.0007
2	93	98.540	30.6916	0.3115
3	35	30.623	19.1581	0.6256
4+	8	8.706	0.4984	0.0573
			$\chi_0^2 \approx$	1.018

Our hypotheses are:
H_0: the rockets can be modeled by a Poisson random variable.
H_1: the rockets cannot be modeled by a Poisson random variable.

Classical approach: The critical value, with df $= 5 - 1 = 4$, is $\chi^2_{0.05} = 9.488$. Since the test statistic is not in the critical region ($\chi^2_0 < \chi^2_{0.05}$), we do not reject H_0.

P-value approach: Using the chi-square table, we find the row that corresponds to 4 degrees of freedom. The value of 1.018 is less than 1.064, which has an area under the chi-square distribution of 0.90 to the right. Therefore, we have P-value > 0.90 [Tech: P-value = 0.907]. Since P-value $> \alpha$, we do not reject the null hypothesis.

Conclusion: There is not sufficient evidence to conclude that the distribution of rocket hits is different from a Poisson distribution. That is, the rocket hits do appear to be modeled by a Poisson random variable.

12.2 Tests for Independence and the Homogeneity of Proportions

1. The chi-square test for independence is a test of a single population and is to determine if there is an association between two characteristics of that population. The chi-square test for homogeneity is a test to determine if two distinct populations are the same with respect to a single characteristic, i.e. if that characteristic is exhibited by the same percentage of the two populations. The procedures and assumptions for the two tests are the same.

3. **(a)** $\chi^2_0 = \sum \dfrac{(O_i - E_i)^2}{E_i} = \dfrac{(34 - 36.26)^2}{36.26} + \dfrac{(43 - 44.63)^2}{44.63} + \cdots + \dfrac{(17 - 20.89)^2}{20.89} = 1.701$

(b) Classical Approach:
There are 2 rows and 3 columns, so df $= (2-1)(3-1) = 2$ and the critical value is $\chi^2_{0.05} = 5.991$. The test statistic, 1.701, is less than the critical value so we do not reject H_0.

P-value Approach:
There are 2 rows and 3 columns, so we find the P-value using $(2-1)(3-1) = 2$ degrees of freedom. The P-value is the area under the chi-square distribution with 2 degrees of freedom to the right of $\chi^2_0 = 1.701$. Using Table VII, we find the row that corresponds to 2 degrees of freedom. The value of 1.701 is less than 4.605, which has an area under the chi-square distribution of 0.10 to the right. Therefore, we have P-value > 0.10 [Tech: P-value = 0.427]. Since P-value $> \alpha$, we do not reject H_0.

Conclusion:
There is not sufficient evidence, at the $\alpha = 0.05$ level of significance, to conclude that the two variables are dependent. We conclude that X and Y are not related.

5. The hypotheses are $H_0 : p_1 = p_2 = p_3$ and $H_1 :$ at least one proportion differs from the others.

   The expected counts are calculated as $\dfrac{\text{(row total)} \cdot \text{(column total)}}{\text{(table total)}} = \dfrac{229 \cdot 120}{363} \approx 75.702$ (for the first cell) and so on. The observed and expected counts are shown in the following table.

	Category 1	Category 2	Category 3	Total
Success	76	84	69	229
	(75.702)	(78.857)	(74.441)	
Failure	44	41	49	134
	(44.298)	(46.143)	(43.559)	
Total	120	125	118	363

Since none of the expected counts is less than 5, the requirements of the test are satisfied.

The test statistic is $\chi_0^2 = \sum \dfrac{(O_i - E_i)^2}{E_i} = \dfrac{(76 - 75.702)^2}{75.702} + \cdots + \dfrac{(49 - 43.559)^2}{43.559} = 1.989$.

Classical Approach:

df $= (2-1)(3-1) = 2$ so the critical value is $\chi_{0.01}^2 = 9.210$. The test statistic, 1.989, is less than the critical value so we do not reject H_0.

P-value Approach:

Using Table VII, we find the row that corresponds to 2 degrees of freedom. The value of 1.989 is less than 4.605, which has an area under the chi-square distribution of 0.10 to the right. Therefore, we have P-value > 0.10 [Tech: P-value $= 0.370$]. Since P-value $> \alpha$, we do not reject H_0.

Conclusion:

There is not sufficient evidence, at the $\alpha = 0.01$ level of significance, to conclude that at least one of the proportions is different from the others.

7. (a) The expected counts are calculated as $\dfrac{\text{(row total)} \cdot \text{(column total)}}{\text{(table total)}} = \dfrac{199 \cdot 150}{380} \approx 78.553$ (for the first cell) and so on, giving the following table of observed and expected counts:

Sexual Activity	Family Structure				Total
	Both Parents	**One Parent**	**Parent/ Stepparent**	**Nonparental Guardian**	
Had intercourse	64	59	44	32	199
	(78.553)	(52.368)	(41.895)	(26.184)	
Did not have	86	41	36	18	181
	(71.447)	(47.632)	(38.105)	(23.816)	
Total	150	100	80	50	380

   (b) All expected frequencies are greater than 5 so all requirements for a chi-square test are satisfied. That is, all expected frequencies are greater than or equal to 1, and no more than 20% of the expected frequencies are less than 5.

   (c) $\chi_0^2 = \sum \dfrac{(O_i - E_i)^2}{E_i} = \dfrac{(64 - 78.553)^2}{78.553} + \cdots + \dfrac{(18 - 23.816)^2}{23.816} \approx 10.358$ [Tech: 10.357]

(d) H_0 : the row and column variables are independent

H_1 : the row and column variables are dependent

$df = (2-1)(4-1) = 3$ so the critical value is $\chi^2_{0.05} = 7.815$.

Classical Approach:
The test statistic is 10.358 which is greater than the critical value so we reject H_0.

P-value Approach:
Using Table VII, we find the row that corresponds to 2 degrees of freedom. The value of 10.358 is greater than 9.348, which has an area under the chi-square distribution of 0.025 to the right. Therefore, we have P-value < 0.025 [Tech: P-value = 0.016]. Since P-value < α, we reject H_0.

Conclusion:
There is sufficient evidence, at the $\alpha = 0.05$ level of significance, to conclude that family structure and sexual activity are dependent.

(e) The biggest difference between observed and expected occurs under the family structure in which both parents are present. Fewer females were sexually active than was expected when both parents were present. This means that having both parents present seems to have an impact on whether the child is sexually active.

(f) The conditional frequencies and bar chart show that sexual activity varies by family structure.

Sexual Intercourse	Both Parents	One Parent	Parent/Stepparent	Nonparent Guardian
Yes	$\frac{64}{150} \approx 0.427$	$\frac{59}{100} = 0.590$	$\frac{44}{80} = 0.550$	$\frac{32}{50} = 0.640$
No	$\frac{86}{150} \approx 0.573$	$\frac{41}{100} = 0.410$	$\frac{36}{80} = 0.450$	$\frac{18}{50} = 0.360$

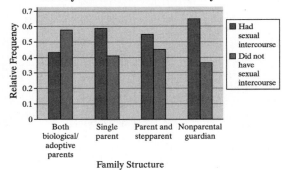

Family Structure and Sexual Activity

9. (a) The expected counts are calculated as $\dfrac{\text{(row total)} \cdot \text{(column total)}}{\text{(table total)}} = \dfrac{634 \cdot 551}{1996} \approx 175.017$ (for the first cell) and so on, giving the following table:

Happiness	Health				Total
	Excellent	Good	Fair	Poor	
Very Happy	271	261	82	20	634
	(175.017)	(295.718)	(128.642)	(34.622)	
Pretty Happy	247	567	231	53	1098
	(303.105)	(512.143)	(222.791)	(59.961)	
Not Too Happy	33	103	92	36	264
	(72.878)	(123.138)	(53.567)	(14.417)	
Total	551	931	405	109	1996

All expected frequencies are greater than 5 so the requirements for a chi-square test are satisfied.

$$\chi_0^2 = \sum \frac{(O_i - E_i)^2}{E_i} = \frac{(271 - 175.017)^2}{175.017} + \cdots + \frac{(36 - 14.417)^2}{14.417} \approx 182.173 \quad \text{[Tech: 182.174]}$$

df $= (3-1)(4-1) = 6$ so the critical value is $\chi_{0.05}^2 = 12.592$.

H_0 : the row and column variables are independent

H_1 : the row and column variables are dependent

Classical Approach:
The test statistic is 182.173 which is greater than the critical value so we reject H_0.

P-value Approach:
Using Table VII, we find the row that corresponds to 6 degrees of freedom. The value of 182.173 is greater than 18.548, which has an area under the chi-square distribution of 0.005 to the right. Therefore, we have P-value < 0.005 [Tech: P-value < 0.001]. Since P-value < α, we reject H_0.

Conclusion:
There is sufficient evidence, at the $\alpha = 0.05$ level of significance, to conclude that health and happiness are related.

(b) The conditional frequencies and bar chart show that the level of happiness varies by health status.

Happiness	Health			
	Excellent	Good	Fair	Poor
Very Happy	$\dfrac{271}{551} \approx 0.492$	$\dfrac{261}{931} \approx 0.280$	$\dfrac{82}{405} \approx 0.202$	$\dfrac{20}{109} \approx 0.183$
Pretty Happy	$\dfrac{247}{551} \approx 0.448$	$\dfrac{567}{931} \approx 0.609$	$\dfrac{231}{405} \approx 0.570$	$\dfrac{53}{109} \approx 0.486$
Not Too Happy	$\dfrac{33}{551} \approx 0.60$	$\dfrac{103}{931} \approx 0.111$	$\dfrac{92}{405} \approx 0.227$	$\dfrac{36}{109} \approx 0.330$

Health and Happiness

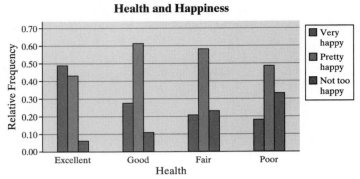

(c) The proportion of individuals who are "very happy" is much higher for individuals in "excellent" health than any other health category. Further, the proportion of individuals who are "not too happy" is much lower for individuals in "excellent" health compared to the other health categories. The level of happiness seems to decline as health status declines.

11. (a) The expected counts are calculated as $\dfrac{\text{(row total)} \cdot \text{(column total)}}{\text{(table total)}} = \dfrac{474 \cdot 393}{1054} \approx 176.738$ (for the first cell) and so on, giving the following table:

Years of Education	Smoking Status			Total
	Current	**Former**	**Never**	
< 12	178	88	208	474
	(176.738)	(96.688)	(200.573)	
12	137	69	143	349
	(130.130)	(71.191)	(147.679)	
13–15	44	25	44	113
	(42.134)	(23.050)	(47.816)	
16 or more	34	33	51	118
	(43.998)	(24.070)	(49.932)	
Total	393	215	446	1054

All expected frequencies are greater than 5 so the requirements for a chi-square test are satisfied.

$$\chi_0^2 = \sum \frac{(O_i - E_i)^2}{E_i} = \frac{(178 - 176.738)^2}{176.738} + \cdots + \frac{(51 - 49.932)^2}{49.932} \approx 7.803$$

df $= (4-1)(3-1) = 6$ so the critical value is $\chi_{0.05}^2 = 12.592$.

H_0 : the row and column variables are independent

H_1 : the row and column variables are dependent

Classical Approach:
The test statistic is 7.803 which is less than the critical value so we do not reject H_0.

P-value Approach:
Using Table VII, we find the row that corresponds to 6 degrees of freedom. The value of 7.803 is less than 10.645, which has an area under the chi-square distribution of 0.10 to the right. Therefore, we have P-value > 0.10 [Tech: P-value = 0.253]. Since P-value > α, we do not reject H_0.

Conclusion:
There is not sufficient evidence, at the $\alpha = 0.05$ level of significance, to conclude that smoking status and years of education are associated. That is, it does not appear that years of education plays a role in determining smoking status.

(b) The conditional frequencies and bar chart show that the distribution of smoking status is similar at all levels of education. This supports the result from part (a).

Years of Education	Current	Former	Never
< 12	$\dfrac{178}{474} \approx 0.376$	$\dfrac{88}{474} \approx 0.186$	$\dfrac{208}{474} \approx 0.439$
12	$\dfrac{137}{349} \approx 0.393$	$\dfrac{69}{349} \approx 0.198$	$\dfrac{143}{349} \approx 0.410$
13–15	$\dfrac{44}{113} \approx 0.389$	$\dfrac{25}{113} \approx 0.221$	$\dfrac{44}{113} \approx 0.389$
16 or more	$\dfrac{34}{118} \approx 0.288$	$\dfrac{33}{118} \approx 0.280$	$\dfrac{51}{118} \approx 0.432$

Smoking Status vs. Level of Education

13. (a) The expected counts are calculated as $\dfrac{(\text{row total}) \cdot (\text{column total})}{(\text{table total})} = \dfrac{1028 \cdot 385}{1437} \approx 275.421$ (for the first cell) and so on, giving the following table:

Opinion	Highest Degree					Total
	Less Than High School	High School	Junior College	Bachelor	Graduate	
Male	302	551	29	100	46	1028
	(275.421)	(537.250)	(37.915)	(122.330)	(55.084)	
Female	83	200	24	71	31	409
	(109.579)	(213.750)	(15.085)	(48.670)	(21.916)	
Total	385	751	53	171	77	1437

All expected frequencies are greater than 5 so all requirements for a chi-square test are satisfied.

$$\chi_0^2 = \sum \frac{(O_i - E_i)^2}{E_i} = \frac{(302 - 275.421)^2}{275.421} + \cdots + \frac{(31 - 21.916)^2}{21.916} \approx 37.198$$

df $= (2-1)(5-1) = 4$ so the critical value is $\chi_{0.05}^2 = 9.488$.

H_0 : $p_{\text{LTHS}} = p_{\text{HS}} = p_{\text{JC}} = p_{\text{B}} = p_{\text{G}}$

H_1 : at least one of the proportions is different from the others

Classical Approach:
The test statistic is 37.198 which is greater than the critical value so we reject H_0.

P-value Approach:
Using Table VII, we find the row that corresponds to 4 degrees of freedom. The value of 37.198 is greater than 14.860, which has an area under the chi-square distribution of 0.005 to the right. Therefore, we have *P*-value < 0.005 [Tech: *P*-value < 0.001]. Since *P*-value < α, we reject H_0.

Conclusion:
There is sufficient evidence, at the $\alpha = 0.05$ level of significance, to conclude that at least one of the proportions is different from the others. That is, the evidence suggests that the proportion of individuals who feel everyone has an equal opportunity to obtain a quality education in the United States is different for at least one level of education.

(b) The conditional frequencies and bar chart show that the distribution of opinion varies by education. Higher proportions of individuals with either a "high school" education or "less than a high school" education seem to believe that everyone has an equal opportunity to obtain a quality education. Individuals whose highest degree is from a junior college appear to agree least with the statement.

Opinion	Less than High School	High School	Junior College	Bachelor	Graduate
Yes	$\frac{302}{385} \approx 0.784$	0.734	0.547	0.585	0.597
No	$\frac{83}{385} \approx 0.216$	0.266	0.453	0.415	0.403

Equal Opportunity for Education

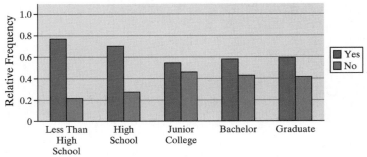

15. (a) The expected counts are calculated as $\dfrac{\text{(row total)} \cdot \text{(column total)}}{\text{(table total)}} = \dfrac{67 \cdot 217}{1108} \approx 13.122$ (for the first cell)

and so on, giving the following table:

Side Effect	Treatment					Total
	Placebo	Celebrex 50 mg	Celebrex 100 mg	Celebrex 200 mg	Naproxen 500 mg	
Ulcer	5	8	7	13	34	67
	(13.122)	(14.089)	(13.727)	(13.364)	(12.699)	
No Ulcer	212	225	220	208	176	1041
	(203.878)	(218.911)	(213.273)	(207.636)	(197.301)	
Total	217	233	227	221	210	1108

All expected frequencies are greater than 5 so all requirements for a chi-square test are satisfied.

$$\chi_0^2 = \sum \frac{(O_i - E_i)^2}{E_i} = \frac{(5 - 13.122)^2}{13.122} + \cdots + \frac{(176 - 197.301)^2}{197.301} \approx 49.701 \text{ [Tech: 49.703]}.$$

$df = (2-1)(5-1) = 4$ so the critical value is $\chi^2_{0.01} = 13.277$.

$H_0 : p_{\text{placebo}} = p_{50 \text{ mg}} = p_{100 \text{ mg}} = p_{200 \text{ mg}} = p_{\text{Naproxen}}$

H_1 : at least one of the proportions is not equal to the rest

<u>Classical Approach:</u>
The test statistic is 49.701 which is greater than the critical value so we reject H_0.

<u>P-value Approach:</u>
Using Table VII, we find the row that corresponds to 4 degrees of freedom. The value of 49.701 is greater than 14.860, which has an area under the chi-square distribution of 0.005 to the right. Therefore, we have P-value < 0.005 [Tech: P-value < 0.001]. Since P-value < α, we reject H_0.

<u>Conclusion:</u>
There is sufficient evidence, at the $\alpha = 0.01$ level of significance, to conclude that at least one of the proportions is different from the others. That is, the evidence suggests that the incidence of ulcers varies by treatment.

(b) The conditional frequencies and bar chart show that the percentage of subjects suffering ulcers varies by treatment. This supports the conclusion from part (a).

	Placebo	**Celebrex 50 mg**	**Celebrex 100 mg**	**Celebrex 200 mg**	**Naproxen 500 mg**
Ulcer	$\frac{5}{217} \approx 0.023$	$\frac{8}{233} \approx 0.034$	$\frac{7}{227} \approx 0.031$	$\frac{13}{221} \approx 0.059$	$\frac{34}{210} \approx 0.162$
No Ulcer	$\frac{212}{217} \approx 0.977$	$\frac{225}{233} \approx 0.966$	$\frac{220}{227} \approx 0.969$	$\frac{208}{221} \approx 0.941$	$\frac{176}{210} \approx 0.838$

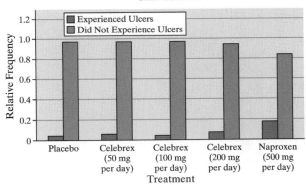

17. (a) Since there were no individuals who gave "career" as a reason for dropping, we have omitted that category from the analysis.

Gender	Drop Reason			Total
	Personal	**Work**	**Course**	
Female	5	3	13	21
	(4.62)	(6.72)	(9.66)	
Male	6	13	10	29
	(6.38)	(9.28)	(13.34)	
Total	11	16	23	50

(b) The expected counts are calculated as $\dfrac{(\text{row total})\cdot(\text{column total})}{(\text{table total})} = \dfrac{21\cdot 11}{50} = 4.62$ (for the first cell) and

so on (included in the table from part (a)). All expected frequencies are greater than 1 and only one (out of six) expected frequency is less than 5. All requirements for a chi-square test are satisfied.

$$\chi_0^2 = \sum \frac{(O_i - E_i)^2}{E_i} = \frac{(5-4.62)^2}{4.62} + \cdots + \frac{(10-13.34)^2}{13.34} \approx 5.595$$

$df = (2-1)(3-1) = 2$ so the critical value is $\chi_{0.10}^2 = 4.605$.

Classical Approach:
The test statistic is 5.595 which is greater than the critical value so we reject H_0.

P-value Approach:
Using Table VII, we find the row that corresponds to 2 degrees of freedom. The value of 5.595 is greater than 4.605, which has an area under the chi-square distribution of 0.10 to the right. Therefore, we have P-value < 0.10 [Tech: P-value = 0.061]. Since P-value < α, we reject H_0.

Conclusion:
There is sufficient evidence, at the $\alpha = 0.1$ level of significance, to conclude that gender and drop reason are dependent. Females are more likely to drop because of the course, while males are more likely to drop because of work.

(c) The conditional frequencies and bar chart show that the distribution of genders varies by reason for dropping. This supports the results from part (b).

	Personal	**Work**	**Course**
Female	$\dfrac{5}{11} \approx 0.455$	$\dfrac{3}{16} \approx 0.188$	$\dfrac{13}{23} \approx 0.565$
Male	$\dfrac{6}{11} \approx 0.545$	$\dfrac{13}{16} \approx 0.813$	$\dfrac{10}{23} \approx 0.435$

19. (a) The population being studied is healthy adult women aged 45 years or older. The sample consists of the 39,876 women in the study.

(b) The response variable is whether or not the subject had a cardiovascular event (such as a heart attack or stroke). This is a qualitative variable because the values serve to simply classify the subjects.

(c) There are two treatments: 100 mg of aspirin and a placebo

(d) Because the subjects are all randomly assigned to the treatments, this is a completely randomized design.

(e) Randomization controls for any other explanatory variables because individuals affected by these lurking variables should be equally dispersed between the two treatment groups.

(f) $H_0 : p_1 = p_2$ vs. $H_1 : p_1 \neq p_2$

$$\hat{p} = \frac{x_1 + x_2}{n_1 + n_2} = \frac{477 + 522}{19{,}934 + 19{,}942} = \frac{999}{39{,}876} \approx 0.025 \text{, } \hat{p}_1 = \frac{477}{19{,}934} \approx 0.024 \text{, } \hat{p}_2 = \frac{522}{19{,}942} \approx 0.026$$

The test statistic is $z_0 = \dfrac{\hat{p}_1 - \hat{p}_2}{\sqrt{\hat{p}(1-\hat{p})}\sqrt{\dfrac{1}{n_1} + \dfrac{1}{n_2}}} = \dfrac{0.024 - 0.026}{\sqrt{0.025(0.975)}\sqrt{\dfrac{1}{19{,}934} + \dfrac{1}{19{,}942}}} \approx -1.28$ [Tech: -1.44]

Classical Approach:

The critical values for $\alpha = 0.05$ are $-z_{0.025} = -1.96$ and $z_{0.025} = 1.96$.

The test statistic is -1.28, which is not in the critical region so we do not reject H_0.

P-value Approach:

$$
\begin{aligned}
P - \text{value} &= 2 \cdot P\big(Z > |-1.28|\big) \\
&= 2 \cdot P(Z > 1.28) \\
&= 2(0.1003) \\
&= 0.2006 \text{ [Tech: 0.1511]}
\end{aligned}
$$

Therefore, we have P-value $> \alpha = 0.05$ and we do not reject H_0.

Conclusion:

There is not sufficient evidence, at the $\alpha = 0.05$ level of significance, to conclude that a difference exists between the proportions of cardiovascular events in the aspirin group versus the placebo group.

(g) The expected counts are calculated as $\dfrac{(\text{row total}) \cdot (\text{column total})}{(\text{table total})} = \dfrac{999 \cdot 19{,}934}{39{,}876} \approx 499.400$ (for the first cell) and so on.

	aspirin	placebo	Total
cardiovascular event	477 (499.400)	522 (499.600)	999
no event	19,457 (19,434.600)	19,420 (19,442.400)	38,877
Total	19,934	19,942	39,876

All expected frequencies are greater than 1 and only one (out of six) expected frequency is less than 5. All requirements for a chi-square test are satisfied.

$$\chi_0^2 = \sum \frac{(O_i - E_i)^2}{E_i} = \frac{(477 - 499.400)^2}{499.400} + \cdots + \frac{(19{,}420 - 19{,}442.400)^2}{19{,}442.400} \approx 2.061$$

$\text{df} = (2-1)(2-1) = 1$ so the critical value is $\chi_{0.05}^2 = 3.841$.

Classical Approach:

The test statistic is 2.061 which is less than the critical value so we do not reject H_0.

P-value Approach:

Using Table VII, we find the row that corresponds to 1 degree of freedom. The value of 2.061 is less than 2.706, which has an area under the chi-square distribution of 0.10 to the right. Therefore, we have P-value > 0.10 [Tech: P-value = 0.151]. Since P-value $> \alpha$, we do not reject H_0.

Conclusion:

There is not sufficient evidence, at the $\alpha = 0.05$ level of significance, to conclude that a difference exists between the proportions of cardiovascular events in the aspirin group versus the placebo group.

(h) Using technology in part (f), the test statistic to four decimal places is $z_0 = -1.4355$ which gives $(z_0)^2 \approx 2.061 = \chi_0^2$. Our conclusion is that for comparing two proportions, $z_0^2 = \chi_0^2$.

Chapter 12 Review Exercises

1. Using $\alpha = 0.05$, we want to test

   H_0 : The wheel is balanced vs. H_1 : The wheel is not balanced

   If the wheel is balanced, each slot is equally likely. Thus, we would expect the proportion of red to be

   $\frac{18}{38} = \frac{9}{19}$, the proportion of black to be $\frac{18}{38} = \frac{9}{19}$, and the proportion of green to be $\frac{2}{18} = \frac{1}{19}$. There are 500

   spins. To determine the expected number of each color, multiply 500 by the given proportions for each color.

   For example, the expected count of red would be $500\left(\frac{9}{19}\right) \approx 236.842$. We summarize the observed and

   expected counts in the following table:

	O_i	E_i	$(O_i - E_i)^2$	$\dfrac{(O_i - E_i)^2}{E_i}$
Red	233	236.842	14.7610	0.0623
Black	237	236.842	0.0250	0.0001
Green	30	26.316	13.5719	0.5157
			$\chi_0^2 \approx$	0.578

   Since all the expected cell counts are greater than or equal to 5, the requirements for the goodness-of-fit test are satisfied.

   Classical approach: The critical value, with df = 3 − 1 = 2, is $\chi_{0.05}^2 = 5.991$. Since the test statistic is not in

   the critical region ($\chi_0^2 < \chi_{0.05}^2$), we do not reject the null hypothesis.

   P-value approach: Using the chi-square table, we find the row that corresponds to 2 degrees of freedom. The value of 0.578 is less than 4.605, which has an area under the chi-square distribution of 0.10 to the right. Therefore, we have P-value > 0.10 [Tech: 0.749]. Since P-value > α , we do not reject the null hypothesis.

   Conclusion: There is not enough evidence, at the $\alpha = 0.05$ level of significance, to conclude that the wheel is out of balance.

2. Using $\alpha = 0.05$, we want to test

   H_0 : The data follow the expected distribution

   H_1 : The data do not follow the expected distribution

   There are 77 series to consider. To determine the expected counts, multiply 77 by the given percentage for each number of games in the series. For example, the expected count for a 4 game series would be $77(0.125) = 9.625$. We summarize the observed and expected counts in the following table:

	O_i	E_i	$(O_i - E_i)^2$	$\dfrac{(O_i - E_i)^2}{E_i}$
4	15	9.625	28.8906	3.0016
5	15	19.25	18.0625	0.9383
6	17	24.0625	49.8789	2.0729
7	30	24.0625	35.2539	1.4651
			$\chi_0^2 \approx$	7.478

   Since all the expected cell counts are greater than or equal to 5, the requirements for the goodness-of-fit test are satisfied.

   Classical approach: The critical value, with df = 4 − 1 = 3, is $\chi_{0.05}^2 = 7.815$. Since the test statistic is not in

   the critical region ($\chi_0^2 < \chi_{0.05}^2$), we do not reject the null hypothesis.

P-value approach: Using the chi-square table, we find the row that corresponds to 4 degrees of freedom. The value of 7.478 is less than 9.488, which has an area under the chi-square distribution of 0.05 to the right. Therefore, we have *P*-value > 0.05 [Tech: *P*-value = 0.058]. Since *P*-value > α, we do not reject the null hypothesis.

Conclusion: There is not enough evidence, at the $\alpha = 0.05$ level of significance, to conclude that the teams playing in the World Series have not been evenly matched. That is, the evidence suggests that the teams have been evenly matched. On the other hand, the *P*-value is suggestive of an issue. In particular, there are fewer six-game series than we would expect. Perhaps the team that is down "goes all out" in game 6, trying to force game 7.

3. (a) The expected counts are calculated as $\dfrac{\text{(row total)}\cdot\text{(column total)}}{\text{(table total)}} = \dfrac{499\cdot 325}{1316} \approx 123.233$ (for the first cell) and so on, giving the following table:

	Class			Total
	First	**Second**	**Third**	
Survived	203	118	178	499
	(123.233)	(108.066)	(267.701)	
Did Not Survive	122	167	528	817
	(201.767)	(176.934)	(438.299)	
Total	325	285	706	1316

All expected frequencies are greater than 5 so all requirements for a chi-square test are satisfied. That is, all expected frequencies are greater than or equal to 1, and no more than 20% of the expected frequencies are less than 5.

$$\chi_0^2 = \sum \frac{(O_i - E_i)^2}{E_i} = \frac{(203-123.333)^2}{123.333} + \cdots + \frac{(528-438.299)^2}{438.299} \approx 132.882 \ \ [\text{Tech: } 133.052]$$

H_0: survival status and social class are independent

H_1: survival status and social class are dependent

df $= (2-1)(3-1) = 2$ so the critical value is $\chi_{0.05}^2 = 5.991$.

Classical Approach:
The test statistic is 132.882 which is greater than the critical value so we reject H_0.

P-value Approach:
Using Table VII, we find the row that corresponds to 2 degrees of freedom. The value of 132.882 is greater than 10.597, which has an area under the chi-square distribution of 0.005 to the right. Therefore, we have *P*-value < 0.005 [Tech: *P*-value < 0.001]. Since *P*-value < α, we reject H_0.

Conclusion:
There is sufficient evidence, at the $\alpha = 0.05$ level of significance, to conclude that survival status and social class are dependent.

(b) The conditional distribution and bar graph support the conclusion that a relationship exists between survival status and social class. Individuals with higher-class tickets survived in greater proportions than those with lower-class tickets.

	Class		
	First	**Second**	**Third**
Survived	$\frac{203}{325} \approx 0.625$	$\frac{118}{285} \approx 0.414$	$\frac{178}{706} \approx 0.252$
Did Not Survive	$\frac{122}{325} \approx 0.375$	$\frac{167}{285} \approx 0.586$	$\frac{528}{706} \approx 0.748$

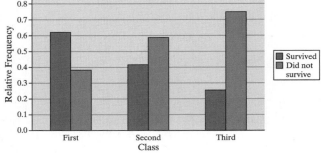

Titanic

4. The expected counts are calculated as $\frac{(\text{row total}) \cdot (\text{column total})}{(\text{table total})} = \frac{1279 \cdot 40}{4953} \approx 10.329$ (for the first cell) and so on, giving the following table:

Less than	Class					Total
H.S. degree	**22-27**	**28-32**	**33-36**	**37-42**	**43+**	
Yes	14	34	140	1010	81	1279
	(10.329)	(25.565)	(124.724)	(1043.755)	(74.628)	
No	26	65	343	3032	208	3674
	(29.671)	(73.435)	(358.276)	(2998.245)	(214.372)	
Total	40	99	483	4042	289	4953

All expected frequencies are greater than 5 so all requirements for a chi-square test are satisfied. That is, all expected frequencies are greater than or equal to 1, and no more than 20% of the expected frequencies are less than 5.

$$\chi_0^2 = \sum \frac{(O_i - E_i)^2}{E_i} = \frac{(14 - 10.329)^2}{10.329} + \cdots + \frac{(208 - 214.372)^2}{214.372} \approx 10.238 \quad [\text{Tech: } 10.239]$$

H_0: gestation period and completing high school are independent

H_1: gestation period and completing high school are dependent

df $= (2-1)(5-1) = 4$ so the critical value is $\chi_{0.05}^2 = 9.488$.

Classical Approach:
The test statistic is 10.238 which is greater than the critical value so we reject H_0.

P-value Approach:
Using Table VII, we find the row that corresponds to 4 degrees of freedom. The value of 10.238 is greater than 9.488, which has an area under the chi-square distribution of 0.05 to the right. Therefore, we have *P*-value < 0.05 [Tech: *P*-value < 0.037]. Since *P*-value < α, we reject H_0.

Conclusion:
There is sufficient evidence, at the $\alpha = 0.05$ level of significance, to conclude that gestation period and completing high school are dependent.

5. **(a)** The expected counts are calculated as $\dfrac{(\text{row total}) \cdot (\text{column total})}{(\text{table total})} = \dfrac{1976 \cdot 700}{2334} \approx 592.631$ (for the first cell) and so on, giving the following table:

	Political Affiliation			Total
	Republican	**Independent**	**Democrat**	
Important	644	662	670	1976
	(592.631)	(691.685)	(691.685)	
Not Important	56	155	147	358
	(107.369)	(125.315)	(125.315)	
Total	700	817	817	2334

All expected frequencies are greater than 5 so all requirements for a chi-square test are satisfied. That is, all expected frequencies are greater than or equal to 1, and no more than 20% of the expected frequencies are less than 5.

$$\chi_0^2 = \sum \frac{(O_i - E_i)^2}{E_i} = \frac{(644 - 592.631)^2}{592.631} + \cdots + \frac{(147 - 125.315)^2}{125.315} \approx 41.767$$

$H_0 : p_R = p_I = p_D$
$H_1 :$ at least one proportion is different from the others

$df = (2-1)(3-1) = 2$ so the critical value is $\chi_{0.05}^2 = 5.991$.

Classical Approach:
The test statistic is 41.767 which is greater than the critical value so we reject H_0.

P-value Approach:
Using Table VII, we find the row that corresponds to 2 degrees of freedom. The value of 41.767 is greater than 10.597, which has an area under the chi-square distribution of 0.005 to the right. Therefore, we have P-value < 0.005 [Tech: P-value < 0.001]. Since P-value < α, we reject H_0.

Conclusion:
There is sufficient evidence, at the $\alpha = 0.05$ level of significance, to conclude that at least one proportion is different from the others. That is, the evidence suggests that the proportion of adults who feel morality is important when deciding how to vote is different for at least one political affiliation.

(b) The conditional distribution and bar graph support the conclusion that the proportion of adults who feel morality is important when deciding how to vote is different for at least one political affiliation. It appears that a higher proportion of Republicans feel that morality is important when deciding how to vote than for Democrats or Independents.

	Political Affiliation		
	Republican	**Independent**	**Democrat**
Important	$\dfrac{644}{700} = 0.920$	$\dfrac{662}{817} \approx 0.810$	$\dfrac{670}{817} \approx 0.820$
Not Important	$\dfrac{56}{700} = 0.080$	$\dfrac{155}{817} \approx 0.190$	$\dfrac{147}{817} \approx 0.180$

Moral Values of Candidates

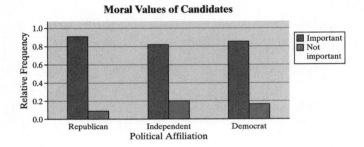

Chapter 12 Test

1. Using $\alpha = 0.01$, we want to test

    H_0 : The dice are fair   vs. H_1 : The dice are loaded

    If the dice are fair, the sum of the two dice will follow the given distribution. There are 400 rolls. To determine the expected number of times for each sum, multiply 400 by the given relative frequency for each

    sum. For example, the expected count for the sum 2 would be $400\left(\dfrac{1}{36}\right) \approx 11.111$. We summarize the

    observed and expected counts in the following table:

Sum	Observed Count (O_i)	Expected Rel. Freq.	Expected Count (E_i)	$(O_i - E_i)^2$	$(O_i - E_i)^2 / E_i$
2	16	1/36	11.111	23.902	2.1512
3	23	2/36	22.222	0.605	0.0272
4	31	3/36	33.333	5.443	0.1633
5	41	4/36	44.444	11.861	0.2669
6	62	5/36	55.556	41.525	0.7474
7	59	6/36	66.667	58.783	0.8817
8	59	5/36	55.556	11.861	0.2135
9	45	4/36	44.444	0.309	0.0070
10	34	3/36	33.333	0.445	0.0133
11	19	2/36	22.222	10.381	0.4672
12	11	1/36	11.111	0.012	0.0011
				$\chi_0^2 \approx$	4.940

Since all the expected cell counts are greater than or equal to 5, the requirements for the goodness-of-fit test are satisfied.

Classical approach: The critical value, with df $= 11 - 1 = 10$, is $\chi_{0.01}^2 = 23.209$. Since the test statistic is not in the critical region ($\chi_0^2 < \chi_{0.01}^2$), we do not reject the null hypothesis.

P-value approach: Using the chi-square table, we find the row that corresponds to 10 degrees of freedom. The value of 4.940 is less than 15.987, which has an area under the chi-square distribution of 0.10 to the right. Therefore, we have P-value > 0.10 [Tech: 0.895]. Since P-value $> \alpha$, we do not reject the null hypothesis.

Conclusion: There is not enough evidence, at the $\alpha = 0.01$ level of significance, to conclude that the dice are loaded. The data indicate that the dice are fair.

2. Using $\alpha = 0.1$, we want to test

H_0 : Educational attainment in the U.S. is the same as in 2000   vs.

H_1 : Educational attainment is different today than in 2000.

There are 500 Americans in the sample. To determine the expected number attaining each degree level, multiply 500 by the given relative frequency for each level. For example, the expected count for "not a high school graduate" would be $500(0.158) = 79$. We summarize the observed and expected counts in the table:

Observed Count (O_i)	Expected Rel. Freq.	Expected Count (E_i)	$(O_i - E_i)^2$	$(O_i - E_i)^2 / E_i$
72	0.158	79	49.00	0.6203
159	0.331	165.5	42.25	0.2553
85	0.176	88	9.00	0.1023
44	0.078	39	25.00	0.6410
92	0.170	85	49.00	0.5765
48	0.087	43.5	20.25	0.4655
			$\chi_0^2 \approx$	2.661

Since all the expected cell counts are greater than or equal to 5, the requirements for the goodness-of-fit test are satisfied.

<u>Classical approach:</u> The critical value, with df $= 6 - 1 = 5$, is $\chi_{0.10}^2 = 9.236$. Since the test statistic is not in the critical region ($\chi_0^2 < \chi_{0.10}^2$), we do not reject the null hypothesis.

<u>P-value approach:</u> Using the chi-square table, we find the row that corresponds to 5 degrees of freedom. The value of 2.661 is less than 9.236, which has an area under the chi-square distribution of 0.10 to the right. Therefore, we have P-value > 0.10 [Tech: 0.752]. Since P-value $> \alpha$, we do not reject the null hypothesis.

<u>Conclusion:</u> There is not enough evidence, at the $\alpha = 0.1$ level of significance, to conclude that the distribution of educational attainment has changed since 2000. That is, the data indicate that educational attainment has not changed.

3. (a) The expected counts are calculated as $\dfrac{\text{(row total)} \cdot \text{(column total)}}{\text{(table total)}} = \dfrac{1997 \cdot 547}{2712} \approx 402.787$ (for the first cell) and so on, giving the following table:

Race	Region				Total
	Not HS Graduate	**HS Graduate**	**Some College**	**Graduated College**	
White	421	520	656	400	1997
	(402.787)	(452.123)	(679.657)	(462.432)	
Black	56	57	158	28	299
	(60.307)	(67.694)	(101.761)	(69.237)	
Asian or Pacific Islander	13	8	11	40	72
	(14.522)	(16.301)	(24.504)	(16.673)	
Hispanic	38	17	68	101	224
	(45.180)	(50.714)	(76.236)	(51.870)	
Other	19	12	30	59	120
	(24.204)	(27.168)	(40.841)	(27.788)	
Total	547	614	923	628	2712

All expected frequencies are greater than 5 so all requirements for a chi-square test are satisfied. That is, all expected frequencies are greater than or equal to 1, and no more than 20% of the expected frequencies are less than 5.

$$\chi_0^2 = \sum \frac{(O_i - E_i)^2}{E_i} = \frac{(421 - 402.787)^2}{402.787} + \cdots + \frac{(59 - 27.788)^2}{27.788} \approx 240.871 \quad [\text{Tech: } 240.873]$$

H_0 : Race and Region are independent

H_1 : Race and Region are dependent

df $= (5-1)(4-1) = 12$ so the critical value is $\chi_{0.05}^2 = 21.026$.

<u>Classical Approach:</u>
The test statistic is 240.871 which is greater than the critical value so we reject H_0 .

<u>*P*-value Approach:</u>
Using Table VII, we find the row that corresponds to 12 degrees of freedom. The value of 240.871 is greater than 28.300, which has an area under the chi-square distribution of 0.005 to the right. Therefore, we have *P*-value < 0.005 [Tech: *P*-value < 0.001]. Since *P*-value < α , we reject H_0 .

<u>Conclusion:</u>
There is sufficient evidence, at the $\alpha = 0.05$ level of significance, to conclude that race and region are dependent. That is, there appears to be a link between race and region in the United States.

(b) This summary supports the conclusion in part (a) by showing that the proportion for each racial category is not the same within each region of the United States. For example, whites make up a larger proportion in the Midwest than they do in any other region. Blacks make up a larger proportion in the South than they do in any other region. Hispanics make up a larger proportion in the West than they do any other region.

	Political Affiliation			
	Northwest	**Midwest**	**South**	**West**
White	$\frac{421}{547} \approx 0.770$	$\frac{520}{614} \approx 0.847$	$\frac{656}{923} \approx 0.711$	$\frac{400}{628} \approx 0.637$
Black	$\frac{56}{547} \approx 0.102$	$\frac{57}{614} \approx 0.093$	$\frac{158}{923} \approx 0.171$	$\frac{28}{628} \approx 0.045$
Asian or Pacific Islander	$\frac{13}{547} \approx 0.024$	$\frac{8}{614} \approx 0.013$	$\frac{11}{923} \approx 0.012$	$\frac{40}{628} \approx 0.064$
Hispanic	$\frac{38}{547} \approx 0.069$	$\frac{17}{614} \approx 0.028$	$\frac{68}{923} \approx 0.074$	$\frac{101}{628} \approx 0.161$
Other	$\frac{19}{547} \approx 0.035$	$\frac{12}{614} \approx 0.020$	$\frac{30}{923} \approx 0.033$	$\frac{59}{628} \approx 0.094$

Race Versus Region of the United States

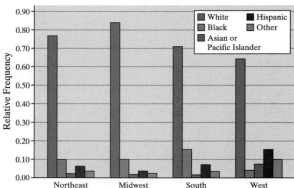

4. The expected counts are calculated as $\dfrac{(\text{row total}) \cdot (\text{column total})}{(\text{table total})} = \dfrac{344 \cdot 550}{1510} \approx 125.298$ (for the first cell) and so on, giving the following table:

	Education			Total
	High School or Less	**Some College**	**College Graduate**	
Enjoy	143	102	99	344
	(125.298)	(105.934)	(112.768)	
Do not Enjoy	407	363	396	1166
	(424.702)	(359.066)	(382.232)	
Total	550	465	495	1510

All expected frequencies are greater than 5 so all requirements for a chi-square test are satisfied. That is, all expected frequencies are greater than or equal to 1, and no more than 20% of the expected frequencies are less than 5.

$$\chi_0^2 = \sum \frac{(O_i - E_i)^2}{E_i} = \frac{(143 - 125.298)^2}{125.298} + \cdots + \frac{(396 - 382.232)^2}{382.232} \approx 5.605$$

$H_0: \ p_{\text{H}} = p_{\text{C}} = p_{\text{G}}$

$H_1:$ at least one proportion is different from the others

df $= (2-1)(3-1) = 2$ so the critical value is $\chi_{0.05}^2 = 5.991$.

<u>Classical Approach:</u>
The test statistic is 5.605 which is less than the critical value so we do not reject H_0.

<u>P-value Approach:</u>
Using Table VII, we find the row that corresponds to 2 degrees of freedom. The value of 5.605 is less than 5.991, which has an area under the chi-square distribution of 0.05 to the right. Therefore, we have P-value > 0.05 [Tech: P-value $= 0.061$]. Since P-value $> \alpha$, we do not reject H_0.

<u>Conclusion:</u>
There is not sufficient evidence, at the $\alpha = 0.05$ level of significance, to conclude that at least one proportion is different from the others. That is, the evidence suggests that the proportions of individuals who enjoy gambling do not differ among the levels of education.

5. The expected counts are calculated as $\dfrac{(\text{row total}) \cdot (\text{column total})}{(\text{table total})} = \dfrac{3179 \cdot 137}{8874} \approx 49.079$ (for the first cell) and so on, giving the following table:

	Education							Total
	Almost Daily	**Several Times a Week**	**Several Times a Month**	**Once a Month**	**Several Times a Year**	**Once a Year**	**Never**	
Smoker	80	409	294	362	433	336	1265	3179
	(49.079)	(271.902)	(241.094)	(298.412)	(360.387)	(323.847)	(1634.280)	
Nonsmoker	57	350	379	471	573	568	3297	5695
	(87.921)	(487.098)	(431.906)	(534.588)	(645.613)	(580.153)	(2927.720)	
Total	137	759	673	833	1006	904	4562	8874

All expected frequencies are greater than 5 so all requirements for a chi-square test are satisfied. That is, all expected frequencies are greater than or equal to 1, and no more than 20% of the expected frequencies are less than 5.

$$\chi_0^2 = \sum \frac{(O_i - E_i)^2}{E_i} = \frac{(80 - 49.079)^2}{49.079} + \cdots + \frac{(3297 - 2927.720)^2}{2927.720} \approx 330.803$$

H_0 : The distribution of bar visits is the same for smokers and nonsmokers

H_1 : The distribution of bar visits is different for smokers and nonsmokers

df $= (2-1)(7-1) = 6$ so the critical value is $\chi_{0.05}^2 = 12.592$.

Classical Approach:
The test statistic is 330.803 which is greater than the critical value so we reject H_0 .

P-value Approach:
Using Table VII, we find the row that corresponds to 6 degrees of freedom. The value of 330.803 is greater than 18.548, which has an area under the chi-square distribution of 0.005 to the right. Therefore, we have P-value < 0.005 [Tech: P-value < 0.001]. Since P-value < α , we reject H_0 .

Conclusion:
There is sufficient evidence, at the $\alpha = 0.05$ level of significance, to conclude that the distributions of time spent in bars differ between smokers and nonsmokers.

To determine if smokers tend to spend more time in bars than nonsmokers, we can examine the conditional distribution of time spent in bars by smoker status.

	Education						
	Almost Daily	**Several Times a Week**	**Several Times a Month**	**Once a Month**	**Several Times a Year**	**Once a Year**	**Never**
Smoker	$\frac{80}{137} \approx 0.584$	$\frac{409}{759} \approx 0.539$	$\frac{294}{673} \approx 0.437$	$\frac{362}{833} \approx 0.435$	$\frac{433}{1006} \approx 0.430$	$\frac{336}{904} \approx 0.372$	$\frac{1265}{4562} \approx 0.277$
Nonsmoker	$\frac{57}{137} \approx 0.416$	$\frac{350}{759} \approx 0.461$	$\frac{379}{673} \approx 0.563$	$\frac{471}{833} \approx 0.565$	$\frac{573}{1006} \approx 0.570$	$\frac{568}{904} \approx 0.628$	$\frac{3297}{4562} \approx 0.723$

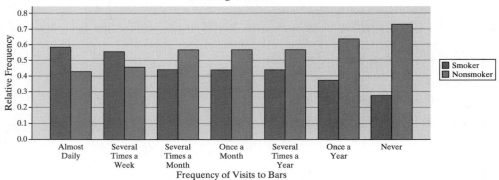

Based on the conditional distribution and the bar graph, it appears that smokers tend to spend more time in bars than nonsmokers.

Chapter 13
Comparing Three or More Means

13.1 Comparing Three or More Means (One-Way Analysis of Variance)

1. analysis of variance.

3. The mean square due to treatment estimate of σ^2 is a weighted average of the squared deviations of each sample mean from the grand mean of all the samples. The mean square due to error estimate of σ^2 is the weighted average of the sample variances.

5.

Source of Variation	Sum of Squares	Degrees of Freedom	Mean Squares	F - Test Statistic
Treatment	387	2	$\dfrac{387}{2} = 193.5$	$\dfrac{193.5}{297.852} \approx 0.650$
Error	8042	27	$\dfrac{8042}{27} \approx 297.852$	
Total	8429	29		

7. $\overline{x} = \dfrac{10 \cdot 40 + 10 \cdot 42 + 10 \cdot 44}{10 + 10 + 10} = \dfrac{1260}{30} = 42$; $\text{MST} = \dfrac{10(40-42)^2 + 10(42-42)^2 + 10(44-42)^2}{3-1} = \dfrac{80}{2} = 40$;

$\text{MSE} = \dfrac{(10-1) \cdot 48 + (10-1) \cdot 31 + (10-1) \cdot 25}{30-3} = \dfrac{936}{27} \approx 34.667$;

$F_0 = \dfrac{40}{34.667} \approx 1.154$

9. $\overline{x}_1 = \dfrac{28 + 23 + 30 + 27}{4} = \dfrac{108}{4} = 27$; $\overline{x}_2 = \dfrac{22 + 25 + 17 + 23}{4} = \dfrac{87}{4} = 21.75$; $\overline{x}_3 = \dfrac{25 + 24 + 19 + 30}{4} = \dfrac{98}{4} = 24.5$;

$\overline{x} = \dfrac{28 + 23 + 30 + \ldots + 19 + 30}{12} = \dfrac{293}{12} \approx 24.417$;

$s_1^2 = \dfrac{(28-27)^2 + (23-27)^2 + (30-27)^2 + (27-27)^2}{4-1} = \dfrac{26}{3} \approx 8.667$;

$s_2^2 = \dfrac{(22-21.75)^2 + (25-21.75)^2 + (17-21.75)^2 + (23-21.75)^2}{4-1} = \dfrac{34.75}{3} \approx 11.583$;

$s_3^2 = \dfrac{(25-24.5)^2 + (24-24.5)^2 + (19-24.5)^2 + (30-24.5)^2}{4-1} = \dfrac{61}{3} \approx 20.333$;

$\text{MST} = \dfrac{4(27-24.417)^2 + 4(21.75-24.417)^2 + 4(24.5-24.417)^2}{3-1} \approx \dfrac{55.167}{2} \approx 27.584$;

$\text{MSE} = \dfrac{(4-1) \cdot 8.667 + (4-1) \cdot 11.583 + (4-1) \cdot 20.333}{12-3} = \dfrac{121.749}{9} \approx 13.528$;

$F_0 = \dfrac{27.584}{13.528} \approx 2.04$.

11. (a) The hypotheses are $H_0 : \mu_{\text{sludge plot}} = \mu_{\text{spring disk}} = \mu_{\text{no till}}$ versus H_1 : at least one mean differs from the others.

(b) (1) Each sample must be a simple random sample.
(2) The three samples must be independent of each other.
(3) The samples must come from normally distributed populations.
(4) The populations must have equal variances.

(c) Since P-value $= 0.007$ is smaller than the $\alpha = 0.05$ level of significance, we reject H_0. There is sufficient evidence to conclude that at least one plot type has a mean number of plants that differs from the other means.

(d) Yes, the boxplots support the results obtained in part (c). The boxplots indicate that significantly more plants are growing in the spring disk plot than in the other plot types.

(e) The needed summary statistics are: $\bar{x}_{\text{sludge plot}} \approx 28.333$, $\bar{x}_{\text{spring disk}} = 33$, $\bar{x}_{\text{no till}} = 28.5$, $\bar{x} \approx 29.944$,

$s^2_{\text{sludge plot}} \approx 7.867$, $s^2_{\text{spring disk}} = 3.2$, $s^2_{\text{no till}} = 6.7$, $n_{\text{sludge plot}} = n_{\text{spring disk}} = n_{\text{no till}} = 6$, $n = 18$, and $k = 3$.
Thus,

$$\text{MST} = \frac{6(28.333 - 29.994)^2 + 6(33 - 29.994)^2 + 6(28.5 - 29.994)^2}{3 - 1} \approx \frac{84.114}{2} \approx 42.057,$$

$$\text{MSE} = \frac{(6-1) \cdot 7.867 + (6-1) \cdot 3.2 + (6-1) \cdot 6.7}{18 - 3} = \frac{88.835}{15} \approx 5.922, \text{ and } F_0 = \frac{42.057}{5.922} \approx 7.10.$$

13. (a) The hypotheses are $H_0 : \mu_M = \mu_T = \mu_W = \mu_R = \mu_F$ versus H_1 : at least one mean differs from the others.

(b) (1) Each sample must be a simple random sample.
(2) The five samples must be independent of each other.
(3) The samples must come from normally distributed populations.
(4) The populations must have equal variances.

(c) Since P-value < 0.001 is smaller than the $\alpha = 0.01$ level of significance, we reject H_0. There is sufficient evidence to conclude that at least one weekday has a mean number of births that differs from the other weekdays.

(d) Yes, from the boxplots it appears that Monday has fewer births than other weekdays.

(e) The needed summary statistics are: $\bar{x}_M \approx 10,696.3$, $\bar{x}_T \approx 12,237.1$, $\bar{x}_W \approx 11,586.9$, $\bar{x}_R \approx 11,897.9$,

$\bar{x}_F \approx 12,002.9$, $\bar{x} = 11,684.2$, $s^2_M = 279,042.5$, $s^2_T \approx 633,537.8$, $s^2_W \approx 136,085.3$, $s^2_R \approx 252,402.7$,

$s^2_F \approx 166,186.1$ $n_M = n_T = n_W = n_R = n_F = 8$, $n = 40$, and $k = 5$. Thus,

$$\text{MST} = \frac{8(10,696.3 - 11,684.2)^2 + 8(12,237.1 - 11,684.2)^2 + \ldots + 8(12,002.9 - 11,684.2)^2}{5 - 1}$$

$$\approx \frac{11,506,795.9}{4} \approx 2,876,698.8,$$

$$\text{MSE} = \frac{(8-1) \cdot 279,042.5 + (8-1) \cdot 633,537.8 + \ldots + (8-1) \cdot 116,186.1}{40 - 5} = \frac{10,270,780.8}{35} \approx 293,450.9, \text{ and}$$

$$F_0 = \frac{2,876,698.8}{293,450.9} \approx 9.80.$$

(f) From the boxplots, Monday appears to have significantly fewer births than the other days.

15. (a) The hypotheses are $H_0 : \mu_{\text{Financial}} = \mu_{\text{Energy}} = \mu_{\text{Utilities}}$ versus H_1 : at least one mean differs from the others.

(b) (1) Each sample is a simple random sample.
(2) The three samples are independent of each other.
(3) The samples are from normally distributed populations.
(4) $s_{\text{Financial}} \approx 5.12$, $s_{\text{Energy}} \approx 4.87$, and $s_{\text{Utilities}} \approx 4.53$. The largest sample standard deviation is less than twice the smallest sample standard deviation, so the requirement that the populations have equal variances is satisfied.

(c) Using a TI-84 Plus, we obtain the following output:

Since P-value ≈ 0.150 is larger than the $\alpha = 0.05$ level of significance, we do not reject H_0. By hand we obtain $F_0 = 2.08 < F_{0.05,2,21} \approx F_{0.05,2,20} = 3.49$, so we do not reject H_0. There is not sufficient evidence to conclude that at least one mean is different from the others.

(d)

17. (a) The hypotheses are $H_0 : \mu_{\text{Large Fam. Car}} = \mu_{\text{Pass. Van}} = \mu_{\text{Mid. Util. Veh.}}$ versus H_1 : at least one mean differs from the others.

(b) (1) Each sample is a simple random sample.
(2) The three samples are independent of each other.
(3) The samples are from normally distributed populations.
(4) The largest sample standard deviation is less than twice the smallest sample standard deviation ($4.53 < 2 \cdot 3.53$), so the requirement that the populations have equal variances is satisfied.

(c) Using a TI-84 Plus, we obtain the following output:

Since P-value ≈ 0.178 is larger than the $\alpha = 0.01$ level of significance, we do not reject H_0. By hand we obtain $F_0 = 1.91 < F_{0.01,2,18} \approx F_{0.01,2,20} = 5.85$, so we do not reject H_0. There is not sufficient evidence to conclude that at least one mean is different from the others.

(d)

19. **(a)** The hypotheses are $H_0 : \mu_{\text{Alaska}} = \mu_{\text{Florida}} = \mu_{\text{Texas}}$ versus H_1 : at least one mean differs from the others.

(b) (1) Each sample is a simple random sample.
(2) The three samples are independent of each other.
(3) The samples are from normally distributed populations.
(4) The largest sample standard deviation is less than twice the smallest sample standard deviation ($0.397 < 2 \cdot 0.252$), so the requirement that the populations have equal variances is satisfied.

(c) Using a TI-84 Plus, we obtain the following output:

```
One-way ANOVA       One-way ANOVA
 F=5.810954298      ↑ MS=.583755556
 p=.0135328894       Error
 Factor               df=15
  df=2                SS=1.50686667
  SS=1.16751111       MS=.100457778
↓ MS=.583755556      Sxp=.31695075
```

Since P-value ≈ 0.014 is less than $\alpha = 0.05$, we reject H_0. By hand we obtain

$F_0 = 5.81 > F_{0.05, 2, 15} = 3.68$, so we reject H_0. There is sufficient evidence to conclude that at least one mean is different from the others.

(d)

21. **(a)** The hypotheses are $H_0 : \mu_{67-0-301} = \mu_{67-0-400} = \mu_{67-0-353}$ versus H_1 : at least one mean differs from the others.

(b) $s_{67-0-301} \approx 107.8$, $s_{67-0-400} \approx 385.0$, $s_{67-0-353} \approx 195.8$. Since the largest standard deviation is greater than twice the smallest standard deviation (i.e., since $385.0 > 2 \cdot 107.8$), the rule of thumb for meeting the ANOVA requirement of equal variances is violated, and so we cannot perform an ANOVA.

23. **(a)** $s_{\text{control}} = \sqrt{21.6} \approx 4.6$, $s_{\text{RAP}} = \sqrt{13.0} \approx 3.6$, and $s_{\text{headache}} = \sqrt{8.4} \approx 2.9$

(b) The problem states that children were stratified in two age groups: 4 to 11 years, and 12 to 18 years. That is, the children in the study were separated by a common characteristic, age. Thus, the researchers used stratified sampling.

(c) $H_0 : \mu_1 = \mu_2$ versus $H_1 : \mu_1 \neq \mu_2$. The level of significance is $\alpha = 0.05$. Since the sample size of both groups is 70, we use $n_1 - 1 = 69$ degrees of freedom. The test statistic is

$$t_0 = \frac{(\bar{x}_1 - \bar{x}_2) - (\mu_1 - \mu_2)}{\sqrt{\dfrac{s_1^2}{n_1} + \dfrac{s_2^2}{n_2}}} = \frac{(11.7 - 9.0) - 0}{\sqrt{\dfrac{21.6}{70} + \dfrac{13.0}{70}}} \approx 3.84 .$$

Classical approach: Since this is a two-tailed test with 69 degrees of freedom, the critical values are $\pm t_{0.025} = \pm 1.994$. Since the test statistic $t_0 \approx 3.84$ is more than the critical value $t_{0.025} = 1.994$ (i.e., since the test statistic falls within the critical region), we reject H_0.

P-value approach: The *P*-value for this two-tailed test is the area under the *t*-distribution with 69 degrees of freedom to the right of $t_0 = 3.84$ plus the area to the left of -3.84. From the *t*-distribution table in the row corresponding to 70 degrees of freedom (since there is no entry for 69), 3.84 falls above 3.435 whose right-tail area is 0.0005. We must double this value in order to get the total area in both tails: 0.001. So, *P*-value < 0.001. (Using technology, we find *P*-value = 0.000191.) Because the *P*-value is less than the level of significance $\alpha = 0.05$, we reject H_0.

Conclusion: There is sufficient evidence at the $\alpha = 0.05$ level of significance to conclude that the population mean CBCL scores are different.

(d) It is not necessary to check the normality assumption because the samples are all considered to be large samples (sample size greater than 30).

(e) $\bar{x} = \dfrac{11.7 + 9.0 + 12.4}{3} = 11.03$

$MST = \dfrac{70(11.7 - 11.03)^2 + 70(9.0 - 11.03)^2 + 70(12.4 - 11.03)^2}{3 - 1} = \dfrac{451.269}{2} \approx 225.635$,

$MSE = \dfrac{(70-1) \cdot 21.6 + (70-1) \cdot 13.0 + (70-1) \cdot 8.4}{210 - 3} = \dfrac{2967}{207} \approx 14.333$, and $F_0 = \dfrac{225.635}{14.333} \approx 15.74$.

Since $F_0 = 15.74 > F_{0.05,2,207} \approx F_{0.05,2,200} = 3.04$, we reject H_0. There is sufficient evidence to conclude that at least one mean CBCL scores is different from the others.

(f) No; the result from part (e) only indicates that *at least one* mean is different. It does not indicate which pairs are different nor even how many. The result from part (c) indicates that the mean scores for the control group and the RAP group are different, but additional pairs could also be significantly different.

13.2 Post Hoc Tests on One-Way Analysis of Variance

1. The purpose of multiple comparison tests is to determine which means differ significantly. We use them when the results of a one-way ANOVA lead us to reject the null hypothesis, telling us that at least one of the means is different.

3. (a) We look in the table of critical values for Tukey's Test with $\alpha = 0.05$ and where the row corresponding to $\nu = 10$ intersects with the column corresponding to $k = 3$ to find the critical value $q_{0.05,10,3} = 3.877$.

(b) We look in the table of critical values for Tukey's Test with $\alpha = 0.05$ and where the row corresponding to $\nu = 24$ intersects with the column corresponding to $k = 5$ to find the critical value $q_{0.05,24,5} = 4.166$.

(c) We look in the table of critical values for Tukey's Test with $\alpha = 0.05$ and where the row corresponding to $\nu = 30$ (since there is no row in the table for $\nu = 32$) intersects with the column corresponding to $k = 8$ to find the critical value $q_{0.05,32,8} \approx q_{0.05,30,8} = 4.602$.

(d) For this test, $k = 5$ and $\nu = n - k = 65 - 5 = 60$, so we look in the table with $\alpha = 0.05$ and where the row corresponding to $\nu = 60$ intersects with the column corresponding to $k = 5$ to find the critical value $q_{0.05,60,5} = 3.977$.

5. Arrange the sample means in ascending order: $\overline{x}_2 = 9.1$, $\overline{x}_1 = 9.5$, $\overline{x}_3 = 18.1$. Now $n = 5 \cdot 3 = 15$ and $k = 3$, so $\nu = n - k = 12$ and the critical value is $q_{0.05,12,3} = 3.773$.

We first test $H_0 : \mu_3 = \mu_2$. The test statistic is $q_0 = \dfrac{\overline{x}_3 - \overline{x}_2}{\sqrt{\frac{s^2}{2} \cdot \left(\frac{1}{n_3} + \frac{1}{n_2} \right)}} = \dfrac{18.1 - 9.1}{\sqrt{\frac{26.2}{2} \cdot \left(\frac{1}{5} + \frac{1}{5} \right)}} \approx 3.932$. Since this test

statistic is greater than $q_{0.05,12,3} = 3.773$, we reject $H_0 : \mu_3 = \mu_2$.

Next, we test $H_0 : \mu_3 = \mu_1$. The test statistic is $q_0 = \dfrac{\overline{x}_3 - \overline{x}_1}{\sqrt{\frac{s^2}{2} \cdot \left(\frac{1}{n_3} + \frac{1}{n_1} \right)}} = \dfrac{18.1 - 9.5}{\sqrt{\frac{26.2}{2} \cdot \left(\frac{1}{5} + \frac{1}{5} \right)}} \approx 3.757$. Since this test

statistic is less than $q_{0.05,12,3} = 3.773$, we do not reject $H_0 : \mu_3 = \mu_1$.

Finally, we test $H_0 : \mu_1 = \mu_2$. The test statistic is $q_0 = \dfrac{\overline{x}_1 - \overline{x}_2}{\sqrt{\frac{s^2}{2} \cdot \left(\frac{1}{n_1} + \frac{1}{n_2} \right)}} = \dfrac{9.5 - 9.1}{\sqrt{\frac{26.2}{2} \cdot \left(\frac{1}{5} + \frac{1}{5} \right)}} \approx 0.175$. Since this test

statistic is less than $q_{0.05,12,3} = 3.773$, we do not reject $H_0 : \mu_1 = \mu_2$.

The results of the test are ambiguous since we have that $\mu_1 = \mu_3$ and $\mu_1 = \mu_2$, but $\mu_2 \neq \mu_3$. That is, we have: $\underline{\mu_2 \quad \mu_1 \quad \mu_3}$.

7. **(a)** The hypotheses are $H_0 : \mu_1 = \mu_2 = \mu_3$ versus H_1 : at least one mean differs from the others. Using MINITAB, we obtain the following output:

    One-way ANOVA: Sample 1, Sample 2, Sample 3
    ```
    Source  DF      SS      MS      F      P
    Factor   2    630.3   315.2   9.72   0.002
    Error   15    486.5    32.4
    Total   17   1116.8
    ```
    Since P-value ≈ 0.002 is less than the $\alpha = 0.05$ level of significance, we reject H_0. There is sufficient evidence to conclude that at least one mean differs from the others.

    (b) Using MINITAB, we obtain the following output:

    Tukey 95% Simultaneous Confidence Intervals: All Pairwise Comparisons
    ```
    Individual confidence level = 97.97%
    ```

    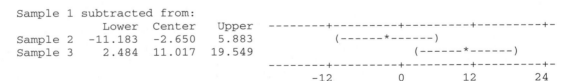

    ```
    Sample 1 subtracted from:
              Lower   Center   Upper
    Sample 2 -11.183  -2.650   5.883
    Sample 3   2.484  11.017  19.549
    ```

    ```
    Sample 2 subtracted from:
              Lower   Center   Upper
    Sample 3  5.134   13.667  22.199
    ```

    From the output we have:

    $-11.183 < \mu_2 - \mu_1 < 5.883$. This interval contains 0, so we do not reject $H_0 : \mu_1 = \mu_2$.

    $2.484 < \mu_3 - \mu_1 < 19.549$. This interval does not contain 0, so we reject $H_0 : \mu_1 = \mu_3$.

    $5.134 < \mu_3 - \mu_2 < 22.199$. This interval does not contain 0, so we reject $H_0 : \mu_2 = \mu_3$.

    Thus, we have $\mu_1 = \mu_2 \neq \mu_3$. That is, we have: $\underline{\mu_1 \quad \mu_2 \quad \mu_3}$.

(c) The boxplots shown to the right support the result of parts (a) and (b).

9. From the provided output, we have:

$1.021 < \mu_{\text{spring disk}} - \mu_{\text{sludge plot}} < 8.313$. This interval does not contain 0, so we reject $H_0 : \mu_{\text{sludge plot}} = \mu_{\text{spring disk}}$

$-3.479 < \mu_{\text{no till}} - \mu_{\text{sludge plot}} < 3.813$. This interval contains 0, so we do not reject $H_0 : \mu_{\text{sludge plot}} = \mu_{\text{no till}}$.

$-8.146 < \mu_{\text{no till}} - \mu_{\text{spring disk}} < -0.854$. This interval does not contain 0, so we reject $H_0 : \mu_{\text{spring disk}} = \mu_{\text{no till}}$.

Thus, we have $\mu_{\text{sludge plot}} = \mu_{\text{no till}} \neq \mu_{\text{spring disk}}$. That is, we have: $\underline{\mu_{\text{sludge plot}} \quad \mu_{\text{no till}}} \quad \underline{\mu_{\text{spring disk}}}$.

From the pairwise comparisons, it appears that the spring disk method has a greater mean number of plants. Therefore, we should recommend spring-disk method of planting.

11. We use MINITAB to obtain the following output:

Tukey 95% Simultaneous Confidence Intervals: All Pairwise Comparisons
Individual confidence level = 97.97%

Alaska subtracted from:

	Lower	Center	Upper
Florida	-0.7015	-0.2267	0.2482
Texas	-0.0849	0.3900	0.8649

```
 --------+---------+---------+---------+-
          (-------*-------)
                       (------*-------)
 --------+---------+---------+---------+-
     -0.60      0.00      0.60      1.20
```

Florida subtracted from:

	Lower	Center	Upper
Texas	0.1418	0.6167	1.0915

```
 --------+---------+---------+---------+-
                         (-------*-------)
 --------+---------+---------+---------+-
     -0.60      0.00      0.60      1.20
```

From the output, we have:

$-0.7015 < \mu_{\text{Florida}} - \mu_{\text{Alaska}} < 0.2482$. This interval contains 0, so we do not reject $H_0 : \mu_{\text{Alaska}} = \mu_{\text{Florida}}$.

$-0.0849 < \mu_{\text{Texas}} - \mu_{\text{Alaska}} < 0.8649$. This interval contain 0s, so we do not reject $H_0 : \mu_{\text{Alaska}} = \mu_{\text{Texas}}$.

$0.1418 < \mu_{\text{Texas}} - \mu_{\text{Florida}} < 1.0915$. This interval does not contain 0, so we reject $H_0 : \mu_{\text{Florida}} = \mu_{\text{Texas}}$.

The results are ambiguous. We have that $\mu_{\text{Alaska}} = \mu_{\text{Florida}}$ and $\mu_{\text{Alaska}} = \mu_{\text{Texas}}$, but $\mu_{\text{Florida}} \neq \mu_{\text{Texas}}$. That is, we have: $\underline{\mu_{\text{Florida}} \quad \underline{\mu_{\text{Alaska}} \quad \mu_{\text{Texas}}}}$.

13. (a) The hypotheses are $H_0 : \mu_{\text{Placebo}} = \mu_{\text{Rit 10}} = \mu_{\text{Rit 17.5}} = \mu_{\text{Add 7.5}} = \mu_{\text{Add 12.5}}$ versus H_1 : at least one mean differs from the others. Using MINITAB, we obtain the following output:

One-way ANOVA: Placebo, Ritalin 10 m, Ritalin 17.5, Adderall 7.5, Adderall 12.

Source	DF	SS	MS	F	P
Factor	4	7297	1824	5.75	0.001
Error	30	9519	317		
Total	34	16816			

Since P-value $= 0.001 < \alpha = 0.05$ level of significance, we reject H_0. There is sufficient evidence to support the hypothesis that at least one of the mean abilities to follow rules is different from the others.

(b) Using MINITAB, we obtain the following output:

```
Tukey 95% Simultaneous Confidence Intervals:  All Pairwise Comparisons
Individual confidence level = 99.31%

Placebo subtracted from:

           Lower   Center   Upper   -----+---------+---------+---------+----
Rit 10      0.25    27.86   55.46                      (---------*--------)
Rit 17.5    1.11    28.71   56.32                      (---------*--------)
Add 7.5     4.11    31.71   59.32                       (---------*--------)
Add 12.5   16.40    44.00   71.60                           (---------*--------)
                                    -----+---------+---------+---------+----
                                       -30        0        30       60


Rit 10 subtracted from:

           Lower   Center   Upper   -----+---------+---------+---------+----
Rit 17.5  -26.75     0.86   28.46          (---------*--------)
Add 7.5   -23.75     3.86   31.46          (--------*---------)
Add 12.5  -11.46    16.14   43.75              (--------*---------)
                                    -----+---------+---------+---------+----
                                       -30        0        30       60


Rit 17.5 subtracted from:

           Lower   Center   Upper   -----+---------+---------+---------+----
Add 7.5   -24.60     3.00   30.60          (--------*--------)
Add 12.5  -12.32    15.29   42.89              (--------*--------)
                                    -----+---------+---------+---------+----
                                       -30        0        30       60


Add 7.5 subtracted from:

           Lower   Center   Upper   -----+---------+---------+---------+----
Add 12.5  -15.32    12.29   39.89              (--------*--------)
                                    -----+---------+---------+---------+----
                                       -30        0        30       60
```

From the output, we have:

$0.25 < \mu_{\text{Rit 10}} - \mu_{\text{Placebo}} < 55.46$. This interval does not contain 0, so we reject $H_0 : \mu_{\text{Placebo}} = \mu_{\text{Rit 10}}$.

$1.11 < \mu_{\text{Rit 17.5}} - \mu_{\text{Placebo}} < 56.32$. This interval does not contain 0, so we reject $H_0 : \mu_{\text{Placebo}} = \mu_{\text{Rit 17.5}}$.

$4.11 < \mu_{\text{Add 7.5}} - \mu_{\text{Placebo}} < 59.32$. This interval does not contain 0, so we reject $H_0 : \mu_{\text{Placebo}} = \mu_{\text{Add 7.5}}$.

$16.40 < \mu_{\text{Add 12.5}} - \mu_{\text{Placebo}} < 71.60$. This interval does not contain 0, so we reject $H_0 : \mu_{\text{Placebo}} = \mu_{\text{Add 12.5}}$.

$-26.75 < \mu_{\text{Rit 17.5}} - \mu_{\text{Rit 10}} < 28.46$. This interval contains 0, so we do not reject $H_0 : \mu_{\text{Rit 10}} = \mu_{\text{Rit 17.5}}$.

$-23.75 < \mu_{\text{Add 7.5}} - \mu_{\text{Rit 10}} < 31.46$. This interval contains 0, so we do not reject $H_0 : \mu_{\text{Rit 10}} = \mu_{\text{Add 7.5}}$.

$-11.5 < \mu_{\text{Add 12.5}} - \mu_{\text{Rit 10}} < 43.75$. This interval contains 0, so we do not reject $H_0 : \mu_{\text{Rit 10}} = \mu_{\text{Add 12.5}}$.

$-24.60 < \mu_{\text{Add 7.5}} - \mu_{\text{Rit 17.5}} < 30.60$. This interval contains 0, so we do not reject $H_0 : \mu_{\text{Rit 17.5}} = \mu_{\text{Add 7.5}}$.

$-12.32 < \mu_{\text{Add 12.5}} - \mu_{\text{Rit 17.5}} < 42.89$. This interval contains 0, so we do not reject $H_0 : \mu_{\text{Rit 17.5}} = \mu_{\text{Add 12.5}}$.

$-15.32 < \mu_{\text{Add 12.5}} - \mu_{\text{Add 7.5}} < 39.89$. This interval contains 0, so we do not reject $H_0 : \mu_{\text{Add 7.5}} = \mu_{\text{Add 12.5}}$.

Thus, we have that $\mu_{\text{Placebo}} \neq \mu_{\text{Rit 10}} = \mu_{\text{Rit 17.5}} = \mu_{\text{Add 7.5}} = \mu_{\text{Add 12.5}}$. That is, we have:

$\underline{\mu_{\text{Placebo}}}$ $\mu_{\text{Rit 10}}$ $\mu_{\text{Rit 17.5}}$ $\mu_{\text{Add 7.5}}$ $\mu_{\text{Add 12.5}}$. The mean ability to follow rules of a child taking the placebo is less than the mean ability to follow rules of a child taking any of the four drug treatments.

(c) The boxplots shown to the right support the results of parts (a) and (b).

15. (a) The hypotheses are $H_0 : \mu_{\text{nonsmokers}} = \mu_{\text{light smokers}} = \mu_{\text{heavy smokers}}$ versus H_1 : at least one mean differs from the others. Using MINITAB, we obtain the following output:

One-way ANOVA: nonsmoker, light smoker, heavy smoker
```
Source   DF      SS      MS      F       P
Factor    2   1494.9   747.4   19.62   0.000
Error    18    685.7    38.1
Total    20   2180.6
```
Since *P*-value < 0.001 is less than the $\alpha = 0.05$ level of significance, we reject H_0. There is sufficient evidence to support the hypothesis that at least one of the mean heart rates is different from the others.

(b) Using MINITAB, we obtain the following output:

Tukey 95% Simultaneous Confidence Intervals: All Pairwise Comparisons
```
Individual confidence level = 98.00%

nonsmoker subtracted from:
       Lower   Center   Upper   ----+---------+---------+---------+-----
light  3.578   12.000   20.422                    (------*------)
heavy  12.150  20.571   28.993                         (------*------)
                                ----+---------+---------+---------+-----
                                  -12        0        12        24

light subtracted from:
       Lower   Center   Upper   ----+---------+---------+---------+-----
heavy  0.150   8.571    16.993                  (------*------)
                                ----+---------+---------+---------+-----
                                  -12        0        12        24
```

From the output, we have:

$3.578 < \mu_{\text{light smokers}} - \mu_{\text{nonsmokers}} < 20.422$. This interval does not contain 0, so we reject

$H_0 : \mu_{\text{nonsmokers}} = \mu_{\text{light smokers}}$.

$12.150 < \mu_{\text{heavy smokers}} - \mu_{\text{nonsmokers}} < 28.993$. This interval does not contain 0, so we reject

$H_0 : \mu_{\text{nonsmokers}} = \mu_{\text{heavy smokers}}$.

$0.150 < \mu_{\text{heavy smokers}} - \mu_{\text{light smokers}} < 16.993$. This interval does not contain 0, so we reject

$H_0 : \mu_{\text{light smokers}} = \mu_{\text{heavy smokers}}$.

Thus, we have that $\mu_{\text{nonsmokers}} \neq \mu_{\text{light smokers}} \neq \mu_{\text{heavy smokers}}$. That is, we have:

$\underline{\mu_{\text{nonsmokers}} \quad \mu_{\text{light smokers}} \quad \mu_{\text{heavy smokers}}}$. It appears that the mean heart rate of nonsmokers is lower than the mean heart rate of smokers and that the mean heat rate of light smokers is lower than the mean heart rate of heavy smokers.

(c) The boxplots shown to the right support the result of parts (a) and (b).

Heavy smokers

Light smokers

Nonsmokers

Heart Rate (per minute)

17. (a) The hypotheses are $H_0 : \mu_{skim} = \mu_{mixed} = \mu_{whole}$ versus H_1 : at least one mean differs from the others. Using MINITAB, we obtain the following output:

One-way ANOVA: Skim Milk, Mixed Milk, Whole Milk

```
Source   DF      SS      MS      F      P
Factor    2  117596   58798  70.51  0.000
Error    21   17511     834
Total    23  135107
```

Since P-value < 0.001 is less than the $\alpha = 0.05$ level of significance, we reject H_0. There is sufficient evidence to support the hypothesis that at least one of the mean calcium intake levels is different from the others.

(b) Using MINITAB, we obtain the following output:

Tukey 95% Simultaneous Confidence Intervals: All Pairwise Comparisons
Individual confidence level = 98.00%

Skim Milk subtracted from:

```
              Lower   Center   Upper    ---------+---------+---------+---------+
Mixed Milk   108.65   145.00  181.35                                 (---*--)
Whole Milk   -43.10    -6.75   29.60                    (--*---)
                                        ---------+---------+---------+---------+
                                             -100        0       100       200
```

Mixed Milk subtracted from:

```
              Lower   Center    Upper   ---------+---------+---------+---------+
Whole Milk  -188.10  -151.75  -115.40   (---*--)
                                        ---------+---------+---------+---------+
                                             -100        0       100       200
```

From the output, we have:

$108.65 < \mu_{mixed} - \mu_{skim} < 181.35$. This interval does not contain 0, so we reject $H_0 : \mu_{skim} = \mu_{mixed}$.

$-43.10 < \mu_{whole} - \mu_{skim} < 29.60$. This interval contains 0, so we do not reject $H_0 : \mu_{skim} = \mu_{whole}$.

$-188.10 < \mu_{whole} - \mu_{mixed} < -115.40$. This interval does not contain 0, so we reject $H_0 : \mu_{mixed} = \mu_{whole}$.

Thus, we have that $\mu_{mixed} \neq \mu_{skim} = \mu_{whole}$. That is, we have: $\underline{\mu_{mixed}}$ $\underline{\mu_{skim}\ \mu_{whole}}$.

It appears that the mean calcium intake of children who drank mixed milk is different from the mean calcium intakes of children who drank skim or whole milk.

(c) The boxplots shown to the right support the result of parts (a) and (b).

Skim milk

Mixed milk

Whole milk

Data

19. (a) This was a cohort study because the researchers identified the participants to be included in the study and then followed them over a period of time.

(b) $\bar{x}_{HS} = \dfrac{2.6 + 4.1 + \ldots + 5.5}{10} = 4.05$, $\bar{x}_S = \dfrac{3.1 + 3.8 + \ldots + 4.1}{10} = 4.2$, $\bar{x}_{NS} = \dfrac{4.7 + 4.2 + \ldots + 5.1}{10} = 4.2$,

$\bar{x}_{OD} = \dfrac{5.0 + 7.3 + \ldots + 6.0}{10} = 5.8$, $\bar{x}_{NR} = \dfrac{4.5 + 5.5 + \ldots + 4.7}{10} = 4.65$

The means do not appear to be equal.

(c) $s_{HS} = \sqrt{\dfrac{(2.6 - 4.05)^2 + \ldots + (5.5 - 4.05)^2}{10 - 1}} \approx 0.991$, $s_S = \sqrt{\dfrac{(3.1 - 4.2)^2 + \ldots + (4.1 - 4.2)^2}{10 - 1}} \approx 0.554$,

$s_{NS} = \sqrt{\dfrac{(4.7 - 4.2)^2 + \ldots + (5.1 - 4.2)^2}{10 - 1}} \approx 1.099$, $s_{OD} = \sqrt{\dfrac{(5.0 - 5.8)^2 + \ldots + (6.0 - 5.8)^2}{10 - 1}} \approx 0.921$,

$s_{NR} = \sqrt{\dfrac{(4.5 - 4.65)^2 + \ldots + (4.7 - 4.65)^2}{10 - 1}} \approx 0.566$

Since the largest sample standard deviation is less than twice the smallest sample standard deviation $(1.099 < 2 \cdot 0.554)$, the population variances can be assumed to be equal.

(d) For 95% confidence, $\alpha / 2 = 0.025$. With 9 degrees of freedom, $t_{0.025} = 2.262$. Then:

Lower bound $= \bar{x} - t_{0.025} \cdot \dfrac{s}{\sqrt{n}} = 4.05 - 2.262 \cdot \dfrac{0.991}{\sqrt{10}} \approx 4.05 - 0.71 = 3.34$

Upper bound $= \bar{x} + t_{0.025} \cdot \dfrac{s}{\sqrt{n}} = 4.05 + 2.262 \cdot \dfrac{0.991}{\sqrt{10}} \approx 4.05 + 0.71 = 4.76$

(e) There are $_5C_2 = \dfrac{5!}{2!(5-2)!} = \dfrac{5!}{2!3!} = \dfrac{5 \cdot 4}{2} = 10$ possible pairwise comparisons among treatment levels.

(f) Assuming all the tests are independent, we obtain $P(\text{Type I error}) = 1 - (1 - \alpha)^k$

$$= 1 - (1 - 0.05)^{10}$$
$$= 1 - (0.95)^{10}$$
$$= 1 - 0.5987$$
$$= 0.4013$$

(g) The hypotheses are $H_0 : \mu_{HS} = \mu_S = \mu_{NS} = \mu_{OD} = \mu_{NR}$ versus H_1 : at least one mean differs from the others. Using MINITAB, we obtain the following output:

One-way ANOVA: HS, S, NS, OD, NR

Source	DF	SS	MS	F	P
Factor	4	20.630	5.157	7.03	0.000
Error	45	33.010	0.734		
Total	49	53.640			

Since P-value < 0.001 is less than the $\alpha = 0.05$ level of significance, we reject H_0. There is sufficient evidence to support the hypothesis that at least one mean time to degree completion is different from the others.

Using MINITAB, we obtain the following output:

```
Tukey 95% Simultaneous Confidence Intervals:  All Pairwise Comparisons
Individual confidence level = 99.33%

HS subtracted from:
       Lower   Center   Upper   --------+---------+---------+---------+-
S    -0.9388   0.1500  1.2388                (------*------)
NS   -0.9388   0.1500  1.2388                (------*------)
OD    0.6612   1.7500  2.8388                       (-------*------)
NR   -0.4888   0.6000  1.6888                   (------*------)
                               --------+---------+---------+---------+-
                                    -1.5       0.0       1.5       3.0

S subtracted from:
       Lower   Center   Upper   --------+---------+---------+---------+-
NS   -1.0888   0.0000  1.0888               (------*------)
OD    0.5112   1.6000  2.6888                     (-------*------)
NR   -0.6388   0.4500  1.5388                  (------*------)
                               --------+---------+---------+---------+-
                                    -1.5       0.0       1.5       3.0

NS subtracted from:
       Lower   Center   Upper   --------+---------+---------+---------+-
OD    0.5112   1.6000  2.6888                     (-------*------)
NR   -0.6388   0.4500  1.5388                  (------*------)
                               --------+---------+---------+---------+-
                                    -1.5       0.0       1.5       3.0

OD subtracted from:
       Lower   Center   Upper   --------+---------+---------+---------+-
NR   -2.2388  -1.1500 -0.0612          (------*-------)
                               --------+---------+---------+---------+-
                                    -1.5       0.0       1.5       3.0
```

From the output, we have:

$-0.9388 < \mu_S - \mu_{HS} < 1.2388$. This interval contains 0, so we do not reject $H_0 : \mu_S = \mu_{HS}$.

$-0.9388 < \mu_{NS} - \mu_{HS} < 1.2388$. This interval contains 0, so we do not reject $H_0 : \mu_{NS} = \mu_{HS}$.

$0.6612 < \mu_{OD} - \mu_{HS} < 2.8388$. This interval does not contain 0, so we reject $H_0 : \mu_{OD} = \mu_{HS}$.

$-0.4888 < \mu_{NR} - \mu_{HS} < 1.6888$. This interval contains 0, so we do not reject $H_0 : \mu_{NR} = \mu_{HS}$.

$-1.0888 < \mu_{NS} - \mu_S < 1.0888$. This interval contains 0, so we do not reject $H_0 : \mu_{NS} = \mu_S$.

$0.5112 < \mu_{OD} - \mu_S < 2.6888$. This interval does not contain 0, so we reject $H_0 : \mu_{OD} = \mu_S$.

$-0.6388 < \mu_{NR} - \mu_S < 1.5388$. This interval contains 0, so we do not reject $H_0 : \mu_{NR} = \mu_S$.

$0.5112 < \mu_{OD} - \mu_{NS} < 2.6888$. This interval does not contain 0, so we reject $H_0 : \mu_{OD} = \mu_{NS}$.

$-0.6388 < \mu_{NR} - \mu_{NS} < 1.5388$. This interval contains 0, so we do not reject $H_0 : \mu_{NR} = \mu_{NS}$.

$-2.2388 < \mu_{NR} - \mu_{OD} < -0.0612$. This interval does not contain 0, so we reject $H_0 : \mu_{NR} = \mu_{OD}$.

Thus, we have that $\mu_{HS} = \mu_S = \mu_{NS} = \mu_{NR} \neq \mu_{OD}$. That is, we have: $\underline{\mu_{HS} \quad \mu_S \quad \mu_{NS} \quad \mu_{NR}} \quad \underline{\mu_{OD}}$.

The mean time to degree completion appears to be the same for all initial institution types except for those with an open-door policy. It appears that students starting at an institution with an open-door policy take longer to complete their bachelor's degree.

13.3 The Randomized Complete Block Design

1. In a completely randomized design, the researcher manipulates a single factor fixed at $k \geq 2$ levels and randomly distributes the experimental units among the levels. The researcher may be able to control some factors so they remain fixed. In a randomized complete block design, there is a factor whose level cannot be fixed. So the researcher partitions the experimental units according to this factor, forming blocks which may form homogeneous groups. Within each block, the experimental units are assigned to one of the treatments.

3. *Randomized* means that the order in which the treatment is applied within each block is random. *Complete* refers to the fact that each block gets every treatment.

5. A researcher can deal with explanatory variables by:
   (1) Controlling their levels so that they remain fixed throughout the experiment or setting them at fixed levels, and
   (2) Randomizing so that any effects that cannot be identified or controlled are minimized.

7. **(a)** The treatment *P*-value = 0.013 is sufficient evidence against the null hypothesis that the population means are equal. The researcher will conclude that at least one population mean is different from the others.

   (b) The mean square error is MSE = 17.517.

   (c) The *P*-values comparing treatment 3 with treatments 1 and 2 are small (less than 0.05) while the *P*-value comparing treatments 1 and 2 is large. Therefore, we conclude that means 1 and 2 are the same, but mean 3 is different. That is, we conclude $\mu_1 = \mu_2 \neq \mu_3$ or $\underline{\mu_1 \quad \mu_2} \quad \underline{\mu_3}$.

9. **(a)** The treatment *P*-value = 0.108 is not sufficient evidence against the null hypothesis that the population means are equal. In other words, the researcher does not reject the null hypothesis that population means are equal.

   (b) The mean square error is MSE = 0.944.

   (c) We would only use Tukey's test if the null hypothesis from part (a) had been rejected.

11. **(a)** Using MINITAB, we obtain the following output:

```
Two-way ANOVA: Response versus Block, Treatment
Source       DF      SS     MS      F       P
Block         4   5.836  1.459  10.13   0.003
Treatment     2   4.468  2.234  15.51   0.002
Error         8   1.152  0.144
Total        14  11.456
```

Since the treatment *P*-value = 0.002, we reject the null hypothesis that the population means are equal. Thus, we conclude that at least one mean is different from the others.

   (b) Using MINITAB, we obtain the following output:

```
Tukey 95.0% Simultaneous Confidence Intervals
Response Variable Response
All Pairwise Comparisons among Levels of Treatment

Treatment = 1  subtracted from:

Treatment    Lower   Center    Upper   --------+---------+---------+--------
2           -1.606   -0.920  -0.2344            (---------*---------)
3           -1.986   -1.300  -0.6144   (--------*---------)
                                       --------+---------+---------+--------
                                         -1.40     -0.70      0.00
```

```
Treatment = 2  subtracted from:

Treatment   Lower   Center   Upper   --------+---------+---------+--------
3          -1.066  -0.3800  0.3056              (---------*--------)
                                      --------+---------+---------+--------
                                          -1.40      -0.70      0.00
```

Tukey Simultaneous Tests
Response Variable Response
All Pairwise Comparisons among Levels of Treatment

```
Treatment = 1  subtracted from:

            Difference    SE of              Adjusted
Treatment   of Means    Difference  T-Value   P-Value
2            -0.920       0.2400     -3.833    0.0123
3            -1.300       0.2400     -5.417    0.0016

Treatment = 2  subtracted from:

            Difference    SE of              Adjusted
Treatment   of Means    Difference  T-Value   P-Value
3            -0.3800      0.2400     -1.583    0.3066
```

The interval for Treatment 2 minus Treatment 1 does not include 0 and the *P*-value = 0.0123 for this comparison, which is small, so reject the null hypothesis that $\mu_1 = \mu_2$. Likewise, the interval for Treatment 3 minus Treatment 1 does not include 0 and the *P*-value = 0.0016, which is small, so we reject the null hypothesis that $\mu_1 = \mu_3$. Finally, the interval for Treatment 3 minus Treatment 2 does include 0 and the *P*-value = 0.3066, which is not small, so we do not reject the null hypothesis that $\mu_2 = \mu_3$. Thus, the Tukey's tests indicate that the means of Treatments 2 and 3 are equal, but the mean of Treatment 1 is different. That is, we conclude: $\mu_1 \neq \mu_2 = \mu_3$ or $\underline{\mu_1} \quad \underline{\mu_2 \quad \mu_3}$.

(c) The boxplots shown to the right support the results found in parts (a) and (b).

13. (a) The sample standard deviations for the provided data are $s_{87\text{ Octane}} \approx 1.240$, $s_{89\text{ Octane}} \approx 1.439$, and $s_{92\text{ Octane}} \approx 1.225$. Since the largest standard deviation is less than twice the smallest, $1.439 < 2(1.225)$, the requirement for equal population variances is satisfied.

(b) Using MINITAB, we obtain the following output:

```
Two-way ANOVA: Miles per Gallon versus Car and Driver, Octane level
Source          DF     SS       MS      F       P
Car and Driver   5   25.0161  5.00322  92.84  0.000
Octane level     2    1.1411  0.57056  10.59  0.003
Error           10    0.5389  0.05389
Total           17   26.6961
```

Since the treatment (octane level) *P*-value = 0.003, there is sufficient evidence to conclude that the mean number of miles per gallon is different among the three octane levels at the $\alpha = 0.05$ level of significance.

(c) Using MINITAB, we obtain the following output:

```
Tukey 95.0% Simultaneous Confidence Intervals
Response Variable Miles
All Pairwise Comparisons among Levels of Octane level

Octane level = 1   subtracted from:

Octane
level    Lower    Center   Upper      +---------+---------+---------+------
2       -0.3344   0.03333  0.4010     (----------*---------)
3        0.1823   0.55000  0.9177                     (----------*---------)
                                       +---------+---------+---------+------
                                     -0.35      0.00      0.35      0.70

Octane level = 2   subtracted from:

Octane
level    Lower   Center   Upper       +---------+---------+---------+------
3        0.1490  0.5167   0.8844                   (----------*---------)
                                       +---------+---------+---------+------
                                     -0.35      0.00      0.35      0.70


Tukey Simultaneous Tests
Response Variable Miles per Gallon
All Pairwise Comparisons among Levels of Octane level

Octane level = 1   subtracted from:

Octane  Difference    SE of               Adjusted
level   of Means   Difference  T-Value    P-Value
2        0.03333     0.1340    0.2487     0.9666
3        0.55000     0.1340    4.1037     0.0055

Octane level = 2   subtracted from:

Octane  Difference    SE of               Adjusted
level   of Means   Difference  T-Value    P-Value
3        0.5167      0.1340    3.855      0.0081
```

The interval for 89 octane minus 87 octane includes 0 and the *P*-value = 0.9666 for this comparison, which is not small, so we do not reject the null hypothesis that $\mu_{87} = \mu_{89}$. The interval for 92 octane minus 87 octane does not include 0 and the *P*-value = 0.0055, which is small, so we reject the null hypothesis that $\mu_{87} = \mu_{92}$. Finally, the interval for 92 octane minus 89 octane does not include 0 and the *P*-value = 0.0081, which is small, so we reject the null hypothesis that $\mu_{89} = \mu_{92}$. We conclude

$\mu_{87} = \mu_{89} \neq \mu_{92}$ or $\underline{\mu_{87}\ \ \mu_{89}}\ \ \mu_{92}$

(d) The Tukey's tests indicate that the mean gasoline mileages for 87 octane and 89 octane are equal, but the mean gasoline mileage for 92 octane is different. In fact, since the confidence intervals for 92 octane minus 87 octane and 92 octane minus 87 octane contain all positive values, we can conclude that the mean gasoline mileage for 92 octane is higher than for 87 octane and 89 octane.

15. (a) The sample standard deviations for the provided data are $s_{\text{Jeep}} \approx 448$, $s_{\text{Saturn}} \approx 453$, $s_{\text{Toyota}} \approx 830$, and $s_{\text{Hyundai}} \approx 614$. Since the largest standard deviation is less than twice the smallest, $830 < 2(448)$, the requirement for equal population variances is satisfied.

(b) Using MINITAB, we obtain the following output:

Two-way ANOVA: Cost versus Location, Brand

Source	DF	SS	MS	F	P
Location	3	3404468	1134823	10.09	0.003
Brand	3	926488	308829	2.75	0.105
Error	9	1012007	112445		
Total	15	5342964			

Since the treatment (brand) *P*-value = 0.105, there is not sufficient evidence to conclude that the mean cost of repair is different among the car brands at the $\alpha = 0.05$ level of significance.

(c) Since the null hypothesis was not rejected, there is no need to do a post hoc analysis.

17. (a) The sample standard deviations for the provided data are $s_{\text{lift-off}-1} \approx 5.65$, $s_{\text{return}+1} \approx 9.97$, and $s_{+1\text{ month}} \approx 9.00$. Since the largest standard deviation is less than twice the smallest, $9.97 < 2(5.65)$, the requirement for equal population variances is satisfied.

(b) Using MINITAB, we obtain the following output:

Two-way ANOVA: Water Consumption versus Rat, Day

Source	DF	SS	MS	F	P
Rat	5	861.78	172.356	8.62	0.002
Day	2	635.44	317.722	15.89	0.001
Error	10	199.89	19.989		
Total	17	1697.11			

Since the treatment (day) *P*-value = 0.001, there is sufficient evidence to conclude that the mean water consumption of rats is different among the experiment levels at the $\alpha = 0.05$ level of significance.

(c) Using MINITAB with Treatment 1 representing Lift-off Minus 1, Treatment 2 representing Return Plus 1, and Treatment 3 representing Return Plus 1 Month, we obtain the following output:

```
Tukey 95.0% Simultaneous Confidence Intervals
Response Variable Water Consumption
All Pairwise Comparisons among Levels of Day

Day = 1  subtracted from:

Day    Lower   Center   Upper    -+---------+---------+---------+-----
2    -0.9152   6.167    13.25    (---------*---------)
3     7.4181  14.500    21.58              (---------*---------)
                                 -+---------+---------+---------+-----
                                0.0       7.0      14.0      21.0

Day = 2  subtracted from:

Day  Lower  Center   Upper    -+---------+---------+---------+-----
3    1.251   8.333    15.42       (---------*---------)
                                -+---------+---------+---------+-----
                               0.0       7.0      14.0      21.0

Tukey Simultaneous Tests
Response Variable Water Consumption
All Pairwise Comparisons among Levels of Day

Day = 1  subtracted from:
      Difference      SE of              Adjusted
Day   of Means     Difference  T-Value   P-Value
2        6.167        2.581      2.389    0.0885
3       14.500        2.581      5.617    0.0006

Day = 2  subtracted from:
      Difference      SE of              Adjusted
Day   of Means     Difference  T-Value   P-Value
3        8.333        2.581      3.228    0.0225
```

The interval for Return Plus 1 minus Lift-off minus 1 includes 0 and the P-value = 0.0885 for this comparison, which is not small, so we do not reject the null hypothesis that $\mu_{\text{lift-off}-1} = \mu_{\text{return}+1}$. The interval for Return Plus 1 Month minus Lift-off Minus 1 does not include 0 and the P-value = 0.0006, which is small, so we reject the null hypothesis that $\mu_{\text{lift-off}-1} = \mu_{\text{return}+1 \text{ month}}$. Finally, the interval for Return Plus 1 Month minus Return Plus 1 does not include 0 and the P-value = 0.0225, which is small, so we reject the null hypothesis that $\mu_{\text{return}+1} = \mu_{\text{return}+1 \text{ month}}$. Thus, we conclude that

$$\mu_{\text{lift-off}-1} = \mu_{\text{return}+1} \neq \mu_{\text{return}+1 \text{ month}} \quad \text{or} \quad \underline{\mu_{\text{lift-off}-1} \quad \mu_{\text{return}+1}} \quad \underline{\mu_{\text{return}+1 \text{ month}}} \,.$$

19. (a) Using MINITAB, we obtain the following output:

Two-way ANOVA: Wait Time versus Day, Procedure

```
Source      DF      SS        MS        F       P
Day          6   6538.75   1089.79   249.50   0.000
Procedure    1      3.02      3.02     0.69   0.438
Error        6     26.21      4.37
Total       13   6567.97
```

Since the treatment (procedure) P-value = 0.438, there is not sufficient evidence to conclude that the mean wait time is different for the two loading procedures at the $\alpha = 0.05$ level of significance.

(b)

Observation	1	2	3	4	5	6	7
Wait time before new procedure	11.6	25.9	20.0	38.2	57.3	32.1	81.8
Wait time after new procedure	10.7	28.3	19.2	35.9	59.2	31.8	75.3
$d_i =$ after − before	−0.9	2.4	−0.8	−2.3	1.9	−0.3	−6.5

Using technology, $\bar{d} \approx -0.93$ minutes and $s_d \approx 2.96$ minutes, where we measure differences as wait time after new procedure minus wait time before new procedure. The hypotheses are $H_0 : \mu_d = 0$ versus $H_1 : \mu_d \neq 0$. The level of significance is $\alpha = 0.05$. The test statistic is $t_0 = \dfrac{\bar{d}}{s_d / \sqrt{n}} = \dfrac{-0.93}{2.96 / \sqrt{7}} \approx -0.83$.

Classical approach: Since this is a two-tailed test with 6 degrees of freedom, the critical values are $\pm t_{0.025} = \pm 2.447$. Since the test statistic $t_0 \approx -0.83$ falls between the critical values $\pm t_{0.025} = \pm 2.447$ (i.e., since the test statistic does not fall within the critical region), we do not reject H_0.

P-value approach: The P-value for this two-tailed test is the area under the t-distribution with 6 degrees of freedom to the left of the test statistic $t_0 = -0.83$ plus the area to the right of $t_0 = 0.83$. From the t-distribution table in the row corresponding to 6 degrees of freedom, 0.83 falls between 0.718 and 0.906 whose right-tail areas are 0.25 and 0.20, respectively. We must double these values in order to get the total area in both tails: 0.50 and 0.40. So, $0.40 < P\text{-value} < 0.50$. (Using technology, we find P-value $\approx$ 0.438.) Because the P-value is greater than the level of significance $\alpha = 0.05$, we do not reject H_0.

Conclusion: There is not sufficient evidence at the $\alpha = 0.05$ level of significance to conclude that the new loading procedure changed the wait time.

(c) The P-values for the two approaches are both 0.438. Yes, it appears that the method presented in this section is a generalization of the matched-pair t-test.

13.4   Two-Way Analysis of Variance

1. In a completely randomized design, the researcher examines a single factor fixed at k levels and randomly distributes the experimental units among the levels. In a randomized complete block design, there is a factor whose level cannot be fixed. So the researcher partitions the experimental units according to this factor, forming blocks. The experimental units are then randomly assigned to the blocks, being sure each block has each treatment. In a factorial design, the researcher has two variables A and B, called factor A and factor B. The factors are fixed, so factor A has a levels and factor B has b levels. The experimental units are then uniformly distributed among the cells formed by the factors, and the response variable is measured at each of the ab cells.

3. An interaction plot graphically displays the mean response for each treatment (combination of factor levels). The plots are useful for illustrating potential interaction between the factors.

5. The plots are relatively parallel, so there is no significant interaction among the factors.

7. The plots cross, so there is significant interaction among the factors.

9. **(a)** Main effect of factor $A = \dfrac{20+14+1+3}{4} - \dfrac{12+8+6+7}{4} = 1.25$;

   Main effect of factor $B = \dfrac{6+7+1+3}{4} - \dfrac{12+8+20+14}{4} = -9.25$.

   Interaction effects: If factor B is fixed at low and factor A changes from low to high, the response

   variable increases by $\dfrac{20+14}{2} - \dfrac{12+8}{2} = 7$ units. On the other hand, if factor B is fixed at high and factor

   A changes from low to high, the response variable increases by $\dfrac{1+3}{2} - \dfrac{6+7}{2} = -4.5$ (decreases by 4.5)

   units. So the change in the response variable for factor A depends on the level of factor B, meaning there is an interaction effect between factors A and B. If factor A is fixed at low and factor B changes from

   low to high, the response variable increases by $\dfrac{6+7}{2} - \dfrac{12+8}{2} = -3.5$ (decreases by 3.5) units. On the

   other hand, if factor A is fixed at high and factor B changes from low to high, the response variable

   increases by $\dfrac{3+1}{2} - \dfrac{20+14}{2} = -15$ (decreases by 15) units. So the change in the response variable for

   factor B depends on the level of factor A, meaning there is an interaction effect between factors A and B.

   (b)

   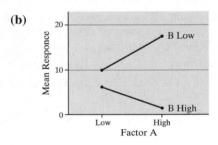

11. **(a)** Since the interaction P-value is 0.836, which is large, there is no evidence of an interaction effect.

    (b) Since the P-value for factor A is 0.003, which is small, there is evidence of a difference in means from factor A. Since the P-value for factor B is < 0.001, which is small, there is also evidence of a difference in means from factor B.

    (c) MSE $= 45.3$

13. **(a)** Since the interaction *P*-value is < 0.001, which is small, there is evidence of an interaction effect.

(b) In part (a), we found evidence of an interaction effect. Therefore, we do not consider main effects because they could be misleading.

(c) MSE = 31.9

15. Using MINITAB with the provided data, we obtain the following output:

Two-way ANOVA: Response versus Factor A, Factor B

Source	DF	SS	MS	F	P
Factor A	1	30.15	30.150	0.32	0.579
Factor B	2	912.90	456.452	4.84	0.021
Interaction	2	6.80	3.402	0.04	0.965
Error	18	1699.27	94.404		
Total	23	2649.13			

(a) Since the interaction *P*-value is 0.965, which is large, there is no evidence of an interaction effect.

(b) Since the *P*-value for factor A is 0.579, which is large, there is no evidence of a difference in the means from factor A. Since the *P*-value for factor B is 0.021, which is small, there is evidence of a difference in the means from factor B.

(c) The lines in the interaction plot are close to parallel, supporting the conclusion that there is no interaction effect.

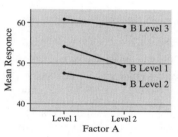

(d) The Tukey's test for factor A is unnecessary since there is no evidence for difference in means. For factor B, we use MINITAB to perform Tukey's test. The results are:

```
Tukey 95.0% Simultaneous Confidence Intervals
Response Variable Response
All Pairwise Comparisons among Levels of Factor B

Factor B = 1   subtracted from:

Factor B   Lower   Center   Upper    --+---------+---------+---------+----
2         -18.55   -6.150    6.251   (-------*-------)
3          -3.53    8.875   21.276                 (-------*-------)
                                     --+---------+---------+---------+----
                                     -15         0        15        30

Factor B = 2   subtracted from:

Factor B   Lower   Center   Upper    --+---------+---------+---------+----
3          2.624   15.03    27.43               (-------*-------)
                                     --+---------+---------+---------+----
                                     -15         0        15        30

Tukey Simultaneous Tests
Response Variable Response
All Pairwise Comparisons among Levels of Factor B

Factor B = 1   subtracted from:

           Difference    SE of              Adjusted
Factor B   of Means    Difference  T-Value  P-Value
2            -6.150       4.858     -1.266    0.4317
3             8.875       4.858      1.827    0.1893

Factor B = 2   subtracted from:

           Difference    SE of              Adjusted
Factor B   of Means    Difference  T-Value  P-Value
3            15.03        4.858      3.093    0.0165
```

The interval for $B_2 - B_1$ includes 0 and the *P*-value = 0.4317 for this comparison, which is large, so we do not reject the null hypothesis that $B_1 = B_2$. Likewise, the interval for $B_3 - B_1$ includes 0 and the *P*-value = 0.1893, which is large, so we do not reject the null hypothesis that $B_1 = B_3$. Finally, the interval for $B_3 - B_2$ does not include 0 and the *P*-value = 0.0165, which is small, so we reject the null hypothesis that $B_2 = B_3$. Thus, the Tukey's test for factor B is ambiguous. It shows $B_1 = B_2$ and $B_1 = B_3$ but $B_2 \neq B_3$. That is, $\underline{B_2 \quad B_1} \quad \underline{B_1 \quad B_3}$.

17. (a) This is a 2×3 factorial design with 2 replications within each cell.

(b) Using MINITAB with the provided data, we obtain the following output:

Two-way ANOVA: Cholesterol versus Gender, Age

Source	DF	SS	MS	F	P
Gender	1	310.08	310.08	2.51	0.164
Age	2	2177.17	1088.58	8.81	0.016
Interaction	2	628.17	314.08	2.54	0.159
Error	6	741.50	123.58		
Total	11	3856.92			

Since the interaction *P*-value is 0.159, which is large, there is no evidence of an interaction effect.

(c) Since the *P*-value for gender is 0.164, which is large, there is no evidence of a difference in the means for the two genders. Since the *P*-value for age is 0.016, which is small, there is evidence of a difference in the means for the three age groups.

(d) The plots are not parallel, so there is significant interaction among the factors. This result contradicts the result found in part (b).

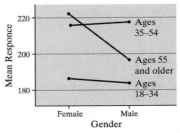

(e) The Tukey's test for gender is unnecessary since there is no evidence for difference in means for the two genders. For age, we use MINITAB to perform Tukey's test:

```
Tukey 95.0% Simultaneous Confidence Intervals
Response Variable Cholesterol
All Pairwise Comparisons among Levels of Age

Age = 1  subtracted from:

Age   Lower    Center  Upper    ---+---------+---------+---------+---
2     7.3765   31.50   55.62                  (---------*--------)
3     0.1265   24.25   48.37              (---------*--------)
                                ---+---------+---------+---------+---
                                 -25        0        25        50

Age = 2  subtracted from:

Age   Lower    Center  Upper    ---+---------+---------+---------+---
3     -31.37   -7.250  16.87    (---------*----------)
                                ---+---------+---------+---------+---
                                 -25        0        25        50
```

```
Tukey Simultaneous Tests
Response Variable Cholesterol
All Pairwise Comparisons among Levels of Age

Age = 1  subtracted from:

       Difference    SE of              Adjusted
Age    of Means   Difference  T-Value   P-Value
2        31.50      7.861      4.007     0.0166
3        24.25      7.861      3.085     0.0490
```

```
Age = 2   subtracted from:
        Difference      SE of                  Adjusted
Age     of Means     Difference   T-Value      P-Value
3         -7.250        7.861      -0.9223       0.6473
```

The interval for $\mu_{35-54} - \mu_{18-34}$ does not include 0 and the P-value = 0.0166 for this comparison, which is small, so we reject the null hypothesis that $\mu_{18-34} = \mu_{35-54}$. Likewise, the interval for $\mu_{55+} - \mu_{18-34}$ does not include 0 and the P-value = 0.0490, which is small, so we reject the null hypothesis that $\mu_{18-34} = \mu_{55+}$. Finally, the interval for $\mu_{55+} - \mu_{35-54}$ includes 0 and the P-value = 0.6473, which is large, so we do not reject the null hypothesis that $\mu_{35-54} = \mu_{55+}$. Thus, the Tukey's test for age shows $\mu_{18-34} \neq \mu_{35-54} = \mu_{55+}$. That is, $\underline{\mu_{18-34}}$ $\underline{\mu_{35-54}\ \ \mu_{55+}}$.

19. (a) Using technology, we find the sample standard deviation of the data within each cell: $s_{3.75/301} \approx 311$, $s_{3.75/400} \approx 320$, $s_{3.75/353} \approx 257$, $s_{4/301} \approx 298$, $s_{4/400} \approx 250$, $s_{4/353} \approx 168$, $s_{5/301} \approx 236$, $s_{5/400} \approx 222$, and $s_{5/353} \approx 267$. The largest standard deviation (320) is less than twice the smallest standard deviation (168), so the requirement of equal population variances is satisfied. [$320 < 2(168)$.]

(b) Using MINITAB with the provided data, we obtain the following output:

Two-way ANOVA: Strength versus Slump, Mixture

Source	DF	SS	MS	F	P
Slump	2	1234813	617406	8.95	0.002
Mixture	2	1824169	912084	13.23	0.000
Interaction	4	256415	64104	0.93	0.469
Error	18	1241100	68950		
Total	26	4556496			

Since the interaction P-value is 0.469, which is large, there is no evidence of an interaction effect.

(c) Since the P-value for slump is 0.002, which is small, there is sufficient evidence of difference in the means for the three types of slump. Since the P-value for mixture is < 0.001, which is small, there is evidence of a difference in the means for the three mixtures.

(d)

(e) We use MINITAB to perform Tukey's test. The results are:

```
Tukey 95.0% Simultaneous Confidence Intervals
Response Variable Strength
All Pairwise Comparisons among Levels of Slump

Slump = 1   subtracted from:

Slump   Lower    Center   Upper  --------+---------+---------+--------
2       -634.3   -318.3    -2.4          (---------*----------)
3       -835.4   -519.4   -203.5  (----------*---------)
                                  --------+---------+---------+--------
                                       -600      -300       0

Slump = 2   subtracted from:

Slump   Lower    Center   Upper  --------+---------+---------+--------
3       -517.1   -201.1   114.9              (---------*----------)
                                  --------+---------+---------+--------
                                       -600      -300       0
```

Tukey Simultaneous Tests
Response Variable Strength
All Pairwise Comparisons among Levels of Slump

Slump = 1   subtracted from:

Slump	Difference of Means	SE of Difference	T-Value	Adjusted P-Value
2	-318.3	123.8	-2.572	0.0481
3	-519.4	123.8	-4.196	0.0015

Slump = 2   subtracted from:

Slump	Difference of Means	SE of Difference	T-Value	Adjusted P-Value
3	-201.1	123.8	-1.625	0.2612

Tukey 95.0% Simultaneous Confidence Intervals
Response Variable Strength
All Pairwise Comparisons among Levels of Mixture

Mixture = 1   subtracted from:

```
Mixture    Lower   Center   Upper   ---+---------+---------+---------+---
2         320.691  636.7    952.6                      (------*-----)
3          -2.086  313.9    629.9               (-----*------)
                                    ---+---------+---------+---------+---
                                    -500        0        500       1000
```

Mixture = 2   subtracted from:

```
Mixture    Lower   Center   Upper   ---+---------+---------+---------+---
3         -638.8  -322.8   -6.803   (------*-----)
                                    ---+---------+---------+---------+---
                                    -500        0        500       1000
```

Tukey Simultaneous Tests
Response Variable Strength
All Pairwise Comparisons among Levels of Mixture

Mixture = 1   subtracted from:

Mixture	Difference of Means	SE of Difference	T-Value	Adjusted P-Value
2	636.7	123.8	5.143	0.0002
3	313.9	123.8	2.536	0.0517

Mixture = 2   subtracted from:

Mixture	Difference of Means	SE of Difference	T-Value	Adjusted P-Value
3	-322.8	123.8	-2.608	0.0448

Tukey's test for Slump: The interval for $\mu_4 - \mu_{3.75}$ does not include 0 and the *P*-value = 0.0481 for this comparison, which is small, so we reject the null hypothesis that $\mu_{3.75} = \mu_4$. Next, the interval for $\mu_5 - \mu_{3.75}$ does not include 0 and the *P*-value = 0.0015, which is small, so we reject the null hypothesis that $\mu_{3.75} = \mu_5$. Finally, the interval for $\mu_5 - \mu_4$ includes 0 and the *P*-value = 0.2612, which is large, so we do not reject the null hypothesis that $\mu_4 = \mu_5$. Thus, we have $\mu_{3.75} \neq \mu_4 = \mu_5$ or $\underline{\mu_{3.75}} \quad \underline{\mu_4 \quad \mu_5}$.

That is, the Tukey's test indicates that the mean 28-day strength for slump 3.75 is significantly different from the mean 28-day strength for slump 4 and slump 5.

Tukey's test for Mixture: The interval for $\mu_{400} - \mu_{301}$ does not include 0 and the P-value = 0.0002, which is small, so we reject the null hypothesis that $\mu_{301} = \mu_{400}$. Next, the interval for $\mu_{353} - \mu_{301}$ includes 0 and the P-value = 0.0517, which is not small, so we do not reject the null hypothesis that $\mu_{301} = \mu_{353}$. Finally, the interval for $\mu_{353} - \mu_{400}$ does not include 0 and the P-value = 0.0448, which is small, so we reject the null hypothesis that $\mu_{400} = \mu_{353}$. Thus, we have $\mu_{301} = \mu_{353} \neq \mu_{400}$ or

$\underline{\mu_{301} \quad \mu_{353}} \quad \mu_{400}$. That is, the Tukey's test indicates that the mean 28-day strength of mixture 67-0-400 is significantly different from the mean 28-day strength of the other two mixtures.

21. (a) Using technology, we find the sample standard deviation the data within each cell:

$s_{\text{specialty / Chicago}} \approx 4.00$, $s_{\text{specialty / Bolingbrook}} \approx 3.20$, $s_{\text{specialty / Peoria}} \approx 3.63$, $s_{\text{general / Chicago}} \approx 2.50$,

$s_{\text{general / Bolingbrook}} \approx 3.91$,, and $s_{\text{general / Peoria}} \approx 2.52$. The largest standard deviation (4.00) is less than twice the smallest standard deviation (2.50), so the requirement of equal population variances is satisfied. [$4.00 < 2(2.50)$.]

(b) Using MINITAB with the provided data, we obtain the following output:

Two-way ANOVA: Cost versus Type, Location

Source	DF	SS	MS	F	P
Type	1	0.720	0.7200	0.06	0.804
Location	2	31.126	15.5630	1.39	0.287
Interaction	2	3.573	1.7867	0.16	0.854
Error	12	134.560	11.2134		
Total	17	169.980			

Since the interaction P-value is 0.854, which is large, there is no evidence of an interaction effect.

(c) Since the P-value for type is 0.804, which is large, there is no evidence of a difference in the means for the type of service center. Since the P-value for location is 0.287, which is large, there is no evidence of a difference in the means for the locations.

(d)

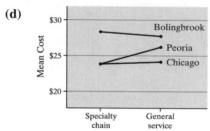

(e) Since there were no significant main effects, there is no need to do Tukey's tests.

Chapter 13 Review Exercises

1. (a) We look in the table of critical values for Tukey's Test with $\alpha = 0.05$ and where the row corresponding to $\nu = 16$ intersects with the column corresponding to $k = 7$ to find the critical value $q_{0.05,16,7} = 4.741$.

(b) We look in the table of critical values for Tukey's Test with $\alpha = 0.05$ and where the row corresponding to $\nu = 30$ intersects with the column corresponding to $k = 6$ to find the critical value $q_{0.05,30,6} = 4.302$.

(c) We look in the table of critical values for Tukey's Test with $\alpha = 0.01$ and where the row corresponding to $\nu = 40$ (since there is no row in the table for $\nu = 42$) intersects with the column corresponding to $k = 4$ to find the critical value $q_{0.01,42,4} \approx q_{0.01,40,4} = 4.696$.

(d) For this test, $k = 6$ and $\nu = n - k = 46 - 6 = 40$, so we look in the table with $\alpha = 0.05$ and where the row corresponding to $\nu = 40$ intersects with the column corresponding to $k = 6$ to find the critical value $q_{0.05,40,6} = 4.232$.

2. Arrange the sample means in ascending order: $\bar{x}_4 = 47$, $\bar{x}_2 = 64$, $\bar{x}_3 = 70$, $\bar{x}_1 = 78$. Now $n = 31 \cdot 4 = 124$ and $k = 4$, so $\nu = n - k = 120$ and the critical value is $q_{0.05,120,4} = 3.685$. We first test $H_0 : \mu_1 = \mu_4$. The

test statistic is $q_0 = \dfrac{\bar{x}_1 - \bar{x}_4}{\sqrt{\frac{s^2}{2} \cdot \left(\frac{1}{n_1} + \frac{1}{n_4} \right)}} = \dfrac{78 - 47}{\sqrt{\frac{373}{2} \cdot \left(\frac{1}{31} + \frac{1}{31} \right)}} \approx 8.937$. Since this test statistic is greater than

$q_{0.05,120,4} = 3.685$, we reject $H_0 : \mu_1 = \mu_4$.

Second, we test $H_0 : \mu_1 = \mu_2$. The test statistic is $q_0 = \dfrac{\bar{x}_1 - \bar{x}_2}{\sqrt{\frac{s^2}{2} \cdot \left(\frac{1}{n_1} + \frac{1}{n_2} \right)}} = \dfrac{78 - 64}{\sqrt{\frac{373}{2} \cdot \left(\frac{1}{31} + \frac{1}{31} \right)}} \approx 4.036$. Since this

test statistic is greater than $q_{0.05,120,4} = 3.685$, we reject $H_0 : \mu_1 = \mu_2$.

Third, we test $H_0 : \mu_1 = \mu_3$. The test statistic is $q_0 = \dfrac{\bar{x}_1 - \bar{x}_3}{\sqrt{\frac{s^2}{2} \cdot \left(\frac{1}{n_1} + \frac{1}{n_3} \right)}} = \dfrac{78 - 70}{\sqrt{\frac{373}{2} \cdot \left(\frac{1}{31} + \frac{1}{31} \right)}} \approx 2.306$. Since this test

statistic is less than $q_{0.05,120,4} = 3.685$, we do not reject $H_0 : \mu_1 = \mu_3$.

Fourth, we test $H_0 : \mu_3 = \mu_4$. The test statistic is $q_0 = \dfrac{\bar{x}_3 - \bar{x}_4}{\sqrt{\frac{s^2}{2} \cdot \left(\frac{1}{n_3} + \frac{1}{n_4} \right)}} = \dfrac{70 - 47}{\sqrt{\frac{373}{2} \cdot \left(\frac{1}{31} + \frac{1}{31} \right)}} \approx 6.631$. Since this test

statistic is greater than $q_{0.05,120,4} = 3.685$, we reject $H_0 : \mu_3 = \mu_4$.

Fifth, we test $H_0 : \mu_3 = \mu_2$. The test statistic is $q_0 = \dfrac{\bar{x}_3 - \bar{x}_2}{\sqrt{\frac{s^2}{2} \cdot \left(\frac{1}{n_3} + \frac{1}{n_2} \right)}} = \dfrac{70 - 64}{\sqrt{\frac{373}{2} \cdot \left(\frac{1}{31} + \frac{1}{31} \right)}} \approx 1.730$. Since this test

statistic is less than $q_{0.05,120,4} = 3.685$, we do not reject $H_0 : \mu_3 = \mu_2$.

Finally, we test $H_0 : \mu_2 = \mu_4$. The test statistic is $q_0 = \dfrac{\bar{x}_2 - \bar{x}_4}{\sqrt{\frac{s^2}{2} \cdot \left(\frac{1}{n_2} + \frac{1}{n_4} \right)}} = \dfrac{64 - 47}{\sqrt{\frac{373}{2} \cdot \left(\frac{1}{31} + \frac{1}{31} \right)}} \approx 4.901$. Since this

test statistic is greater than $q_{0.05,120,4} = 3.685$, we reject $H_0 : \mu_2 = \mu_4$.

The results of the test are ambiguous since we have that $\mu_1 \neq \mu_2$, but $\mu_1 = \mu_3 \neq \mu_4$ and $\mu_2 = \mu_3 \neq \mu_4$.

3. **(a)** The hypotheses are: $H_0 : \mu_{\text{organic}} = \mu_{\text{mineral}} = \mu_{\text{surface}}$ versus H_1 : At least one of the means is different from the others.

   (b) (1) Each of the 3 samples must be a simple random sample.
   (2) The three samples must be independent of each other.
   (3) The samples must come from normally distributed populations.
   (4) The populations must have equal variances.

   (c) Since P-value < 0.001 is smaller than the $\alpha = 0.01$ level of significance, the researcher will reject the null hypothesis and conclude that the mean concentration of dissolved carbon is different in at least one of the collection areas.

   (d) Answers will vary. One possibility follows: Yes, the boxplots support the results obtained in part (c). The boxplots indicate a lower concentration of dissolved carbon in ground water obtained in mineral soil than from either surface water or ground water obtained from organic soil.

   (e) The needed summary statistics are: $\bar{x}_{\text{organic}} \approx 14.867$, $\bar{x}_{\text{mineral}} \approx 10.027$, $\bar{x}_{\text{surface}} \approx 13.045$, $\bar{x} \approx 12.345$, $s_{\text{organic}}^2 \approx 39.015$, $s_{\text{mineral}}^2 = 24.789$, $s_{\text{surface}}^2 \approx 15.419$, $n_{\text{organic}} = 31$, $n_{\text{mineral}} = 47$, and $n_{\text{surface}} = 44$, and $n = 122$, and $k = 3$. Thus,

$$\text{MST} = \frac{31(14.867 - 12.345)^2 + 47(10.027 - 12.345)^2 + 44(13.045 - 12.345)^2}{3 - 1} \approx 235.636,$$

$$MSE = \frac{(31-1) \cdot 39.015 + (47-1) \cdot 24.789 + (44-1) \cdot 15.419}{122-3} \approx 24.990 \text{, and}$$

$$F = \frac{MST}{MSE} = \frac{235.636}{24.990} \approx 9.429$$

(f) From the MINITAB output we have: $-7.588 < \mu_{\text{mineral}} - \mu_{\text{organic}} < -2.092$;

$-4.606 < \mu_{\text{surface}} - \mu_{\text{organic}} < 0.964$; and $0.527 < \mu_{\text{surface}} - \mu_{\text{mineral}} < 5.510$.

The first interval does not contain 0, so we reject $H_0 : \mu_{\text{organic}} = \mu_{\text{mineral}}$ .. The second interval contains 0, so we do not reject $H_0 : \mu_{\text{organic}} = \mu_{\text{surface}}$. The third interval does not contain 0, so we reject

$H_0 : \mu_{\text{mineral}} = \mu_{\text{surface}}$. Thus, we have $\mu_{\text{organic}} = \mu_{\text{surface}} \neq \mu_{\text{mineral}}$ or $\underline{\mu_{\text{organic}} \quad \mu_{\text{surface}}} \quad \mu_{\text{mineral}}$. Thus, the Tukey's test indicates that the mean dissolved carbon is equal in the surface water and in the ground water taken from organic soil, but it differs from the mean dissolved carbon concentration in ground water taken from mineral soil.

4. From the MINITAB output we have: $-6.914 < \mu_{\text{select structural}} - \mu_{\text{no. 1}} < 13.064$;

$1.936 < \mu_{\text{select structural}} - \mu_{\text{no. 2}} < 21.914$; $-2.989 < \mu_{\text{select structural}} - \mu_{\text{below grade}} < 16.989$;

$-1.139 < \mu_{\text{no. 1}} - \mu_{\text{no. 2}} < 18.839$; $-6.064 < \mu_{\text{no. 1}} - \mu_{\text{below grade}} < 13.914$; and

$-14.914 < \mu_{\text{no. 2}} - \mu_{\text{below grade}} < 5.064$. The first interval contains 0, so we do not reject $H_0 : \mu_1 = \mu_2$ .. The second interval does not contain 0, so we reject $H_0 : \mu_1 = \mu_3$. The third interval contains 0, so we do not reject $H_0 : \mu_1 = \mu_4$. The fourth interval contains 0, so we do not reject $H_0 : \mu_2 = \mu_3$. The fifth interval contains 0, so we do not reject $H_0 : \mu_2 = \mu_4$. The sixth interval contains 0, so we do not reject $H_0 : \mu_3 = \mu_4$. Thus, Tukey's test is ambiguous. It indicates that $\mu_{\text{select structural}} \neq \mu_{\text{no. 2}}$, but that

$\mu_{\text{select structural}} = \mu_{\text{no. 1}} = \mu_{\text{no. 2}} = \mu_{\text{below grade}}$ or $\underline{\mu_{\text{select structural}} \quad \mu_{\text{no. 1}} \quad \mu_{\text{below grade}}} \quad \mu_{\text{no. 2}}$.

5. **(a)** The hypotheses are $H_0 : \mu_{\text{front}} = \mu_{\text{middle}} = \mu_{\text{back}}$ versus H_1 : at least one mean differs from the others. Using MINITAB, we obtain the following output:

One-way ANOVA: Front, Middle, Back
```
Source   DF    SS      MS      F      P
Factor    2   2.350   1.175   7.50   0.003
Error    27   4.229   0.157
Total    29   6.580
```

The P-value is 0.003, so we reject the null hypothesis and conclude that at least one of the means is different.

(b) Using MINITAB, we obtain the following output:

Tukey 95% Simultaneous Confidence Intervals
All Pairwise Comparisons

Individual confidence level = 98.04%

Front subtracted from:
```
          Lower    Center   Upper    --+---------+---------+---------+-------
Middle   -0.9779  -0.5386  -0.0993     (--------*--------)
Back     -1.0760  -0.6367  -0.1974   (--------*--------)
                                     --+---------+---------+---------+-------
                                     -1.00     -0.50     0.00      0.50
```

Middle subtracted from:
```
         Lower    Center   Upper     --+---------+---------+---------+-------
Back    -0.5374  -0.0981   0.3412                 (--------*--------)
                                     --+---------+---------+---------+-------
                                     -1.00     -0.50     0.00      0.50
```

From the output we have: $-0.9779 < \mu_{middle} - \mu_{front} < -0.0993$; $-1.0760 < \mu_{back} - \mu_{front} < -0.1974$; and $-0.5374 < \mu_{back} - \mu_{middle} < 0.3412$.

The first interval does not contain 0, so we reject $H_0 : \mu_{front} = \mu_{middle}$. Likewise, the second interval does not contain 0, so we reject $H_0 : \mu_{front} = \mu_{back}$. Finally, the third interval contains 0, so we do not reject $H_0 : \mu_{middle} = \mu_{back}$. Thus, we have $\mu_{front} \neq \mu_{middle} = \mu_{back}$ or $\underline{\mu_{front}} \quad \underline{\mu_{middle} \quad \mu_{back}}$. That is, there is evidence that students who choose seats in the front of the mathematics classroom have a different mean GPA from that of students choosing seats in the middle or the back of the classroom.

(c) The boxplots shown to the right support the result of parts (a) and (b), indicating that GPAs of students who sit in front are generally higher than students who sit in the middle and back.

6. (a) Using MINITAB, we obtain the following output:

Two-way ANOVA: Distance versus Person, Catapult

Source	DF	SS	MS	F	P
Person	3	45.688	15.2292	16.01	0.001
Catapult	3	105.688	35.2292	37.03	0.000
Error	9	8.563	0.9514		
Total	15	159.938			

Since the treatment (catapult) P-value < 0.001, we reject the null hypothesis that the population means are equal. Thus, we conclude that at least one mean is different from the others.

(b) Using MINITAB , we obtain the following output:

```
Tukey 95.0% Simultaneous Confidence Intervals
Response Variable Distance
All Pairwise Comparisons among Levels of Catapult

Catapult = 1  subtracted from:

Catapult  Lower   Center  Upper    ------+---------+---------+---------+
2         -1.656  0.5000  2.656    (-------*------)
3          1.844  4.0000  6.156             (------*-------)
4          4.094  6.2500  8.406                      (------*------)
                                   ------+---------+---------+---------+
                                       0.0       3.0       6.0       9.0

Catapult = 2  subtracted from:

Catapult  Lower   Center  Upper    ------+---------+---------+---------+
3         1.344   3.500   5.656             (-------*------)
4         3.594   5.750   7.906                    (------*------)
                                   ------+---------+---------+---------+
                                       0.0       3.0       6.0       9.0

Catapult = 3  subtracted from:

Catapult  Lower    Center  Upper   ------+---------+---------+---------+
4         0.09438  2.250   4.406          (------*-------)
                                   ------+---------+---------+---------+
                                       0.0       3.0       6.0       9.0
```

```
Tukey Simultaneous Tests
Response Variable Distance
All Pairwise Comparisons among Levels of Catapult
Catapult = 1   subtracted from:

            Difference      SE of            Adjusted
Catapult    of Means     Difference  T-Value  P-Value
2              0.5000       0.6897    0.7249   0.8847
3              4.0000       0.6897    5.7996   0.0012
4              6.2500       0.6897    9.0618   0.0000

Catapult = 2   subtracted from:

            Difference      SE of            Adjusted
Catapult    of Means     Difference  T-Value  P-Value
3              3.500        0.6897    5.075    0.0031
4              5.750        0.6897    8.337    0.0001

Catapult = 3   subtracted from:

            Difference      SE of            Adjusted
Catapult    of Means     Difference  T-Value  P-Value
4              2.250        0.6897    3.262    0.0405
```

The interval for $\mu_2 - \mu_1$ includes 0 and the P-value = 0.8847 for this comparison, which is large, so we do not reject the null hypothesis that $\mu_1 = \mu_2$. The interval for $\mu_3 - \mu_1$ does not include 0 and the P-value = 0.0012, which is small, so we reject the null hypothesis that $\mu_1 = \mu_3$. The interval for $\mu_4 - \mu_1$ does not include 0 and the P-value < 0.0001, which is small, so we reject the null hypothesis that $\mu_4 = \mu_1$. The interval for $\mu_3 - \mu_2$ does not include 0 and the P-value = 0.0031, which is small, so we reject the null hypothesis that $\mu_2 = \mu_3$. The interval for $\mu_4 - \mu_2$ does not include 0 and the P-value = 0.0001, which is small, so we reject the null hypothesis that $\mu_2 = \mu_4$. Finally, the interval for $\mu_4 - \mu_3$ does not include 0 and the P-value = 0.0405, which is small, so we reject the null hypothesis that $\mu_3 = \mu_4$. Thus, the Tukey's tests indicate that the means of Mark IV V2 and Balloon Ballista are equal, but the means of Hydrolaunch and Waterworks II are different. That is, we conclude: $\mu_1 = \mu_2 \neq \mu_3 \neq \mu_4$

or $\underline{\mu_1 \ \mu_2} \quad \underline{\mu_3} \quad \underline{\mu_4}$.

(c) The boxplots shown to the right support the results found in parts (a) and (b).

7. (a) Using technology, we find the sample standard deviation the data within each cell:
$s_{M/\text{one 8 hr}} \approx 3.42$, $s_{M/\text{two 4 hr}} \approx 2.92$, $s_{M/\text{two 2 hr}} \approx 2.39$, $s_{F/\text{one 8 hr}} \approx 4.72$, $s_{F/\text{two 4 hr}} \approx 2.70$, and $s_{F/\text{two 2 hr}} \approx 3.87$. The largest standard deviation (4.72) is less than twice the smallest standard deviation (2.39), so the requirement of equal population variances is satisfied. [$4.72 < 2(2.39)$.]

(b) Using MINITAB with the provided data, we obtain the following output:

Two-way ANOVA: Score versus Gender, Program

Source	DF	SS	MS	F	P
Gender	1	20.83	20.833	1.77	0.196
Program	2	890.60	445.300	37.90	0.000
Interaction	2	24.07	12.033	1.02	0.374
Error	24	282.00	11.750		
Total	29	1217.50			

Since the interaction *P*-value is 0.374, which is large, there is no evidence of an interaction effect.

(c) Since the *P*-value for gender is 0.196, which is large, there is no evidence of a significant difference in the means for gender. Since we have the program *P*-value program < 0.001, which is small, there is evidence of a significant difference in the means for type of driving program. The null hypothesis

$H_0 : \mu_{8 \text{ hour}} = \mu_{4 \text{ hour}} = \mu_{2 \text{ hour}}$ is rejected.

(d) The lines in the interaction plot shown to the right are close to parallel, supporting the conclusion that there is no interaction effect.

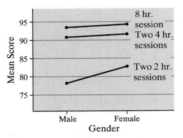

(e) The Tukey's test for gender is unnecessary since there is no evidence of a difference in means. For type of program, with Program 1 = One 8-hour session, Program 2 = Two 4-hour session, and Program 3 = Two 2-hour sessions, we use MINITAB to perform Tukey's test. The results are:

Tukey 95.0% Simultaneous Confidence Intervals
Response Variable Score
All Pairwise Comparisons among Levels of Program

Program = 1 subtracted from:

```
Program   Lower   Center   Upper      ---+---------+---------+---------+---
2         -6.03   -2.20    1.626                           (-------*------)
3         -16.33  -12.50   -8.674      (-------*-------)
                                       ---+---------+---------+---------+---
                                       -15.0     -10.0     -5.0      0.0
```

Program = 2 subtracted from:

```
Program   Lower   Center   Upper      ---+---------+---------+---------+---
3         -14.13  -10.30   -6.474         (------*-------)
                                       ---+---------+---------+---------+---
                                       -15.0     -10.0     -5.0      0.0
```

Tukey Simultaneous Tests
Response Variable Score
All Pairwise Comparisons among Levels of Program

Program = 1 subtracted from:

Program	Difference of Means	SE of Difference	T-Value	Adjusted P-Value
2	-2.20	1.533	-1.435	0.3394
3	-12.50	1.533	-8.154	0.0000

Program = 2 subtracted from:

Program	Difference of Means	SE of Difference	T-Value	Adjusted P-Value
3	-10.30	1.533	-6.719	0.0000

The interval for $\mu_{\text{two 4 hr}} - \mu_{\text{one 8 hr}}$ includes 0 and the *P*-value = 0.3394 for this comparison, which is large, so we do not reject the null hypothesis that $\mu_{\text{one 8 hr}} = \mu_{\text{two 4 hr}}$. The interval for $\mu_{\text{two 2 hr}} - \mu_{\text{one 8 hr}}$ does not include 0 and the *P*-value < 0.0001, which is small, so we reject the null hypothesis that $\mu_{\text{one 8 hr}} = \mu_{\text{two 2 hr}}$. Finally, the interval for $\mu_{\text{two 2 hr}} - \mu_{\text{two 4 hr}}$ does not include 0 and the *P*-value < 0.0001, which is small, so we reject the null hypothesis that $\mu_{\text{two 4 hr}} = \mu_{\text{two 2 hr}}$. Thus, we have

$\mu_{\text{one 8 hr}} = \mu_{\text{two 4 hr}} \neq \mu_{\text{two 2 hr}}$ or $\underline{\mu_{\text{one 8 hr}} \quad \mu_{\text{two 4 hr}}} \quad \mu_{\text{two 2 hr}}$. That is, the Tukey's test indicates that the mean score after taking one 8-hour defensive driving class is equal to the mean score after taking two 4-hour classes, but that the mean score after taking two 2-hour classes is different.

Chapter 13 Test

1. (a) The hypotheses are $H_0 : \mu_{\text{nonsmokers}} = \mu_{\text{light smokers}} = \mu_{\text{heavy smokers}}$ versus $H_1 :$ at least one mean differs from the others. Using MINITAB, we obtain the following output:

One-way ANOVA: Machine I, Machine II, Machine III, Machine IV

Source	DF	SS	MS	F	P
Factor	3	3.637	1.212	5.81	0.004
Error	24	5.011	0.209		
Total	27	8.649			

Since *P*-value = 0.004 is less than the $\alpha = 0.05$ level of significance, we reject H_0. There is sufficient evidence to support the hypothesis that at least one of the means is different from the others.

(b) Using MINITAB, we obtain the following output:

Tukey 95% Simultaneous Confidence Intervals   All Pairwise Comparisons
Individual confidence level = 98.90%

Machine I subtracted from:

	Lower	Center	Upper
Machine II	-1.4307	-0.7571	-0.0836
Machine III	-0.4736	0.2000	0.8736
Machine IV	-0.7450	-0.0714	0.6022

```
             ------+---------+---------+---------+---
Machine II      (------*------)
Machine III             (------*------)
Machine IV          (-----*------)
             ------+---------+---------+---------+---
               -1.0       0.0       1.0       2.0
```

Machine II subtracted from:

	Lower	Center	Upper
Machine III	0.2836	0.9571	1.6307
Machine IV	0.0121	0.6857	1.3593

```
------+---------+---------+---------+---
                            (------*-----)
                          (------*------)
------+---------+---------+---------+---
   -1.0       0.0       1.0       2.0
```

Machine III subtracted from:

	Lower	Center	Upper
Machine IV	-0.9450	-0.2714	0.4022

```
------+---------+---------+---------+--
            (-----*------)
------+---------+---------+---------+--
   -1.0       0.0       1.0       2.0
```

From the output, we have:

$-1.4307 < \mu_{\text{II}} - \mu_{\text{I}} < -0.0836$. This interval does not contain 0, so we reject $H_0 : \mu_{\text{I}} = \mu_{\text{II}}$.

$-0.4736 < \mu_{\text{III}} - \mu_{\text{I}} < 0.8736$. This interval contains 0, so we do not reject $H_0 : \mu_{\text{I}} = \mu_{\text{III}}$.

$-0.7450 < \mu_{\text{IV}} - \mu_{\text{I}} < 0.6022$. This interval contains 0, so we do not reject $H_0 : \mu_{\text{I}} = \mu_{\text{IV}}$.

$0.2836 < \mu_{\text{III}} - \mu_{\text{II}} < 1.6307$. This interval does not contain 0, so we reject $H_0 : \mu_{\text{II}} = \mu_{\text{III}}$.

$0.0121 < \mu_{IV} - \mu_{II} < 1.3593$. This interval does not contain 0, so we reject $H_0 : \mu_{II} = \mu_{IV}$.

$-0.9450 < \mu_{IV} - \mu_{III} < 0.4022$. This interval contains 0, so we do not reject $H_0 : \mu_{III} = \mu_{IV}$.

Thus, we have that $\mu_I = \mu_{III} = \mu_{IV} \neq \mu_{II}$. That is, we have: $\underline{\mu_I \quad \mu_{III} \quad \mu_{IV}} \quad \underline{\mu_{II}}$. It appears that the mean amount dispensed by Machine II is less than the mean amount dispensed by the other machines.

(c) The boxplots shown to the right support the result of parts (a) and (b).

2. (a) The hypotheses are $H_0 : \mu_{large\ car} = \mu_{passenger\ van} = \mu_{utility\ van}$ versus H_1 : At least one of the mean femur forces is different from the others.

(b) Assuming that the samples were randomly chosen and are independent, the normal probability plots indicate the populations are normally distributed. Now, $s_{large\ car} \approx 1.441$, $s_{passenger\ van} \approx 2.277$, and $s_{utility\ van} \approx 1.218$. Thus, the requirement of equal variances is satisfied, since the largest standard deviation is less than twice the smallest standard deviation. $[2.277 < 2(1.218) = 2.456.]$

(c) Using MINITAB, we obtain the following output:

One-way ANOVA: Large Family Cars, Passenger Vans, Midsize Utility Vehicles

Source	DF	SS	MS	F	P
Factor	2	47.74	23.87	8.19	0.003
Error	18	52.45	2.91		
Total	20	100.19			

Since P-value $= 0.003$ is smaller than the $\alpha = 0.05$ level of significance, so we reject H_0. There is sufficient evidence to conclude that at least one of the mean femur forces differs from the others.

(d)

3. (a) Using MINITAB, we obtain the following output:

Two-way ANOVA: Tensile Strength versus Thickness, Type

Source	DF	SS	MS	F	P
Thickness	3	97239906	32413302	28.27	0.001
Type	2	23628163	11814081	10.30	0.011
Error	6	6880388	1146731		
Total	11	127748456			

Since the treatment (type) P-value $= 0.011$, we reject the null hypothesis that the population means are equal. Thus, we conclude that at least one mean is different from the others.

(b) Using MINITAB , we obtain the following output:

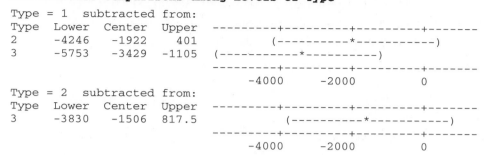

```
Tukey 95.0% Simultaneous Confidence Intervals
Response Variable Tensile Strength
All Pairwise Comparisons among Levels of Type

Type = 1   subtracted from:
Type  Lower   Center   Upper    ---------+---------+---------+-------
2     -4246   -1922    401               (----------*-----------)
3     -5753   -3429    -1105    (-----------*----------)
                                ---------+---------+---------+-------
                                    -4000      -2000        0

Type = 2   subtracted from:
Type  Lower   Center   Upper    ---------+---------+---------+-------
3     -3830   -1506    817.5             (----------*-----------)
                                ---------+---------+---------+-------
                                    -4000      -2000        0
```

```
Tukey Simultaneous Tests
Response Variable Tensile Strength
All Pairwise Comparisons among Levels of Type

Type = 1   subtracted from:
       Difference      SE of              Adjusted
Type   of Means    Difference  T-Value    P-Value
2        -1922        757.2    -2.539     0.0972
3        -3429        757.2    -4.528     0.0095

Type = 2   subtracted from:
       Difference      SE of              Adjusted
Type   of Means    Difference  T-Value    P-Value
3        -1506        757.2    -1.989     0.1956
```

The interval for $\mu_2 - \mu_1$ includes 0 and the *P*-value = 0.0972 for this comparison, which is large, so we do not reject the null hypothesis that $\mu_1 = \mu_2$. The interval for $\mu_3 - \mu_1$ does not include 0 and the *P*-value = 0.0095, which is small, so we reject the null hypothesis that $\mu_1 = \mu_3$. Finally, the interval for $\mu_3 - \mu_2$ includes 0 and the *P*-value = 0.1956, which is large, so we do not reject the null hypothesis that $\mu_2 = \mu_3$. Thus, the Tukey's test is ambiguous. It indicates that $\mu_3 \neq \mu_1 = \mu_2$, but $\mu_2 = \mu_3$. That is: $\underline{\mu_1 \quad \mu_2} \quad \mu_3$.

(c)

4. (a) Using technology, we find the sample standard deviation the data within each cell:
$s_{M/20\text{-}34} \approx 0.46$, $s_{M/35\text{-}49} \approx 0.48$, $s_{M/50\text{-}64} \approx 0.53$, $s_{F/20\text{-}34} \approx 0.31$, $s_{F/35\text{-}49} \approx 0.28$, and $s_{F/50\text{-}64} \approx 0.32$.
The largest standard deviation (0.53) is less than twice the smallest standard deviation (0.28), so the requirement of equal population variances is satisfied. [$0.53 < 2(0.28) = 0.56$.]

(b) Using MINITAB with the provided data, we obtain the following output:

Two-way ANOVA: Learning Time versus Gender, Age

Source	DF	SS	MS	F	P
Gender	1	0.05042	0.050417	0.30	0.589
Age	2	0.51083	0.255417	1.53	0.243
Interaction	2	0.27083	0.135417	0.81	0.459
Error	18	2.99750	0.166528		
Total	23	3.82958			

Since the interaction P-value is 0.459, which is larger than $\alpha = 0.05$, there is no evidence of an interaction effect.

(c) Since the P-value for gender is 0.589, which is large, there is no evidence of a difference in the means for gender. Since the P-value for age is 0.243, which is also large, there is no evidence of a difference in the means for the three age groups.

(d) The lines in the interaction plot shown to the right are not very parallel, contradicting the conclusion that there is no interaction effect.

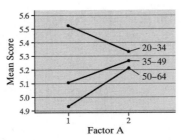

(e) The Tukey's test is unnecessary since there is no evidence of a difference in means.

5. (a)

(b) The plot indicates a significant interaction effect since the graphs for the two levels of gender are not parallel.

(c) Since there is a significant interaction effect, any conclusion about main effects would be confounded.

Chapter 14

Inference on the Least-Squares Regression Model and Multiple Regression

Section 14.1

1. There are two requirements for conducting inference on the least-squares regression model. First, for each particular value of the explanatory variable x, the corresponding responses in the population have a mean that depends linearly on x. Second, the response variables are normally distributed with mean $\mu_{y|x} = \beta_0 + \beta_1 x$ and standard deviation σ.

   The first requirement is tested by examining a plot of the residuals against the explanatory variable. If the plot shows any discernible pattern, then the linear model is inappropriate. The normality requirement is checked by examining a normal probability plot of the residuals.

3. The y-coordinates on the regression line represent the mean value of the response variable for any given value of the explanatory variable.

5. If we do not reject $H_0 : \beta_1 = 0$ then the best estimate for the value of the response variable for any value of the explanatory variable is the sample mean of y. That is, $\overline{y}$.

7. (a) Using technology, we get: $\beta_0 \approx b_0 = -2.3256$ and $\beta_1 \approx b_1 = 2.0233$.

   (b) We calculate $\hat{y} = 2.0233x - 2.3256$ to generate the following table:

x	y	$\hat{y}$	$y - \hat{y}$	$(y - \hat{y})^2$	$(x - \overline{x})^2$
3	4	3.74	0.26	0.0654	5.76
4	6	5.77	0.23	0.0541	1.96
5	7	7.79	−0.79	0.6252	0.16
7	12	11.84	0.16	0.0265	2.56
8	14	13.86	0.14	0.0195	6.76
			Sum:	0.7907	17.20

   $$s_e = \sqrt{\frac{\Sigma(y - \hat{y})^2}{n - 2}} = \sqrt{\frac{0.7907}{5 - 2}} \approx 0.5134 \text{ is}$$
   the point estimate for σ.

   (c) $s_{b_1} = \dfrac{s_e}{\sqrt{\Sigma(x - \overline{x})^2}} = \dfrac{0.5134}{\sqrt{17.20}} \approx 0.1238$

   (d) We have the hypotheses $H_0 : \beta_1 = 0$ versus $H_1 : \beta_1 \neq 0$. The test statistic is

   $$t_0 = \frac{b_1}{s_{b_1}} = \frac{2.0233}{0.1238} = 16.343 \text{ [Tech: 16.344]}.$$

   Classical approach: The level of significance is $\alpha = 0.05$. Since this is a two-tailed test with $n - 2 = 5 - 2 = 3$ degrees of freedom, the critical values are $\pm t_{0.025} = \pm 3.182$. Since $t_0 = 16.343 > t_{0.025} = 3.182$ (the test statistic falls in a critical region), we reject H_0.

   P-value approach: The P-value for this two-tailed test is the area under the t-distribution with 3 degrees of freedom to the right of $t_0 = 16.343$ [Tech: 16.344], plus the area to the left of $-t_0 = -16.343$ [Tech: −16.344]. From the t-distribution table in the row corresponding to 3 degrees of freedom, 16.343 falls to the right of 12.924, whose right-tail area is 0.0005. We must double this value in order to get the total area in both tails: 0.001. So, P-value < 0.001. [Tech: P-value = 0.0005]. Because the P-value is less than the level of significance $\alpha = 0.05$, we reject H_0.

   Conclusion: There is sufficient evidence to conclude that a linear relationship exists between x and y.

9. (a) Using technology, we get: $\beta_0 \approx b_0 = 1.2$ and $\beta_1 \approx b_1 = 2.2$.

   (b) We calculate $\hat{y} = 2.2x + 1.2$ to generate the following table:

x	y	$\hat{y}$	$y - \hat{y}$	$(y - \hat{y})^2$	$(x - \overline{x})^2$
−2	−4	−3.2	−0.8	0.64	4
−1	0	−1	1.	1	1
0	1	1.2	−0.2	0.04	0
1	4	3.4	0.6	0.36	1
2	5	5.6	−0.6	0.36	4
			Sum	2.40	10

   $$s_e = \sqrt{\frac{\Sigma(y - \hat{y})^2}{n - 2}} = \sqrt{\frac{2.4}{5 - 2}} = 0.8944 \text{ is the}$$
   point estimate for σ.

(c) $s_{b_1} = \dfrac{s_e}{\sqrt{\Sigma(x-\overline{x})^2}} = \dfrac{0.8944}{\sqrt{10}} = 0.2828$

(d) The hypotheses are $H_0 : \beta_1 = 0$ versus $H_1 : \beta_1 \neq 0$. The test statistic is

$$t_0 = \dfrac{b_1}{s_{b_1}} = \dfrac{2.2}{0.2828} = 7.778 \ [\text{Tech: } 7.779].$$

Classical approach: The level of significance is $\alpha = 0.05$. Since this is a two-tailed test with $n-2 = 5-2 = 3$ degrees of freedom, the critical values are $\pm t_{0.025} = \pm 3.182$. Since $t_0 = 7.778 > t_{0.025} = 3.182$ (the test statistic falls in a critical region), we reject H_0.

P-value approach: The P-value for this two-tailed test is the area under the t-distribution with 3 degrees of freedom to the right of $t_0 = 7.778$ [Tech: 7.779], plus the area to the left of $-t_0 = -7.778$ [Tech: –7.779]. From the t-distribution table in the row corresponding to 3 degrees of freedom, 7.778 falls between 7.453 and 10.215, whose right-tail areas are 0.0025 and 0.001, respectively. We must double these values in order to get the total area in both tails: 0.005 and 0.002. So, $0.002 < P\text{-value} < 0.005$. [Tech: P-value = 0.0044]. Because the P-value is less than the level of significance $\alpha = 0.05$, we reject H_0.

Conclusion: There is sufficient evidence to conclude that a linear relationship exists between x and y.

11. (a) Using technology, we get: $\beta_0 \approx b_0 = 116.6$ and $\beta_1 \approx b_1 = -0.72$.

(b) We calculate $\hat{y} = -0.72x + 116.6$ to generate the following table:

x	y	$\hat{y}$	$y - \hat{y}$	$(y - \hat{y})^2$	$(x - \overline{x})^2$
20	100	102.2	–2.2	4.84	400
30	95	95	0	0	100
40	91	87.8	3.2	10.24	0
50	83	80.6	2.4	5.76	100
60	70	73.4	–3.4	11.56	400
			Sum	32.40	1000

$s_e = \sqrt{\dfrac{\Sigma(y-\hat{y})^2}{n-2}} = \sqrt{\dfrac{32.4}{5-2}} = 3.2863$ is the point estimate for σ.

(c) $s_{b_1} = \dfrac{s_e}{\sqrt{\Sigma(x-\overline{x})^2}} = \dfrac{3.2863}{\sqrt{1000}} = 0.1039$

(d) The hypotheses are $H_0 : \beta_1 = 0$ versus $H_1 : \beta_1 \neq 0$. The test statistic is

$$t_0 = \dfrac{b_1}{s_{b_1}} = \dfrac{-0.72}{0.1039} = -6.929 \ [\text{Tech: } -6.928]$$

Classical approach: The level of significance is $\alpha = 0.05$. Since this is a two-tailed test with $n-2 = 5-2 = 3$ degrees of freedom, the critical values are $\pm t_{0.025} = \pm 3.182$. Since $t_0 = -6.929 < -t_{0.025} = -3.182$ (the test statistic falls in a critical region), we reject H_0.

P-value approach: The P-value for this two-tailed test is the area under the t-distribution with 3 degrees of freedom to the left of $t_0 = -6.929$ [Tech: –6.928], plus the area to the right of 6.929 [Tech: 6.928]. From the t-distribution table in the row corresponding to 3 degrees of freedom, 6.929 falls between 5.841 and 7.453, whose right-tail areas are 0.005 and 0.0025, respectively. We must double these values in order to get the total area in both tails: 0.01 and 0.005. So, $0.005 < P\text{-value} < 0.01$. [Tech: P-value = 0.0062]. Because the P-value is less than the level of significance $\alpha = 0.05$, we reject H_0.

Conclusion: There is sufficient evidence to conclude that a linear relationship exists between x and y.

13. (a) Using technology, we get:
$\beta_0 \approx b_0 = 12.4932$ and $\beta_1 \approx b_1 = 0.1827$.

(b) Using technology, we calculate $\hat{y} = 0.1827x + 12.4932$ to generate the table:

x	y	$\hat{y}$	$y - \hat{y}$	$(y - \hat{y})^2$	$(x - \overline{x})^2$
27.75	17.5	17.5640	–0.0640	0.0041	1.6782
24.5	17.1	16.9701	0.1299	0.0169	3.8202
25.5	17.1	17.1528	–0.0528	0.0028	0.9112
26	17.3	17.2442	0.0558	0.0031	0.2066
25	16.9	17.0615	–0.1615	0.0261	2.1157
27.75	17.6	17.5640	0.0360	0.0013	1.6782
26.5	17.3	17.3356	–0.0356	0.0013	0.0021
27	17.5	17.4269	0.0731	0.0053	0.2975
26.75	17.3	17.3813	–0.0813	0.0066	0.0873
26.75	17.5	17.3813	0.1187	0.0141	0.0873
27.5	17.5	17.5183	–0.0183	0.0003	1.0930
			Sum	0.0819	11.9773

$$s_e = \sqrt{\frac{\Sigma(y-\hat{y})^2}{n-2}} = \sqrt{\frac{0.0819}{11-2}} = 0.0954$$

(c) A normal probability plot shows that the residuals are approximately normally distributed.

(d) $s_{b_1} = \dfrac{s_e}{\sqrt{\Sigma(x-\overline{x})^2}} = \dfrac{0.0954}{\sqrt{11.9773}} = 0.0276$

(e) The hypotheses are $H_0 : \beta_1 = 0$ versus $H_1 : \beta_1 \neq 0$. The test statistic is

$$t_0 = \frac{b_1}{s_{b_1}} = \frac{0.1827}{0.0276} = 6.620 \quad [\text{Tech: } 6.630]$$

Classical approach: The level of significance is $\alpha = 0.01$. Since this is a two-tailed test with $n - 2 = 11 - 2 = 9$ degrees of freedom, the critical values are $\pm t_{0.005} = \pm 3.250$. Since $t_0 = 6.620 > t_{0.005} = 3.250$ (the test statistic falls in a critical region), we reject H_0.

P-value approach: The P-value for this two-tailed test is the area under the t-distribution with 9 degrees of freedom to the right of $t_0 = 6.620$ [Tech: 6.630], plus the area to the left of $-t_0 = -6.620$ [Tech: −9.630]. From the t-distribution table in the row corresponding to 9 degrees of freedom, 6.620 falls to the right of 4.781, whose right-tail area is 0.0005. We must double this value in order to get the total area in both tails: 0.001. So, P-value < 0.001. [Tech: P-value = 0.0001]. Because the P-value is less than the level of significance $\alpha = 0.01$, we reject H_0.

Conclusion: There is sufficient evidence to conclude that a linear relationship exists between height and head circumference.

(f) For a 95% confidence interval we use $t_{\alpha/2} = t_{0.025} = 2.262$:

Lower bound:
$$b_1 - t_{\alpha/2} \cdot s_{b_1} = 0.1827 - 2.262 \cdot 0.0276$$
$$= 0.1203 \quad [\text{Tech: } 0.1204]$$

Upper bound:
$$b_1 + t_{\alpha/2} \cdot s_{b_1} = 0.1827 + 2.262 \cdot 0.0276$$
$$= 0.2451$$

(g) We use the regression equation to obtain a good estimate of the child's head circumference if the child's height is 26.5 inches:
$$\hat{y} = 0.1827x + 12.4932$$
$$= 0.1827 \cdot 26.5 + 12.4932$$
$$\approx 17.34 \text{ inches}$$

15. (a) Using technology, we get:
$$\beta_0 \approx b_0 = 2675.6 \text{ and } \beta_1 \approx b_1 = 0.6764$$

(b) Using technology, we calculate $\hat{y} = 0.6764x + 2675.6$ to generate the table:

x	y	$\hat{y}$	$y - \hat{y}$	$(y-\hat{y})^2$	$(x-\overline{x})^2$
2300	4070	4231.32	−161.32	26025.6	338724
3390	5220	4968.62	251.38	63191.6	258064
2430	4640	4319.26	320.74	102874.9	204304
2890	4620	4630.41	−10.41	108.4	64
3330	4850	4928.04	−78.04	6089.5	200704
2480	4120	4353.08	−233.08	54326.2	161604
3380	5020	4961.86	58.14	3380.7	248004
2660	4890	4474.84	415.16	172361.9	49284
2620	4190	4447.78	−257.78	66449.7	68644
3340	4630	4934.80	−304.80	92902.8	209764
			Sum	587711.3	1739160

$$s_e = \sqrt{\frac{\Sigma(y-\hat{y})^2}{n-2}} = \sqrt{\frac{587711.3}{10-2}} = 271.04$$

(c) A normal probability plot shows that the residuals are normally distributed.

(d) $s_{b_1} = \dfrac{s_e}{\sqrt{\Sigma(x-\overline{x})^2}} = \dfrac{271.04}{\sqrt{1,739,160}} = 0.2055$

(e) The hypotheses are $H_0 : \beta_1 = 0$ versus $H_1 : \beta_1 \neq 0$. The test statistic is

$$t_0 = \frac{b_1}{s_{b_1}} = \frac{0.6764}{0.2055} = 3.291.$$

Classical approach: The level of significance is $\alpha = 0.05$. Since this is a two-tailed test with $n - 2 = 10 - 2 = 8$ degrees of freedom, the critical values are $\pm t_{0.025} = \pm 2.306$. Since $t_0 = 3.291 > t_{0.025} = 2.306$ (the test statistic falls in a critical region), we reject H_0.

P-value approach: The P-value for this two-tailed test is the area under the t-distribution with 8 degrees of freedom to the right of $t_0 = 3.291$, plus the area to the left of $-t_0 = -3.291$. From the t-distribution table in the row corresponding to 8 degrees of freedom, 3.291 falls between 2.896 and 3.355, whose right-tail areas are 0.01 and 0.005, respectively. We must double these values in order to get the total area in both tails: 0.02 and 0.01. So, $0.01 < P\text{-value} < 0.02$. [Tech: P-value = 0.0110]. Because the P-value is less than the level of significance $\alpha = 0.05$, we reject H_0.

Conclusion: There is sufficient evidence to conclude that a linear relationship exists between the 7-day and 28-day strength of this type of concrete.

(f) For a 95% confidence interval we use $t_{\alpha/2} = t_{0.025} = 2.306$:

Lower bound:
$$b_1 - t_{\alpha/2} \cdot s_{b_1} = 0.6764 - 2.306 \cdot 0.2055$$
$$= 0.2025$$

Upper bound:
$$b_1 + t_{\alpha/2} \cdot s_{b_1} = 0.6764 + 2.306 \cdot 0.2055$$
$$= 1.1503$$

(g) We use the regression equation to calculate the mean 28-day strength of this concrete if the 7-day strength is 3000 psi:
$$\hat{y} = 0.6764x + 2675.6$$
$$= 0.6764 \cdot 3000 + 2675.6$$
$$\approx 4704.8 \text{ psi}$$

17. (a) Using technology, we get:
$\beta_0 \approx b_0 = 1.2286$ and $\beta_1 \approx b_1 = 0.7622$.

(b) Using technology, we calculate $\hat{y} = 0.7622x + 1.2286$ to generate the table at the top of the next column. Using the results from the table, we obtain:

$$s_e = \sqrt{\frac{\Sigma(y - \hat{y})^2}{n - 2}} = \sqrt{\frac{93.9456}{11 - 2}} = 3.2309$$

x	y	$\hat{y}$	$y - \hat{y}$	$(y - \hat{y})^2$	$(x - \bar{x})^2$
4.33	3.28	4.5289	−1.2489	1.5598	23.6903
3.25	5.09	3.7058	1.3843	1.9163	14.3434
−1.78	0.54	−0.1281	0.6681	0.4464	1.5444
−3.20	2.88	−1.2104	4.0904	16.7314	7.0901
1.29	2.69	2.2118	0.4782	0.2287	3.3389
3.58	7.41	3.9573	3.4527	11.9211	16.9519
1.48	−4.83	2.3567	−7.1867	51.6487	4.0694
−4.40	−2.38	−2.1251	−0.2549	0.0650	14.9207
−0.86	2.37	0.5731	1.7969	3.2288	0.1042
−6.12	−4.27	−3.4361	−0.8338	0.6952	31.1668
−3.48	−3.77	−1.4239	−2.3461	5.5042	8.6596
			Sum	93.9456	125.8797

(c) A normal probability plot shows that the residuals are approximately normally distributed.

(d) $s_{b_1} = \dfrac{s_e}{\sqrt{\Sigma(x - \bar{x})^2}} = \dfrac{3.2309}{\sqrt{125.8797}} = 0.2880$

(e) The hypotheses are $H_0 : \beta_1 = 0$ versus $H_1 : \beta_1 \neq 0$. The test statistic is
$$t_0 = \frac{b_1}{s_{b_1}} = \frac{0.7622}{0.2880} = 2.647 .$$

Classical approach: The level of significance is $\alpha = 0.1$. Since this is a two-tailed test with $n - 2 = 11 - 2 = 9$ degrees of freedom, the critical values are $\pm t_{0.05} = \pm 1.833$. Since $t_0 = 2.647 > t_{0.05} = 1.833$ (the test statistic falls in a critical region), we reject H_0.

P-value approach: The P-value for this two-tailed test is the area under the t-distribution with 9 degrees of freedom to the right of $t_0 = 2.647$, plus the area to the left of $-t_0 = -2.647$. From the t-distribution table in the row corresponding to 9 degrees of freedom, 2.647 falls between 2.398 and 2.821, whose right-tail areas are 0.02 and 0.01, respectively. We must double these values in order to get the total area in both tails: 0.04 and 0.02. So, $0.02 < P\text{-value} < 0.04$. [Tech: P-value = 0.0266]. Because the P-value is less than the level of significance $\alpha = 0.10$, we reject H_0.

Conclusion: There is sufficient evidence to conclude that a linear relationship exists between the rate of return of the S&P 500 index and the rate of return of UTX.

(f) For a 90% confidence interval we use
$t_{\alpha/2} = t_{0.05} = 1.833$:

Lower bound:
$b_1 - t_{\alpha/2} \cdot s_{b_1} = 0.7622 - 1.833 \cdot 0.2880$

$= 0.2343$

Upper bound:
$b_1 + t_{\alpha/2} \cdot s_{b_1} = 0.7622 + 1.833 \cdot 0.2880$

$= 1.2901$ [Tech: 1.2900]

(g) We can use the regression equation to calculate the mean rate of return:
$\hat{y} = 0.7622x + 1.2286$

$= 0.7622(3.25) + 1.2286 \approx 3.706\%$

19. (a) Using technology, we get:
$\beta_0 \approx b_0 = 163.9988$ and $\beta_1 \approx b_1 = 63.1624$.

(b) Using technology, we calculate
$\hat{y} = 63.1624x + 163.9988$ to generate the table:

x	y	$\hat{y}$	$y - \hat{y}$	$(y - \hat{y})^2$	$(x - \bar{x})^2$
18	1390	1300.9220	89.0780	7934.8901	16
21	1340	1490.4092	−150.4092	22622.9274	1
27	1910	1869.3836	40.6164	1649.6919	25
18	1150	1300.9220	−150.9220	22777.4501	16
20	1360	1427.2468	−67.2468	4522.1321	4
25	1780	1743.0588	36.9412	1364.6523	9
25	1590	1743.0588	−153.0588	23426.9963	9
19	1470	1364.0844	105.9156	11218.1143	9
17	1190	1237.7596	−47.7596	2280.9794	25
28	1770	1932.5460	−162.5460	26421.2021	36
30	2290	2058.8708	231.1292	53420.7071	64
20	1660	1427.2468	232.7532	54174.0521	4
18	1480	1300.9220	179.0780	32068.9301	16
22	1370	1553.5716	−183.5716	33698.5323	0
			Sum	297581.2576	234

$$s_e = \sqrt{\frac{\Sigma(y - \hat{y})^2}{n - 2}} = \sqrt{\frac{297581.2576}{14 - 2}}$$

$$= 157.4752$$

(c) A normal probability plot shows that the residuals are approximately normally distributed.

(d) $s_{b_1} = \dfrac{s_e}{\sqrt{\Sigma(x - \bar{x})^2}} = \dfrac{157.4752}{\sqrt{234}} = 10.2945$

(e) The hypotheses are $H_0 : \beta_1 = 0$ versus $H_1 : \beta_1 \neq 0$. The test statistic is
$$t_0 = \frac{b_1}{s_{b_1}} = \frac{63.1624}{10.2945} = 6.136 .$$

Classical approach: The level of significance is $\alpha = 0.01$. Since this is a two-tailed test with $n - 2 = 14 - 2 = 12$ degrees of freedom, the critical values are $\pm t_{0.005} = \pm 3.055$. Since $t_0 = 6.136 > t_{0.005} = 3.055$ (the test statistic falls in a critical region), we reject H_0 .

P-value approach: The P-value for this two-tailed test is the area under the t-distribution with 12 degrees of freedom to the right of $t_0 = 6.136$, plus the area to the left of $-t_0 = -6.136$. From the t-distribution table in the row corresponding to 12 degrees of freedom, 6.136 falls to the right of 4.318, whose right-tail area is 0.0005. We must double this value in order to get the total area in both tails: 0.001. So, P-value < 0.001. [Tech: P-value = 0.0001]. Because the P-value is less than the level of significance $\alpha = 0.01$, we reject H_0 .

Conclusion: There is sufficient evidence to conclude that a linear relationship exists between ACT scores and SAT scores.

(f) For a 95% confidence interval we use
$t_{\alpha/2} = t_{0.025} = 2.179$:

Lower bound:
$b_1 - t_{\alpha/2} \cdot s_{b_1} = 63.1624 - 2.179 \cdot 10.2945$

$= 40.731$

Upper bound:
$b_1 + t_{\alpha/2} \cdot s_{b_1} = 63.1624 + 2.179 \cdot 10.2945$

$= 85.594$

(g) We can use the regression equation to calculate the mean energy expenditure:
$\hat{y} = 63.1624x + 163.9988$

$= 63.1624(26) + 163.9988 \approx 1806.2$

21. (a) No linear relationship appears to exist between calories and sugar:

(b) Using technology, we obtain
$\hat{y} = -0.0146x + 17.8675$.

(c) We use the equation from part (b) to generate the table:

x	y	$\hat{y}$	$y - \hat{y}$	$(y - \hat{y})^2$	$(x - \bar{x})^2$
180	10	15.2395	−5.2395	27.4524	1327.0408
200	18	14.9475	3.0525	9.3178	269.8980
210	14	14.8015	−0.8015	0.6424	41.3265
220	20	14.6555	5.3445	28.5637	12.7551
220	0	14.6555	−14.6555	214.7837	12.7551
230	28	14.5095	13.4905	181.9936	184.1837
240	2	14.3635	−12.3635	152.8561	555.6122
270	20	13.9255	6.0745	36.8996	2869.8980
320	2	13.1955	−11.1955	125.3392	10727.0408
110	10	16.2615	−6.2615	39.2064	11327.0408
180	12	15.2395	−3.2395	10.4944	1327.0408
200	22	14.9475	7.0525	49.7378	269.8980
220	24	14.6555	9.3445	87.3197	12.7551
230	24	14.5095	9.4905	90.0696	184.1837
			Sum	1054.6761	29121.4286

$$s_e = \sqrt{\frac{\Sigma(y - \hat{y})^2}{n - 2}} = \sqrt{\frac{1054.6761}{14 - 2}} = 9.3749$$

(d) A normal probability plot shows that the residuals are approximately normally distributed.

(e) $s_{b_1} = \dfrac{s_e}{\sqrt{\Sigma(x - \bar{x})^2}} = \dfrac{9.3749}{\sqrt{29121.4286}} = 0.0549$

(f) The hypotheses are $H_0 : \beta_1 = 0$ versus $H_1 : \beta_1 \neq 0$. The test statistic is

$$t_0 = \frac{b_1}{s_{b_1}} = \frac{-0.0146}{0.0549} = -0.266 \ [\text{Tech:} -0.265].$$

<u>Classical approach</u>: The level of significance is $\alpha = 0.01$. Since this is a two-tailed test with $n - 2 = 14 - 2 = 12$ degrees of freedom, the critical values are $\pm t_{0.005} = \pm 3.055$. Since $t_0 = -0.266 > t_{0.005} = -3.055$ (the test statistic does not fall in a critical region), we do not reject H_0.

<u>P-value approach</u>: The P-value for this two-tailed test is the area under the t-distribution with 12 degrees of freedom to the left of $t_0 = -0.266$, plus the area to the right of 0.266. From the t-distribution table in the row corresponding to 12 degrees of freedom, 0.266 falls to the left of 0.695, whose right-tail area is 0.25. We must double this value in order to get the total area in both tails: 0.50. So, P-value > 0.50. [Tech: P-value = 0.7955]. Because the P-value is greater than the level of significance $\alpha = 0.01$, we do not reject H_0.

<u>Conclusion</u>: A linear relationship does not exist between the number of calories per serving and the number of grams per serving in high-protein and moderate-protein energy bars.

(g) For a 95% confidence interval we use $t_{\alpha/2} = t_{0.025} = 2.179$:
Lower bound:
$$b_1 - t_{\alpha/2} \cdot s_{b_1} = -0.0146 - 2.179 \cdot 0.0549$$
$$= -0.1342 \ [\text{Tech:} -0.1343]$$
Upper bound:
$$b_1 + t_{\alpha/2} \cdot s_{b_1} = -0.0146 + 2.179 \cdot 0.0549$$
$$= 0.1050 \ [\text{Tech: } 0.1051]$$

(h) We do not recommend using the least-squares regression line to predict the sugar content of the energy bars because we did not reject the null hypothesis, which means there is no linear correlation between calories per serving and number of grams per serving. A good estimate of for the sugar content would be $\bar{y} = 14.7$ grams.

23. (a) Using technology, we get
$\hat{y} = 2{,}018{,}994\,x - 3{,}771{,}542{,}830$.

(b) The hypotheses are $H_0 : \beta_1 = 0$ versus
$H_1 : \beta_1 \neq 0$. For a test at the $\alpha = 0.01$ level
of significance with $n - 2 = 11 - 2 = 9$
degrees of freedom, the critical values are
$\pm t_{0.005} = \pm 3.250$. Using technology we find
the test statistic to be $t_0 = 21.812$, and we
find the P-value to be P-value < 0.0001 .

Classical approach: Since $t_0 = 21.812 >$
$t_{0.005} = 3.250$ (the test statistic falls within
the critical region), we reject H_0 .

P-value approach: Since P-value < 0.0001
$< \alpha = 0.01$ we reject H_0 .

Conclusion: A linear relationship exists
between year and population of the U.S.

(c)

(d)

(e) A linear model is not appropriate. The
scatter plot seems to show some curvature,
and there is a pattern in the residual plot.

(f) The moral is that the inferential procedures
may lead us to believe that a linear relation
exists between the two variables even
though diagnostic tools (such as residual
plots) indicate that a linear model is
inappropriate.

25. (a)
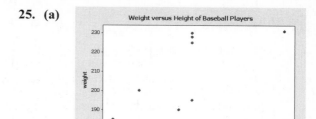

(b) Using technology, we get the least-squares
equation: $\hat{y} = 4.0432\,x - 92.9811$.

The hypotheses are $H_0 : \beta_1 = 0$ versus
$H_1 : \beta_1 \neq 0$. For a test at the $\alpha = 0.05$ level
of significance with $n - 2 = 9 - 2 = 7$
degrees of freedom, the critical values
$\pm t_{0.025} = \pm 2.365$. Using technology we find
that the test statistic to be $t_0 = 2.563$, and
we find P-value $= 0.0374$.

Classical approach: Since $t_0 = 2.563 >$
$t_{0.025} = 2.365$ (the test statistic falls within
the critical region), we reject H_0 .

P-value approach: Since P-value $= 0.0374$
$< \alpha = 0.05$ we reject H_0 .

Conclusion: A linear relationship exists
between the height and weight of baseball
players.

(c) With Randy Johnson removed, we obtain
the equation: $\hat{y} = 5.8067\,x - 221.2133$.

The hypotheses are $H_0 : \beta_1 = 0$ versus
$H_1 : \beta_1 \neq 0$. For a test at the $\alpha = 0.05$ level
of significance with $n - 2 = 8 - 2 = 6$
degrees of freedom, the critical values
$\pm t_{0.025} = \pm 2.447$. Using technology we find
that the test statistic to be $t_0 = 2.139$, and
we find P-value $= 0.0763$.

Classical approach: Since $t_0 = 2.139 <$
$t_{0.025} = 2.447$ (the test statistic does not fall
in the critical region), we do not reject H_0 .

P-value approach: Since P-value $= 0.0763$
$> \alpha = 0.05$ we do not reject H_0 .

Conclusion: A linear relationship does not
exist between the height and weight of
baseball players.

Since Randy Johnson has sufficient effect on
the relationship to change the outcome of
the hypothesis test, he is an influential
observation.

Section 14.2

1. A confidence interval is an interval constructed about the predicted value of y, at a given level of x, and is used to measure the accuracy of the predicted mean response of all the individuals in the population at that level. A prediction interval is an interval constructed about the predicted value of y and is used to measure the accuracy of a single individual's predicted value at a given level for x. Since there is more variability in individuals than in means, for the same level of confidence, the prediction interval will always be wider than the confidence interval.

3. (a) From Problem 7 in Section 14.1, the least-squares regression equation is $\hat{y} = 2.0233x - 2.3256$. So, the predicted mean value of y when $x = 7$ is $\hat{y} = 2.0233(7) - 2.3256 \approx 11.8$.

   (b) From Problem 7 in Section 14.1, we have that $n = 5$, $s_e = 0.5134$, $\bar{x} = 5.4$, and $\sum(x - \bar{x})^2 = 17.20$. For a 95% confidence interval with $n - 2 = 3$ degrees of freedom, we use $t_{0.025} = 3.182$.

   Lower Bound:

   $$\hat{y} - t_{\alpha/2} \cdot s_e \sqrt{\frac{1}{n} + \frac{(x^* - \bar{x})^2}{\sum(x - \bar{x})^2}}$$

   $$= 11.8 - 3.182 \cdot 0.5134 \cdot \sqrt{\frac{1}{5} + \frac{(7 - 5.4)^2}{17.20}}$$

   $$\approx 10.8 \text{ [Tech: 10.9]}$$

   Upper Bound:

   $$\hat{y} + t_{\alpha/2} \cdot s_e \sqrt{\frac{1}{n} + \frac{(x^* - \bar{x})^2}{\sum(x - \bar{x})^2}}$$

   $$= 11.8 + 3.182 \cdot 0.5134 \cdot \sqrt{\frac{1}{5} + \frac{(7 - 5.4)^2}{17.20}}$$

   $$\approx 12.8$$

   (c) The predicted value of y when $x = 7$ is also $\hat{y} = 11.8$.

   (d) Lower Bound:

   $$\hat{y} - t_{\alpha/2} \cdot s_e \sqrt{1 + \frac{1}{n} + \frac{(x^* - \bar{x})^2}{\sum(x - \bar{x})^2}}$$

   $$= 11.8 - 3.182 \cdot 0.5134 \cdot \sqrt{1 + \frac{1}{5} + \frac{(7 - 5.4)^2}{17.20}}$$

   $$\approx 9.9$$

Upper Bound:

$$\hat{y} + t_{\alpha/2} \cdot s_e \sqrt{1 + \frac{1}{n} + \frac{(x^* - \bar{x})^2}{\sum(x - \bar{x})^2}}$$

$$= 11.8 + 3.182 \cdot 0.5134 \cdot \sqrt{1 + \frac{1}{5} + \frac{(7 - 5.4)^2}{17.20}}$$

$$\approx 13.7$$

   (e) In (a) and (b) we are predicting the mean response (a point estimate and an interval estimate) for the population of individuals with $x = 7$. In (c) and (d) we are predicting the response (a point estimate and an interval estimate) for a single individual with $x = 7$.

5. (a) From Problem 9 in Section 14.1, the least-squares regression equation is $\hat{y} = 2.2x + 1.2$. The predicted mean value of y when $x = 1.4$ is $\hat{y} = 2.2(1.4) + 1.2 = 4.3$.

   (b) From Problem 9 in Section 14.1, we have that $n = 5$, $s_e = 0.8944$, $\bar{x} = 0$, and $\sum(x - \bar{x})^2 = 10$. For a 95% confidence interval with $n - 2 = 3$ degrees of freedom, we use $t_{0.025} = 3.182$.

   Lower Bound:

   $$\hat{y} - t_{\alpha/2} \cdot s_e \sqrt{\frac{1}{n} + \frac{(x^* - \bar{x})^2}{\sum(x - \bar{x})^2}}$$

   $$= 4.3 - 3.182 \cdot 0.8944 \cdot \sqrt{\frac{1}{5} + \frac{(1.4 - 0)^2}{10}}$$

   $$\approx 2.5$$

   Upper Bound:

   $$\hat{y} + t_{\alpha/2} \cdot s_e \sqrt{\frac{1}{n} + \frac{(x^* - \bar{x})^2}{\sum(x - \bar{x})^2}}$$

   $$= 4.3 + 3.182 \cdot 0.8944 \cdot \sqrt{\frac{1}{5} + \frac{(1.4 - 0)^2}{10}}$$

   $$\approx 6.1$$

   (c) The predicted value of y when $x = 1.4$ is also $\hat{y} = 4.3$.

   (d) Lower Bound:

   $$\hat{y} - t_{\alpha/2} \cdot s_e \sqrt{1 + \frac{1}{n} + \frac{(x^* - \bar{x})^2}{\sum(x - \bar{x})^2}}$$

   $$= 4.3 - 3.182 \cdot 0.8944 \cdot \sqrt{1 + \frac{1}{5} + \frac{(1.4 - 0)^2}{10}}$$

   $$\approx 0.9$$

Upper Bound:

$$\hat{y} + t_{\alpha/2} \cdot s_e \sqrt{1 + \frac{1}{n} + \frac{(x^* - \bar{x})^2}{\Sigma(x - \bar{x})^2}}$$

$$= 4.3 + 3.182 \cdot 0.8944 \cdot \sqrt{1 + \frac{1}{5} + \frac{(1.4 - 0)^2}{10}}$$

$$\approx 7.7$$

7. (a) From Problem 13 in Section 14.1, the least-squares regression equation is $\hat{y} = 0.1827x + 12.4932$. So, the predicted mean value of y when $x = 25.75$ is $\hat{y} = 0.1827(25.75) + 12.4932 = 17.20$ inches.

(b) From Problem 13 in Section 14.1, we have that $n = 11$, $s_e = 0.0954$, $\bar{x} = 26.4545$, and $\Sigma(x - \bar{x})^2 \approx 11.9773$. For a 95% confidence interval with $n - 2 = 9$ degrees of freedom, we use $t_{0.025} = 2.262$.

Lower Bound:

$$\hat{y} - t_{\alpha/2} \cdot s_e \sqrt{\frac{1}{n} + \frac{(x^* - \bar{x})^2}{\Sigma(x - \bar{x})^2}}$$

$$= 17.20 - 2.262 \cdot 0.0954 \cdot \sqrt{\frac{1}{11} + \frac{(25.75 - 26.4545)^2}{11.9773}}$$

$$\approx 17.12 \text{ inches}$$

Upper Bound:

$$\hat{y} + t_{\alpha/2} \cdot s_e \sqrt{\frac{1}{n} + \frac{(x^* - \bar{x})^2}{\Sigma(x - \bar{x})^2}}$$

$$= 17.20 + 2.262 \cdot 0.0954 \cdot \sqrt{\frac{1}{11} + \frac{(25.75 - 26.4545)^2}{11.9773}}$$

$$\approx 17.28 \text{ inches}$$

(c) The predicted value of y when $x = 25.75$ is also $\hat{y} = 17.20$ inches.

(d) Lower Bound:

$$\hat{y} - t_{\alpha/2} \cdot s_e \sqrt{1 + \frac{1}{n} + \frac{(x^* - \bar{x})^2}{\Sigma(x - \bar{x})^2}}$$

$$= 17.20 - 2.262 \cdot 0.0954 \cdot \sqrt{1 + \frac{1}{11} + \frac{(25.75 - 26.4545)^2}{11.9773}}$$

$$\approx 16.97 \text{ inches}$$

Upper Bound:

$$\hat{y} + t_{\alpha/2} \cdot s_e \sqrt{1 + \frac{1}{n} + \frac{(x^* - \bar{x})^2}{\Sigma(x - \bar{x})^2}}$$

$$= 17.20 + 2.262 \cdot 0.0954 \cdot \sqrt{1 + \frac{1}{11} + \frac{(25.75 - 26.4545)^2}{11.9773}}$$

$$\approx 17.43 \text{ inches}$$

(e) In (a) and (b) we are predicting the mean head circumference (a point estimate and an interval estimate) for the population of all children who are 25.75 inches tall. In (c) and (d) we are predicting the head circumference (a point estimate and an interval estimate) of a single child who is 25.75 inches tall.

9. (a) From Problem 15 in Section 14.1, the least-squares regression equation is $\hat{y} = 0.6764x + 2675.6$. So, the predicted mean value of y when $x = 2550$ is $\hat{y} = 0.6764 \cdot 2550 + 2675.6 = 4400.4$ psi.

(b) From Problem 15 in Section 14.1, we have that $n = 10$, $s_e = 271.04$, $\bar{x} = 2882$, and $\Sigma(x - \bar{x})^2 = 1,739,160$. For a 95% confidence interval with $n - 2 = 8$ degrees of freedom, we use $t_{0.025} = 2.306$.

Lower Bound:

$$\hat{y} - t_{\alpha/2} \cdot s_e \sqrt{\frac{1}{n} + \frac{(x^* - \bar{x})^2}{\Sigma(x - \bar{x})^2}}$$

$$= 4400.4 - 2.306 \cdot 271.04 \cdot \sqrt{\frac{1}{10} + \frac{(2550 - 2882)^2}{1,739,160}}$$

$$\approx 4147.8 \text{ psi}$$

Upper Bound:

$$\hat{y} + t_{\alpha/2} \cdot s_e \sqrt{\frac{1}{n} + \frac{(x^* - \bar{x})^2}{\Sigma(x - \bar{x})^2}}$$

$$= 4400.4 + 2.306 \cdot 271.04 \cdot \sqrt{\frac{1}{10} + \frac{(2550 - 2882)^2}{1,739,160}}$$

$$\approx 4653.0 \text{ [Tech: 4653.1] psi}$$

(c) The predicted value of y when $x = 2550$ is also $\hat{y} = 4400.4$ psi.

(d) Lower Bound:

$$\hat{y} - t_{\alpha/2} \cdot s_e \sqrt{1 + \frac{1}{n} + \frac{(x^* - \bar{x})^2}{\Sigma(x - \bar{x})^2}}$$

$$= 4400.4 - 2.306 \cdot 271.04 \cdot \sqrt{1 + \frac{1}{10} + \frac{(2550 - 2882)^2}{1,739,160}}$$

$$\approx 3726.3 \text{ psi}$$

Upper Bound:

$$\hat{y} + t_{\alpha/2} \cdot s_e \sqrt{1 + \frac{1}{n} + \frac{(x^* - \bar{x})^2}{\Sigma(x - \bar{x})^2}}$$

$$= 4400.4 + 2.306 \cdot 271.04 \cdot \sqrt{1 + \frac{1}{10} + \frac{(2550 - 2882)^2}{1,739,160}}$$

$$\approx 5074.5 \text{ [Tech: 5074.6] psi}$$

(e) In (a) and (b) we are predicting the mean 28-day strength (a point estimate and an interval estimate) for the population of all concrete that has a 7-day strength of 2550 psi. In (c) and (d) we are predicting the 28-day strength (a point estimate and an interval estimate) for a single batch of concrete with 7-day strength of 2550 psi.

11. (a) From Problem 17 in Section 14.1, the least-squares regression equation is $\hat{y} = 0.7622x + 1.2286$. So, the predicted mean value of y when $x = 4.2$ is $\hat{y} = 0.7622 \cdot 4.2 + 1.2286 = 4.430\%$.

(b) From Problem 17 in Section 14.1, we have that $n = 11$, $s_e = 3.2309$, $\overline{x} \approx -0.5373$, and $\sum(x-\overline{x})^2 = 125.8797$. For a 90% confidence interval with $n - 2 = 9$ degrees of freedom, we use $t_{0.05} = 1.833$.

Lower Bound:

$$\hat{y} - t_{\alpha/2} \cdot s_e \sqrt{\frac{1}{n} + \frac{(x^*-\overline{x})^2}{\sum(x-\overline{x})^2}}$$

$$= 4.430 - 1.833 \cdot 3.2309 \cdot \sqrt{\frac{1}{11} + \frac{(4.2 - (-0.5373))^2}{125.8797}}$$

$$\approx 1.357\%$$

Upper Bound:

$$\hat{y} + t_{\alpha/2} \cdot s_e \sqrt{\frac{1}{n} + \frac{(x^*-\overline{x})^2}{\sum(x-\overline{x})^2}}$$

$$= 4.430 + 1.833 \cdot 3.2309 \cdot \sqrt{\frac{1}{11} + \frac{(4.2 - (-0.5373))^2}{125.8797}}$$

$$\approx 7.503\%$$

(c) The predicted value of y when $x = 4.2$ is also $\hat{y} = 4.430\%$.

(d) Lower Bound:

$$\hat{y} - t_{\alpha/2} \cdot s_e \sqrt{1 + \frac{1}{n} + \frac{(x^*-\overline{x})^2}{\sum(x-\overline{x})^2}}$$

$$= 4.430 - 1.833 \cdot 3.2309 \cdot \sqrt{1 + \frac{1}{11} + \frac{(4.2 - (-0.5373))^2}{125.8797}}$$

$$\approx -2.242\%$$

Upper Bound:

$$\hat{y} + t_{\alpha/2} \cdot s_e \sqrt{1 + \frac{1}{n} + \frac{(x^*-\overline{x})^2}{\sum(x-\overline{x})^2}}$$

$$= 4.430 + 1.833 \cdot 3.2309 \cdot \sqrt{1 + \frac{1}{11} + \frac{(4.2 - (-0.5373))^2}{125.8797}}$$

$$\approx 11.102\%$$

(e) Although the predicted rates of return in parts (a) and (c) are the same, the intervals are different because the distribution of the mean rate of return, part (a), has less variability than the distribution of the individual rate of return of a particular S&P 500 rate of return, part (c).

13. (a) From Problem 19 in Section 14.1, the least-squares regression equation is $\hat{y} = 63.1642x + 163.9988$. So, the predicted mean value of y when $x = 23$ is given by $\hat{y} = 63.1642 \cdot 23 + 163.9988 \approx 1617$.

(b) From Problem 19 in Section 14.1, we have that $n = 14$, $s_e = 157.4572$, $\overline{x} = 22$, and $\sum(x-\overline{x})^2 = 234$. For a 95% confidence interval with $n - 2 = 12$ degrees of freedom, we use $t_{0.025} = 2.179$.

Lower Bound:

$$\hat{y} - t_{\alpha/2} \cdot s_e \sqrt{\frac{1}{n} + \frac{(x^*-\overline{x})^2}{\sum(x-\overline{x})^2}}$$

$$= 1617 - 2.179 \cdot 157.4752 \cdot \sqrt{\frac{1}{14} + \frac{(23-22)^2}{234}}$$

$$\approx 1523 \ [\text{Tech: } 1522]$$

Upper Bound:

$$\hat{y} + t_{\alpha/2} \cdot s_e \sqrt{\frac{1}{n} + \frac{(x^*-\overline{x})^2}{\sum(x-\overline{x})^2}}$$

$$= 1617 + 2.179 \cdot 157.4752 \cdot \sqrt{\frac{1}{14} + \frac{(23-22)^2}{234}}$$

$$\approx 1711$$

(c) The predicted value of y when $x = 23$ is also 1617.

(d) Lower Bound:

$$\hat{y} - t_{\alpha/2} \cdot s_e \sqrt{1 + \frac{1}{n} + \frac{(x^*-\overline{x})^2}{\sum(x-\overline{x})^2}}$$

$$= 1617 - 2.179 \cdot 157.4752 \cdot \sqrt{1 + \frac{1}{14} + \frac{(23-22)^2}{234}}$$

$$\approx 1261$$

Upper Bound:

$$\hat{y} + t_{\alpha/2} \cdot s_e \sqrt{1 + \frac{1}{n} + \frac{(x^*-\overline{x})^2}{\sum(x-\overline{x})^2}}$$

$$= 1617 + 2.179 \cdot 157.4752 \cdot \sqrt{1 + \frac{1}{14} + \frac{(23-22)^2}{234}}$$

$$\approx 1973$$

(e) Although the predicted SAT scores in parts (a) and (c) are the same, the intervals are different because the distribution of the mean SAT scores, part (a), has less variability than the distribution of individual SAT scores, part (c).

15. (a) In Problem 21 of Section 14.1, we showed that no linear relationship exists between sugar content and calories in high-protein and moderate protein energy bars. Therefore, it does not make sense to construct either a confidence interval or prediction interval based on the least-squares regression equation.

(b) Since no linear relation exists between x and y, we use the techniques of Section 9.2 on the y-data to construct the confidence interval. The summary statistics are $n = 14$, $\bar{y} \approx 14.71$ grams and $s_y \approx 9.03$ grams. For 95% confidence with $n - 1 = 13$, we use $t_{0.025} = 2.160$.

Lower Bound:

$$\bar{y} - t_{0.025} \cdot \frac{s_y}{\sqrt{n}} = 14.71 - 2.160 \cdot \frac{9.03}{\sqrt{14}}$$

$$\approx 9.5 \text{ grams}$$

Upper Bound:

$$\bar{y} + t_{0.025} \cdot \frac{s_y}{\sqrt{n}} = 14.71 + 2.160 \cdot \frac{9.03}{\sqrt{14}}$$

$$\approx 19.9 \text{ grams}$$

17. (a)

(b) $n = 9$, $\sum x_i = 454$, $\sum y_i = 24,100$, $\sum x_i^2 = 23,356$, $\sum y_i^2 = 78,190,000$, $\sum x_i y_i = 1,281,800$

$$r = \cfrac{\sum x_i y_i - \cfrac{\sum x_i \sum y_i}{n}}{\sqrt{\left(\sum x_i^2 - \cfrac{(\sum x_i)^2}{n}\right)}\sqrt{\left(\sum y_i^2 - \cfrac{(\sum y_i)^2}{n}\right)}}$$

$$= \cfrac{1,281,800 - \cfrac{(454)(24,100)}{9}}{\sqrt{\left(23,356 - \cfrac{(454)^2}{9}\right)}\sqrt{\left(78,190,000 - \cfrac{(24,100)^2}{9}\right)}}$$

$$\approx 0.839$$

(c) Looking to Table II, the critical value for a correlation coefficient when $n = 9$ is 0.666. Since $0.839 > 0.666$, we conclude that a linear relation exists between size and price.

(d) Using technology, we obtain the regression equation: $\hat{y} = 145.4990x - 4661.8395$

(e) For each inch added to the size of a plasma television, the price will increase by about $145.50.

(f) It is not reasonable to interpret the intercept. It would indicate the price of a 0-inch plasma television, which makes no sense.

(g) We compute the coefficient of determination: $r^2 = (0.839)^2 \approx 0.704$. About 70.4% of the variability in price is explained by the variability in size.

(h) $\beta_0 \approx b_0 = -4661.8395$; $\beta_1 \approx b_1 = 145.4990$

(i) We use the least-squares regression equation $\hat{y} = 145.4990x - 4661.8395$ to complete the table that follows:

x	y	$\hat{y}$	$y - \hat{y}$	$(y - \hat{y})^2$	$(x - \bar{x})^2$
60	4000	4068.1005	−68.1005	4637.6781	91.3095
60	3600	4068.1005	−468.101	219118.0781	91.3095
58	5000	3777.1025	1222.898	1495478.2955	57.0871
50	3200	2613.1105	586.8895	344439.2852	0.1975
50	1800	2613.1105	−813.111	661148.6852	0.1975
50	1600	2613.1105	−1013.11	1026392.8852	0.1975
42	1900	1449.1185	450.8815	203294.1270	71.3079
42	1700	1449.1185	250.8815	62941.5270	71.3079
42	1300	1449.1185	−149.119	22236.3270	71.3079
			Sum	4039686.8885	454.2222

$$s_e = \sqrt{\frac{\sum(y - \hat{y})^2}{n - 2}} = \sqrt{\frac{4039686.8885}{9 - 2}}$$

$$= 759.6697$$

(j) A normal probability plot shows that the residuals are approximately normally distributed.

(k) $s_{b_1} = \dfrac{s_e}{\sqrt{\Sigma(x-\bar{x})^2}} = \dfrac{759.6697}{\sqrt{454.2222}} = 35.6443$

(l) The hypotheses are $H_0 : \beta_1 = 0$ versus $H_1 : \beta_1 \neq 0$. The test statistic is

$t_0 = \dfrac{b_1}{s_{b_1}} = \dfrac{145.4990}{35.6443} = 4.082$.

Classical approach: The level of significance is $\alpha = 0.05$. Since this is a two-tailed test with $n - 2 = 7$ degrees of freedom, the critical values are $\pm t_{0.025} = \pm 2.365$. Since $t_0 = 4.082 > t_{0.025} = 2.365$ (the test statistic falls in a critical region), we reject H_0.

P-value approach: The *P*-value for this two-tailed test is the area under the *t*-distribution with 7 degrees of freedom to the right of $t_0 = 4.082$, plus the area to the left of $-t_0 = -4.082$. From the *t*-distribution table in the row corresponding to 7 degrees of freedom, 4.082 falls between 4.029 and 4.785, whose right-tail areas are 0.0025 and 0.001, respectively. We must double these values in order to get the total area in both tails: 0.005 and 0.002. So, $0.002 < P\text{-value} < 0.005$. [Tech: *P*-value = 0.0047]. Because the *P*-value is less than the level of significance $\alpha = 0.05$, we reject H_0.

Conclusion: There is sufficient evidence to conclude that a linear relationship exists between the size and price of plasma televisions.

(m) For a 95% confidence interval we use $t_{\alpha/2} = t_{0.025} = 2.365$:

Lower bound:

$b_1 - t_{\alpha/2} \cdot s_{b_1} = 145.4990 - 2.365 \cdot 35.6443$
$= 61.2002$

Upper bound:

$b_1 + t_{\alpha/2} \cdot s_{b_1} = 145.4990 + 2.365 \cdot 35.6443$
$= 229.7978$

(n) The predicted mean value of y when $x = 50$ is $\hat{y} = 145.4990 \cdot 50 - 4661.8395 \approx \$2613.1$.

(o) For a 95% confidence interval with $n - 2 = 7$ degrees of freedom, we use $t_{0.025} = 2.365$.

Lower Bound:

$\hat{y} - t_{\alpha/2} \cdot s_e \sqrt{\dfrac{1}{n} + \dfrac{(x^* - \bar{x})^2}{\Sigma(x - \bar{x})^2}}$

$= 2613.1 - 2.365 \cdot 759.6697 \cdot \sqrt{\dfrac{1}{9} + \dfrac{(50 - 50.4444)^2}{454.2222}}$

$\approx \$2013.1$

Upper Bound:

$\hat{y} + t_{\alpha/2} \cdot s_e \sqrt{\dfrac{1}{n} + \dfrac{(x^* - \bar{x})^2}{\Sigma(x - \bar{x})^2}}$

$= 2613.1 + 2.365 \cdot 759.6697 \cdot \sqrt{\dfrac{1}{9} + \dfrac{(50 - 50.4444)^2}{454.2222}}$

$\approx \$3213.1$

(p) The price of the Samsung/HP-T5064 is below average price for 50-inch plasma televisions. Its price, $1600, is below the low end of the confidence interval found in part (o).

(q) The predicted value of y when $x = 50$ is also $\hat{y} = \$2,613.1$.

(r) Lower Bound:

$\hat{y} - t_{\alpha/2} \cdot s_e \sqrt{1 + \dfrac{1}{n} + \dfrac{(x^* - \bar{x})^2}{\Sigma(x - \bar{x})^2}}$

$= 2613.1 - 2.365 \cdot 759.6697 \cdot \sqrt{1 + \dfrac{1}{9} + \dfrac{(50 - 50.4444)^2}{454.2222}}$

$\approx \$718.9$

Upper Bound:

$\hat{y} + t_{\alpha/2} \cdot s_e \sqrt{1 + \dfrac{1}{n} + \dfrac{(x^* - \bar{x})^2}{\Sigma(x - \bar{x})^2}}$

$= 2613.1 + 2.365 \cdot 759.6697 \cdot \sqrt{1 + \dfrac{1}{9} + \dfrac{(50 - 50.4444)^2}{454.2222}}$

$\approx \$4507.3$

(s) Answers will vary. Possible lurking variables: brand name, picture quality, and audio quality.

Section 14.3

1. A correlation matrix is a matrix that shows the linear correlation among all variables under consideration in a multiple linear regression model.

3. If the null hypothesis is rejected, we conclude that at least one of the explanatory variables is linearly related to the response variable. It is very possible that one or more of the $\beta_i = 0$ even though the null hypothesis is rejected.

5. The difference between R^2 and adjusted R^2 is that R^2 becomes larger as more explanatory variables are added to the regression model, regardless of their contribution to the explanation of the response variable. Adjusted R^2 compensates for this ability to artificially increase R^2 by simply adding more explanatory variables. It modifies R^2 based on the sample size, n, and the number of explanatory variables, k. Adjusted R^2 actually decreases if the additional variable does little to explain the variation in the response variable. So, the adjusted R^2 is better for determining whether an additional explanatory variable should be added to the model.

7. **(a)** The slope coefficient of x_1 is 3. This indicates that $\hat{y}$ will increase 3 units, on average, for every one unit increase in x_1, provided x_2 remains constant. The slope coefficient of x_2 is -4. This indicates that $\hat{y}$ will decrease 4 units, on average, for every one unit increase in x_2, provided x_1 remains constant.

   (b) $\hat{y} = 5 + 3x_1 - 4x_2$
   $\hat{y} = 5 + 3(10) - 4x_2$
   $\hat{y} = 35 - 4x_2$

(c) $\hat{y} = 5 + 3x_1 - 4x_2$
$\hat{y} = 5 + 3(15) - 4x_2$
$\hat{y} = 50 - 4x_2$

(d) $\hat{y} = 5 + 3x_1 - 4x_2$
$\hat{y} = 5 + 3(20) - 4x_2$
$\hat{y} = 65 - 4x_2$

(e) Changing the value of x_1 has the effect of changing the y-intercept of the graph of the regression line.

9. **(a)** $R^2_{adj} = 1 - \left(\dfrac{n-1}{n-k-1}\right)\left(1 - R^2\right)$

   $= 1 - \left(\dfrac{25-1}{25-3-1}\right)\left(1 - 0.653\right)$

   $= 0.603$

   (b) $F = \dfrac{R^2}{1 - R^2} \cdot \dfrac{n - (k+1)}{k}$

   $= \dfrac{0.653}{1 - 0.653} \cdot \dfrac{25 - (3+1)}{3}$

   $= 13.173$

   (c) $R^2_{adj} = 1 - \left(\dfrac{25-1}{25-4-1}\right)\left(1 - 0.665\right) = 0.598$

   Do not add the new variable. Its addition reduces the proportion of variance explained by the model.

11. **(a)** **Correlations: x1, x2, x3, y**

```
            x1          x2          x3
x2    -0.460
x3     0.183    -0.012
y     -0.748     0.788     0.258
```

Because the correlations among the explanatory variables are all between -0.7 and 0.7, multicollinearity does not appear to be a problem.

(b) **Regression Analysis: y versus x1, x2, x3**

```
The regression equation is
y = 7.96 - 0.104 x1 + 0.936 x2 + 0.115 x3


Predictor         Coef    SE Coef        T        P
Constant        7.9647     0.5407    14.73    0.000
x1             -0.10354    0.02080    -4.98    0.003
x2              0.9359     0.1999     4.68    0.003
x3              0.11524    0.03188    3.61    0.011


S = 0.245127    R-Sq = 94.0%    R-Sq(adj) = 91.0%

Analysis of Variance

Source            DF        SS         MS        F        P
Regression         3    5.6485     1.8828    31.33    0.000
Residual Error     6    0.3605     0.0601
Total              9    6.0090
```

The regression equation is $\hat{y} = 7.9647 - 0.1035x_1 + 0.9359x_2 + 0.1152x_3$.

(c) From the output in part (b), our test statistic is $F_0 = 31.33$. Since P-value < 0.001, we reject the null hypothesis and conclude that at least one of the explanatory variables is linearly related to the response variable.

(d) From the output in part (b), conclude the following:

Reject $H_0 : \beta_1 = 0$ since $t_{\beta 1} = -4.98$ and P-value $= 0.003 < \alpha = 0.05$.

Reject $H_0 : \beta_2 = 0$ since $t_{\beta 2} = 4.68$ and P-value $= 0.003 < \alpha = 0.05$

Reject $H_0 : \beta_3 = 0$ since $t_{\beta 3} = 3.61$ and P-value $= 0.011 < \alpha = 0.05$

13. **(a)** **Correlations: x1, x2, x3, x4, y**

```
            x1          x2          x3          x4
x2    -0.184
x3     0.182    -0.106
x4    -0.166    -0.558     0.291
y     -0.046     0.932    -0.063    -0.750
```

Because the correlations among the explanatory variables are all between -0.7 and 0.7, multicollinearity does not appear to be a problem.

(b) **Regression Analysis: y versus x1, x2, x3, x4**

```
The regression equation is
y = 279 + 0.07 x1 + 1.15 x2 + 0.390 x3 - 2.94 x4

Predictor       Coef    SE Coef       T       P
Constant      279.02      61.34    4.55   0.003
x1             0.068      1.159    0.06   0.955
x2            1.1460     0.1546    7.41   0.000
x3            0.3897     0.2639    1.48   0.183
x4           -2.9378     0.8235   -3.57   0.009

S = 13.1177    R-Sq = 95.9%    R-Sq(adj) = 93.6%

Analysis of Variance
Source              DF        SS       MS       F       P
Regression           4   28152.4   7038.1   40.90   0.000
Residual Error       7    1204.5    172.1
Total               11   29356.9
```

The regression equation is $\hat{y} = 279.02 + 0.068x_1 + 1.1460x_2 + 0.3897x_3 - 2.9378x_4$.

$F_0 = 40.90$ and P-value < 0.001 so we reject the null hypothesis and conclude that at least one of the explanatory variables is linearly related to the response variable.

Based on the regression output, we see that x_1 and x_3 have slope coefficients that are not significantly different from zero (P-value > 0.05).

(c) The variable x_1 has the highest P-value. After removing this variable, the regression equation becomes:
$\hat{y} = 282.33 + 1.1426x_2 + 0.3941x_3 - 2.9570x_4$
The new test statistic is $F_0 = 62.29$ and P-value < 0.001, so the model is still significant.

The variable x_3 has a slope coefficient that is not significantly different from zero. Removing this variable gives $\hat{y} = 281.62 + 1.1584x_2 - 2.6267x_4$
The new test statistic is $F_0 = 76.89$ and P-value < 0.001 so the model is still significant. Both of the explanatory variables have slope coefficients that are significantly different from zero.
Reject $H_0 : \beta_2 = 0$ since $t_{\beta_2} = 7.89$ and P-value $< 0.001 < \alpha = 0.05$

Reject $H_0 : \beta_4 = 0$ since $t_{\beta_4} = -3.54$ and P-value $= 0.006 < \alpha = 0.05$

(d)

None of the residual plots show any discernible pattern, and the boxplot does not show any outliers. Therefore, the linear model is appropriate.

(e) $\hat{y} = 281.62 + 1.1584x_2 - 2.6267x_4$

$= 281.62 + 1.1584(35.6) - 2.6267(29)$

$= 246.7$

(f)

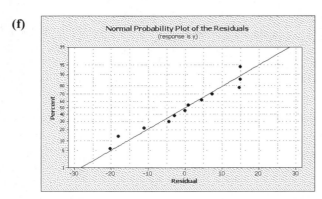

Because the points are roughly linear (i.e. lie within the confidence bounds), it is reasonable to conclude that the residuals are normally distributed. Therefore, it is reasonable to construct confidence and prediction intervals.

(g) 95% Confidence Interval: Lower Bound: 237.45; Upper Bound: 255.91

95% Prediction Interval: Lower Bound: 214.93; Upper Bound: 278.43

15. (a) Yes, the *P*-value is very small so we would reject the null hypothesis and conclude that at least one of the explanatory variables is linearly related to the response variable.

(b) $b_1 = -0.87$: A one point increase in the computer confidence variable reduces the computer anxiety measurement by 0.87 point, on average, provided all other explanatory variables remain constant.

$b_2 = -0.51$: A one point increase in the computer knowledge variable reduces the computer anxiety measurement by 0.51 point, on average, provided all other explanatory variables remain constant.

$b_3 = -0.45$: A one point increase in the computer liking variable reduces the computer anxiety measurement by 0.45 point, on average, provided all other explanatory variables remain constant.

$b_4 = 0.33$: A one point increase in the trait anxiety variable increases the computer anxiety measurement by 0.33 point, on average, provided all other explanatory variables remain constant.

Yes, the slope coefficients of the model are all reasonable.

(c) $\hat{y} = 84.04 - 0.87(25) - 0.51(19) - 0.45(20) + 0.33(43) = 57.79$

(d) $R^2 = 0.69$ indicates that 69% of the variance in computer anxiety is explained by this regression model.

(e) This statement means that the requirements of a multiple linear regression model were checked and were met.

17. (a) Regression Analysis: Wind Chill versus Air Temp, Wind Speed

```
The regression equation is
Wind Chill = - 12.5 + 1.33 Air Temp - 0.414 Wind Speed

Predictor        Coef   SE Coef       T       P
Constant     -12.5091    0.9194  -13.61   0.000
Air Temp      1.33408   0.04024   33.16   0.000
Wind Speed   -0.41430   0.03365  -12.31   0.000

S = 1.62942    R-Sq = 99.3%   R-Sq(adj) = 99.1%
```

The regression equation is $\hat{y} = -12.5091 + 1.3341x_1 - 0.4143x_2$.

(b)

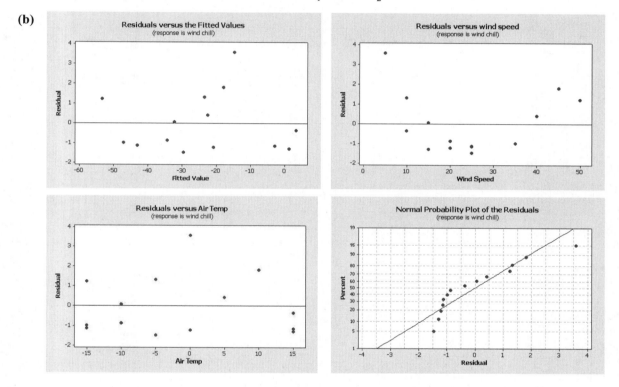

There is a pattern in the residual plot, so we conclude that the linear model is not appropriate.

19. (a) Correlations: Slump, 7-day psi, 28-day psi

```
                Slump    7-day psi
7-day psi      -0.460
28-day psi     -0.753        0.737
```

Because the correlation between Slump and 7-day strength is between -0.7 and 0.7, multicollinearity does not appear to be a problem.

(b) **Regression Analysis: 28-day psi versus Slump, 7-day psi**

```
The regression equation is
28-day psi = 3891 - 296 Slump + 0.552 7-day psi

Predictor     Coef   SE Coef      T       P
Constant    3890.5     828.8   4.69   0.002
Slump       -295.9     109.9  -2.69   0.027
7-day psi   0.5523    0.2171   2.54   0.034

S = 276.505    R-Sq = 76.0%    R-Sq(adj) = 70.1%

Analysis of Variance
Source           DF        SS       MS      F      P
Regression        2   1942115   971058  12.70  0.003
Residual Error    8    611639    76455
Total            10   2553755
```

The regression equation is $\hat{y} = 3890.5 - 295.9x_1 + 0.5523x_2$.

(c)

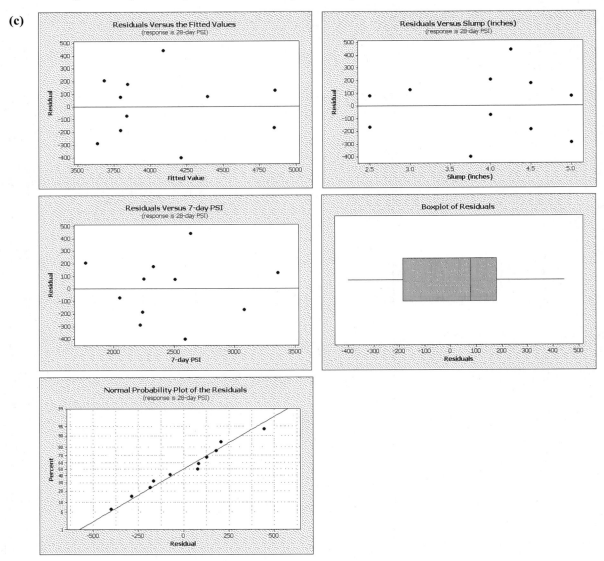

None of the residual plots show any discernible pattern, and the boxplot does not show any outliers. Therefore, the linear model is appropriate.

(d) $b_1 = -295.9$: A one unit increase in slump of the concrete decreases the 28-day strength by 295.9 pounds per square inch, on average, provided the other variable (7-day strength) remains constant.

$b_2 = 0.5523$: A one pound per square inch increase in the 7-day strength increases the 28-day strength by 0.5523 pounds per square inch, on average, provided the other variable (slump) remains constant.

(e) $R^2 = 0.760$ means that 76.0% of the variance in 28-day strength is explained by the model.

$R^2_{adj} = 0.701$ (or 70.1%) modifies the value of R^2 taking the sample size and number of explanatory variables into account.

(f) From part (b), we have $F_0 = 12.70$ and P-value $= 0.003 < \alpha = 0.05$. Therefore, we reject the null hypothesis and conclude that at least one of the explanatory variables is linearly related to the 28-day strength of the concrete.

(g) From part (b), we have the following:
Reject $H_0 : \beta_1 = 0$ since $t_{\beta 1} = -2.69$ and P-value $= 0.027 < \alpha = 0.05$.
Reject $H_0 : \beta_2 = 0$ since $t_{\beta 2} = 2.54$ and P-value $= 0.034 < \alpha = 0.05$.

(h) $\hat{y} = 3890.5 - 295.9(3.5) + 0.552(2450) = 4207.3$ psi

(i) $\hat{y} = 3890.5 - 295.9(3.5) + 0.552(2450) = 4207.3$ psi (the prediction is the same for an individual as it is for the mean response, given particular values for the explanatory variables).

(j) 95% Confidence Interval: Lower Bound: 3988.6 psi; Upper Bound: 4427.2 psi; We are 95% confident that the mean 28-day strength of all concrete for which slump is 3.5 inches and 7-day strength is 2450 psi is between 3988.6 psi and 4427.2 psi.

95% Prediction Interval: Lower Bound: 3533.6 psi; Upper Bound: 4882.2 psi; We are 95% confident that the mean 28-day strength of a specific sample of concrete for which slump is 3.5 inches and 7-day strength is 2450 psi is between 3533.6 psi and 4882.2 psi.

21. (a) **Correlations: Square Footage, Bedrooms, Baths, Asking Price**

	Square Foota	Bedrooms	Baths
Bedrooms	0.267		
Baths	0.274	0.565	
Asking Price	0.665	0.409	0.626

Because the correlations among the explanatory variables are all between -0.7 and 0.7, multicollinearity does not appear to be a problem.

(b) **Regression Analysis: Asking Price versus Square Footage, Bedrooms, Baths**

```
The regression equation is
Asking Price = 25.6 + 0.0719 Square Footage - 0.8 Bedrooms + 55.3 Baths

Predictor          Coef   SE Coef      T      P
Constant          25.63     98.19   0.26  0.800
Square Footage  0.07193   0.02761   2.61  0.028
Bedrooms          -0.81     27.28  -0.03  0.977
Baths             55.35     27.46   2.02  0.075

S = 46.2441    R-Sq = 65.6%    R-Sq(adj) = 54.1%

Analysis of Variance

Source           DF     SS     MS     F      P
Regression        3  36675  12225  5.72  0.018
Residual Error    9  19247   2139
Total            12  55921
```

The regression equation is $\hat{y} = 25.63 + 0.0719x_1 - 0.81x_2 + 55.35x_3$.

(c) From the output in (b), we get $F_0 = 5.72$ and P-value $= 0.018 < \alpha = 0.05$. We reject the null hypothesis and conclude that at least one of the explanatory variables is linearly associated with asking price.

(d) From part (b), we have the following:

Square Footage: Reject $H_0 : \beta_1 = 0$ since $t_0 = 2.61$ and P-value $= 0.028 < \alpha = 0.05$

Bedrooms: Do not reject $H_0 : \beta_2 = 0$ since $t_0 = -0.03$ and P-value $= 0.977 > \alpha = 0.05$

Baths: Do not reject $H_0 : \beta_3 = 0$ since $t_0 = 2.02$ and P-value $= 0.075 > \alpha = 0.05$

We conclude that the square footage is linearly related to asking price for a house.

(e) The variable x_2, bedrooms, has the highest P-value. Removing this variable, we get:

Regression Analysis: Asking Price versus Square Footage, Baths

```
The regression equation is
Asking Price = 24.1 + 0.0718 Square Footage + 54.9 Baths


Predictor          Coef   SE Coef     T      P
Constant          24.09     79.18  0.30  0.767
Square Footage  0.07181   0.02593  2.77  0.020
Baths             54.92     22.08  2.49  0.032


S = 43.8732    R-Sq = 65.6%    R-Sq(adj) = 58.7%


Analysis of Variance


Source          DF     SS     MS     F      P
Regression       2  36673  18336  9.53  0.005
Residual Error  10  19249   1925
Total           12  55921
```

The new regression equation is: $\hat{y} = 24.09 + 0.0718x_1 + 54.92x_3$

Square Footage: Reject $H_0 : \beta_1 = 0$ since $t_0 = 2.77$ and P-value $= 0.020 < \alpha = 0.05$

Bathrooms: Reject $H_0 : \beta_3 = 0$ since $t_0 = 2.49$ and P-value $= 0.032 < \alpha = 0.05$

We conclude that the square footage and number of bathrooms are both linearly related to asking price.

(f)

None of the residual plots show any discernible pattern, the boxplot does not show any outliers, and the normal probability plot looks fairly linear. Therefore, the linear model is appropriate.

(g) $b_1 = 0.0718$: If the house increases by 1 square foot, then the asking price will increase by $71.80, on average, assuming that the number of bathrooms remains constant.

$b_3 = 54.9$: For each bathroom added to the house, the asking price will increase by $54,900, on average, assuming that the square footage remains constant.

(h) 95% Confidence Interval: Lower Bound: 365.1; Upper Bound: 429.1; We are 95% confident that the mean asking price for all 2900-square-foot houses with 3 bathrooms is between $365,100 and $429,100.

95% Prediction Interval: Lower Bound: 294.2; Upper Bound: 499.9; We are 95% confident that the asking price of a randomly chosen 2900-square-foot house with 3 bathrooms is between $294,200 and $499,900.

23. (a) Correlations: Engine, Weight, HP, MPG

```
         Engine   Weight       HP
Weight    0.764
HP        0.920    0.558
MPG      -0.597   -0.390   -0.735
```

Yes, because the correlation between engine size and weight and the correlation between engine size and horse power are larger than 0.7, there is reason to be concerned about multicollinearity.

(b) **Regression Analysis: MPG versus Engine, Weight, HP**

```
The regression equation is
MPG = 35.8 + 4.80 Engine - 0.00394 Weight - 0.0658 HP

Predictor        Coef    SE Coef       T       P
Constant       35.827      9.212    3.89   0.004
Engine          4.803      3.361    1.43   0.187
Weight      -0.003945   0.003847   -1.03   0.332
HP           -0.06583    0.02636   -2.50   0.034

S = 1.46624    R-Sq = 62.6%    R-Sq(adj) = 50.1%

Analysis of Variance

Source          DF        SS       MS       F       P
Regression       3    32.344   10.781    5.01   0.026
Residual Error   9    19.349    2.150
Total           12    51.692
```

The regression equation is $\hat{y} = 35.827 + 4.803x_1 - 0.00394x_2 - 0.06583x_3$.

(c) From the output in (b), we get $F_0 = 5.01$ and P-value $= 0.026 < \alpha = 0.05$. We reject the null hypothesis and conclude that at least one of the explanatory variables is linearly associated with gasoline mileage.

(d) From part (b), we have the following:

Engine Size: Do not reject $H_0 : \beta_1 = 0$ since $t_0 = 1.43$ and P-value $= 0.187 > \alpha = 0.05$.

Curb Weight: Do not reject $H_0 : \beta_2 = 0$ since $t_0 = -1.03$ and P-value $= 0.332 > \alpha = 0.05$.

Horsepower: Reject $H_0 : \beta_3 = 0$ since $t_0 = -2.50$ and P-value $= 0.034 < \alpha = 0.05$.

We conclude that horsepower is linearly associated with gas mileage, but curb weight and engine size are not linearly associated with gas mileage.

(e) Removing the explanatory variable curb weight, x_2, we obtain:

Regression Analysis: MPG versus Engine, HP

```
The regression equation is
MPG = 26.8 + 2.15 Engine - 0.0503 HP

Predictor      Coef    SE Coef       T       P
Constant     26.805      2.740    9.78   0.000
Engine        2.147      2.147    1.00   0.341
HP         -0.05032    0.02164   -2.32   0.042

S = 1.47003    R-Sq = 58.2%    R-Sq(adj) = 49.8%

Analysis of Variance

Source          DF        SS       MS       F       P
Regression       2    30.082   15.041    6.96   0.013
Residual Error  10    21.610    2.161
Total           12    51.692
```

The new regression equation is $\hat{y} = 26.805 + 2.147x_1 - 0.05032x_3$.

Testing the explanatory variables in this new model, we find:

Engine Size: Do not reject $H_0 : \beta_1 = 0$ since $t_0 = 1.00$ and P-value $= 0.341 > \alpha = 0.05$.

Horsepower: Reject $H_0 : \beta_3 = 0$ since $t_0 = -2.32$ and P-value $= 0.042 < \alpha = 0.05$.

We conclude that horsepower is linearly associated with gas mileage, but engine size is not.

Removing the explanatory variable engine size, x_1, we obtain:

Regression Analysis: MPG versus HP

```
The regression equation is
MPG = 28.7 - 0.0304 HP

Predictor        Coef    SE Coef      T       P
Constant       28.723      1.956   14.68   0.000
HP           -0.030399   0.008457   -3.59   0.004

S = 1.47000   R-Sq = 54.0%   R-Sq(adj) = 49.8%

Analysis of Variance

Source           DF      SS      MS       F      P
Regression        1   27.922  27.922   12.92  0.004
Residual Error   11   23.770   2.161
Total            12   51.692
```

The new regression equation is $\hat{y} = 28.723 - 0.0304x_3$.

(f)

None of the residual plots show any discernible pattern, the boxplot does not show any outliers, and the normal probability plot looks roughly linear. Therefore, the linear model is appropriate.

(g) $b_3 = -0.0304$: If the horsepower is increased by 1, the gas mileage will decrease b 0.0304 miles per gallon, on average.

(h) 95% Confidence Interval: Lower Bound: 21.622; Upper Bound: 23.665; We are 95% confident that the mean gas mileage for all automobiles with 200 horsepower is between 21.622 and 23.665 miles per gallon.

95% Prediction Interval: Lower Bound: 19.251; Upper Bound: 26.036; We are 95% confident that the gas mileage for a randomly chosen automobile with 200 horsepower is between 19.251 and 26.036 miles per gallon.

25. (a)

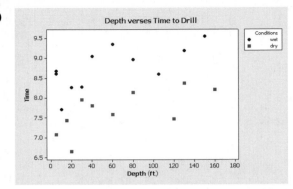

(b) Regression Analysis: Time versus Depth, Conditions

```
The regression equation is
Time = 8.35 + 0.00682 Depth - 1.14 Conditions

Predictor          Coef    SE Coef      T       P
Constant         8.3511     0.1597  52.30   0.000
Depth           0.006816   0.001766   3.86   0.001
Conditions       -1.1359     0.1787  -6.36   0.000

S = 0.407631    R-Sq = 74.2%    R-Sq(adj) = 71.3%

Analysis of Variance
Source           DF        SS       MS       F       P
Regression        2     8.5814   4.2907   25.82   0.000
Residual Error   18     2.9909   0.1662
Total            20    11.5723
```

The regression equation is $\hat{y} = 8.3511 + 0.006816x_1 - 1.1359x_2$.

(c) For the explanatory variable x_1, depth, $t_0 = 3.86$ and the P-value $= 0.001$. So, we reject the null hypothesis and conclude that there is a significant linear relation between depth and time needed to drill another 5 feet. For the explanatory variable x_2, condition, $t_0 = -6.36$ and the P-value < 0.001. So, we reject the null hypothesis and conclude that there is a significant linear relation between condition and time needed to drill another 5 feet.

(d) 95% Confidence Interval: Lower Bound: 7.598; Upper Bound: 8.196; We are 95% confident that the mean time it takes to drill an additional 5 feet is between 7.598 minutes and 8.196 minutes.

95% Prediction Interval: Lower Bound: 6.990; Upper Bound: 8.804; We are 95% confident that for a randomly chosen drilling, it will take between 6.990 minutes and 8.804 minutes to drill an additional 5 feet.

Chapter 14 Review Exercises

1. The least-squares regression model is $y_i = \beta_1 x_i + \beta_0 + \varepsilon_i$. The requirements to perform inference on the least-squares regression line are (1) for any particular values of the explanatory variable x, the mean of the corresponding responses in the population depends linearly on x, and (2) the response variables, y_i, are normally distributed with mean $\mu_{y|x} = \beta_1 x + \beta_0$ and the standard deviation σ.

We verify these requirements by checking to see that the residuals are normally distributed, with mean 0 and constant variance σ^2 and that the residuals are independent. We do this by constructing residual plots and a normal probability plot of the residuals.

2. (a) Using technology, we get: $\beta_0 \approx b_0 = 3.8589$ and $\beta_1 \approx b_1 = -0.1049$. Substituting $x = 5$ into the least-squares regression equation, we get: $\hat{y} = -0.1049(5) + 3.8589 \approx 3.334$.

So, the mean GPA of students who choose to sit in the fifth row is 3.334.

(b) We calculate $\hat{y} = -0.1049x + 3.8589$ to generate the table:

x	y	$\hat{y}$	$y - \hat{y}$	$(y - \hat{y})^2$	$(x - \overline{x})^2$
1	4	3.754	0.2460	0.0605	21.6960
2	3.35	3.649	−0.2991	0.0895	13.3802
2	3.5	3.649	−0.1491	0.0222	13.3802
2	3.67	3.649	0.0209	0.0004	13.3802
2	3.75	3.649	0.1009	0.0102	13.3802
3	3.37	3.544	−0.1742	0.0303	7.0644
3	3.62	3.544	0.0758	0.0057	7.0644
4	2.35	3.439	−1.0893	1.1866	2.7486
4	2.71	3.439	−0.7293	0.5319	2.7486
4	3.75	3.439	0.3107	0.0965	2.7486
5	3.1	3.334	−0.2344	0.0549	0.4328
5	3.22	3.334	−0.1144	0.0131	0.4328
5	3.36	3.334	0.0256	0.0007	0.4328
5	3.58	3.334	0.2456	0.0603	0.4328
5	3.67	3.334	0.3356	0.1126	0.4328
5	3.69	3.334	0.3556	0.1265	0.4328
5	3.72	3.334	0.3856	0.1487	0.4328
5	3.84	3.334	0.5056	0.2556	0.4328
6	2.35	3.230	−0.8795	0.7735	0.1170
6	2.63	3.230	−0.5995	0.3594	0.1170
6	3.15	3.230	−0.0795	0.0063	0.1170
6	3.69	3.230	0.4605	0.2121	0.1170
6	3.71	3.230	0.4805	0.2309	0.1170
7	2.88	3.125	−0.2446	0.0598	1.8012
7	2.93	3.125	−0.1946	0.0379	1.8012
7	3	3.125	−0.1246	0.0155	1.8012
7	3.21	3.125	0.0854	0.0073	1.8012
7	3.53	3.125	0.4054	0.1643	1.8012
7	3.74	3.125	0.6154	0.3787	1.8012
7	3.75	3.125	0.6254	0.3911	1.8012
7	3.9	3.125	0.7754	0.6012	1.8012
8	2.3	3.020	−0.7197	0.5180	5.4854
8	2.54	3.020	−0.4797	0.2301	5.4854
8	2.61	3.020	−0.4097	0.1679	5.4854
9	2.71	2.915	−0.2048	0.0419	11.1696
9	3.74	2.915	0.8252	0.6810	11.1696
9	3.75	2.915	0.8352	0.6976	11.1696
11	1.71	2.705	−0.9950	0.9900	28.5380
			Sum:	9.3709	194.5526

$$s_e = \sqrt{\frac{\Sigma(y - \hat{y})^2}{n - 2}} = \sqrt{\frac{9.3709}{38 - 2}} \approx 0.5102$$

(c) A normal probability plot shows that the residuals are approximately normally distributed.

(d) $\quad s_{b_1} = \dfrac{s_e}{\sqrt{\Sigma(x - \overline{x})^2}}$

$$= \frac{0.5102}{\sqrt{194.5526}} \approx 0.0366$$

(e) We have the hypotheses $H_0 : \beta_1 = 0$ versus $H_1 : \beta_1 \neq 0$. The test statistic is

$$t_0 = \frac{b_1}{s_{b_1}} = \frac{-0.1049}{0.0366} = -2.866 \ [\text{Tech:} -2.868]$$

Classical approach: The level of significance is $\alpha = 0.05$. Since this is a two-tailed test with $n - 2 = 38 - 2 = 36$ degrees of freedom, the critical values are $\pm t_{0.025} = \pm 2.028$. Since $t_0 = -2.866 < -t_{0.025} = -2.028$ (the test statistic falls in a critical region), we reject H_0.

P-value approach: The P-value for this two-tailed test is the area under the t-distribution with 36 degrees of freedom to the left of $t_0 = -2.866$ [Tech: −2.868], plus the area to the right of 2.866 [Tech: 2.868]. From the t-distribution table in the row corresponding to 36 degrees of freedom, 2.866 falls between 2.719 and 2.990, whose right-tail areas are 0.005 and 0.0025, respectively. We must double these values in order to get the total area in both tails: 0.01 and 0.005. So, $0.005 < P$-value < 0.01. [Tech: P-value = 0.0069]. Because the P-value is less than the level of significance $\alpha = 0.05$, we reject H_0.

Conclusion: There is sufficient evidence to conclude that a linear relationship exists between the row chosen by students on the first day of class and their cumulative GPAs.

(f) For a 95% confidence interval with d.f. = 36, we use $t_{0.025} = 2.028$:

Lower bound:
$$b_1 - t_{\alpha/2} \cdot s_{b_1} = -0.1049 - 2.028 \cdot 0.0366$$
$$= -0.1791$$

Upper bound:
$$b_1 + t_{\alpha/2} \cdot s_{b_1} = -0.1049 + 2.028 \cdot 0.0366$$
$$= -0.0307$$

(g) Lower Bound:
$$\hat{y} - t_{\alpha/2} \cdot s_e \sqrt{\frac{1}{n} + \frac{(x^* - \overline{x})^2}{\Sigma(x - \overline{x})^2}}$$
$$= 3.334 - 2.028 \cdot 0.5102 \cdot \sqrt{\frac{1}{38} + \frac{(5 - 5.6579)^2}{194.5526}}$$
$$\approx 3.159$$

Upper Bound:
$$\hat{y} + t_{\alpha/2} \cdot s_e \sqrt{\frac{1}{n} + \frac{(x^* - \overline{x})^2}{\Sigma(x - \overline{x})^2}}$$
$$= 3.334 + 2.028 \cdot 0.5102 \cdot \sqrt{\frac{1}{38} + \frac{(5 - 5.6579)^2}{194.5526}}$$
$$\approx 3.509$$

(h) The predicted GPA of a randomly selected student who chose a seat in the fifth row is also $\hat{y} = -0.1049(5) + 3.8589 \approx 3.334$.

(i) Lower Bound:
$$\hat{y} - t_{\alpha/2} \cdot s_e \sqrt{1 + \frac{1}{n} + \frac{(x^* - \overline{x})^2}{\Sigma(x - \overline{x})^2}}$$
$$= 3.334 - 2.028 \cdot 0.5102 \cdot \sqrt{1 + \frac{1}{38} + \frac{(5 - 5.6579)^2}{194.5526}}$$
$$\approx 2.285$$

Upper Bound:
$$\hat{y} + t_{\alpha/2} \cdot s_e \sqrt{1 + \frac{1}{n} + \frac{(x^* - \overline{x})^2}{\Sigma(x - \overline{x})^2}}$$
$$= 3.334 + 2.028 \cdot 0.5102 \cdot \sqrt{1 + \frac{1}{38} + \frac{(5 - 5.6579)^2}{194.5526}}$$
$$\approx 4.383 \quad \text{[Tech: 4.384]}$$

(j) Although the predicted GPAs in parts (a) and (b) are the same, the intervals are different because the distribution of the mean GPA, part (a), has less variability than the distribution of individual GPAs, part (h).

3. (a) Using technology, we get $\beta_0 \approx b_0 = -399.2$ and $\beta_1 \approx b_1 = 2.5315$. Substituting $x = 900$ into the least-squares regression equation, we get $\hat{y} = -399.2 + 2.5315(900) \approx 1879.2$

[Tech: 1879.1]. So, the mean rent of a 900-square-foot apartment in Queens is \$1879.2.

(b) Using technology, we calculate $\hat{y} = -399.2 + 2.5315x$ to generate the table:

x	y	$\hat{y}$	$y - \hat{y}$	$(y - \hat{y})^2$	$(x - \overline{x})^2$
500	650	866.55	−216.55	46,893.9	160,000
588	1215	1089.32	125.678	15,795	97,344
1000	2000	2132.3	−132.3	17,503.3	10,000
688	1655	1342.47	312.528	97,673.8	44,944
825	1250	1689.29	−439.29	19,2974	5625
1259	2700	2787.96	−87.959	7736.7	128,881
650	1200	1246.28	−46.275	2141.38	62,500
560	1250	1018.44	231.56	53620	115,600
1073	2350	2317.1	32.9005	1082.44	29,929
1452	3300	3276.54	23.462	550.465	304,704
1305	3100	2904.41	195.593	38,256.4	164,025
				474,227	1,123,552

$$s_e = \sqrt{\frac{\Sigma(y - \hat{y})^2}{n - 2}} = \sqrt{\frac{474,227}{11 - 2}} \approx 229.547$$

(c) A normal probability plot shows that the residuals are normally distributed.

(d) $s_{b_1} = \dfrac{s_e}{\sqrt{\Sigma(x - \overline{x})^2}} = \dfrac{229.547}{\sqrt{1,123,552}} = 0.2166$

(e) We have the hypotheses $H_0 : \beta_1 = 0$ versus $H_1 : \beta_1 \neq 0$. The test statistic is
$$t_0 = \frac{b_1}{s_{b_1}} = \frac{2.5315}{0.2166} = 11.687$$

Classical approach: The level of significance is $\alpha = 0.05$. Since this is a two-tailed test with $n - 2 = 11 - 2 = 9$ degrees of freedom, the critical values are $\pm t_{0.025} = \pm 2.262$. Since $t_0 = 11.687 > t_{0.025} = 2.262$ (the test statistic falls in a critical region), we reject H_0.

P-value approach: The P-value for this two-tailed test is the area under the t-distribution with 9 degrees of freedom to the right of $t_0 = 11.687$, plus the area to the left of $-t_0 = -11.687$. From the t-distribution table in the row corresponding to 9 degrees of freedom, 11.687 falls to the right of 4.781, whose right-tail area is 0.0005. We must double this value in order to get the total area in both tails: 0.001. So, P-value < 0.001. [Tech: P-value < 0.0001]. Because the P-value is less than the level of significance $\alpha = 0.05$, we reject H_0.

Conclusion: There is sufficient evidence to conclude that a linear relationship exists between the square footage of an apartment in Queens, New York, and the monthly rent.

(f) For a 95% confidence interval we use $t_{0.025} = 2.262$:

Lower Bound:
$$b_1 - t_{\alpha/2} \cdot s_{b_1} = 2.5315 - 2.262(0.2166)$$
$$= 2.0416$$

Upper Bound:
$$b_1 + t_{\alpha/2} \cdot s_{b_1} = 2.5315 + 2.262(0.2166)$$
$$= 3.0214$$

(g) We have $n = 11$, $s_e = 229.547$, $\bar{x} = 900$ and $\sum(x-\bar{x})^2 = 1,123,552$. For a 90% confidence interval with $n-2 = 9$ degrees of freedom, we use $t_{0.05} = 1.833$.

Lower Bound:
$$\hat{y} - t_{\alpha/2} \cdot s_e \sqrt{\frac{1}{n} + \frac{(x^*-\bar{x})^2}{\sum(x-\bar{x})^2}}$$
$$= 1879.2 - 1.833(229.547)\sqrt{\frac{1}{11} + \frac{(900-900)^2}{1,123,552}}$$
$$= \$1752.3 \quad [\text{Tech: } \$1752.2]$$

Upper Bound:
$$\hat{y} + t_{\alpha/2} \cdot s_e \sqrt{\frac{1}{n} + \frac{(x^*-\bar{x})^2}{\sum(x-\bar{x})^2}}$$
$$= 1879.2 + 1.833(229.547)\sqrt{\frac{1}{11} + \frac{(900-900)^2}{1,123,552}}$$
$$= \$2006.1 \quad [\text{Tech: } \$2006.0]$$

(h) The predicted value of y is also $\hat{y} = \$1879.2$ [Tech: 1879.1].

(i) Lower Bound:
$$\hat{y} - t_{\alpha/2} \cdot s_e \sqrt{1 + \frac{1}{n} + \frac{(x^*-\bar{x})^2}{\sum(x-\bar{x})^2}}$$
$$= 1879.2 - 1.833(229.547)\sqrt{1 + \frac{1}{11} + \frac{(900-900)^2}{1,123,552}}$$
$$= \$1439.7 \quad [\text{Tech: } \$1439.6]$$

Upper Bound:
$$\hat{y} + t_{\alpha/2} \cdot s_e \sqrt{1 + \frac{1}{n} + \frac{(x^*-\bar{x})^2}{\sum(x-\bar{x})^2}}$$
$$= 1879.2 + 1.833(229.547)\sqrt{1 + \frac{1}{11} + \frac{(900-900)^2}{1,123,552}}$$
$$= \$2318.7 \quad [\text{Tech: } \$2318.6]$$

(j) Although the predicted rents in parts (a) and (h) are the same, the intervals are different because the distribution of the means, in part (a), has less variability than the distribution of the individuals, in part (h).

4. (a) Using technology, we get $\hat{y} = -2054.6x + 20976.2$.

(b) The hypotheses are $H_0 : \beta_1 = 0$, $H_1 : \beta_1 \neq 0$. The level of significance is $\alpha = 0.05$. Since this is a two-tailed test with $n-2 = 12-2 = 10$ degrees of freedom, the critical values are $\pm t_{0.025} = \pm 2.228$. Using technology we obtain the test statistic $t_0 = -10.111$ and P-value < 0.0001.

Classical Approach: Since $t_0 = -10.111 < -t_{0.025} = -2.228$ (the test statistic falls within critical region), we reject H_0.

P-value Approach: Since P-value $< 0.0001 < \alpha = 0.05$, we reject H_0.

Conclusion: There is sufficient evidence to conclude that a linear relationship exists between age of used Chevy Camaros and their prices.

(c)

(d) The plot of residuals shows a pattern which indicates that a linear model is not appropriate. The moral is to perform graphical diagnostic tests along with the inferential procedures before drawing any conclusions regarding the model.

5.   (a)   **Correlations: Midterm, Project, Course Grade**

```
                 Midterm        Project
Project          -0.076
Course Grade      0.704          0.526
```

No; the correlation between midterm and project is between −0.7 and 0.7, multicollinearity does not appear to be a problem.

(b)   **Regression Analysis: Course Grade versus Midterm, Project**

```
The regression equation is
Course Grade = - 91.3 + 1.52 Midterm + 0.441 Project

Predictor    Coef    SE Coef      T       P
Constant    -91.29     30.77   -2.97   0.021
Midterm      1.5211    0.3148   4.83   0.002
Project      0.4407    0.1170   3.77   0.007

S = 3.99690   R-Sq = 83.3%   R-Sq(adj) = 78.5%

Analysis of Variance
Source          DF       SS       MS       F       P
Regression       2   558.27   279.14   17.47   0.002
Residual Error   7   111.83    15.98
Total            9   670.10
```

The regression equation is $\hat{y} = -91.29 + 1.5211x_1 + 0.4407x_2$.

(c)

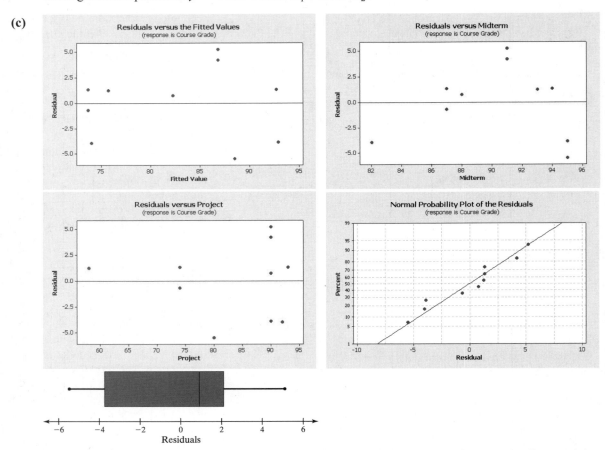

None of the residual plots show any discernible pattern, the boxplot does not show any outliers, and the normal probability plot looks fairly linear. Therefore, the linear model is appropriate.

(d) $b_1 = 1.5211$: This indicates that when midterm grade increases by 1 point, the course grade increases by 1.5211 points, on average, assuming the project grade remains constant.

$b_2 = 0.4407$ **:** This indicates that when the project grade increases by 1 point, the course grade increases by 0.4407 point, on average, assuming the midterm grade remains constant.

(e) $R^2 = 0.833$; 83.3% of the variance in course grade is explained by the regression model. $R^2_{adj} = 0.785$ modifies the value of R^2 taking the sample size and number of explanatory variables into account.

(f) From the output in (b), we get $F_0 = 17.47$ and P-value = $0.002 < \alpha = 0.05$. We reject the null hypothesis and conclude that at least one of the explanatory variables is linearly associated with course grade.

(g) From part (b), we get the following:
Midterm: Reject $H_0 : \beta_1 = 0$ since $t_0 = 4.83$ and P-value = $0.002 < \alpha = 0.05$
Project: Reject $H_0 : \beta_2 = 0$ since $t_0 = 3.77$ and P-value = $0.007 < \alpha = 0.05$
Both explanatory variables have slope coefficients that are significantly different than zero.

(h) $\hat{y} = -91.29 + 1.5211x_1 + 0.4407x_2 = -91.29 + 1.5211(85) + 0.4407(75) \approx 71.1$
The predicted mean final course grade is 71.1.

(i) 95% Confidence Interval: Lower Bound: 69.00; Upper Bound: 82.04; We are 95% confident that the mean course grade of all students who score an 83 on their midterm and a 92 on their project is between 69 and 82.

95% Prediction Interval: Lower Bound: 64.04; Upper Bound: 87.00; We are 95% confident that the final course grade of a randomly selected student who scored an 83 on their midterm and a 92 on their project is between 64 and 87.

6. (a) Correlations: Protein, Fat, Carbohydrates, Calories

```
                Calories      Protein          Fat
Protein          -0.730
Fat              -0.064       -0.261
Carbohydrates     0.801       -0.787       -0.541
```

Yes; because the correlation between carbohydrate and protein is smaller than -0.7 , multicollinearity may be a problem.

(b) Regression Analysis: Calories versus Protein, Fat, Carbohydrates

```
The regression equation is
Calories = 111 - 0.24 Protein + 5.39 Fat + 2.96 Carbohydrates

Predictor         Coef   SE Coef       T       P
Constant        111.24     78.15    1.42   0.192
Protein         -0.239     1.546   -0.15   0.881
Fat              5.388     1.890    2.85   0.021
Carbohydrates   2.9558    0.7944    3.72   0.006

S = 6.65682    R-Sq = 83.5%    R-Sq(adj) = 77.3%

Analysis of Variance
Source           DF        SS       MS       F       P
Regression        3   1794.41   598.14   13.50   0.002
Residual Error    8    354.51    44.31
Total            11   2148.92
```

The regression equation is $\hat{y} = 111.24 - 0.239x_1 + 5.388x_2 + 2.9558x_3$.

(c) From the output in (b), we get $F_0 = 13.50$ and P-value = $0.002 < \alpha = 0.05$. We reject the null hypothesis and conclude that at least one of the explanatory variables is linearly associated with calories.

(d) From part (b), we get the following:

Protein: Do not reject $H_0 : \beta_1 = 0$ since $t_0 = -0.15$ and P-value $= 0.881 > \alpha = 0.05$

Fat: Reject $H_0 : \beta_2 = 0$ since $t_0 = 2.85$ and P-value $= 0.021 < \alpha = 0.05$

Carbs: Reject $H_0 : \beta_3 = 0$ since $t_0 = 3.72$ and P-value $= 0.006 < \alpha = 0.05$

The explanatory variable x_1, protein, is not significantly different from 0, so we should remove it from the model.

(e) After removing x_2, protein, we get the following:

Regression Analysis: Calories versus Fat, Carbohydrates

```
The regression equation is
Calories = 101 + 5.48 Fat + 3.05 Carbohydrates

Predictor          Coef  SE Coef      T      P
Constant         101.20    41.13   2.46  0.036
Fat               5.481    1.692   3.24  0.010
Carbohydrates    3.0536   0.4543   6.72  0.000

S = 6.28549   R-Sq = 83.5%   R-Sq(adj) = 79.8%

Analysis of Variance
Source           DF        SS      MS      F      P
Regression        2   1793.35  896.68  22.70  0.000
Residual Error    9    355.57   39.51
Total            11   2148.92
```

The new regression equation is $\hat{y} = 101.20 + 5.481x_2 + 3.0536x_3$.

Both explanatory variables have slope coefficients that are significantly different than zero.

(f)

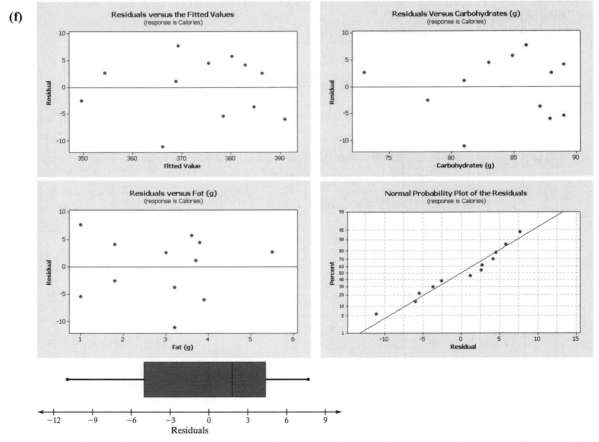

None of the residual plots show any discernible pattern, the boxplot does not show any outliers, and the normal probability plot looks fairly linear. Therefore, the linear model is appropriate.

(g) $b_2 = 5.481$: This indicates that when the fat content increases by 1 gram, the cereal increases by 5.481 calories, on average, assuming carbohydrate content remains constant.

$b_3 = 3.0536$: This indicates that when carbohydrate content increases by 1 gram, the cereal increases by 3.0536 calories, on average, assuming the fat content remains constant.

(h) $R^2 = 0.835$; 83.5% of the variance in calories is explained by the regression model. $R^2_{adj} = 0.798$ modifies the value of R^2 taking the sample size and number of explanatory variables into account.

(i) 95% Confidence Interval:
Lower Bound: 388.08; Upper Bound: 407.81
We are 95% confident that the mean calories in 100 grams of cereal containing 4 grams of fat and 90 grams of carbohydrates will be between 388.08 and 407.81.

95% Prediction Interval:
Lower Bound: 380.64; Upper Bound: 415.25
We are 95% confident that 100 grams of a randomly selected cereal containing 4 grams of fat and 90 grams of carbohydrates contains between 380.64 and 415.25 calories.

Chapter 14 Test

1. The requirements to perform inference on the least-squares regression line are (1) for any particular value of the explanatory variable x, the mean of the corresponding responses in the population depends linearly on x, and (2) the response variables, y_i, are normally distributed with mean $\mu_{y|x} = \beta_1 x + \beta_0$ and standard deviation σ.

2. **(a)** Using technology, we get:
$\beta_0 \approx b_0 = -0.3091$ and $\beta_1 \approx b_1 = 0.2119$.

We use the regression equation to calculate the mean number of chirps at 80.2°F:
$\hat{y} = 0.2119 \cdot 80.2 - 0.3091 = 16.69$

(b) Using technology, we calculate
$\hat{y} = 0.2119x - 0.3091$ to generate the table:

x	y	$\hat{y}$	$y - \hat{y}$	$(y - \hat{y})^2$	$(x - \bar{x})^2$
88.6	20.0	18.467	1.533	2.349	73.274
93.3	19.8	19.463	0.337	0.113	175.828
80.6	17.1	16.772	0.328	0.108	0.314
69.7	14.7	14.462	0.238	0.057	106.916
69.4	15.4	14.398	1.002	1.003	113.210
79.6	15.0	16.560	−1.560	2.434	0.194
80.6	16.0	16.772	−0.772	0.596	0.314
76.3	14.4	15.861	−1.461	2.134	13.988
71.6	16.0	14.865	1.135	1.289	71.234
84.3	18.4	17.556	0.844	0.712	18.148
75.2	15.5	15.628	−0.128	0.016	23.426
82.0	17.1	17.069	0.031	0.001	3.842
83.3	16.2	17.344	−1.144	1.309	10.628
82.6	17.2	17.196	0.004	0.000	6.554
83.5	17.0	17.387	−0.387	0.149	11.972
			Sum	12.270	629.836

$$s_e = \sqrt{\frac{\sum(y - \hat{y})^2}{n - 2}} = \sqrt{\frac{12.270}{15 - 2}} = 0.9715$$

(c) A normal probability plot shows that the residuals are normally distributed.

(d) $s_{b_1} = \dfrac{s_e}{\sqrt{\sum(x - \bar{x})^2}} = \dfrac{0.9715}{\sqrt{629.836}} = 0.0387$

(e) The hypotheses are $H_0 : \beta_1 = 0$ versus $H_1 : \beta_1 \neq 0$. The test statistic is
$$t_0 = \frac{b_1}{s_{b_1}} = \frac{0.2119}{0.0387} = 5.475 .$$

Classical approach: The level of significance is $\alpha = 0.05$. Since this is a two-tailed test with $n - 2 = 15 - 2 = 13$ degrees of freedom, the critical values are $\pm t_{0.025} = \pm 2.160$. Since $t_0 = 5.475 > t_{0.025} = 2.160$ (the test statistic falls in a critical region), we reject H_0 .

P-value approach: The *P*-value for this two-tailed test is the area under the *t*-distribution with 13 degrees of freedom to the right of $t_0 = 5.475$, plus the area to the left of $-t_0 = -5.475$. From the *t*-distribution table in the row corresponding to 13 degrees of freedom, 5.475 falls to the right of 4.221, whose right-tail area is 0.0005. We must double this value in order to get the total area in both tails: 0.001. So, *P*-value < 0.001 [Tech: 0.0001]. Because the *P*-value is less than the level of significance $\alpha = 0.05$, we reject H_0.

Conclusion: There is sufficient evidence to conclude that a linear relationship exists between temperature and the chirps per second.

(f) For a 95% confidence interval we use $t_{0.025} = 2.160$:

Lower Bound:
$$b_1 - t_{\alpha/2} \cdot s_{b_1} = 0.2119 - 2.160 \cdot 0.0387$$
$$= 0.1283$$

Upper Bound:
$$b_1 + t_{\alpha/2} \cdot s_{b_1} = 0.2119 + 2.160 \cdot 0.0387$$
$$= 0.2955$$

(g) We have $n = 15$, $s_e = 0.9715$, $\bar{x} = 80.04$ and $\sum(x - \bar{x})^2 = 629.836$. For a 90% confidence interval with df $= n - 2 = 13$, we use $t_{0.05} = 1.771$:

Lower Bound:
$$\hat{y} - t_{\alpha/2} \cdot s_e \sqrt{\frac{1}{n} + \frac{(x^* - \bar{x})^2}{\sum(x - \bar{x})^2}}$$
$$= 16.69 - 1.771 \cdot 0.9715 \cdot \sqrt{\frac{1}{15} + \frac{(80.2 - 80.04)^2}{629.836}}$$
$$= 16.24 \text{ chirps}$$

Upper Bound:
$$\hat{y} + t_{\alpha/2} \cdot s_e \sqrt{\frac{1}{n} + \frac{(x^* - \bar{x})^2}{\sum(x - \bar{x})^2}}$$
$$= 16.69 + 1.771 \cdot 0.9715 \cdot \sqrt{\frac{1}{15} + \frac{(80.2 - 80.04)^2}{629.836}}$$
$$= 17.13 \text{ chirps}$$

(h) The predicted value of y is also $\hat{y} = 16.69$ chirps.

(i) Lower Bound:
$$\hat{y} - t_{\alpha/2} \cdot s_e \sqrt{1 + \frac{1}{n} + \frac{(x^* - \bar{x})^2}{\sum(x - \bar{x})^2}}$$
$$= 16.69 - 1.771 \cdot 0.9715 \cdot \sqrt{1 + \frac{1}{15} + \frac{(80.2 - 80.04)^2}{629.836}}$$
$$= 14.91 \text{ chirps}$$

Upper Bound:
$$\hat{y} + t_{\alpha/2} \cdot s_e \sqrt{1 + \frac{1}{n} + \frac{(x^* - \bar{x})^2}{\sum(x - \bar{x})^2}}$$
$$= 16.69 + 1.771 \cdot 0.9715 \cdot \sqrt{1 + \frac{1}{15} + \frac{(80.2 - 80.04)^2}{629.836}}$$
$$= 18.47 \quad [\text{Tech: } 18.46] \text{ chirps}$$

(j) Although the predicted numbers of chirps in parts (a) and (h) are the same, the intervals are different because the distribution of the means, part (a), has less variability than the distribution of the individuals, part (h).

3. (a) Using technology we get: $\beta_0 \approx b_0 = 29.705$ and $\beta_1 \approx b_1 = 2.6351$. We use the regression equation to calculate the mean height: $\hat{y} = 2.6351(7) + 29.705 = 48.15$ inches.

(b) Using technology, we calculate $\hat{y} = 2.6351x + 29.705$ to generate the table:

x	y	$\hat{y}$	$y - \hat{y}$	$(y - \hat{y})^2$	$(x - \bar{x})^2$
2	36.1	34.975	1.125	1.266	14.823
2	34.2	34.975	−0.775	0.601	14.823
2	31.1	34.975	−3.875	15.015	14.823
3	36.3	37.610	−1.310	1.716	8.123
3	39.5	37.610	1.890	3.572	8.123
4	41.5	40.245	1.255	1.575	3.423
4	38.6	40.245	−1.645	2.706	3.423
5	45.6	42.880	2.720	7.397	0.722
5	44.8	42.880	1.920	3.686	0.722
5	44.6	42.880	1.720	2.958	0.722
6	49.8	45.515	4.285	18.359	0.023
7	43.2	48.150	−4.950	24.506	1.323
7	47.9	48.150	−0.250	0.063	1.323
8	51.4	50.785	0.615	0.378	4.623
8	48.3	50.785	−2.485	6.177	4.623
8	50.9	50.785	0.115	0.013	4.623
9	52.2	53.420	−1.220	1.490	9.923
9	51.3	53.420	−2.120	4.496	9.923
10	55.6	56.056	−0.456	0.208	17.223
10	59.5	56.056	3.444	11.864	17.223
			Sum	108.045	140.550

$$s_e = \sqrt{\frac{\sum(y - \hat{y})^2}{n-2}} = \sqrt{\frac{108.045}{20-2}} = 2.450$$

(c) A normal probability plot shows that the residuals are normally distributed.

(d) $s_{b_1} = \dfrac{s_e}{\sqrt{\sum(x-\overline{x})^2}} = \dfrac{2.450}{\sqrt{140.550}} = 0.2067$

(e) The hypotheses are $H_0 : \beta_1 = 0$ versus $H_1 : \beta_1 \neq 0$. The test statistic is

$$t_0 = \frac{b_1}{s_{b_1}} = \frac{2.6351}{0.2067} = 12.751.$$

Classical approach: The level of significance is $\alpha = 0.05$. Since this is a two-tailed test with $n - 2 = 20 - 2 = 18$ degrees of freedom, the critical values are $\pm t_{0.025} = \pm 2.101$. Since $t_0 = 12.751 > t_{0.025} = 2.101$ (the test statistic falls in a critical region), we reject H_0.

P-value approach: The P-value for this two-tailed test is the area under the t-distribution with 18 degrees of freedom to the right of $t_0 = 12.751$, plus the area to the left of $-t_0 = -12.751$. From the t-distribution table in the row corresponding to 18 degrees of freedom, 12.751 falls to the right of 3.922, whose right-tail area is 0.0005. We must double this value in order to get the total area in both tails: 0.001. So, P-value < 0.001 [Tech: P-value < 0.0001]. Because the P-value is less than the level of significance $\alpha = 0.05$, we reject H_0.

Conclusion: There is sufficient evidence to conclude that a linear relationship exists between boys' ages and heights.

(f) For a 95% confidence interval we use $t_{0.025} = 2.101$.
Lower Bound:
$$b_1 - t_{\alpha/2} \cdot s_{b_1} = 2.6351 - 2.101 \cdot 0.2067$$
$$= 2.2008$$
Upper Bound:
$$b_1 + t_{\alpha/2} \cdot s_{b_1} = 2.6351 + 2.101 \cdot 0.2067$$
$$= 3.0694$$

(g) We have $n = 20$, $s_e = 2.450$, $\overline{x} = 5.85$ and $\sum(x-\overline{x})^2 = 629.836$. For a 90% confidence interval with $n - 2 = 18$ degrees of freedom, we use $t_{0.05} = 1.734$.

Lower Bound:
$$\hat{y} - t_{\alpha/2} \cdot s_e \sqrt{\frac{1}{n} + \frac{(x^* - \overline{x})^2}{\sum(x-\overline{x})^2}}$$
$$= 48.15 - 1.734 \cdot 2.450 \cdot \sqrt{\frac{1}{20} + \frac{(7-5.85)^2}{140.550}}$$
$$= 47.11 \text{ [Tech: 47.12] inches}$$
Upper Bound:
$$\hat{y} + t_{\alpha/2} \cdot s_e \sqrt{\frac{1}{n} + \frac{(x^* - \overline{x})^2}{\sum(x-\overline{x})^2}}$$
$$= 48.15 + 1.734 \cdot 2.450 \cdot \sqrt{\frac{1}{20} + \frac{(7-5.85)^2}{140.550}}$$
$$= 49.19 \text{ inches}$$

(h) The predicted value of y is also $\hat{y} = 48.15$ inches.

(i) Lower Bound:
$$\hat{y} - t_{\alpha/2} \cdot s_e \sqrt{1 + \frac{1}{n} + \frac{(x^* - \overline{x})^2}{\sum(x-\overline{x})^2}}$$
$$= 48.15 - 1.734 \cdot 2.450 \cdot \sqrt{1 + \frac{1}{20} + \frac{(7-5.85)^2}{140.550}}$$
$$= 43.78 \text{ inches}$$
Upper Bound:
$$\hat{y} + t_{\alpha/2} \cdot s_e \sqrt{1 + \frac{1}{n} + \frac{(x^* - \overline{x})^2}{\sum(x-\overline{x})^2}}$$
$$= 48.15 + 1.734 \cdot 2.450 \cdot \sqrt{1 + \frac{1}{20} + \frac{(7-5.85)^2}{140.550}}$$
$$= 52.52 \text{ inches}$$

(j) Although the predicted heights in parts (a) and (h) are the same, the intervals are different because the distribution of the means, part (a), has less variability than the distribution of the individuals, part (h).

4. (a) Using technology we get: $\beta_0 \approx b_0 = 67.388$ and $\beta_1 \approx b_1 = -0.2632$.

(b) Using technology, we calculate
$\hat{y} = -0.2632x + 67.388$ to generate the table:

x	y	$\hat{y}$	$y - \hat{y}$	$(y - \hat{y})^2$	$(x - \bar{x})^2$
15	65	63.440	1.560	2.433	598.81
16	60	63.177	−3.177	10.093	550.87
28	58	60.019	−2.019	4.076	131.57
61	60	51.334	8.666	75.101	463.52
53	46	53.439	−7.439	55.344	183.04
43	66	56.071	9.929	98.582	12.46
16	56	63.177	−7.177	51.509	550.87
25	75	60.808	14.192	201.403	209.40
28	46	60.019	−14.019	196.527	131.57
34	45	58.440	−13.440	180.627	29.93
37	58	57.650	0.350	0.122	6.10
41	70	56.597	13.403	179.627	2.34
43	73	56.071	16.929	286.587	12.46
49	45	54.492	−9.492	90.099	90.81
53	60	53.439	6.561	43.042	183.04
61	56	51.334	4.666	21.772	463.52
68	30	49.492	−19.492	379.924	813.93
			Sum	1876.869	4434.24

$$s_e = \sqrt{\frac{\sum (y - \hat{y})^2}{n-2}} = \sqrt{\frac{1876.869}{17-2}} = 11.1859$$

(c) A normal probability plot shows that the residuals are normally distributed.

(d) $s_{b_1} = \dfrac{s_e}{\sqrt{\sum (x - \bar{x})^2}} = \dfrac{11.1859}{\sqrt{4434.24}} = 0.1680$

(e) The hypotheses are $H_0 : \beta_1 = 0$ versus $H_1 : \beta_1 \neq 0$. The test statistic is
$$t_0 = \frac{b_1}{s_{b_1}} = \frac{-0.2632}{0.1680} = -1.567 .$$

<u>Classical approach</u>: The level of significance is $\alpha = 0.05$. Since this is a two-tailed test with $n - 2 = 17 - 2 = 15$ degrees of freedom, the critical values are $\pm t_{0.025} = \pm 2.131$. Since $t_0 = -1.567 > -t_{0.025} = -2.131$ (the test statistic does not fall in a critical region), we do not reject H_0.

<u>P-value approach</u>: The P-value for this two-tailed test is the area under the t-distribution with 15 degrees of freedom to the left of $t_0 = -1.567$, plus the area to the right of 1.567. From the t-distribution table in the row corresponding to 15 degrees of freedom, 1.567 falls between 1.341 and 1.753, whose right-tail areas are 0.10 and 0.05, respectively. We must double these values in order to get the total area in both tails: 0.20 and 0.10. So, $0.10 < P\text{-value} < 0.20$ [Tech: P-value = 0.1380]. Because the P-value is larger than $\alpha = 0.05$, we do not reject H_0.

<u>Conclusion</u>: There is not sufficient evidence to support that a linear relationship exists between a woman's age and grip strength.

(f) Since there is no significant linear relationship, the best estimate is the sample mean grip strength, which is $\bar{y} = 57$ psi.

5. **(a)** **Correlations: Fat, Protein, Sugar, Carbs, Calories**

```
            Fat     Protein     Sugar      Carbs
Protein    0.507
Sugar      0.412     0.247
Carbs      0.576     0.550      0.671
Calories   0.950     0.661      0.542      0.784
```

Because the correlations among the explanatory variables are all between -0.7 and 0.7, multicollinearity does not appear to be a problem.

(b) **Regression Analysis: Calories versus Fat, Protein, Sugar, Carbs**

```
The regression equation is
Calories = - 11.9 + 9.08 Fat + 4.47 Protein + 1.49 Sugar + 3.52 Carbs


Predictor    Coef   SE Coef       T       P
Constant   -11.90     14.20   -0.84   0.463
Fat        9.0765    0.2421   37.49   0.000
Protein    4.4676    0.5935    7.53   0.005
Sugar       1.488     1.112    1.34   0.273
Carbs      3.5202    0.3035   11.60   0.001


S = 4.39186    R-Sq = 99.9%    R-Sq(adj) = 99.8%


Analysis of Variance

Source           DF      SS      MS       F       P
Regression        4   89342   22336  1157.97   0.000
Residual Error    3      58      19
Total             7   89400
```

The regression equation is $\hat{y} = -11.90 + 9.0765x_1 + 4.4676x_2 + 1.488x_3 + 3.5202x_4$.

(c) From the output in (b), we get $F_0 = 1157.97$ and P-value $< 0.001 < \alpha = 0.05$. We reject the null hypothesis and conclude that at least one of the explanatory variables is linearly associated with calories.

(d) From part (b), we have the following:

Fat: Reject $H_0: \beta_1 = 0$ since $t_0 = 37.49$ and P-value $< 0.001 < \alpha = 0.05$.

Protein: Reject $H_0: \beta_2 = 0$ since $t_0 = 7.53$ and P-value $= 0.005 < \alpha = 0.05$.

Sugar: Do not reject $H_0: \beta_3 = 0$ since $t_0 = 1.34$ and P-value $= 0.273 > \alpha = 0.05$.

Carbohydrates: Reject $H_0: \beta_4 = 0$ since $t_0 = 11.60$ and P-value $= 0.001 < \alpha = 0.05$.

We conclude that the explanatory variables fat, protein, and carbohydrates all have a significant linear relation with the response variable calories. We also conclude that the explanatory variable sugar is not linearly related to fat. So, we remove x_3, sugar, from the model.

(e) Removing the variable x_3, fat, we obtain:

Regression Analysis: Calories versus Fat, Protein, Carbs

```
The regression equation is
Calories = - 4.4 + 9.11 Fat + 4.29 Protein + 3.77 Carbs

Predictor    Coef   SE Coef      T       P
Constant    -4.42     14.29   -0.31   0.772
Fat        9.1099    0.2635   34.57   0.000
Protein    4.2937    0.6338    6.77   0.002
Carbs      3.7672    0.2637   14.28   0.000

S = 4.80656   R-Sq = 99.9%   R-Sq(adj) = 99.8%

Analysis of Variance

Source           DF      SS      MS        F       P
Regression        3   89308   29769  1288.54   0.000
Residual Error    4      92      23
Total             7   89400
```

The new regression equation is $\hat{y} = -4.42 + 9.1099x_1 + 4.2937x_2 + 3.7672x_4$

Fat: Reject $H_0 : \beta_1 = 0$ since $t_0 = 34.57$ and P-value $< 0.001 < \alpha = 0.05$.

Protein: Reject $H_0 : \beta_2 = 0$ since $t_0 = 6.77$ and P-value $= 0.002 < \alpha = 0.05$.

Carbohydrates: Reject $H_0 : \beta_4 = 0$ since $t_0 = 14.28$ and P-value $< 0.001 < \alpha = 0.05$.

We conclude that the remaining explanatory variables fat, protein, and carbohydrates all have a significant linear relation with the response variable calories. We believe that this is our best model.

(f)

None of the residual plots show any discernible pattern, the boxplot does not show any outliers, and the normal probability plot looks fairly linear. Therefore, the linear model is appropriate.

(g) $b_1 = 9.1099$: If fat is increased by one gram, calories will increase by 9.1099 calories, on average, assuming that protein and carbohydrates remain constant.

$b_2 = 4.2937$: If protein is increased by one gram, calories will increase by 4.2937 calories, on average, assuming that fat and carbohydrates remain constant.

$b_4 = 3.7672$: If carbohydrates are increased by one gram, calories will increase by 3.7672 calories, on average, assuming that fat and protein remain constant.

(h) $R^2 = 99.9\%$; Virtually all (99.9%) of the variation in calories is explained by the least-squares regression model.

$R^2_{adj} = 99.8\%$ modifies the value of R^2 taking the sample size and number of explanatory variables into account.

(i) 95% Confidence Interval: Lower Bound: 656.16; Upper Bound: 668.17; We are 95% confident that the mean number of calories for all fast-food cheeseburgers with 38 grams of fat, 29 grams of protein, and 52 grams of carbohydrates is between 656.16 and 668.17 calories.

95% Prediction Interval: Lower Bound: 647.53; Upper Bound: 676.80; We are 95% confident that the number of calories for a randomly selected fast-food cheeseburger with 38 grams of fat, 29 grams of protein, and 52 grams of carbohydrates is between 647.53 and 676.80 calories.

Chapter 15

Nonparametric Statistics

15.1 An Overview of Nonparametric Statistics

1. Parametric procedures are based on some assumptions about the distribution of the population that is being examined, and typically involve the estimation of a population parameter, such as the mean or variance of the population. Nonparametric methods, perhaps better called distribution-free methods, make no assumptions about the underlying population distributions and usually do not involve estimating any population parameters, although they may be used to perform inference about population parameters such as the population median.

3. The power of a test is the probability of making a Type II error, i.e. the error of not rejecting the null hypothesis when the alternative hypothesis is in fact correct.

5. Disadvantages of nonparametric procedures include:
   1. Nonparametric tests are often misused; that is, they are often used when a more powerful parametric method is appropriate.
   2. Nonparametric tests are typically less powerful than parametric tests.
   3. Nonparametric tests are typically less efficient than parametric tests.

15.2 Runs Test for Randomness

1. A phenomenon is random if individual outcomes are unpredictable (follow no particular pattern), although there may be an overall pattern in the long run. A run is an unbroken sequence of similar events, items, or symbols that is followed by an event, item, or symbol that is mutually exclusive from the first event, item, or symbol.

3. (a) $n = 14$, $n_1 = 6$, $n_2 = 8$ and $r = 6$.

   (b) Since $n_1 \leq 20$ and $n_2 \leq 20$, the critical values are 3 and 12, from Table X.

   (c) The test statistic is $r = 6$, which is not in the critical region, so we do not reject H_0. There is not enough evidence to conclude that the sequence is not random.

5. (a) $n = 19$, $n_1 = 9$, $n_2 = 10$ and $r = 5$.

   (b) Since $n_1 \leq 20$ and $n_2 \leq 20$, the critical values are 5 and 16, from Table X.

   (c) The test statistic is $r = 5$, which is in the critical region and so we reject H_0. There is enough evidence to conclude that the sequence is not random.

7. $n = 20$, $n_1 = 9$ (two-seam fastballs), $n_2 = 11$ (split-finger fastballs), and $r = 12$. Since $n_1 \leq 20$ and $n_2 \leq 20$, the critical values are 6 and 16, from Table X. The test statistic is $r = 12$, which is not in the critical region, so we do not reject H_0. There is not enough evidence at the $\alpha = 0.05$ level of significance to conclude that Okajima's pitches are not random.

9. $n = 14$, $n_1 = 8$ (on-time flights), $n_2 = 6$ (late arrivals) and $r = 7$. Since $n_1 \leq 20$ and $n_2 \leq 20$, the critical values are 3 and 12, from Table X. The test statistic is $r = 7$, which is not in the critical region, so we do not reject H_0. There is not enough evidence at the $\alpha = 0.05$ level of significance to conclude that the arrival status of Flight 0426 is not random.

11. $n = 20$, $n_1 = 14$ (accepted bottles), $n_2 = 6$ (rejected bottles) and $r = 4$. Since $n_1 \leq 20$ and $n_2 \leq 20$, the critical values are 5 and 14, from Table X. The test statistic is $r = 4$, which is in the critical region, so we reject H_0. There is enough evidence at the $\alpha = 0.05$ level of significance to conclude that the filling machine is not random in the way that it over- or under-fills bottles.

13. $n = 45$, $n_1 = 18$ (positive changes), $n_2 = 27$ (negative changes) and $r = 23$. Since $n_2 > 20$ the critical values are $\pm z_{0.025} = \pm 1.96$. To compute the test statistic, we calculate

$$\mu_r = \frac{2n_1 n_2}{n} + 1 = \frac{2 \cdot 18 \cdot 27}{45} + 1 = 22.6,$$

$$\sigma_r = \sqrt{\frac{2n_1 n_2 (2n_1 n_2 - n)}{n^2 (n-1)}},$$

$$= \sqrt{\frac{2 \cdot 18 \cdot 27 (2 \cdot 18 \cdot 27 - 45)}{45^2 (45-1)}}$$

$$= 3.180$$

and

$$z_0 = \frac{r - \mu_r}{\sigma_r} = \frac{23 - 22.6}{3.180} = 0.13,$$ which is not in the critical region so we do not reject H_0. There is not enough evidence at the $\alpha = 0.05$ level of significance to conclude that stock price fluctuations are not random.

15. **(a)** B B B A A A A A A B B A

(b) $n = 12$, $n_1 = 7$ (A's), $n_2 = 5$ (B's) and $r = 4$. Since $n_1 \leq 20$ and $n_2 \leq 20$, the critical values are 3 and 11, from Table X. The test statistic is $r = 4$, which is not in the critical region so we do not reject H_0. There is not enough evidence at the $\alpha = 0.05$ level of significance to conclude that the sequence of residuals is not random. The data appear to support the assumption.

17. Denoting strengths above 75 with A and those below 75 with B, we obtain the sequence
A A B B B A A A A A A B B B A A
A A B B
This gives us $n = 20$, $n_1 = 12$ (above target), $n_2 = 8$ (below target), $r = 6$. Since $n_1 \leq 20$ and $n_2 \leq 20$, the critical values are 6 and 16, from Table X. The test statistic is $r = 6$, which is in the critical region so we reject H_0. There is enough evidence at the $\alpha = 0.05$ level of significance to conclude that the compression strengths are not random about the target value of 75 psi. It appears the process is no longer in control.

15.3 Inferences about Measures of Central Tendency

1. Small sample efficiency: 0.955; this indicates that a sample size of 1000 would be needed to obtain the same probability of a Type I error as a sample size of 955 in a parametric test. Large sample efficiency: 0.637; this indicates that a sample size of 1000 would be needed to obtain the same probability of a Type I error as a sample size of 637 in a parametric test. Thus, the larger the sample, the more inefficient the sign test becomes.

3. Answers will vary. If the true median $M > M_0$ then we would expect the majority of observations to be greater than M_0 (giving plus signs) and few observations to be less than M_0 (giving minus signs). In this case k is the number of minus signs, and it will be relatively small. So a small value of k is evidence against the null hypothesis and in support of the alternative hypothesis, $H_1 : M > M_0$.

5. Since $H_1 : M < 8$, this is a left-tailed test and k = the number of plus signs = 8. We have $n = 13 + 8 = 21 \leq 25$ and so the critical value, from Table XI, is 6. Since $k > 6$ we do not reject H_0. There is not enough evidence at the 5% level of significance to conclude that the median is less than 8.

7. Since $H_1 : M \neq 100$, this is a two-tailed test and k = the smaller of the number of minus signs and the number of plus signs = 21. We have $n = 21 + 28 = 49 > 25$ and so the critical value, from the normal tables, is $-z_{\alpha/2} = -z_{0.025} = -1.96$. The test statistic is

$$z_0 = \frac{(k + 0.5) - \frac{n}{2}}{\sqrt{n}/2} = \frac{(21 + 0.5) - \frac{49}{2}}{\sqrt{49}/2} = -0.86 > -1.96$$

so we do not reject H_0. There is not enough evidence at the 5% level of significance to conclude that the median differs from 100.

9. We first convert the data to plus and minus signs (and zeros) by comparing the data values to $M_0 = 12$. This gives: $+ + + + - + + -$ $+ + - + + + 0$. There are 11 plus signs and 3 minus signs. Since $H_1 : M > 12$, this is a right-tailed test and k = the number of minus signs = 3. We have $n = 11 + 3 = 14 \leq 25$ and so the critical value, from Table XI, is 3. Since $k \leq 3$ we reject H_0. There is enough evidence at the 5% level of significance to conclude that the median is greater than 12.

11. We record plus signs for values > 5.25 and minus signs for values < 5.25 and count 12 plus signs and 7 minus signs. Since $H_1 : M > 5.25$, this is a right-tailed test and k = the number of minus signs = 7. We have $n = 7 + 12 = 19 \leq 25$ and so the critical value, from Table XI, is 5. Since $k > 5$ we do not reject H_0. There is not enough evidence to conclude that the median pH is greater than 5.25. That is, there is not enough evidence at the 5% level of significance to conclude that the acidity of rain in Glacier National Park has decreased.

13. We record plus signs for values > \$62,000 and minus signs for values < \$62,000 and count 9 plus signs and 2 minus signs (and one 0). Since $H_1 : M > \$62,000$, this is a right-tailed test and k = the number of minus signs = 2. We have $n = 9 + 2 = 11 \leq 25$ and so the critical value, from Table XI, is 2. Since $k \leq 2$ we reject H_0. There is sufficient evidence at the 5% level of significance to conclude that the median income of lawyers who recently graduated from law school is greater than \$62,000.

15. We record plus signs for values > \$800 thousand and minus signs for values < \$800 thousand and count 11 plus signs and 3 minus signs. Since $H_1 : M > 800$, this is a right-tailed test and k = the number of minus signs = 3. We have $n = 11 + 3 = 14 \leq 25$ and so the critical value, from Table XI, is 3. Since $k \leq 3$, we reject H_0. There is sufficient evidence at the 5% level of significance to indicate that the median baseball salary is greater than \$800,000.

17. P-value $= P(X \leq k) = P(X \leq 7)$ calculated from the binomial formula with $n = 19$ and $p = 0.5$. Using technology we get P-value $= 0.1796$.

19. P-value $= P(X \leq k) = P(X \leq 2)$ calculated from the binomial formula with $n = 11$ and $p = 0.5$. Using technology we get P-value $= 0.0327$.

21. Our hypotheses are $H_0 : p = 0.5$ versus $H_1 : p < 0.5$. It was found that 38 of the 100 were women, so we would have 38 plus signs and 62 minus signs. This is a left-tailed test, so k = number of plus signs = 38. Since $n = 100 > 25$, our test statistic is

$$z_0 = \frac{(k + 0.5) - \dfrac{n}{2}}{\dfrac{\sqrt{n}}{2}} = \frac{(38 + 0.5) - \dfrac{100}{2}}{\dfrac{\sqrt{100}}{2}} = -2.3.$$

At the 1% level of significance, our critical value is $-z_{0.01} = -2.33$. Since $z_0 > z_{0.01}$, we do not reject H_0. There is not enough evidence at the 1% level of significance to indicate that less than 50% of gamers are women.

23. Since participants had to agree to be part of the survey, the sample may not be truly random which is required to use the sign test.

15.4 Inferences about the Difference between Two Medians: Dependent Samples

1. If we compute the differences as Population 2 − Population 1, then the researcher believes that $M_D > 0$, where M_D is the median of the differences between the observations in the matched pairs data.

3. Since $H_1 : M_D > 0$, this is a right-tailed test and the test statistic is $T = |T_-| = 16$. The critical value for $n = 12$ and $\alpha = 0.05$ is 17 (from Table XII). Since $T < 17$ we reject H_0. There is enough evidence to conclude that $M_D > 0$.

5. Since $H_1 : M_D < 0$, this is a left-tailed test and the test statistic is $T = T_+ = 33$. The critical value for $n = 15$ and $\alpha = 0.05$ is 30 (from Table XII). Since $T \geq 30$ we do not reject H_0. There is not enough evidence to conclude that $M_D < 0$.

7. Since $H_1 : M_D \neq 0$, this is a two-tailed test and the test statistic is $T = \min(T_+, |T_-|)$ $= \min(50, 121) = 50$. The critical value for $n = 18$ and $\alpha/2 = 0.025$ is 40 (from Table XII). Since $T \geq 40$ we do not reject H_0. There is not enough evidence to conclude that $M_D \neq 0$.

9. Since $H_1 : M_D > 0$, this is a right-tailed test and $T = |T_-| = 300$. The critical value for $n = 40$ and $\alpha = 0.05$ is $-z_\alpha = -z_{0.05} = -1.645$ (since $n > 30$). The test statistic is

$$z_0 = \frac{T - \dfrac{n(n+1)}{4}}{\sqrt{\dfrac{n(n+1)(2n+1)}{24}}} = \frac{300 - \dfrac{40(40+1)}{4}}{\sqrt{\dfrac{40(40+1)(2 \cdot 40 + 1)}{24}}}$$

$$= -1.48 > -1.645$$

so we do not reject H_0. There is not enough evidence to conclude that $M_D > 0$.

11. From the computations in the table below we get $T_+ = 12$, $|T_-| = |-33| = 33$ and $n = 9$.

Before	After	After – Before	Absolute Value	Signed Rank
23.5	19.75	–3.75	3.75	–9
18.5	19.25	0.75	0.75	4.5
21.5	21.75	0.25	0.25	1.5
24	22.5	–1.5	1.5	–7.5
25	25	0	0	0
19.75	19.5	–0.25	0.25	–1.5
35	34.25	–0.75	0.75	–4.5
36.5	35	–1.5	1.5	–7.5
52	51.5	–0.5	0.5	–3
30	31	1	1	6

Since $H_1 : M_D < 0$, this is a left-tailed test and the test statistic is $T = T_+ = 12$. The critical

value for $n = 9$ and $\alpha = 0.05$ is 8 (from Table XII). Since $T \geq 8$ we do not reject H_0. There is not sufficient evidence to indicate that the median waistline has decreased. The exercise regimen does not appear to be effective in reducing waistline.

13. From the computations in the table below we get $T_+ = 5$, $|T_-| = |-16| = 16$ and $n = 6$.

Blue	Red	Red – Blue	Absolute Value	Signed Rank
0.582	0.408	–0.174	0.174	–4
0.481	0.407	–0.074	0.074	–1
0.841	0.542	–0.299	0.299	–6
0.267	0.402	0.135	0.135	3
0.685	0.456	–0.229	0.229	–5
0.450	0.533	0.083	0.083	2

Since $H_1 : M_D \neq 0$, this is a two-tailed test and the test statistic is $T = \min(T_+, |T_-|)$ $= \min(5, 16) = 5$. The critical value for $n = 6$ and $\alpha/2 = 0.025$ is 0 (from Table XII). Since $T \geq 0$ we do not reject H_0. There is not sufficient evidence to indicate that the median reaction times are different for blue versus red.

15. From the computations in the table below we get $T_+ = 31$, $|T_-| = |-5| = 5$ and $n = 8$.

Initial	5 Yrs	5 Yrs – Initial	Absolute Value	Signed Rank
38	52	14	14	8
58	60	2	2	2.5
65	72	7	7	5
74	72	–2	2	–2.5
56	54	–2	2	–2.5
36	48	12	12	7
56	58	2	2	2.5
52	60	8	8	6

Since $H_1 : M_D > 0$, this is a right-tailed test and the test statistic is $T = |T_-| = 5$. The critical value for $n = 8$ and $\alpha = 0.05$ is 5 (from Table XII). Since $T \geq 5$ we do not reject H_0. There is not enough evidence to indicate that the median clarity of the lake has increased.

17. From the computations in the table below we get $T_+ = 20$, $|T_-| = |-35| = 35$ and $n = 10$.

Avis	Hertz	Hertz − Avis	Absolute Value	Signed Rank
67.54	59.14	−8.40	8.40	−10
97.16	97.15	−0.01	0.01	−1
103.43	103.14	−0.29	0.29	−5
38.79	36.63	−2.16	2.16	−9
87.90	87.80	−0.10	0.10	−2
109.86	110.09	0.23	0.23	4
109.35	109.35	0	0	0
82.86	83.03	0.17	0.17	3
112.65	113.32	0.67	0.67	6
116.46	114.53	−1.93	1.93	−8
105.49	107.04	1.55	1.55	7

Since $H_1 : M_D > 0$, this is a right-tailed test and the test statistic is $T = |T_-| = 35$. The critical value for $n = 10$ and $\alpha = 0.05$ is 10 (from Table XII). Since $T \geq 10$ we do not reject H_0. There is not enough evidence to indicate that the median car rental cost for Avis is less than that for Hertz.

19. (a) $H_0 : M_D = 0$ and $H_1 : M_D \neq 0$.

(b) Since P-value $< 0.001 < 0.05 = \alpha$ we should reject H_0 and conclude that space flight affects the red blood cell count of rats.

21. Start by computing the difference between the length of stay for completers referred by employers and 107 days (the median for referrals from the criminal justice system). From the computations in the table below we get $T_+ = 8.5$, $|T_-| = |-57.5| = 57.5$ and $n = 11$.

Employer − Justice	Absolute Value	Signed Rank
−27	27	−11
1	1	1
−12	12	−7
0	0	0
−18	18	−9
−7	7	−4
−22	22	−10
−5	5	−3
8	8	5.5
2	2	2
−8	8	−5.5
−13	13	−8

Since $H_1 : M_D \neq 0$, this is a two-tailed test and the test statistic is $T = \min(T_+, |T_-|)$ $= \min(8.5, 57.5) = 8.5$. The critical value for $n = 11$ and $\alpha/2 = 0.025$ is 10 (from Table XII). Since $T < 10$ we reject H_0. There is enough evidence at the 5% level of significance to conclude that the median length of stay is different for employer referrals than those referred by the criminal justice system.

23. Her results are suspect because the data were not randomly selected.

15.5 Inferences about the Difference between Two Medians: Independent Samples

1. If the two medians are the same, then the sum of the ranks of the observations from population X should be close to the sum of the ranks of the observations from population Y. The test statistic essentially measures the size of the sum of ranks for the sample from population X. If this sum is unusually small, then that is evidence that $M_x < M_y$ and if this sum is unusually large then that is evidence that $M_x > M_y$.

3. The Wilcoxon signed-ranks test is used for dependent samples (matched pairs) while the Mann-Whitney test is used for independent samples.

5. Since $H_1 : M_x \neq M_y$, this is a two-tailed test. As $n_1 = 12$, $n_2 = 15$ and $\alpha = 0.05$, the critical values are $w_{0.025} = 50$ (from Table XIII) and $w_{0.975} = n_1 n_2 - w_{0.025} = 12 \cdot 15 - 50 = 130$. The test statistic is
$$T = S - \frac{n_1(n_1+1)}{2} = 170 - \frac{12 \cdot 13}{2} = 92$$ which is not in the critical region so we do not reject H_0. There is not enough evidence to indicate that $M_x \neq M_y$.

7. Since $H_1 : M_x < M_y$, this is a left-tailed test. As $n_1 = 18$, $n_2 = 16$ and $\alpha = 0.05$, the critical value is $w_{0.05} = 96$ (from Table XIII). The test statistic is

$$T = S - \frac{n_1(n_1+1)}{2} = 210 - \frac{18 \cdot 19}{2} = 39 \text{ which}$$

is less than $w_{0.05} = 96$ and so we reject H_0. There is enough evidence to conclude that $M_x < M_y$.

9. Since $H_1 : M_x > M_y$, this is a right-tailed test. As $n_1 = 15$, $n_2 = 15$ and $\alpha = 0.05$, we first look up $w_{0.05} = 73$ (from Table XIII). Then the critical value is $w_{0.95} = n_1 n_2 - w_{0.05} = 15 \cdot 15 - 73 = 152$. The test statistic is

$$T = S - \frac{n_1(n_1+1)}{2} = 250 - \frac{15 \cdot 16}{2} = 130 \text{ which}$$

is not greater than $w_{0.95} = 152$ so we do not reject H_0. There is not enough evidence to indicate that $M_x > M_y$.

11. Since $H_1 : M_x \neq M_y$, this is a two-tailed test. As $n_1 = 22 > 20$, $n_2 = 25 > 20$ and $\alpha = 0.05$, the critical values are $\pm z_{\alpha/2} = \pm z_{0.025} = \pm 1.96$. We first calculate

$$T = S - \frac{n_1(n_1+1)}{2} = 590 - \frac{22 \cdot 23}{2} = 337 \text{ and}$$

then the test statistic is given by

$$z_0 = \frac{T - \frac{n_1 n_2}{2}}{\sqrt{\frac{n_1 n_2 (n_1 + n_2 + 1)}{12}}} = \frac{337 - \frac{22 \cdot 25}{2}}{\sqrt{\frac{22 \cdot 25 \cdot (22 + 25 + 1)}{12}}}$$
$$= 1.32$$

which is not in the critical region so we do not reject H_0. There is not enough evidence to indicate that $M_x \neq M_y$.

13. From the computations in the table below we have $n_1 = 15$, $n_2 = 13$ and $S = 193.5$.

Women	Men	Women Ranks	Men Ranks
7	12	3.5	11.5
6	20	2	28
10	13	8.5	14.5
17	14	24.5	17.5
9	9	6.5	6.5
13	13	14.5	14.5
14	10	17.5	8.5
15	18	20.5	26.5
18	15	26.5	20.5
15	8	20.5	5
11	13	10	14.5
12	17	11.5	24.5
16	15	23	20.5
7		3.5	
3		1	
	Sum =	193.5	

Since $H_0 : M_W \neq M_M$ and $H_1 : M_W \neq M_M$, this is a two-tailed test. The critical values are $w_{0.025} = 55$ (from Table XIII) and $w_{0.975} = n_1 n_2 - w_{0.025} = 15 \cdot 13 - 55 = 140$. The test statistic is

$$T = S - \frac{n_1(n_1+1)}{2} = 193.5 - \frac{15 \cdot 16}{2} = 73.5$$

which is not in the critical region so we do not reject H_0. There is not enough evidence to indicate that the median sociability score is different for women and men.

15. From the computations in the table below we have $n_1 = 8$, $n_2 = 8$ and $S = 74.5$.

Carpeted	Uncarpeted	Carpeted Ranks	Uncarpeted Ranks
11.8	12.1	10	12
8.2	8.3	4	5
7.1	3.8	2	1
13.0	7.2	13	3
10.8	12.0	8	11
10.1	11.1	6.5	9
14.6	10.1	16	6.5
14.0	13.7	15	14
	Sum =	74.5	

Since $H_1 : M_x > M_y$, this is a right-tailed test. From Table XIII $w_{0.05} = 16$, and so the critical value is $w_{0.95} = n_1 n_2 - w_{0.05} = 8 \cdot 8 - 16 = 48$.

The test statistic is

$$T = S - \frac{n_1(n_1+1)}{2} = 74.5 - \frac{8\cdot 9}{2} = 38.5 \text{ which}$$

is not greater than $w_{0.95} = 48$ and so we do not reject H_0. There is not enough evidence to conclude that the median amount of bacteria in carpeted rooms is greater than that in uncarpeted rooms.

17. From the computations in the table below we have $n_1 = 22$, $n_2 = 20$ and $S = 609$.

Lincoln	Clarendon	Lincoln Ranks	Clarendon Ranks
0.11	0.06	12	2
0.09	0.09	7.5	7.5
0.21	0.14	26.5	18
0.09	0.16	7.5	20.5
0.07	0.03	3	1
0.30	0.12	31.5	14
0.41	0.29	36.5	30
0.33	0.10	33.5	10.5
0.20	0.41	24.5	36.5
0.22	0.08	28	4.5
0.34	0.14	35	18
0.46	0.14	39	18
0.19	0.12	22.5	14
0.67	0.08	41	4.5
0.75	0.09	42	7.5
0.19	0.10	22.5	10.5
0.30	0.21	31.5	26.5
0.33	0.16	33.5	20.5
0.20	0.13	24.5	16
0.42	0.12	38	14
0.25		29	
0.47		40	
	Sum =	609	

Since $H_1: M_x \neq M_y$, this is a two-tailed test. As $n_1 = 22 > 20$ and $\alpha = 0.05$, the critical values are $\pm z_{\alpha/2} = \pm z_{0.025} = \pm 1.96$. We first find $T = S - \frac{n_1(n_1+1)}{2} = 609 - \frac{22\cdot 23}{2} = 356$ and then the test statistic is given by

$$z_0 = \frac{T - \frac{n_1 n_2}{2}}{\sqrt{\frac{n_1 n_2(n_1+n_2+1)}{12}}} = \frac{356 - \frac{22\cdot 20}{2}}{\sqrt{\frac{22\cdot 20\cdot(22+20+1)}{12}}}$$

= 3.43 which is in the critical region so we reject H_0. There is enough evidence to conclude that the median calcium level in rainwater in Lincoln County, Nebraska, differs from the median level in Clarendon County, South Carolina.

19. (a) From the computations in the table below we have $n_1 = 15$, $n_2 = 15$ and $S = 262$.

Prof. A	Prof. B	Prof. A's Ranks	Prof. B's Ranks
A	A	3	3
A	A	3	3
B	A	10	3
B	B	10	10
B	B	10	10
C	B	19.5	10
C	B	19.5	10
C	B	19.5	10
C	B	19.5	10
C	C	19.5	19.5
C	C	19.5	19.5
D	C	26.5	19.5
D	C	26.5	19.5
D	D	26.5	26.5
F	F	29.5	29.5
	Sum =	262	

(b) This is a two-tailed test so $H_1: M_x \neq M_y$. Assuming that $\alpha = 0.05$, the critical values are $w_{0.025} = 65$ (from Table XIII) and $w_{0.975} = n_1 n_2 - w_{0.025} = 15\cdot 15 - 65 = 160$. The test statistic is

$$T = S - \frac{n_1(n_1+1)}{2} = 262 - \frac{15\cdot 16}{2} = 142$$

which is not in the critical region and so we do not reject H_0. There is not enough evidence to support the claim that the median grade differs in the two classes.

15.6 Spearman's Rank-Correlation Test

1. The Spearman rank-correlation test can be performed on qualitative data if the requirements are met and if the data can be ranked. For example, if a restaurant price can be ranked as expensive, moderate, or inexpensive, and the quality of the food can be ranked as good, fair, or poor, then the Spearman rank-correlation test is appropriate. Since Spearman's rank correlation test depends only on the ranks of the data values, as long as the data is ordinal, and so can be ranked, we can perform this test to determine if there is an association between the variables.

3. **(a)** The scatter diagram appears to show a positive linear association.

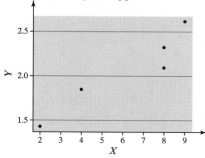

   (b) The table shows the data and their ranks:

X	Y	Rank of X	Rank of Y	d_i	d_i^2
2	1.4	1	1	0	0
4	1.8	2	2	0	0
8	2.1	3.5	3	0.5	0.25
8	2.3	3.5	4	−0.5	0.25
9	2.6	5	5	0	0

$$\sum d_i^2 = 0 + 0 + 0.25 + 0.25 + 0 = 0.5$$

$$r_s = 1 - \frac{6\sum d_i^2}{n(n^2 - 1)} = 1 - \frac{6(0.5)}{5(5^2 - 1)} = 1 - \frac{3}{120} = 0.975$$

   (c) With $n = 5$ and $\alpha = 0.05$ (two tails), the critical values are ± 1.000. Since the test statistic, $r_s = 0.975$, is not greater than 1.000, we do not reject H_0. Note that when $n = 5$, this test is too insensitive to detect an association at the $\alpha = 0.05$ level. However at the $\alpha = 0.10$ level of significance, the critical values are ± 0.900 and since $r_s = 0.975 > 0.900$ we would reject H_0 and conclude that there is an association between X and Y.

5. **(a)** The scatter diagram appears to show a negative linear association.

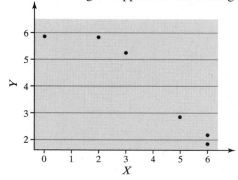

(b) The table shows the data and their ranks:

X	Y	Rank of X	Rank of Y	d_i	d_i^2
0	5.8	1	6	−5	25
2	5.7	2	5	−3	9
3	5.2	3	4	−1	1
5	2.8	4	3	1	1
6	1.9	5.5	1	4.5	20.25
6	2.2	5.5	2	3.5	12.25

$$\sum d_i^2 = 25+9+1+1+20.25+12.25 = 68.5$$

$$r_s = 1 - \frac{6\sum d_i^2}{n(n^2-1)} = 1 - \frac{6(68.5)}{6(6^2-1)} = 1 - \frac{411}{210} \approx -0.957 \text{ [Tech: } r_s = -0.968\text{]}$$

Note: the text formula and technology values are not identical when there are ties among the ranks and the formula is only approximate.

(c) With $n=6$ and $\alpha=0.05$ (two tails), the critical values are ± 0.886. Since the test statistic is less than -0.886, we reject H_0 and conclude that there is an association between X and Y.

7. (a) The table shows the data and their ranks:

% with Bachelor's	Per Capita Income ($)	Rank of X	Rank of Y	d_i	d_i^2
19.0	27,935	1	1	0	0
29.8	38,956	9	9	0	0
24.7	33,236	4	4	0	0
21.2	30,952	2	2	0	0
26.9	32,348	6	3	3	9
27.2	34,397	7	8	−1	1
23.3	33,338	3	5	−2	4
28.3	33,667	8	6	2	4
25.5	34,257	5	7	−2	4

$$\sum d_i^2 = 0+0+0+0+9+1+4+4+4 = 22$$

$$r_s = 1 - \frac{6\sum d_i^2}{n(n^2-1)} = 1 - \frac{6(22)}{9(9^2-1)} = 1 - \frac{132}{720} \approx 0.817 \text{ [Tech: } r_s = 0.817\text{]}$$

As we are testing for a positive association, this is a right-tailed test and with $n=9$ and $\alpha=0.05$ (one tail). The critical value is 0.600. Since the test statistic is greater than 0.600, we reject H_0. There is enough evidence to conclude that per capita personal income and percentage of the population with at least a bachelor's degree are positively associated.

(b) The scatter diagram appears to show a positive linear association, which supports our conclusion.

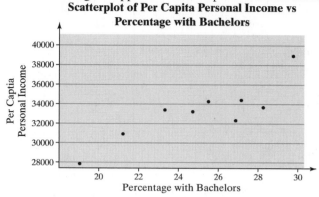

Scatterplot of Per Capita Personal Income vs Percentage with Bachelors

9. (a) The table shows the data and their ranks:

Per Capita Income ($)	Birth Rate	Rank of X	Rank of Y	d_i	d_i^2
39,022	13.8	6	5.5	0.5	0.25
44,077	13.4	8	3	5	25
55,755	14.3	9	8	1	1
39,173	13.8	7	5.5	1.5	2.25
27,897	11.5	1	1	0	0
32,234	14.2	4	7	−3	9
29,515	13.6	2	4	−2	4
31,891	15.7	3	9	−6	36
35,798	12.7	5	2	3	9

$$\sum d_i^2 = 0.25 + 25 + 1 + 2.25 + 0 + 9 + 4 + 36 + 9 = 86.5$$

$$r_s = 1 - \frac{6\sum d_i^2}{n(n^2-1)} = 1 - \frac{6(86.5)}{9(9^2-1)} = 1 - \frac{519}{720} \approx 0.279 \; [\text{Tech: } 0.276]$$

As we are testing for a negative association, this is a left-tailed test and with $n = 9$ and $\alpha = 0.05$ (one tail), the critical value is -0.600. Our test statistic is positive so we do not reject H_0. There is not evidence to conclude that per capita personal income and birth rate are negatively associated.

(b) The scatter diagram appears to show little association, which supports our conclusion.

Scatterplot of Birth Rate vs Per Capita Income

11. (a) The table shows the data and their ranks:

Family Income ($1000)	Commute Time (min)	Rank of X	Rank of Y	d_i	d_i^2
83	27.1	2.5	5	−2.5	6.25
82	17.0	1	1	0	0
86	26.9	4	4	0	0
90	26.0	5	3	2	4
101	32.1	6	6	0	0
83	22.6	2.5	2	0.5	0.25

$$\sum d_i^2 = 6.25 + 0 + 0 + 4 + 0 + 0.25 = 10.5$$

$$r_s = 1 - \frac{6\sum d_i^2}{n(n^2 - 1)} = 1 - \frac{6(10.5)}{6(6^2 - 1)} = 1 - \frac{63}{210} = 0.7 \quad [\text{Tech: } 0.696]$$

As we are testing for a positive association, this is a right-tailed test and with $n = 6$ and $\alpha = 0.10$ (one tail), the critical value is 0.657. Since the test statistic is greater than 0.657, we reject H_0. There is enough evidence to conclude that family income cost and commute time are positively associated.

(b) The scatter diagram perhaps shows a positive association, which supports our conclusion.

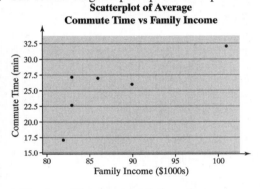

Scatterplot of Average
Commute Time vs Family Income

13. The table shows the data and their ranks:

X	Y	Rank of X	Rank of Y	$d = Y - X$	d^2
2	5	1	1	0	0
3	8	2	2	0	0
4	11	3	3	0	0
5	14	4	4	0	0
6	17	5	5	0	0
				$\sum d^2 =$	0

$r_s = 1.000$, the correlation coefficient of the ranks. X and Y are perfectly positively associated as we can see from the scatter diagram:

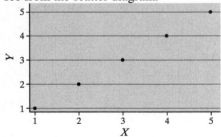

15.7 Kruskal-Wallis Test

1. The sum of ranks for the combined samples will have the value $N(N+1)/2$. If each of the samples comes from the same distribution, then the sum of ranks for sample i should be approximately equal to the fraction n_i/N of the total sum of ranks. Thus the expected value for the sum of ranks of sample i is $n_i(N+1)/2$. The test statistic effectively sums the squared deviations of the ranks in sample i from their expected values. If one of the samples comes from a distribution that differs from that of the others, then its ranks will be unusually small or unusually large and result in a large value for H.

3. **(a)** The tables shows the ranks and the computations based on those ranks:

	X Ranks	Y Ranks	Z Ranks
	6.5	10	4.5
	1.5	12	8
	11	3	1.5
	4.5	6.5	9
R_i	23.5	31.5	23
R_i^2	552.25	992.25	529
n_i	4	4	4

The test statistic is calculated as:

$$H = \frac{12}{N(N+1)}\left[\frac{R_1^2}{n_1}+\cdots+\frac{R_k^2}{n_k}\right]-3(N+1)$$

$$= \frac{12}{12(12+1)}\left[\frac{552.25}{4}+\frac{992.25}{4}+\frac{529}{4}\right]-3(12+1) = 0.875$$

(b) Since there are three populations and each sample size is ≤ 5 we use Table XV and get a critical value of 5.6923 (corresponding to $\alpha = 0.049$).

(c) The test statistic is not greater than the critical value so we do not reject H_0. There is not enough evidence to indicate that the distributions are different.

5. **(a)** H_0: the distributions of the populations are the same.

   H_1: at least one population distribution differs from the others.

(b) The calculations are summarized in the table:

	Monday	**Tuesday**	**Wednesday**	**Thursday**	**Friday**
n_i	8	8	8	8	8
R_i	48	226	144	194.5	207.5
R_i^2	2,304	51,076	20,736	37,830	43,056

$$H = \frac{12}{N(N+1)}\left[\frac{R_1^2}{n_1}+\cdots+\frac{R_k^2}{n_k}\right]-3(N+1)$$

$$= \frac{12}{40(40+1)}\left[\frac{2,304}{8}+\frac{51,076}{8}+\frac{20,736}{8}+\frac{37,830}{8}+\frac{43,056}{8}\right]-3(40+1) = 18.770$$

(c)   Since we are comparing more than 3 populations, we get the critical value from the χ^2 tables. With df $= k - 1 = 5 - 1 = 4$ we find $\chi^2_{0.05} = 9.488$.

(d)   Since the test statistic is greater than the critical value, we reject H_0. There is enough evidence to conclude that the distribution of births varies by day of the week.

(e)   Monday births clearly differ from births on other days of the week.

7.   The table shows the ranks for the data and the calculations based on those ranks:

	Sludge plot	Spring disk	No till	Spring chisel	Great Lakes bt
	1.5	23.5	18	18	10
	6.5	18	3.5	23.5	23.5
	26.5	26.5	13.5	3.5	6.5
	18	30	23.5	10	18
	10	28.5	1.5	21	13.5
	6.5	28.5	13.5	13.5	6.5
R_i	69	155	73.5	89.5	78
R_i^2	4761	24025	5402.25	8010.25	6084
n_i	6	6	6	6	6

$$H = \frac{12}{N(N+1)}\left[\frac{R_1^2}{n_1} + \cdots + \frac{R_k^2}{n_k}\right] - 3(N+1)$$

$$= \frac{12}{30(30+1)}\left[\frac{4761}{6} + \frac{24{,}025}{6} + \frac{5402.25}{6} + \frac{8010.25}{6} + \frac{6084}{6}\right] - 3(30+1) = 10.833$$

Since we are comparing more than 3 populations, we get the critical value from the χ^2 tables. With df $= k - 1 = 5 - 1 = 4$ we find $\chi^2_{0.05} = 9.488$. Since the test statistic is greater than the critical value, we reject H_0. There is enough evidence to support the conclusion that the distribution of corn plant yield varies by plot type.

9. The table shows the ranks for the data and the calculations based on those ranks:

	Simple	**Go/No-Go**	**Choice**
	7	16	15
	11.5	3	11.5
	9	6	13
	4	17	14
	5	10	8
	1	2	18
R_i	37.5	54	79.5
R_i^2	1406.25	2916	6320.25
n_i	6	6	6

$$H = \frac{12}{N(N+1)}\left[\frac{R_1^2}{n_1}+\cdots+\frac{R_k^2}{n_k}\right]-3(N+1)$$

$$= \frac{12}{18(18+1)}\left[\frac{1406.25}{6}+\frac{2916}{6}+\frac{6320.25}{6}\right]-3(18+1) = 5.237$$

Since at least one sample size is greater than 5, we get the critical value from the χ^2 tables. With df $= k-1$ $= 3-1 = 2$ we find $\chi_{0.01}^2 = 9.210$. Since the test statistic is not greater than the critical value, we do not reject H_0. There is not enough evidence to indicate that the distribution of reaction time varies by stimulus.

11. The table shows the ranks for the data and the calculations based on those ranks:

	Large Family	**Passenger Vans**	**Midsize Utility**
	12.5	6	6
	20	3.5	18.5
	3.5	2	17
	8.5	8.5	11
	15	1	15
	15	12.5	21
	10	18.5	6
R_i	84.5	52	94.5
R_i^2	7140.25	2704	8930.25
n_i	7	7	7

$$H = \frac{12}{N(N+1)}\left[\frac{R_1^2}{n_1}+\cdots+\frac{R_k^2}{n_k}\right]-3(N+1)$$

$$= \frac{12}{21(21+1)}\left[\frac{7140.25}{7}+\frac{2704}{7}+\frac{8930.25}{7}\right]-3(21+1) \approx 3.664$$

Since at least one sample size is greater than 5, we get the critical value from the χ^2 tables. With df $= k-1 = 3-1 = 2$ we find $\chi_{0.01}^2 = 9.210$. Since the test statistic is not greater than the critical value, we do not reject H_0. There is not enough evidence to support the conclusion that the distribution of chest compression varies by vehicle category.

Chapter 15 Review Exercises

1. For the given data, $n = 22$, $n_1 = 10$ (number of E's), $n_2 = 12$ (number of W's) and $r = 9$. Since $n_1 \leq 20$ and $n_2 \leq 20$, the critical values are 7 and 17, from Table X. The test statistic is $r = 9$, which is not in the critical region, so we do not reject H_0. There is not enough evidence to conclude that the sequence is not random. It appears that the winning division occurs randomly.

2. We first convert the data to plus and minus signs (and zeros) by comparing the data values to $M_0 = 15$. This gives: $+ + + + + - + + + + + - + - + - + - - -$. There are 13 plus signs and 7 minus signs. Since $H_1 : M > 15$, this is a right-tailed test and k = the number of minus signs = 7. We have $n = 20 \leq 25$ and so the critical value, from Table XI, is 5. Since $k > 5$ we do not reject H_0. There is not enough evidence to indicate that the median number of hours students talk on the phone per week is greater than 15.

3. From the computations in the table below we get $T_+ = 32$, $|T_-| = |-23| = 23$ and $n = 10$.

Height (inches)	Arm Span (inches)	Arm Span – Height	Absolute Value	Signed Rank
59.5	62	2.5	2.5	6
69	65.5	–3.5	3.5	–9.5
77	76	–1	1	–3
59.5	63	3.5	3.5	9.5
74.5	74	–0.5	0.5	–1.5
63	66	3	3	7.5
61.5	61	–0.5	0.5	–1.5
67.5	69	1.5	1.5	4
73	70	–3	3	–7.5
69	71	2	2	5

$T_+ = 32$ and $T_- = -23$. Since $H_1 : M_D \neq 0$, this is a two-tailed test with test statistic $T = \min(T_+, |T_-|) = \min(32, 23) = 23$. The critical value for $n = 10$ and $\alpha/2 = 0.025$ is 8 (from Table XII). Since $T > 8$ we do not reject H_0. There is not enough evidence to conclude that an individual's height and arm span are different.

4. From the computations in the table below we have $n_1 = 16$, $n_2 = 12$ and $S = 216$.

Female	Male	Female Ranks	Male Ranks
0.474	0.541	6	10
0.743	0.659	21	17
0.538	1.050	9	28
0.436	0.880	4	25
0.480	0.577	7	13
0.531	0.752	8	22
0.398	0.849	3	24
0.561	0.544	12	11
0.633	0.464	16	5
0.596	0.675	14	18
0.831	0.626	23	15
0.725	0.393	20	2
0.887		26	

0.905		27	
0.711		19	
0.338		1	
	Sum =	216	

Since $H_1 : M_x \neq M_y$, this is a two-tailed test. The critical values are $w_{0.005} = 42$ (from Table XIII) and

$w_{0.995} = n_1 n_2 - w_{0.005} = 16 \cdot 12 - 42 = 150$. The test statistic is $T = S - \dfrac{n_1(n_1+1)}{2} = 216 - \dfrac{16 \cdot 17}{2} = 80$ which is not

in the critical region so we do not reject H_0. There is not enough evidence to indicate that the median reaction time of males differs from that of females.

5. The table shows the data and their ranks:

Engine Displacement	City mpg	Rank of X	Rank of Y	d_i	d_i^2
3.5	16	9	2.5	6.5	42.25
3.8	17	11.5	4.5	7	49
4.6	15	14.5	1	13.5	182.25
2.2	24	2.5	14	−11.5	132.25
3.5	18	9	7.5	1.5	2.25
2.4	22	5	11.5	−6.5	42.25
2.4	21	5	10	−5	25
2.7	18	7	7.5	−0.5	0.25
3.5	18	9	7.5	1.5	2.25
2.0	24	1	14	−13	169
4.0	17	13	4.5	8.5	72.25
4.6	16	14.5	2.5	12	144
3.8	18	11.5	7.5	4	16
2.2	24	2.5	14	−11.5	132.25
2.4	22	5	11.5	−6.5	42.25

$\sum d_i^2 = 42.25 + 49 + 182.25 + \ldots + 132.25 + 42.25 = 1053.5$

$r_s = 1 - \dfrac{6\sum d_i^2}{n(n^2-1)} = 1 - \dfrac{6(1053.5)}{15(15^2-1)} = 1 - \dfrac{6321}{3360} \approx -0.881$ [Tech: -0.930]

As we are testing for a negative association, this is a left-tailed test and with $n = 15$ and $\alpha = 0.05$ (one tail). The critical value is -0.446. Since the test statistic is less than -0.446, we reject H_0. There is enough evidence to conclude that engine displacement and fuel economy are negatively associated.

6. **(a)** H_0: the distributions of the populations are the same.

   H_1: at least one population distribution differs from the others.

(b) The calculations are summarized in the table:

	Organic	Mineral	Surface
n_i	31	47	44
R_i	2,355.5	2,119.5	3,028.0
R_i^2	5,548,380	4,492,280	9,168,784

$H = \dfrac{12}{N(N+1)} \left[\dfrac{R_1^2}{n_1} + \cdots + \dfrac{R_k^2}{n_k} \right] - 3(N+1)$

$= \dfrac{12}{122(122+1)} \left[\dfrac{5,548,380}{31} + \dfrac{4,492,280}{47} + \dfrac{9,168,784}{44} \right] - 3(122+1) = 17.199$

(c) Since at least one sample size is greater than 5, we get the critical value from the χ^2 tables. With df $= k - 1 = 3 - 1 = 2$ we find $\chi^2_{0.05} = 5.991$.

(d) Since the test statistic is greater than the critical value, we reject H_0. There is enough evidence to support the conclusion that the distribution of concentration of dissolved organic carbon varies by collection area.

(e) The distributions do appear to differ.

7. In general, parametric procedures are based on some assumptions about the distribution of the population that is being examined, and typically involve the estimation of a population parameter, such as the mean or variance of the population. Nonparametric methods, perhaps better called distribution-free methods, make no assumptions about the underlying population distributions and usually do not involve estimating any population parameters, although they may be used to perform inference about population parameters such as the population median. Nonparametric tests generally have fewer assumptions, some have simple computations and some can be applied to qualitative data such as rankings. However, they are typically also less powerful and less efficient than parametric tests. The tests are tabulated below with some of the differences.

Nonparametric Test	Parametric Test	Differences
Runs test for randomness	None	
One-sample sign test	z-Test or t-Test	The nonparametric test is a test of the median and the parametric test is of the mean. The parametric test requires that data come from a normal distribution or the sample size be large.
Wilcoxon Matched-Pairs Signed-Ranks Test	Paired t-Test for Dependent Samples	The nonparametric test is a test of the medians and the parametric test is of the means. The parametric test requires that the differences be normally distributed or the sample size be large. The nonparametric test requires that the differences have a symmetric distribution.
Mann-Whitney Test	Welch's t-Test for Independent Samples	The nonparametric test is a test of the medians and the parametric test is of the means. The parametric test requires that the differences be normally distributed or the sample sizes be large. The nonparametric test requires that the distributions of the two populations be the same, except possibly for the medians.
Spearman's Rank-Correlation Test	Hypothesis Test of the Significance of β_1	The nonparametric test is only for an association and so can also be used on ordinal data whereas the parametric test is only applicable to quantitative data. The parametric test requires that the residuals be normally distributed with constant error variance.
Kruskal-Wallis Test	F-test for One-Way ANOVA	The parametric test requires that the data come from populations that are normally distributed with the same variance and tests if the means are all the same. The nonparametric test requires only that the data can be ranked and tests if the distributions are all the same.

Chapter 15 Test

1. For the given data, $n = 25$, $n_1 = 13$ (number of W's), $n_2 = 12$ (number of E's) and $r = 13$. Since $n_1 \leq 20$ and $n_2 \leq 20$, the critical values are 8 and 19, from Table X. The test statistic is $r = 13$, which is not in the critical region and so we do not reject H_0. There is not enough evidence to indicate that the sequence is not random. It appears that the winning division occurs randomly.

2. We first convert the data to plus and minus signs (and zeros) by comparing the data values to $M_0 = 15$. This gives: $- + - - - - - -$ $- - + - + + - - - - - 0$. There are 4 plus signs and 15 minus signs (and one zero). Since $H_1 : M < 15$, this is a left-tailed test and k = the number of plus signs = 4. We have $n = 4 + 15 = 19 \leq 25$ and so the critical value, from Table XI, is 5. Since $k \leq 5$ we reject H_0. There is enough evidence to conclude that the median number of hours students study per week is less than 15.

3. From the computations in the table below we get $T_+ = 12$, $|T_-| = |-33| = 33$ and $n = 9$.

Before	After	Before – After	Absolute Value	Signed Rank
136	128	8	8	5
120	111	9	9	6.5
129	129	0	0	0
143	148	–5	5	–3.5
115	110	5	5	3.5
113	112	1	1	1.5
89	98	–9	9	–6.5
122	103	19	19	8.5
102	103	–1	1	–1.5
122	103	19	19	8.5

Since $H_1 : M_D > 0$, this is a right-tailed test with test statistic $T = |T_-| = |-11.5| = 11.5$. The critical value for $n = 9$ and $\alpha = 0.05$ is 8 (from Table XII). Since $T \geq 8$ we do not reject H_0. There is not enough evidence to indicate that the median pulse rate has decreased after the exercise program.

4. From the computations in the table below we have $n_1 = 12$, $n_2 = 14$ and $S = 205$.

Texas	Illinois	Texas Ranks	Illinois Ranks
4.69	4.4	17	6.5
5.22	4.4	25.5	6.5
5.1	4.69	23	17
4.76	4.75	21	20
5.22	4.22	25.5	2
4.25	4.63	3	13
4.46	4.64	9	14
5.14	4.45	24	8
4.93	4.54	22	12
4.11	4.49	1	10
4.65	4.35	15	4
4.71	4.36	19	5
	4.69		17
	4.52		11
Sum =		205	

Since $H_1 : M_H > M_C$, this is a right-tailed test. We first look up $w_{0.05} = 52$ (from Table XIII) and then the critical value is $w_{0.95} = n_1 n_2 - w_{0.05} = 12 \cdot 14 - 52 = 116$. The test statistic is
$$T = S - \frac{n_1(n_1 + 1)}{2} = 205 - \frac{12 \cdot 13}{2} = 127$$
which is greater than $w_{0.95} = 116$ and so we reject H_0. There is enough evidence to indicate that the median pH of rain near Houston is greater than that of rain near Chicago.

5. The table shows the data and their ranks:

Temp.	Chirps per sec.	Rank of X	Rank of Y	d_i	d_i^2
88.6	20.0	14	15	−1	1
93.3	19.8	15	14	1	1
80.6	17.1	7.5	10.5	−3	9
69.7	14.7	2	2	0	0
69.4	15.4	1	4	−3	9
79.6	15.0	6	3	3	9
80.6	16.0	7.5	6.5	1	1
76.3	14.4	5	1	4	16
71.6	16.0	3	6.5	−3.5	12.25
84.3	18.4	13	13	0	0
75.2	15.5	4	5	−1	1
82.0	17.1	9	10.5	−1.5	2.25
83.3	16.2	11	8	3	9
82.6	17.2	10	12	−2	4
83.5	17.0	12	9	3	9

$$\sum d_i^2 = 1+1+9+\ldots+4+9 = 83.5$$

$$r_s = 1 - \frac{6\sum d_i^2}{n(n^2-1)} = 1 - \frac{6(83.5)}{15(15^2-1)}$$

$$= 1 - \frac{501}{3360} \approx 0.851 \ [\text{Tech: } 0.850]$$

As we are testing for an association, this is a two-tailed test and with $n=15$ and $\alpha=0.05$ (two tails). The critical values are ± 0.521. Since the test statistic is greater than 0.521, we reject H_0. There is enough evidence to conclude that temperature and number of chirps per second are associated.

6. The table shows the ranks for the data and the calculations based on those ranks:

	Large Family	Passenger Vans	Midsize Utility
	18	16.5	4
	6.5	15	1.5
	12	8	13
	3	19	5
	14	21	1.5
	10	11	9
	16.5	20	6.5
R_i	80	110.5	40.5
R_i^2	6400	12,210.25	1640.25
n_i	7	7	7

$$H = \frac{12}{N(N+1)}\left[\frac{R_1^2}{n_1}+\cdots+\frac{R_k^2}{n_k}\right] - 3(N+1)$$

$$= \frac{12}{21(21+1)}\left[\frac{6,400}{7}+\frac{12,210.25}{7}+\frac{1640.25}{7}\right] - 3(21+1)$$

$$= 9.141$$

Since at least one sample size is greater than 5, we get the critical value from the χ^2 tables.

With df $= k-1 = 3-1 = 2$ we find $\chi_{0.05}^2 = 5.991$. Since the test statistic is greater than the critical value, we reject H_0. There is enough evidence to conclude that the distribution of femur force varies by vehicle category.

7. An efficiency of 0.713 means the nonparametric test would need a sample size of 1000 to achieve the same results as the corresponding parametric test with a sample size of 713.

Appendix B

Lines

1. (a) Slope $= \dfrac{1-0}{2-0} = \dfrac{1}{2}$

(b) If x increases by 2 units, y will increase by 1 unit.

3. (a) Slope $= \dfrac{1-2}{1-(-2)} = \dfrac{-1}{3} = -\dfrac{1}{3}$

(b) If x increases by 3 units, y will decrease by 1 unit.

5. Slope $= \dfrac{y_2 - y_1}{x_2 - x_1} = \dfrac{0-3}{4-2} = \dfrac{-3}{2} = -\dfrac{3}{2}$

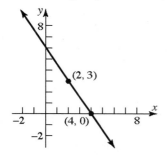

7. Slope $= \dfrac{y_2 - y_1}{x_2 - x_1} = \dfrac{1-3}{2-(-2)} = \dfrac{-2}{4} = -\dfrac{1}{2}$

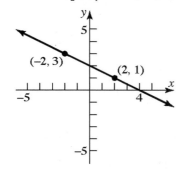

9. Slope $= \dfrac{y_2 - y_1}{x_2 - x_1} = \dfrac{-1-(-1)}{2-(-3)} = \dfrac{0}{5} = 0$

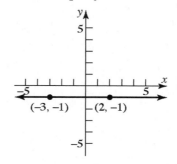

11. Slope $= \dfrac{y_2 - y_1}{x_2 - x_1} = \dfrac{-2-2}{-1-(-1)}$

$= \dfrac{-4}{0}$ undefined.

13. $P = (1, 2); m = 3 = \dfrac{3}{1}$

403

15. $P = (2,4); m = -\dfrac{3}{4} = \dfrac{-3}{4}$

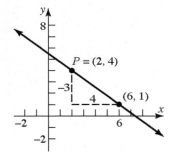

17. $P = (-1,3); m = 0$

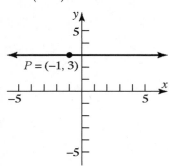

19. $P = (0,3);$ slope undefined

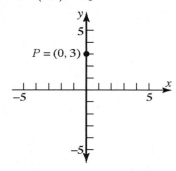

21. Slope $= 3 = \dfrac{3}{1}$

If x increases by 1 unit, then y increases by 3 units.
Answers will vary. Three possible points are:
$x = 1+1 = 2$ and $y = 2+3 = 5$
$\quad (2,5)$
$x = 2+1 = 3$ and $y = 5+3 = 8$
$\quad (3,8)$
$x = 3+1 = 4$ and $y = 8+3 = 11$
$\quad (4,11)$

23. Slope $= -\dfrac{3}{2} = \dfrac{-3}{2}$

If x increases by 2 units, then y decreases by 3 units.
Answers will vary. Three possible points are:
$x = 2+2 = 4$ and $y = -4-3 = -7$
$\quad (4,-7)$
$x = 4+2 = 6$ and $y = -7-3 = -10$
$\quad (6,-10)$
$x = 6+2 = 8$ and $y = -10-3 = -13$
$\quad (8,-13)$

25. Slope $= -2 = \dfrac{-2}{1}$

If x increases by 1 unit, then y decreases by 2 units.
Answers will vary. Three possible points are:
$x = -2+1 = -1$ and $y = -3-2 = -5$
$\quad (-1,-5)$
$x = -1+1 = 0$ and $y = -5-2 = -7$
$\quad (0,-7)$
$x = 0+1 = 1$ and $y = -7-2 = -9$
$\quad (1,-9)$

27. (0, 0) and (2, 1) are points on the line.

Slope $= \dfrac{1-0}{2-0} = \dfrac{1}{2}$

y-intercept is 0; using $y = mx + b$:

$$y = \frac{1}{2}x + 0$$
$$2y = x$$
$$0 = x - 2y$$
$$x - 2y = 0 \text{ or } y = \frac{1}{2}x$$

29. (–1, 3) and (1, 1) are points on the line.

Slope $= \dfrac{1-3}{1-(-1)} = \dfrac{-2}{2} = -1$

Using $y - y_1 = m(x - x_1)$

$$y - 1 = -1(x-1)$$
$$y - 1 = -x + 1$$
$$y = -x + 2$$
$$x + y = 2 \text{ or } y = -x + 2$$

31. $y = 2x + 3$; Slope = 2; y-intercept = 3

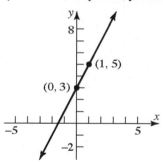

33. $\frac{1}{2}y = x - 1$; $y = 2x - 2$

Slope = 2; y-intercept = –2

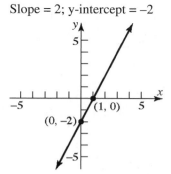

35. $y = \frac{1}{2}x + 2$; Slope$=\frac{1}{2}$; y-intercept = 2

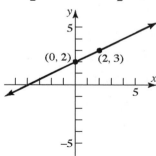

37. $x + 2y = 4$; $2y = -x + 4 \rightarrow y = -\frac{1}{2}x + 2$

Slope $= -\frac{1}{2}$; y-intercept = 2

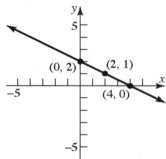

39. $2x - 3y = 6$; $-3y = -2x + 6 \rightarrow y = \frac{2}{3}x - 2$

Slope $= \frac{2}{3}$; y-intercept = –2

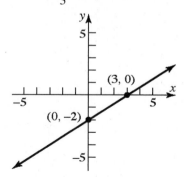

41. $x + y = 1$; $y = -x + 1$

Slope = –1; y-intercept = 1

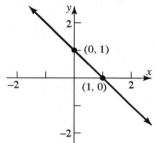

43. $x = -4$; Slope is undefined

y-intercept - none

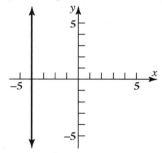

45. $y = 5$; Slope = 0; y-intercept = 5

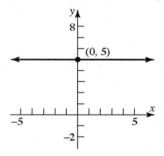

47. $y - x = 0$; $y = x$

Slope = 1; y-intercept = 0

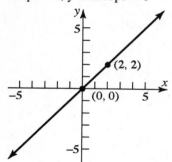

49. $2y - 3x = 0$; $2y = 3x \rightarrow y = \dfrac{3}{2}x$

Slope = $\dfrac{3}{2}$; y-intercept = 0

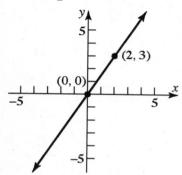